Hans Lyon (Editor)

Theory and Strategy in Histochemistry

A Guide to the Selection and Understanding of Techniques

With the Collaboration of
A. P. Andersen, E. Hasselager, P.-E. Høyer,
M. Møller, P. Prentø, B. van Deurs

With 74 Figures

Springer-Verlag
Berlin Heidelberg NewYork
London Paris Tokyo
HongKong Barcelona Budapest

Dr. Hans Lyon
Københavns Kommunes Hvidovre Hospital
University of Copenhagen
Department of Pathology 134
DK-2650 Hvidovre

ISBN 3-540-19311-1 Springer-Verlag Berlin Heidelberg NewYork
ISBN 0-387-19311-1 Springer-Verlag NewYork Berlin Heidelberg

Data conversion by Danny L. Lewis, Buchproduktion Berlin with TEX
27/3020-543210 – Printed on acid-free paper

Preface

If you want practical information on how to use this book please refer to "Note to the Readers" p. VII.

Histochemistry and cytochemistry are essential tools in biomedical research and routine service laboratories.

Most texts on histochemistry fall into one of two categories:

1. Encyclopaedic texts covering all or nearly all information available on the whole or selected parts of histochemistry.
2. Reviews or surveys of methods found to be useful by the author(s).

While the former category often appeals to the more philosophically inclined reader, direct guidance on the selection of technique may be difficult to find. In contrast, the latter category are often excellent sources for details on how to perform a particular method with a reasonable chance of success. Consideration of the exact mechanism of staining, of possible reasons for failure, and of alternative techniques are, however, frequently lacking.

This book is an introduction to the scientific basis of histochemistry and is intended to provide a background for the selection and development of appropriate methods. It is not a "cook book" and readers expecting exhaustive methodological descriptions will be disappointed.

Although most of the contributors to this book would not describe themselves as histochemists, they have all at some time found it essential to develop a basic understanding of histochemistry. This book contains the information they would have greatly appreciated ready access to at that time.

It is our intention that this book should be useful to:

1. Scientists who would like to develop their understanding of histochemical techniques.
2. Laboratories where histochemical methods are used (frequently or infrequently) in which it should provide assistance in the selection of appropriate techniques.

The dedicated academic may read this book for reasons that are self evident. However, other potential readers may be asking one or more of the following questions:

1. Can the investigation or task I am undertaking be assisted by the application of histochemical techniques?

2. Which techniques should I consider using?
3. What are the problems associated with particular techniques and the investigation of particular cell or tissue components?
4. How should I interpret the results obtained?

We do not claim that all these questions can be directly answered by reference to this text, but these considerations have been foremost in our minds during its preparation. Even though the sequential organization of the book represents a logical approach to the subject, we do not expect readers to start at page 1 and continue right through to the end. We recommend that you refer to the "Note to the Readers" in order to select the sections and sequence most appropriate to your needs.

It is assumed that the reader has a working knowledge of chemistry, basic biochemistry, and histology. It is thus the intention that the text should be suitable for academics, scientific officers and technicians working in pathology, anatomy, biology, and cytology.

We have attempted to strike a balance between the traditional empirical techniques and the modern highly specialized and sophisticated methods. These newer developments together with advances in molecular biology and microscopy have fueled a tremendous increase in our knowledge of cell and tissue biology. Our understanding of the basic substrate for histochemical reactions has therefore altered substantially in recent years and for this reason an introductory review of the chemistry and ultrastructure of cells and extracellular spaces has been included.

All histological and histopathological methods are essentially histochemical in nature. We believe that recognition of this fact leads inevitably to the realization that a clear understanding of the principles set out in this book are the basis for the acceptance of comparable, comprehendable, and even standardizable staining protocols even for the classical empirically derived techniques. Within this context we have considered it inappropriate to include detailed "complete" staining protocols. These are, in any case, notoriously difficult to write with sufficient clarity to enable truly reproducible results. In Appendix A we have, however, described two methods in a way which we believe should make reproducibility possible.

We have elected not to include photomicrographs illustrating the results of histochemical reactions in order to keep the price of the book as low as possible.

The following books were found to be particularly helpful in the preparation of this text, and we therefore wish to acknowledge them specifically: Baker (1966); Bancroft and Stevens (1982); Casselman (1959); Filipe and Lake (1983); Horobin (1982); Lillie and Fullmer (1976); Lillie (1977); Lison (1960); Pearse (1972); (1980); (1985).

We also recommend the following books as sources for detailed descriptions of staining procedures: Bancroft and Cook (1984); Kiernan (1981); Lillie and Fullmer (1976); Lojda et al. (1979); Pearse (1972); (1980); (1985).

Full details of these books are given in the references section.

Hans Lyon

Note to the Readers

The purpose of this note is to assist the readers in getting what they want out of this book as quickly as possible.

We strongly recommend that all readers should familiarize themselves with the *overall* organization of the book as set out in the initial Table of Contents (p. XI).

The sequence of chapters presented reflects one kind of logical progression through the intricacies of histochemistry. We do not necessarily recommend that the book *should* be read in this order. The approach chosen by an individual reader will depend on their specific interests. Chapters 2, 3, 31, and 32 may be considered as appendices to the rest of the text; they will not generally be read in their entirety but should be consulted as required.

Areas of interest may further be found by consulting one of the indexes. These are: a. General index; b. Index of constituents including chemical groups, cells, and tissues detectable by histochemistry; c. Index of dyes including chromogenic reagents, pigments, and stains; d. Index of methods, histochemical and histological. Note that all references are made to Section numbers and not to pages.

Below we have outlined how different readers may wish to approach the book.

1 General Approach

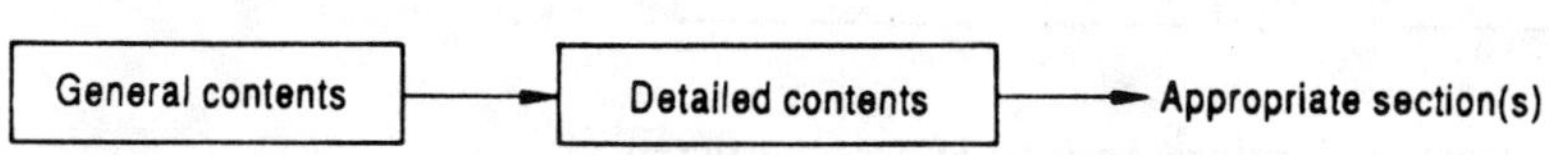

2 Will Histochemistry Be of Help (Task/Investigation)?

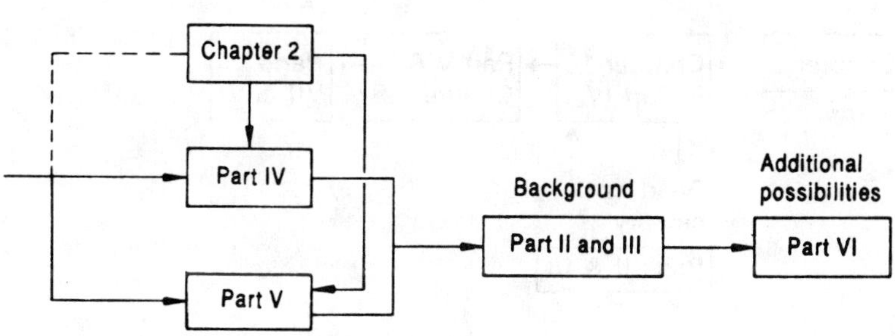

3 Student of Biological Science

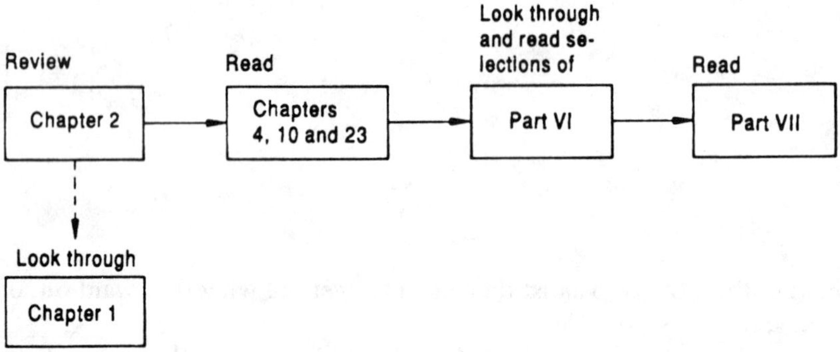

4 Laboratory Technician/Technologist/Scientific Officer

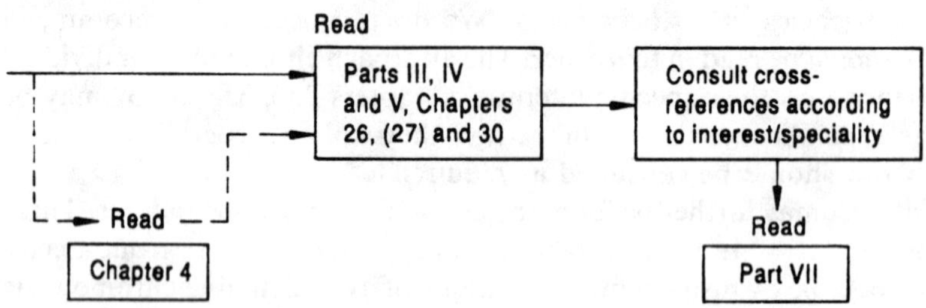

5 Pathologist

a) Should I Use Another Technique?

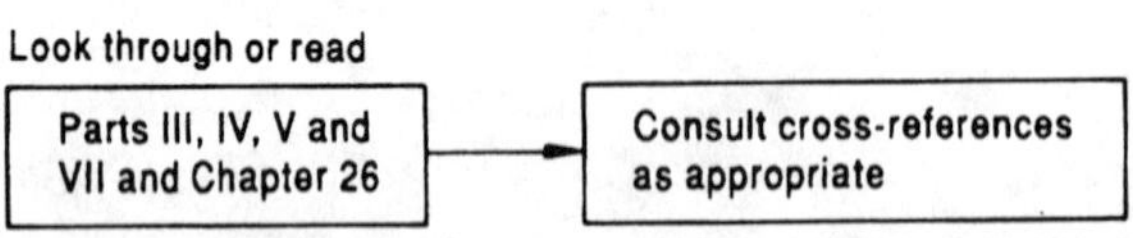

b) Basic Interest in Understanding Histochemistry

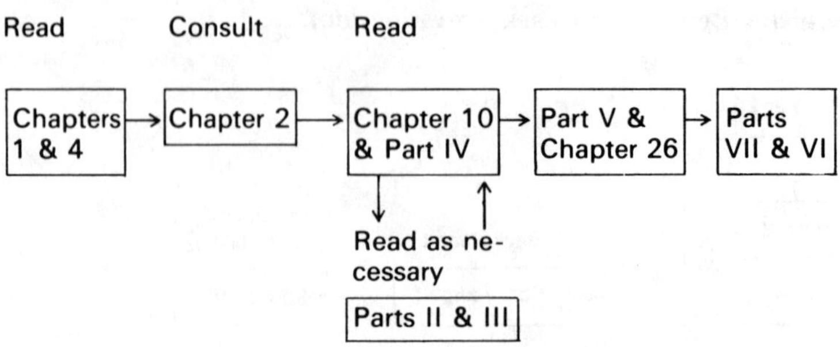

Acknowledgements

The idea for this book is based on a Danish text which was published in 1985 by DSR, Copenhagen, Denmark.

I am indebted to Professor Dietrich H. Wittekind, University of Freiburg, who not only has introduced me to the field of standardization of dyes, stains, and staining methods but who has also consistently encouraged me to prepare the present book.

My thanks are due to all my coauthors for their never failing cooperation and interest in preparing this book.

I gratefully acknowledge the help of Erik Hasselager, Royal Veterinary and Agricultural University, Copenhagen, Denmark, in organizing the text on a personal computer and for supervising and improving the lay-out of tables and figures. I am very much indebted to Palle Jakobsen, Ferrosan Ltd., Denmark, for correcting the formulae and equations. My thanks are also due to Erik Schulte, University of Munich who undertook the task of reading the preliminary drafts of all chapters and not only correcting evident mistakes but also of supplying constructive suggestions for amendments and improvements.

Michael R. Barer has in addition to his valuable contributions to individual chapters carried the burden of correcting the English language throughout for which I am truly grateful.

I wish to thank the Department of Pathology, Hvidovre Hospital, University of Copenhagen and all my colleagues here, past and present, for never failing support and encouragement. I am especially grateful to Professor Per Christoffersen for giving me so excellent working conditions.

The painstaking work of the photographers Susanne Østergaard, Hvidovre Hospital, and Keld Ottosen, The Panum Institute, in preparing the many figures and formulae is gratefully acknowledged.

I am most grateful to my secretary, Ulla Evald, for her never failing interest in converting what at times has been loose thoughts and practically illegible notes into a readable manuscript and who has never complained of having to rewrite the same sections time and time again.

Finally, my thanks are due to the staff of Springer-Verlag, for their encouragement and patience during the preparation of this book.

Hans Lyon

Table of Contents

List of Contributors

Anne Palle Andersen
Novo Nordisk Ltd., Copenhagen, Denmark

Michael R. Barer
Department of Microbiology, University of Newcastle upon Tyne, Newcastle upon Tyne, England

Per Prætorius Clausen
Department of Pathology, University of Odense, Denmark

Bo van Deurs
Department of Medical Anatomy A, The Panum Institute, University of Copenhagen, Denmark

Erik Hasselager
Department of Anatomy and Physiology, Royal Veterinary and Agricultural University, Copenhagen, Denmark

Poul Erik Høyer
Department of Medical Anatomy A, The Panum Institute, University of Copenhagen, Denmark

Astrid K. N. Iversen
Institute of Medical Microbiology, University of Copenhagen, Denmark

Palle Jakobsen
Ferrosan Ltd., Copenhagen, Denmark

Lars Kayser
Department of Medical Anatomy A, The Panum Institute, University of Copenhagen, Denmark

Inger Marie Krogh
Novo Nordisk Ltd., Copenhagen, Denmark

Hans Lyon
Department of Pathology, Hvidovre Hospital, University of Copenhagen, Denmark

Morten Møller
Department of Medical Anatomy B, The Panum Institute, University of Copenhagen, Denmark

Ole William Petersen
Department of Medical Anatomy A, The Panum Institute, University of Copenhagen, Denmark

Poul Prentø
Institute of Cell Biology and Anatomy, University of Copenhagen, Denmark

Erik Schulte
Department of Anatomy, University of Munich, FRG

Jakob Visfeldt
Department of Pathology, Rigshospitalet, University of Copenhagen, Denmark

Mogens Vyberg
Department of Pathology, Aalborg Sygehus, Denmark

Dietrich H. Wittekind
Department of Anatomy II, Albert Ludwig University, Freiburg im Breisgau, FRG

Part 1
General Considerations

1 The Scope of Histochemistry

H. Lyon, M.R. Barer

1.1 Histochemical and Histological Methods

The purpose of histochemical and histological methods is to provide as exact a picture of living tissue as possible. This is exceedingly difficult as any intervention, such as taking a biopsy and preparing it for microscopic examination, effects changes in tissue structure and reactivity. Interpretation of the result therefore requires an understanding of the effects of all aspects of the intervention. One of the primary aims of this book is to give a theoretical background for such understanding.

The discipline of *histochemistry* lies on the boundary between histology and biochemistry. The principal aim of histochemistry is to obtain information regarding the chemical composition and localization of the components of the tissue. As the microscope is the most important single instrument used in histochemistry, the reaction products must be directly visible or be made so. They will in general be coloured.

Aim of Histological Methods. The chief aim of these methods is to visualize and differentiate between tissue components, not to determine the chemical composition.

A division between qualitative and quantitative histochemistry is widely recognized.

1. *Qualitative histochemistry* is concerned with the occurrence and localization of *histochemically demonstrable components*.
2. *Quantitative histochemistry* also assesses the amount of the individual chemical components.

This book is mainly concerned with qualitative histochemistry. Distinction is sometimes made between histochemistry and *cytochemistry*, depending on whether tissues or individual cells are being examined.

H. Lyon (Ed.)
Theory and Strategy in Histochemistry

1.2 The Histochemical Reaction

In principle all reactions that ultimately produce coloured products on tissue sections may be considered as histochemical. In such a reaction it is necessary to understand both the chemical and the histological aspects of the method. The chemical aspects comprise a description of the mechanisms involved, the specificity or selectivity, and the sensitivity. The histological aspects require that the precision with which localization is achieved should be assessed. For a reaction to be designated as histochemical, an account of both chemical and histological aspects must be given. We have adopted the following headings throughout this book.

1.2.1 Mechanism

The following sub-headings have been used: reactive groups, reagents, reaction products, reaction equation, and sources of error.

1.2.2 Selectivity

As tissues contain a vast range of different compounds, it is essential to know which of these may react in the chosen method. If only one chemically well-defined compound reacts, the method is *specific*. In the great majority of cases the method will, however, demonstrate a restricted group of chemical compounds and may then be designated *selective*. Very few methods are specific while the remainder are more or less selective. In a number of cases the selectivity of a method may be increased by inactivating or removing certain reactive groups. These interventions are called respectively *blocking* and *extraction*.

1.2.3 Sensitivity (Detection Limit)

In the assessment of a method it would be useful to know what amount of material and how much of the demonstrable chemical entity within the material are required for the reaction product to be perceived. In quantitative chemistry this can be expressed as the *limit of detection*.

In an ideal world it would be possible to cite the absolute amount of the chemical entity in question that reflects the limit of detection for or sensitivity of the technique applied. In histochemistry the number of variables involved is so large that this is only rarely possible. In this book we have therefore chosen to use the term in the relative sense to indicate the sensitivity of one method compared to others used for demonstrating the same chemical entity. In this way the need for absolute figures is avoided.

It is important to appreciate that relative sensitivity is not a constant property of one method. All steps before and after staining, the exact technical details of staining, the optical system used, the nature of the precipitate, and the substantivity of the reaction product influence the relative potential of one technique to demonstrate a given chemical entity. (The term *substantivity* refers to the ability of the reaction product to attach itself or "stick" to various tissue components, notably proteins. A high substantivity implies low solubility in lipids.)

1.2.4 Localization

This is the central histological consideration which refers to the location of reaction products. Changes in the location of the chemical compound under examination may take place during fixation, dehydration, embedding, sectioning, and during or after the histochemical reaction itself. The compound, one wishes to demonstrate, the intermediate products during the histochemical reaction or the final product may all diffuse. This may result in the compound or reaction product being completely removed from the tissue, diffusely deposited throughout or even specifically bound to other areas away from its site of origin.

According to Grimelius (1968), an assessment of the localization of the reaction product may be made by:

1. *The use of consecutive sections.* The histochemical reaction is performed on one of two adjacent thin sections (1–2μm). The second section is then stained using a general oversight method (Sect.31.2) or a second histochemical procedure whose reaction product distribution is well established.
2. *Double-staining technique.* Two different histochemical reactions are performed consecutively on the same section. It is essential that the two methods used do not affect each other qualitatively or quantitatively. The assessment is made by taking a photograph after the first reaction has been performed and comparing the result with that obtained after the second. If possible, the test should be repeated with the reactions in reverse order.
3. *Restaining technique.* A photograph is taken after the first reaction as above. The section is then destained, a second reaction performed, and the results compared. Interactions between the two reactions should be checked for as with the double staining technique. Comparison is facilitated if thin sections are used. This is a valuable method as it is often possible to perform three or more reactions consecutively with destainings inserted between. Unfortunately, the approach is limited by the almost unavoidable damage to the section during destaining.
4. *Differential count technique.* In this method the number of cells demonstrated by different reactions is compared.

1.2.5 Controls

For the majority of histochemical methods, control reactions are required in order to discriminate between true positive reactions and non-specific reactions. The following control reactions are usually necessary:

1. A positive control reaction performed on a section from another block of the same tissue or another tissue containing the chemical substance under consideration. The substance should be at a concentration close to the relative limit of detection (eliminates false negative reaction).
2. A negative control reaction performed on a positive control tissue block, a consecutive section, or both. Here the control sections are subjected to the same treatment as the test section but a key reagent in the reaction sequence is omitted (eliminates a false positive reaction).
3. Blocking and/or extraction procedures. These assist in determining the selectivity or specificity of the histochemical reaction.

2 The Structural and Chemical Basis for Histochemistry

H. Lyon, B. van Deurs, P. Prentø, E. Hasselager, E. Schulte

The object of histochemistry is to demonstrate tissue and cell components in their native location by using chemically well-defined methods. The fundamental unit for all living organisms is the *cell*. The cell consists of protoplasm which is delimited by the *plasma membrane*. The protoplasm contains the *cell organelles* (e.g. mitochondria and lysosomes) and the *nucleus*. Collections of cells with one or more concerted functions are called *tissues*. A multicellular organism consists of cells, extracellular matrix systems, and fluid systems (blood, lymph, etc.) (see Fig. 2.1). The matrix systems are of fundamental importance for structuring cells into tissues and organs. The *extracellular matrix* is defined as all substances lying outside and between cell surfaces.

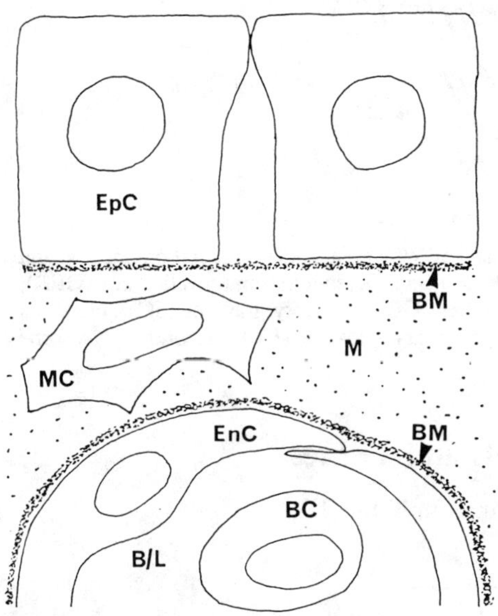

Fig. 2.1. Schematic model for the general organization of a multicellular animal. The epithelial cells (EpC) define the external limits of the organism and rest on a basement membrane (BM), and below this the extracellular matrix (M) is found with various fibre proteins embedded in an amorphous ground substance. Mesenchymal or stromal (MC) cells are also present in the matrix. In addition, blood and lymph vessels (B/L) are present, separated from the matrix by the endothelium (EnC) and a basement membrane. A blood cell (BC) is shown in the vessel.

H. Lyon (Ed.)
Theory and Strategy in Histochemistry

This chapter provides brief descriptions of the chemical and morphological organization of cells and matrix systems accompanied by examples of function. The object is to support and provide a context for the remainder of the text.

In Fig. 2.2 a schematic drawing of a typical mammalian cell is shown.

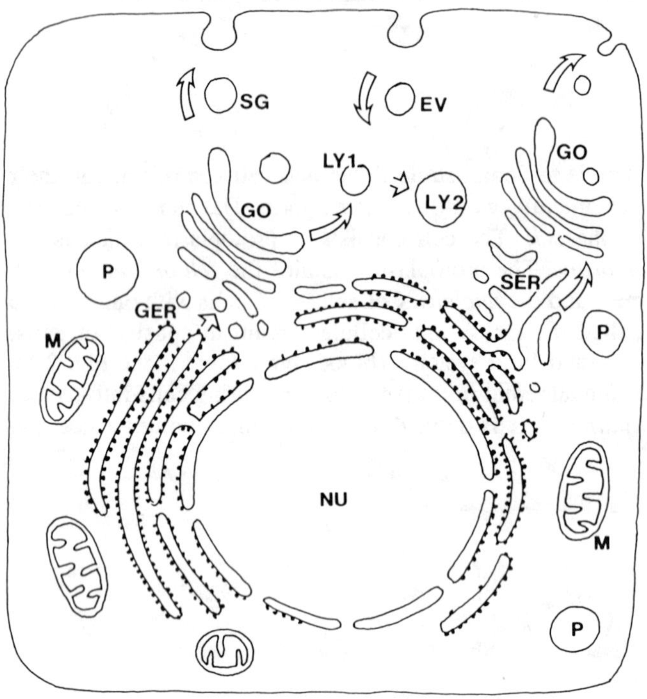

Fig. 2.2. Schematic drawing of a typical mammalian cell with a nucleus (NU), cell organelles, and a centre for the synthesis of glyco- and lipoproteins (granular endoplasmatic reticulum (GER); smooth endoplasmatic reticulum (SER); Golgi complex (GO); secretory granules (SG)), elements of the lysosomal system (endocytotic vacuole (EV); primary lysosome (LY1); secondary lysosome (LY2)), peroxisomes (P), and mitochondria (M).

2.1 Chemical Composition of Cells and Tissues

The chemical composition of cells is outlined in Table 2.1.

Table 2.1. Chemical composition of the cell in % w/w.

Inorganic	(%)	Organic	(%)
Water	75–80	proteins	10–20
		lipids	2–3
		carbohydrates	1–2
		nucleic acids	1–2
Other substances	1–10	other compounds	0.5–1

Although proteins, nucleic acids, carbohydrates, and lipids may occur as pure substances in tissues, they are more frequently found either as molecular complexes or mixed compounds (Table 2.2). For instance, nucleic acids + proteins = nucleoproteins.

Table 2.2. Associations between organic classes in tissue.

	nucleic acids	NUCLEOPROTEINS	
		GLYCOPROTEINS	
	carbohydrates		proteins
GLYCOLIPIDS		PROTEOGLYCANS	
	lipids	LIPOPROTEINS	

2.1.1 Water

Most of the cell is composed of water. It functions as a medium for diffusion and as a solvent for the reactive molecules and ions in the cell (Sect.3.2.1). Physiological processes take place predominantly in the aqueous environment, and water takes part directly in many enzymatic reactions (Chap.24). Up to 5% of the water is "bound" more or less strongly to the other components of the cell, particularly proteins.

Water is an integral part of the structural organization of cells. The formation and maintenance of cell membranes and protein conformation are mainly due to interactions between water and both polar and non-polar residues in macromolecules. This leads to a separation with polar residues remaining solvated in the aqueous phase, and non-polar residues of the protein or membrane becoming buried in the inner, non-aqueous phase. This process is called "hydrophobic stabilization" (Sect.4.5.5).

2.1.2 Other inorganic compounds

Within cells inorganic compounds are found as salts or as components of proteins, carbohydrates, nucleic acids, or lipids. In dissolved form, salts play a key role in regulatory processes such as acid-base and osmotic control. Metals and metal salts are discussed in Chapter 17.

2.1.3 Proteins

Proteins are high molecular weight compounds (macromolecules) which split into amino acids on hydrolysis. Distinction is made between the *primary* structure (sequence of amino acids), the *secondary* structure (geometrical arrangement of the polypeptide chain, e.g. α-helix, β-pleated sheet (Sect.21.7) or triple helix (Sect.2.4.1), *tertiary* structure (three-dimensional structure), and *quaternary* structure (the composition of different polypeptide subunits and possibly non-protein subunits).

A histochemical classification of the *simple* proteins is given in Table 2.3. Several important classes of *conjugated* proteins were listed in Table 2.2; haemoproteins (Sect.18.1.1), other metalloproteins, flavoproteins (Sects.25.1 and 25.4.1), and phosphoproteins (for instance non-histones) should be added to these.

Table 2.3. Classification of proteins found in cells.

Protein type	Examples
Fibre	keratins (also extracellular) myosin actin tubulin
Globular	globulins (also extracellular) globins (conjugated to haem) histones (conjugated to nucleic acids)

With the exception of the terminal amino and carboxyl groups, *the reactive groups* in proteins are represented by the side groups. A large proportion of the side groups are aliphatic hydrocarbons which can take part in the formation of hydrophobic bonds (e.g. with lipids). In the aqueous phase, these groups will tend to orientate themselves towards the inner part of the molecule thereby reducing contact with water to a minimum (cf. Fig. 12.1). In contrast, hydrophilic amino acid radicals are usually found on the outer surface of the protein. At neutral pH both positively (arginyl, lysyl, histidyl) and negatively (asparagyl, glutamyl) charged groups may be present. Further bonding possibilities include: hydrophobic interactions (tyrosyl, tryptophanyl, phenylalanyl), hydrogen bonding (tyrosyl, tryptophanyl, thiol), and covalent bonds (thiol).

2.1.4 Lipids

Lipids are defined as all naturally occurring fat, oil, and wax-like substances, which are soluble in chloroform, but insoluble, or colloid-forming, in water.

Histochemical Classification of Lipids. This has traditionally been distinct from the biochemical classification and remains problematic. Three slightly conflicting approaches have been used. Each recognizes two basic forms but the composition of the resulting categories does not always correspond. Approach I distinguishes between hydrophobic and hydrophilic (Adams, 1965, p.11), approach II recognizes homophasic and heterophasic lipids (Lison, 1960, p.455), while approach III distinguishes between storage lipids and structural lipids according to their distribution and function.

Hydrophobic lipids exist in the lipid phase only and give rise to a high surface tension lipid-water boundary. *Storage lipids* are thus *homophasic* and sharply delimited from the surroundings.

Lipids are classified as *hydrophilic* when, for example, one of the fatty acids in a triglyceride is replaced with a hydrophilic molecule. Such molecules of course still retain both hydrophobic and hydrophilic regions. This gives rise to a polarity which confers the properties of low surface tension, tendency to form micelles in water, and the possibility of organization into the double layers characteristic of biological membranes, hence the term *structural lipids*. In biomembranes the hydrophilic lipids associate with neighbouring molecules and form a complex called *heterophasic lipid*.

Occurrence. Some cell types (fat cells, sebaceous gland cells, cells in the adrenal cortex, and other steroid producing cells, Schwann cells) contain relatively large amounts of lipid. In consequence these tend to be well characterized in terms of composition and function.

Other cell types can show marked fluctuations in the amount of histochemically demonstrable lipid, either synchronous with natural changes in the functional condition of the cells or as a result of pathological changes (e.g. epithelial cells of the lactating mammary gland and the appearance of larger amounts of triglyceride droplets in liver cells in mammals).

Even in cells where lipid is not normally demonstrable using light microscopy, the amount of lipid is considerable (up to 10–20% of the dry weight). Most of this lipid (predominantly phospholipids and cholesterol) is located in the membrane systems of the cell. The rest (including triglycerides) is found dispersively bound to cellular proteins and can become histochemically demonstrable as a result of pathological changes. Sometimes it may be rendered visible by a histochemical demasking process.

2.1.5 Carbohydrates

Carbohydrates are composed of one or several hydroxyaldehydes or hydroxyketones or derivatives of these.

Monosaccharides. These are carbohydrates which cannot be split into lower molecular weight carbohydrates by simple hydrolysis. They are either aldoses, i.e. contain an aldehyde group, or ketoses, i.e. contain a ketone group. They are designated by the number of carbon atoms as trioses, tetroses, pentoses, hexoses, and heptoses. The most important monosaccharides are glucose, mannose, galactose, xylose, arabinose, fucose, ribose, and deoxyribose.

Derivatives of monosaccharides.

a. *Uronic acids* are derived from the simple monosaccharides by substituting the primary hydroxyl group with a carboxyl group. Examples are glucuronic acid, mannuronic acid, and iduronic acid.
b. *Deoxymonosaccharides* are derived by substituting a hydroxyl group with hydrogen. An example is fucose.
c. *Aminomonosaccharides* are derived by substituting a hydroxyl group with an amino group. Examples are glucosamine and galactosamine.
d. *Acetylaminomonosaccharides* are derived from aminomonosaccharides by ester formation between acetic acid and the amino group. An example is N-acetyl glucosamine.
e. *Sulphate esters of monosaccharides* are derived by forming sulphate esters between sulphuric acid and either a hydroxyl group or an amino group in the monosaccharide.
f. *Sialic acids* are a group of naturally occurring N- and O-acyl derivatives of neuraminic acid. Neuraminic acid is a carbohydrate derivative consisting of nine carbon atoms in addition to an amino group and a carboxyl group. The formula for N-acetylneuraminic acid is shown below.

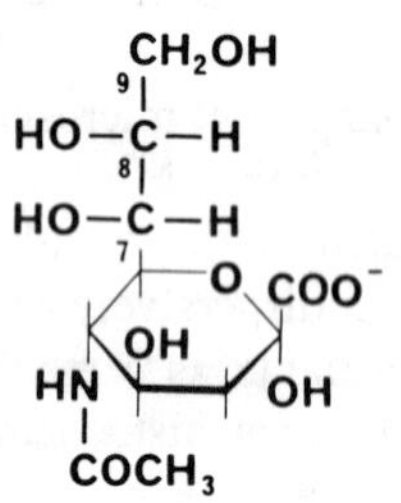

Polysaccharides. These are polymer compounds of monosaccharides and monosaccharide derivatives. They are subdivided into homopolysaccharides and heteropolysaccharides.

A. *Homopolysaccharides* comprise identical repeating monosaccharide or monosaccharide derivative units. Examples include glycogen and chitin which are respectively composed of glucose units bound together by α-1,4-glycoside bonds and N-acetyl-D-glucosamine units bound together by β-1,4-glycoside bonds.
B. *Heteropolysaccharides* or *glycosaminoglycans* (GAG) consist of long nonbranched chains of disaccharide units. The disaccharide component always includes an aminosaccharide. GAG are always charged (polyanionic) due to the presence of sulphate esters and/or uronic acids. Glycosaminoglycans can be *subdivided* as follows:
   1. Polycarboxylates (e.g. hyaluronic acid)
   2. Polycarboxysulphates (e.g. chondroitin-4-sulphate)
   3. Polysulphates (e.g. keratan sulphate)

Examples of acid heteropolysaccharides are given in Table 2.4.

Table 2.4. Examples of acid heteropolysaccharides.

Polysaccharides	Mol. wt (kdal)	Monosaccharides	Bonds in/between disaccharide units	
Hyaluronic acid	4–8,000	D-glucuronic acid, N-acetyl-D-glucosamine	β-1,3	β-1,4
Chondroitin-4-sulphate	5–50	D-glucuronic acid, acetyl-D-galactosamine-4-sulphate	β-1,3	β-1,4
Chondroitin-6-sulphate	5–50	D-glucuronic acid, N-acetyl-D-galactosamine-6-sulphate	β-1,3	β-1,4
Dermatan sulphate	15–40	L-iduronic acid or D-glucuronic acid, N-acetyl-D-galactosamine-4-sulphate	α-1,3	β-1,4
Keratan sulphate	4–19	D-galactose, N-acetyl-D-glucosamine-6-sulphate	β-1,4	β-1,3
Heparan sulphate	5–12	D-glucuronic acid or L-iduronic acid, D-glucosamine-(N-sulphate)-6-sulphate	α-1,4	α-1,4
Heparin	6–25	D-glucuronic acid-2-sulphate or L-iduronic acid-2-sulphate, D-glucosamine-(N-sulphate)-6-sulphate	α-1,4	α-1,4

In biological systems polysaccharides are usually *bound to proteins*. Such protein-carbohydrate complexes are collectively called *mucosubstances*. These compounds are composed of a single polypeptide chain with carbohydrate chains covalently attached to specific side groups. On the basis of their carbohydrate content, mucosubstances are subdivided into 1) *glycoproteins* and 2) *proteoglycans*. In general the protein component dominates in glycoproteins, and the side chains consist of short, frequently branched carbohydrate chains (oligosaccharides). In contrast, the heteropolysaccharide component dominates in proteoglycans (Table 2.5).

The following carbohydrates can be demonstrated histochemically (Chap.22):

I Homoglycans = homopolysaccharides

II Mucosubstances
   1. Proteoglycans
   2. Glycoproteins
      a) sialomucins
      b) sulphomucins
      c) neutral glycoproteins

2.1.6 Nucleic Acids

A *nucleoside* is a pentose (ribose or deoxyribose) to which a purine or pyrimidine base is bound through a N-C1-glycosidic bond, while a *nucleotide* is a phosphoric acid-C5 ester of a nucleoside. *Nucleic acids* are chains of nucleotides bound together through phosphoric acid-C3 ester bonds. If the pentose is deoxyribose,

Table 2.5. Characteristics of glycoproteins and proteoglycans (modified from Pedersen, 1982, p. 251).

Characteristics	Glycoproteins	Proteoglycans
Molecular weight (kdal)	10–1,000	1,000–10,000
Carbohydrate in % w/w of the molecule	1–60	90–95
Number of carbohydrate chains per protein chain	1–50	100–200
Binding between carbohydrate and protein takes place by:		
N-glycoside bond to asparagine	frequent	does not occur
O-glycoside bond to serine	infrequent	frequent
O-glycoside bond to threonine, hydroxyproline or hydroxylysine	less frequent	does not occur
Carbohydrate chains are	often branched	linear
Carbohydrate chains are	short	long
Carbohydrate chains have	no simple monomer sequence	a simple monomer sequence
The carbohydrates contain:		
fucose	less frequently	infrequently
sialic acid	frequently	infrequently
uronic acid	infrequently	frequently
sulphate	less frequently	frequently

deoxyribonucleic acid, DNA, is formed, while, if the pentose is ribose, ribonucleic acid, RNA, is formed.

- *DNA:* deoxyribonucleic acid is a double-stranded molecule forming a double helix held together by hydrogen bonds between the bases. In eukaryotes, DNA is found in the nuclei where it is bound to basic proteins (histones), and in mitochondria.
- *RNA:* ribonucleic acid is predominantly single-stranded. It is transcribed from DNA in the nucleus and transported to the cytoplasm.
- *m-RNA:* messenger RNA, in general, constitutes well under 5% of total cellular RNA and shows a wide range of molecular weights.
- *t-RNA:* transfer RNA constitutes 10–15% of the total cellular RNA, and the molecular weight is approximately 25 kDa.
- *r-RNA:* ribosomal RNA. The molecular weight is between 2,000 and 2,500 kDa. It is found as ribonucleoprotein and ribosomal subunits in the nucleolus and as a substantial constituent (about 40% of the dry mass) of the ribosomes in the cytoplasm.

Ribosomal RNA constitutes at least 80% of the total cellular RNA (nucleoli and cytoplasm) and is the predominant form demonstrated by histochemical methods.

2.1.7 Pigments and Biogenic Amines

Pigments are defined as a group of generally water-insoluble substances which appear coloured or black in unstained tissue sections. Chemically the group is very heterogeneous. Biogenic amines such as serotonin (5-hydroxytryptamine, 5-HT) and the catecholamines adrenaline, noradrenaline, and dopamine, are not pigments in their native form, but react with both formaldehyde and chromate to give chromogenic compounds which have some properties in common with pigments. Discussion of pigments in this text has generally been limited to those found in humans (Table 2.6 and Chap. 18).

Table 2.6 outlines the *classification* of "pigments" and distinguishes between exogenic and endogenic to designate pigments arising outside or within the organism, respectively.

Table 2.6. Classification of "pigments".

I Endogenic
- 1. Haematogenic
  - a. haem
  - b. porphyrins
  - c. haemosiderin
  - d. acid haematins
  - e. bile pigments
  - f. Dubin-Johnson pigment
- 2. Lipofuscins
- 3. Melanins
- 4. Monoamine derived pigments

II Exogenic
- 1. Carotenes
- 2. Carbon
- 3. Certain metal compounds (see Chapters 17 and 18)

2.2 Structure and Function of the Eukaryotic Cell

The eukaryotic cell consists of nucleus and cytoplasm and is delimited by a plasma membrane.

The *nucleus* consists of an apparently amorphous *nucleoplasm* or karyolymph which contains the *chromosomes* and generally one or more *nucleoli*. The chromosomes are usually to a high degree despiralized to an apparently unorganized jumble of chromatin fibres.

The *cytoplasm* contains all the cell organelles (mitochondria, endoplasmic reticulum, Golgi complex, lysosomes, peroxisomes, etc.). These are embedded in the cytoplasmic *matrix* which in addition contains formed elements such as actin filaments, microtubules, etc. Lipid droplets and glycogen granules may also be present. The matrix contains a very large number of different soluble proteins including the enzymes of glycolysis and most of the enzymes and other proteins which take part in protein synthesis. The soluble phase of the cytoplasmic matrix, the *cytosol*, functions as a medium of diffusion or transport for low-molecular soluble substances in the cell.

2.2.1 The Cell Nucleus

In eukaryotic cells (i.e. all cells except bacteria and blue-green algae which are prokaryotes) the genetic material, deoxyribonucleic acid (DNA), is present in a membrane-bound nucleus in the form of chromosomes. One or more nucleoli are present in the nucleus. The nucleolus is essentially a factory for ribosomes, which are later released to the cytoplasm. The nucleus is delimited by the nuclear envelope which consists of an outer and an inner membrane (Sect.2.2.3) separated by the perinuclear space. The outer and inner membrane are connected by a number of nuclear pores where the exchange of materials between cytoplasm and nucleoplasm takes place.

Chromatin. The DNA-containing genetic material in the cell nucleus is called chromatin. Using the light microscope, in most interphase nuclei it is only possible to distinguish between heterochromatin and euchromatin. The *heterochromatin* or condensed chromatin contains all the genes which are inactive in the given cell type. The heterochromatin pattern is thus characteristic for that cell line and, the more heterochromatin, the more restricted the diversity of proteins synthesized. The *euchromatin*, or extended chromatin, is active chromatin. Although it cannot be seen using the light microscope, its presence may be inferred from the relative paucity of heterochromatin in some cell types (e.g. nerve cells). In general, a high activity of gene transcription entails a relatively large nuclear volume and large nucleoli, while a low transcriptional activity entails a relatively small nuclear volume, confluency of heterochromatin granules, and small or virtually absent nucleoli. In some interphase nuclei—and always during nuclear division—it is possible to see the 10–50 nm chromatin fibres condensed into *chromosomes*.

A chromosome is predominantly an aggregate of DNA and histones containing varying amounts and types of non-histone proteins as well as RNA transcripts.

The relative amounts of DNA and histones in chromatin are on a weight basis in the ratio 1:1. The amount of non-histone varies according to cell type and the state of the cell, but is usually found on a weight basis in ratios to DNA of 1:0.5 to 1:1.5. The amount of RNA is small and normally comprises a maximum of around 5% w/w relative to the DNA.

The Molecular Organization of Chromatin. Chromatin consists of DNA in the double helix form and a number of different proteins. Amongst these, the *histones*, which contain a high proportion (up to 25%) of lysine or arginine, are basic in character and are very intimately associated with DNA. The histones have been designated H1, H2A, H2B, H3, and H4. The last four of these interact with each other and with DNA to form the nucleosome fibre, which in turn interacts with histone H1 to form a more compact chromatin fibre.

2.2.2 Ribosomes

Protein synthesis takes place on ribosomes. t-RNA molecules with their associated amino acids recognize complementary nucleotide triplets on m-RNA bound to the ribosome. This process collects amino acids in the correct sequence on the m-RNA. Ribosomes contain the synthetic machinery that catalyzes formation of the polypeptide chain.

Ribosomes are approximately 20 nm in diameter and consist of roughly equal amounts of RNA and protein. Synthesis of proteins can take place on free cytoplasmic ribosomes e.g. haemoglobin synthesis in erythroblasts, or on ribosomes situated on the cytoplasmic surface of the *endoplasmic reticulum* (rough or granular ER, cf. Sect.2.2.4). The polypeptide component of all lysosomal enzymes, secretory products (glandular secretions and certain hormones) as well as most integral membrane proteins (see below) are all formed in the rough ER.

Cells that are very active in protein synthesis have a high content of ribosomes and t-RNA. The nuclear and nucleolar size and the amount of cytoplasmic RNA are also important indices of protein synthesis. The relationships between these indices and the amount of stored protein product may indicate whether a cell is preparing for, actively engaged in, or leaving the synthetic phase of a secretory cycle.

2.2.3 Cellular Membranes

The membranes of all eukaryotic cells have the same basic organization regardless of whether they are plasma membranes (plasmalemma, the membrane surrounding the cell), or membranes of intracellular compartments such as the ER, lysosomes, and mitochondria.

A membrane consists of a continuous 6–9 nm thick *double layer of lipid molecules* orientated with their polar, hydrophilic ends facing outwards. The interior of the membrane is therefore hydrophobic. The lipids are chiefly phosphoglycerides, but also include phosphosphingosides, glycolipids, and sterols, such as cholesterol. While the pattern remains constant for individual membrane types within one species, the relative lipid composition often shows variation between membrane types within species and for the same membrane type between species. Generally, sphingo- and glycolipids and sterols are abundant in plasma membranes (cf. myelin).

Peripheral and *integral proteins* are found in intimate association with the lipid bilayer. The peripheral proteins form no direct links with the hydrophobic part of the membrane, while the integral proteins penetrate this region to a greater or lesser degree. Integral proteins contain one or more non-polar (hydrophobic) amino acid sequences which permit this association to take place.

Membrane proteins of the plasma membrane, and to a lesser extent those associated with other cellular membranes, are *glycosylated*, i.e. they are linked to short side chains of sugars. The carbohydrate portion is outside the membrane proper, and for the plasma membrane, where it forms the cell coat (or glycocalyx), the carbohydrate chains most often terminate with galactose or sialic acid. The plasma membrane is a selectively permeable barrier and the integral proteins play a central role in this function in the form of transport systems such as the Na/K-ATPase (the "Na/K pump"). In addition, membrane proteins are important as antigenic sites (e.g. blood and HLA type determinants) and receptors.

2.2.4 Endoplasmic Reticulum

The plasma membrane, as well as the membranes of secretory granules, lysosomes, and the Golgi complex, all arise from the membranes of the endoplasmic reticulum (ER).

Granular or Rough Endoplasmic Reticiculum. In this ribosomes are attached to the cytoplasmic side of the membrane, and protein synthesis on these ribosomes allows the insertion of integral proteins into the ER membrane and the release of soluble lysosomal or secretory proteins into their appropriate compartments. While the integral proteins destined for the plasma membrane generally do not become functional until they have become fully glycosylated in the Golgi complex (cf. alkaline phosphatase), it is often possible to demonstrate lysosomal enzymes (e.g. acid phosphatase) or secretory proteins (e.g. IgG), while they are still located in the ER. Proteins that remain in the ER itself (e.g. glucose-6-phosphatase, cytochrome P-450) are functional immediately after insertion.

Agranular or Smooth Endoplasmic Reticulum. This is a collective term used by electron microscopists for a number of "cisternal" structures, whose membranes have no attached ribosomes. Some cell types, i.e. liver and steroid hormone synthesizing cells, have a structurally and functionally well-defined agranular ER, which is derived from the granular ER. The smooth ER is involved in lipid synthesis, glycogen metabolism (glucose-6-phosphatase) and detoxification processes. Various kinds of poisoning (e.g. drugs, aromatic hydrocarbons) induce the membrane-bound "mixed function oxidase system" or cytochrome P-450 system leading to an increased specific protein synthesis in the granular ER followed by enormous proliferation of the smooth ER membrane system. This may often be followed light microscopically as an increase in the cytochrome P-450 associated NADPH dehydrogenase activity.

2.2.5 The Golgi Complex

Proteins destined for secretion, for the plasma membrane, for lysosomes, or for the Golgi complex itself, all are delivered to the Golgi complex from the ER. The Golgi complex appears in the electron microscope as a stack of 4–8 cisternae. In the light microscope the complex may sometimes be demonstrated by reactions for enzymes (e.g. alkaline phosphatase, thiamine pyrophosphatase) or glycosylation products (PAS reaction). After histological staining procedures the location of the complex may appear as an unstained or only slightly stained region due to the absence of RNA and the relatively low protein concentration in the organelle.

When newly synthesized polypeptides reach the Golgi region they are processed in a highly specific and vectorial manner through the Golgi stack. This processing includes modifications to the existing glycosylation pattern and further glycosylations with galactose and sialic acid or derivatives of these. The Golgi complex is not only responsible for elaborating the final secretory glycoproteins and mucins, but also for the synthesis of heteroglycans (i.e. hyaluronic acid, chondroitin sulphates).

The finished glycosylated products are either specifically selected for lysosomes (by a mannose-6-phosphate receptor system) or packaged into secretory granules or vesicles and transported to the cell surface. Whether they end up integrated in the plasma membrane or are released from the cell depends on whether or not they have retained an "anchor" in the membrane of the transport vesicle.

2.2.6 Peroxisomes

These organelles vary in size and shape and contain the enzyme catalase (the marker enzyme for cytochemical demonstration of peroxisomes) and one or more H_2O_2-producing oxidases. The function(s) of the peroxisomes is still somewhat unclear, but they are probably involved in lipid metabolism and, like the smooth ER, detoxification (e.g. oxidation of ethanol to acetaldehyde).

2.2.7 Mitochondria

Mitochondria are found in all eukaryotic cells that can utilize O_2, and in most cells they are the main site of ATP production.

Although they are highly pleiomorphic, mitochondria are generally sausage or thread shaped (0.2–0.6 μm wide and one or several μm long). They consist of an outer membrane, which resembles other membranes, and an inner membrane containing more protein than lipid. This inner membrane is folded into the mitochondrial matrix, forming the mitochondrial cristae. Among the many proteins of this inner membrane are the enzymes of the respiratory chain and the ATP synthetase (mitochondrial "ATPase"). Many of the inner membrane enzymes, for instance succinate dehydrogenase and cytochrome oxidase can be histochemically demonstrated (cf. Chap.25). The mitochondrial matrix contains a wide variety of enzymes from the tricarboxylic acid cycle (Krebs cycle or citric acid cycle), fatty acid oxidation, amino acid metabolism, etc. Many of these enzymes are of physiological importance and exhibit a tissue-specific distribution. For instance glutamate dehydrogenase, which is responsible for ammonia formation from amino acids (glutamic acid) has a much higher (histochemical) activity in mammalian liver and intestine than in most other tissues.

Depending on cell type, the number of cristae formed by the mitochondrial inner membrane and the amount of matrix may be more or less abundant. Cells with relatively large ATP requirements (e.g. certain muscle cells, cells of convoluted tubules of the kidney, salt gland cells) have numerous cristae and relatively sparse matrix, while cells which are more involved in intermediary metabolism (e.g. liver cells) may have relatively few cristae, but abundant matrix. This morphological variety reflects differences in the relative and absolute amounts of various mitochondrial proteins and enzymes, which are often detectable by enzyme histochemistry.

Mitochondria are frequently distributed throughout the cytoplasm (e.g. liver cells). However, in some transporting epithelial cells, which show a marked polarity (e.g. tubule cells of the kidney), mitochondria may show a regional localization, typically close to the basal plasma membrane, where the ATP-consuming Na/K-ATP-ase is localized. In heart muscle and cross-striated muscle, mitochondria lie in close proximity to the myofibrils, occasionally forming almost geometrical arrangements (e.g. insect flight muscle).

2.2.8 Endosomes and Lysosomes

In Sects.2.2.2 and 2.2.5 the outward protein traffic of the cell has been outlined, however, cells can also internalize proteins. The process is referred to as *endocytosis* and involves invagination of an area of the plasma membrane which is finally "pinched off" to form an *endocytic vesicle.*

Endocytosis is divided into *phagocytosis* (e.g. a white blood cell ingesting a bacterium) and *pinocytosis* (uptake of cellular solutes). Pinocytosis may be either fluid-phase uptake or receptor-mediated uptake (an adsorptive uptake mechanism, where the molecules in question are bound with a degree of specificity to receptors on the cell surface prior to internalization). Following endocytosis endocytic vesicles fuse to form the *endosome system,* which comprises larger vacuoles as well as smaller vesicles and tubules. These structures continuously undergo the processes of mutual fusion and formation by "pinching off" and may therefore be considered as an interconnected compartment. The endosome system is somehow responsible for the sorting of internalized molecules with respect to their next destination (the cell surface in the case of receptor cycling; lysosomes for substances to be degraded or receptors and ligands to be down-regulated; etc.). Histochemically endosomes may be distinguished from lysosomes by incubating cells with a "marker" of endocytosis at low temperature (18°C in mammals). At this temperature endocytic vesicles do not fuse with lysosomes. These structures may also be differentiated by histochemical demonstration of lysosomal markers such as acid phosphatase.

The endocytotic path involves two main compartments: first endosomes, then lysosomes. A vesicle containing both endocytosed material and lysosomal enzymes is often called a *secondary* lysosome (in contrast to the *primary* lysosome, which is derived directly from the Golgi complex). The secondary lysosomes essentially form a recycling system intermittently receiving new material from endosomes and newly synthesized hydrolytic enzymes from primary lysosomes. How the fusion with primary lysosomes is regulated is as yet unknown.

Secondary lysosomes which are no longer enzymatically active often contain an indigestible residue and are called residual bodies. These may have the character of pigment (e.g. lipofuscin) which can be demonstrated with special cytochemical methods (Sect.18.2.2).

Endocytosis – intracellular digestion – is important from both a basic biological and a cytochemical point of view, since many macromolecules internalized by cells can be detected by enzyme- or immuno-histochemical techniques. Moreover, currently available techniques enable a highly detailed description of the functions and the malfunctions of lysosomes (cf. storage diseases, Sects.31.10.8 and 31.11.2).

The processes described above are called *heterophagy*, literally, "eating" of material "different" from the cell. *Autophagy*, lysosomal degradation of some of the cell's own organelles, also occurs. This may take place as part of the restructuring of a cell, as a result of a cell injury, as an integrated part of morphogenesis (programmed cell death or apoptosis), or sometimes as a response to starvation.

2.2.9 Intracellular Fibre Proteins

The cytoplasmic matrix contains several systems of fibre proteins, which are important in the maintenance and restructuring of cell morphology, intracellular transport, and in cell motility. The intracellular fibre proteins comprise the "cytoskeleton" of the cells. The various elements of the cytoskeleton can be demonstrated using the electron microscope. The most important fibre proteins are tubulins, actins, and the proteins forming the intermediate filaments (e.g. keratins). These fibre proteins constitute a very substantial part of the protein of the cell, actin alone up to more than 10%.

Tubulins. These are globular proteins which associate to form tube-like fibre structures, *microtubules*, which may attain a length of several μm. Microtubules are important in the internal organization of the cell and in establishing "tracks" for the intracellular movement of organelles and granules. The actual movement depends on the interaction between microtubules, actin filaments, and "motor" proteins. The microtubules play a similar role in chromosome movement in nuclear division. In cilia microtubules form the so-called *axoneme* structure, with 9 microtubule doublets surrounding two single microtubules. The presence of the "motor" protein *dynein* on the doublets ("dynein arms") enables these to slide relative to each other leading to movement of the cilium.

Actins. These occur partly in the form of a free globular protein, *G-actin*, partly as a filament protein, *F-actin*, which is a polymer of G-actin in a double helical arrangement. There is a dynamic equilibrium between the two states of actin, in the same way as for free tubulin and microtubules. F-actin comprises the greater part of the thin filaments in the sarcomeres in striated muscle and in smooth muscle cells. Actin occurs in practically all cell types, e.g. epithelial cells and fibroblasts, although it may be of a different subtype (as judged immunocytochemically). In motile cells, and in cultured cells actin is frequently arranged in bundles, the so-called stress fibres, immediately beneath the plasma membrane. Large amounts of F-actin filaments also occur in microvilli where they are arranged parallel to the long axis.

Myosin. This is a large protein molecule which, especially in muscle cells, forms filaments, and which can be specifically demonstrated due to its ATPase-activity (Sects.24.6.4, 31.11.6). Actin and myosin interact in cell movement and in muscle contraction, forming actomyosin.

Intermediate Filaments. Intermediate filaments (IF) are a major fibrous component of the cytoskeleton of almost all cells (except in the early embryo) and have a diameter intermediate between microtubules and F-actin (hence the name "intermediate").

Intermediate filaments fall into several types, each one characteristic for one of the major cell types of the body. *Cytokeratins* are found in both keratinizing and nonkeratinizing epithelia, *neurofilaments* are found in most neurones, *glial fibrillary acidic protein* (GFAP) is found in some types of glial cells, *desmine* predominantly in muscle cells, and *vimentin* in fibroblasts, macrophages, endothelial cells, etc. Most work on intermediate filaments has been done by use of biochemistry and immunohistochemical techniques, as IF are not amenable to histochemical analysis. In fact, except for neurofilaments and cytokeratins ("keratin"), IF are barely detectable by histochemical or histological methods. Only (cyto)keratins will be discussed further because of their role in keratinization and in the formation of desmosomes.

Keratins. Keratins (horn) comprise a group of mechanically and chemically highly resistant fibre proteins found in keratinizing epithelia. They are closely related to the cytokeratins. Keratins occur partly in form of a precursor, *prekeratin,* in keratinocytes. These cells may finally change into dead, cornified cells as a result of a keratinization process (cornification), i.e. a terminal differentiation process which ends in the death of the cell. Keratins are especially found in epidermal cells and derivatives of these, such as nails, horn, and hair.

In epithelial cells which do not cornify, *cytokeratin* is found. Both prekeratin and cytokeratin filaments are frequently termed *tonofilaments*. The tonofilaments "stiffen" the cells and are often connected to *desmosomes* (adhesion structures between neighbouring epithelial cells) and *hemidesmosomes* (adhesion structures between the base of an epithelial cell and the basal lamina).

2.3 The Prokaryotic Cell

This section deals with the structure and chemical composition of bacteria. Only sufficient details to understand the background for the mechanisms of the Gram and Ziehl-Neelsen staining methods will be given (Sect.6.1.6). Bacteria (Fig. 2.3) are prokaryotic cells, i.e. cells without a nucleus, a nuclear membrane, or a mitotic apparatus. The essential DNA of bacteria is found in a single circular molecule (chromosome). Additional, often pathogenicity and antibiotic resistance related functions are coded for on extrachromosomal circular DNA molecules of varying size. These may be present in multiple copies and are known as plasmids. While these structures are not demonstrable histochemically, our recently developed ability to manipulate both eukaryotic and prokaryotic nucleic acids is heavily dependent on plasmids. This technology forms the basis for the development of *in situ nucleic acid hybridization* techniques which can be used to detect and demonstrate genes or smaller base sequences at the cellular level (cf. Sect.20.6).

The bacterial cytoplasm may also contain granules consisting of neutral polymers such as starch, glycogen, or polyphosphate (volutin). There are no mitochondria.

Bacteria have both a plasma membrane and a thick cell wall (Fig. 2.4). The plasma membrane is structurally and functionally very similar to that found in eukaryotic cells. The active transport processes it supports contribute to the development of an inner osmotic pressure well in excess of that found in most fluid environments (between 500 and 2,000 kPa). This would cause bacteria to burst if they were not surrounded by a cell wall with considerable mechanical strength (Fig. 2.3). This strength is due to a large complex polymer, known as peptidoglycan or murein.

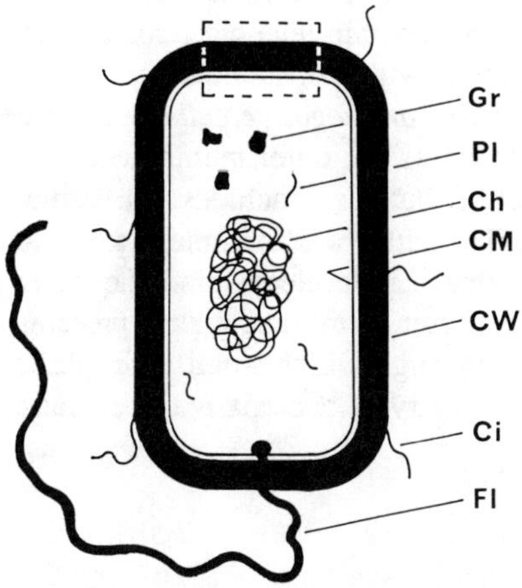

Fig. 2.3. Diagram of a bacterium with granule (Gr), plasmid (Pl), chromosome (Ch), cell membrane (CM), cell wall (CW), cilium (Ci) and flagellum (Fl). The dotted rectangle is expanded in Fig. 2.4.

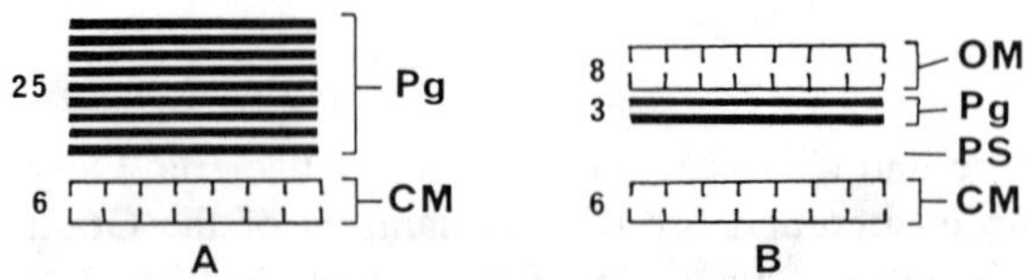

Fig. 2.4. Diagram of bacterial cell wall in Gram-positive (A) and Gram-negative bacteria (B). Peptidoglycan (Pg), cellular membrane (CM), outer membrane (OM), periplasmatic space (PS). The figures indicate the thickness of the layers in nm.

2.3.1 Gram-Positive and Gram-Negative Bacteria

Development and application of the Gram staining method in the latter part of the 19th century fortuitously revealed fundamental properties of bacteria that are valuable both for classification and identification. The difference in staining between Gram positive and Gram negative organisms appear to reflect differences in cell wall structure revealed latterly by electron microscopy.

The peptidoglycan layer is a polymer found in all bacteria (with the exception of mycoplasmas) and consists of alternating molecules of N-acetyl glucosamine (NAG) and N-acetyl muramic acid (NAM) with tetrapeptide side chains (which include D-amino acids) attached to the muramic acid residues. The side chains are bonded together by a further pentapeptide bridge, effectively linking the NAM-NAG polymer. The peptidoglycan layer is thus a single giant molecule. Gram-positive bacteria have a thick peptidoglycan layer which comprises up to 90% of the total cell wall whereas Gram-negative bacteria have a thin layer amounting to between 5 and 20% of the cell wall (Fig. 2.4).

Both groups of bacteria may express proteinaceous and polysaccharide material on their surfaces. In addition *Gram-positive cell walls* contain teichoic acids (Greek *teichos* = wall) (glycerol polymer bonded together with phosphodiester bonds) which form a part of the surface antigen structure. *Gram-negative cell walls* have an additional layer outside the peptidoglycan known as the outer membrane. This layer retains the standard unit membrane structure but also includes substantial quantities of lipopolysaccharide (see Fig. 2.5). The latter are extremely toxic to higher organisms and are referred to as endotoxins. Lipoproteins bind the outer membrane down to the peptidoglycan layer. In the outer membrane matrix proteins are arranged in groups so that they form "pores" through which small hydrophilic molecules may pass. A good survey of the biochemistry of bacteria may be found in Jawetz *et al.* (1989).

2.3.2 Mycobacteria

Mycobacteria are Gram-positive rods which show the property known as acid fastness, i.e. they can withstand decolourization with acid after staining with an

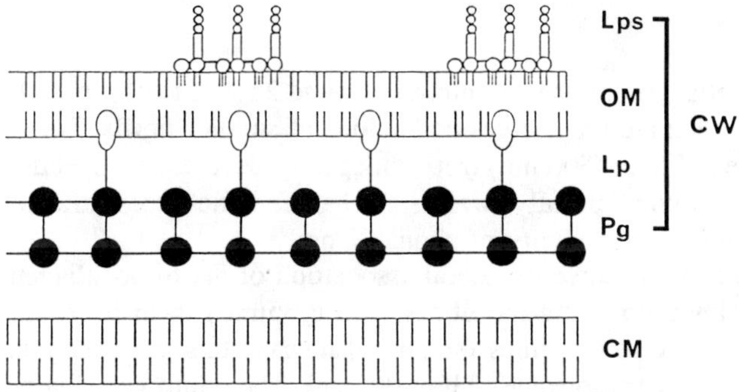

Fig. 2.5. Diagram illustrating the molecular structure of the cell wall (CW) and the cellular membrane (CM) in Gram-negative bacterium. The cell wall is composed of lipopolysaccharide (Lps), outer membrane (OM), lipoprotein (Lp), and peptidoglycan (Pg).

arylmethane dye. Mycobacteria cause chronic diseases such as tuberculosis and leprosy.

The cell wall in mycobacteria is very complex and contains up to 60% lipid. Closest to the cytoplasm one finds murein (a peptidoglycan) just as in other Gram-positive bacteria. The lipids consist of mycolic acids, glycolipids, waxes, mycosides, and phospholipids. The acid fastness is related to the high lipid content and is gradually reduced by differential extraction of the lipids. Bacteria related to mycobacteria also contain mycolic acids and show lesser degrees of acid-fastness (e.g. Nocardia and Corynebacteria).

2.4 The Composition of the Extracellular Matrix

The majority of cells in multicellular organisms are in contact with a complicated network of macromolecules which is permeated with extracellular fluid containing dissolved inorganic and organic compounds. This network forms the extracellular matrix. In addition to functioning as a sort of biological "glue" the matrix plays an active role in the regulation of the adjoining cells by influencing their development, migration, proliferation, form, and metabolic function. In vertebrates the extracellular macromolecules are predominantly proteins (often glycoproteins) and proteoglycans. The macromolecules in the connective tissue are secreted mainly by fibroblasts which are found widespread in the matrix. In Table 2.8 and below a number of important matrix proteins and proteoglycans are reviewed.

2.4.1 Collagens

Collagens are the most abundant proteins in mammals (around 25% of total protein). They usually form fibres which are themselves collections of smaller fibrils. These in turn consist of aggregates of long (300 nm) tropocollagen molecules in a specific arrangement. Electron microscopy usually reveals a characteristic cross striated pattern in collagen fibrils with a periodicity of about 67 nm.

The periodicity results from a staggered lateral association of the tropocollagen molecules. Molecules are lined up along the fibrils with a 35nm gap in between. As the molecular length is close to 4.4 times the period interval this gap accounts for the actual ratio of 5 periods per molecule (335nm/67nm). The 35nm gap is very accessible to the metal stains used in negative staining procedures.

The tropocollagen molecule consists of three polypeptide chains (α-chains) with a unique primary structure. Every third amino acid in the sequence is glycine and the molecule has an unusually high content of proline and hydroxyproline. On the basis of the primary structure, seven different α-chains (see Table 2.8), each with the sequence gly-X-Y repeated 300–350 times, can be distinguished. Due to steric restrictions from proline + hydroxyproline each of the three α-chains forms an extended left-hand helix. In turn, these helices wind around each other to form the greatly extended right-hand triple helix of the tropocollagen molecule. The intimate contact between the three helices is made possible because the glycine residues are always nearest to the centre of the triple helix. This conformation allows the maximum number of intra- and interchain hydrogen bonds, with chain flexibility highly restricted by the proline and hydroxyproline residues.

Collagen structure is further stabilized by covalent cross-links between α-chains and between neighbouring tropocollagen molecules. These cross-links are primarily between lysyl and hydroxylysyl residues. As shown in Table 2.8, it is possible to distinguish between at least five different collagen types depending on which of the seven α-chains take part in the formation of the molecule. The different collagen types contain carbohydrate in highly varying amounts. This carbohydrate, which may be either galactose or glucosylgalactose, is bound by O-glycosidic bonds to hydroxylysyl.

2.4.2 Reticulin

The term reticulin was originally used in relation to reticular fibres in the same way as collagen is used in relation to collagen fibres, i.e. as a term for the protein believed to be characteristic for the fibres. Later, reticulin became a designation for the PAS-positive material in reticular fibrils and basal membranes.

It is now well established that *reticular fibrils* are composed of type III collagen. These fibrils are far thinner and contain more carbohydrate than type I collagen fibrils. These two features are responsible for the histochemical differences between the two fibril types. Characteristically, reticular fibrils are stained black with oxidative argyrophil methods (Sect.8.3.2), red with PAS (Sect.9.2.1), and yellow with

Picrofuchsin (Sect.21.6). In contrast, type I collagen fibrils stain yellow-orange-brown with silver, pink with PAS, and red with Picrofuchsin (Sect.21.6). Specific antibodies directed against the different collagen types are now available so that immunohistochemical identification is possible (Sect.32.5.7).

2.4.3 Basal Membranes

Basal membranes are structures which are found in the boundary between the connective tissue matrix and the adjoining epithelial or endothelial cells (Fig. 2.1), and around muscle cells and fat cells. In addition basal membranes are found between the two cell layers in the glomeruli of the kidney. Light microscopic demonstration can be achieved using the same methods as those applied to reticulin and collagen fibrils (Sect.2.4.2). Examination by electron microscopy reveals an homogeneous, light area called the *lamina lucida* directly basal to the *plasma membrane* of the epithelial or endothelial cell. This consists predominantly of the peripheral portion of the cell surface glycoproteins. Under this lies the *lamina densa* (basal lamina) which is 20–80nm thick (about 300nm in glomeruli). This comprises a network of thin filaments in a finely granular ground substance. Below the basal lamina, reticular fibrils are frequently observed forming a network (reticular lamina) and continuing into the underlying connective tissue. The basal lamina is composed of types IV and V collagen as well as other elements (e.g. laminin and a heparan sulphate proteoglycan).

2.4.4 Elastin

Elastin (Table 2.8) is one of the two components in *elastic fibrils*. The other component consists of "*elastic microfibrils*".

Elastin has a high content of the non-polar amino acids alanine, valine, leucine, and isoleucine, and a considerable amount of tyrosine. The polypeptide chains show random coil configuration (no defined secondary structure) and are bound together by cross links in an *unorganized three-dimensional network*. The cross links in elastin are characteristic and arise between four lysyl groups from four different peptides. The cross links can be isolated from elastin and consist of the for elastin specific amino acids, desmosine and isodesmosine.

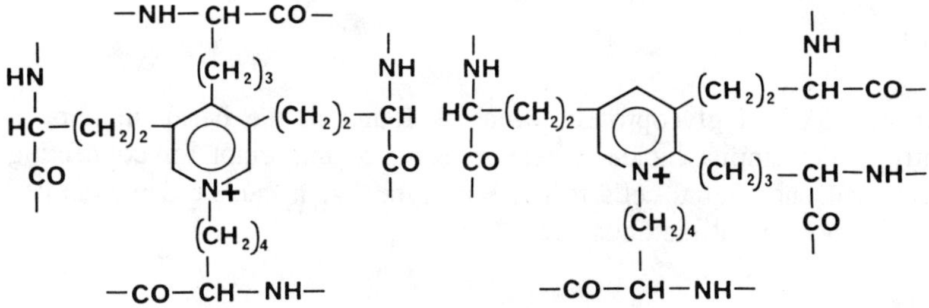

desmosine isodesmosine

Note the central pyridinium ring. The "unorganized" arrangement of elastin is responsible for its elastic properties and also for its autofluorescence. Histochemically, elastic fibres can be demonstrated by their autofluorescence (Sect.30.1) and by a number of dyes (Sect.21.5) which are bound by van der Waal forces (Sect.4.5.4).

The *"elastic" microfibrils* form a more or less continuous layer on the surface of the elastic fibril. They are composed of a glycoprotein with a high content of cystein/cystine and proline, but no hydroxyproline. In oxidative staining methods for elastic fibres the formation of sulphonic acid from cystein/cystine probably plays a role in subsequent dye-binding.

2.4.5 Fibronectin

Fibronectin (Table 2.8) is a glycoprotein which usually occurs as a dimer. As the side groups make cross-linking possible between subunits, a multimer form is also seen. The side groups also provide opportunities for links between fibronectin and other substances. These properties make fibronectin well-suited as an adhesive. For instance, it binds to collagens, fibrin, and proteoglycans. It has been suggested that fibronectin is organized in specific functional domains each of which have particular binding preferences (e.g. to cell surface gangliosides).

Fibronectin occurs in two forms.

1. An *insoluble* form which occurs in close contact with cell surfaces (*pericellular*) and in *matrix* associated with the basal lamina and loose connective tissue.
2. A *soluble* form found in plasma.

Fibronectin takes part in the formation of granulation tissue (an early event in the healing process). Initially fibronectin probably comes from plasma and it appears bound to fibrin. A loose matrix is formed and this serves as a substrate for the migrating fibroblasts and endothelial cells which follow. At first the fibroblasts produce fibronectin, somewhat later, collagen type III, and finally, when the formation of true scar tissue takes place, collagen type I. Analogous processes occur during embryonic development.

Fibronectin can be demonstrated by specific immunohistochemical methods (Sect.32.5.7).

2.4.6 Laminin

Laminin (Table 2.8) is a glycoprotein which is found in the basal membrane, predominantly in the lamina lucida, where it seems to important for connecting epithelial cells and endothelial cells to collagen type IV. It can be demonstrated using immunohistochemical methods (Sect.32.5.7).

2.4.7 Fibrin

Fibrin (Table 2.8) is a fibre protein which takes part in the formation of the fibrin coagulum in healing processes and in clotting. Together with fibronectin (Sect.2.4.5) it forms the matrix that enables the development of granulation tissue.

Fibrin is derived from *fibrinogen* which is a soluble plasma protein (Table 2.8). Histochemical demonstration is achieved with relative selectivity using a number of trichrome methods (Sect.21.6), e.g. Mallory's PTAH and the MSB method. Good results can also be obtained with the DMAB method for tryptophanyl groups (Sects.9.4.5 and 9.4.6).

2.4.8 Proteoglycans

The composition of proteoglycans is shown in Sect.2.1.5 while the occurrence of the participating glycosaminoglycans is shown in Table 2.7. Hyaluronic acid plays a prominent role in the interstitium. It differs from the rest of the glycosaminoglycans in that it comprises a single, very long carbohydrate chain (several thousand disaccharide units), which is not covalently bound to protein. The other glycosaminoglycans have shorter chains (usually 150 disaccharide units) which are always covalently bound to a "core" protein by glycosidic bonds involving the hydroxyl groups of serine and threonine. The proteoglycans often form aggregates with hyaluronic acid in which the proteoglycans, often in very large numbers, are aligned on a central filament of hyaluronic acid. The connection between proteoglycan core protein and hyaluronic acid is non-covalent and is mediated by a special "link" protein (see Fig. 2.6). Proteoglycans may also be bound to matrix proteins such as collagen, elastin, and fibronectin.

Typically the glycosaminoglycan chains show no secondary structure and occupy a very large volume in relation to their mass. They are hydrophilic, bind large quantities of water, and form hydrated gels possessing a large number of negative charges which attract osmotically active cations. Water soluble molecules readily diffuse through the matrix, while cellular movement is retarded to a greater or lesser degree depending on the density. Morphogenetic or inflammatory processes are often accompanied by increased turnover of proteoglycans, notably hyaluronic acid.

Table 2.7. The occurrence of different glycosaminoglycans found in proteoglycans.

Glycosaminoglycan	Occurrence
Hyaluronic acid	connective tissue, skin, vitreus body, cartilage, synovial fluid
Chondroitin-4-sulphate	cartilage, cornea, bone, skin, arteries
Chondroitin-6-sulphate	cornea, bone, skin, arteries
Dermatan sulphate	skin, blood vessels, heart
Keratan sulphate	cartilage, cornea, intervertebral discs
Heparan sulphate	lung, arteries, cell surfaces
Heparin	lung, liver, skin, mast cells

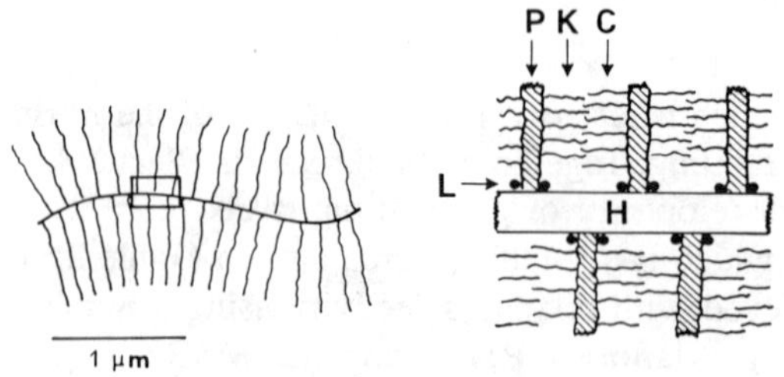

Fig. 2.6. Schematic representation of proteoglycan aggregate. The framed area is expanded on the right. The proteoglycan aggregate consists of chondroitin sulphate (C) and keratin sulphate chains (K) bound covalently to protein chains (P) which are in turn bound non-covalently to a hyaluronic acid chain (H) by special link proteins (L).

Table 2.8. Matrix proteins and their occurrence.

Name	Type	Structure	Occurrence
Collagen		Forms collagen fibres. Consists of tropocollagen units (300 × 1.5 nm), each (consisting of three polypeptide chains (α-chains). Seven different α-chains are known, named α1(I), α1(II), α1(III), α1(IV), α1(V), α2(I), and α2(V).	
	I	Triple helix consists of $[\alpha 1(I)]_2$-α2(I) with broad fibrils. Low content of hydroxylysine and carbohydrate.	Skin, tendon, bone, ligaments, cornea, and parenchymatous organs. Forms approx. 90% of the collagen of the body.
	II	$[\alpha 1(II)]_3$. Thinner fibrils than in type I with a high content of hydroxylysine and carbohydrate.	Hyaline cartilage.
	III	$[\alpha 1(III)]_3$. High content of hydroxyproline. Low content of hydroxylysine and carbohydrate.	Foetal skin, vessel walls, uterus, reticular fibres.
	IV	Precise structure under debate, probably contains procollagen. Possibly $[\alpha 1(IV)]_3$. Very high content of hydroxylysine, high content of carbohydrate.	Basal lamina.
	V	$[\alpha 1(V)]_2\alpha 2(V)$. High content of hydroxylysine and carbohydrate.	Basal lamina.
Elastin		70 kDa glycoprotein with high content of non-polar amino acids alanine, valine, leucine, and isoleucine, and in addition quite a large amount of tyrosine. Elastin forms a non-ordered three-dimensional network, held together by cross-links. The cross-links consist of the specific amino acids desmosine and isodesmosine.	Forms elastic fibres and is especially abundant in the dermis of the skin, the elastic membranes of the blood vessels, and around the alveoli of the lung.

Table 2.8. (*Continued*).

Name	Type	Structure	Occurrence
Elastic microfibrils		Glycoprotein with a high content of cysteine/cystine, and proline, but no hydroxyproline. It is distributed as microfibrils along the surface of the elastic fibres.	Elastic fibres (see elastin).
Fibronectin	Soluble fibronectin = "cold insoluble globulin"	220 kDa glycoprotein which forms a dimer held together by disulphide cross-links at the carboxyl terminals of the two polypeptide subunits.	Blood plasma and other body fluids.
	Insoluble fibronectin = "cell spreading factor"	Usually a dimer, but may occur in a multimer form, when side groups in the polypeptides form cross-links both between the polypeptide chains themselves and to collagen, fibrin, and proteoglycans.	Partly pericellular, in close contact to the cell surface in the form of fibrils, networks, or focal deposits. Partly in the matrix corresponding to loose connective tissue.
Laminin		1,000 kDa asymmetric glycoprotein molecule composed of three 200 kDa A-chains and one 400 kDa B-chain bound together by disulphide bonds.	Basal membranes, especially in lamina lucida.
Fibrinogen		Precursor to fibrin which is composed of three nodules bound together by two rods. Total length 46 nm. Large content of asparagyl, glutamyl, and tyrosyl-O-sulphate.	Soluble protein in blood plasma.
Fibrin		Thrombin acting on fibrinogen splits four arginyl-glycyl peptide bonds and removes four low molecular (10–20 amino acids) polypeptides (fibrinopeptides) with many negative charges. The remainder is a fibrin monomer with about 97% of the amino acids of fibrinogen. The monomers immediately aggregate to the insoluble polymer fibrin.	Takes part in the formation of a coagulum in healing processes and the formation of thrombi. Together with fibronectin it forms the matrix in granulation tissue.
Fibrinoid		Probably a mixture of fibrin and some plasma proteins.	In vessel walls in collagen diseases.

3 Reagents

A.P. Andersen, H. Lyon, P. Jakobsen

A prerequisite for successful and reproducible staining is the use of the correct reagents. The composition, preparation, and storage qualities should be known. A detailed knowledge of the chemical and physical properties of reagents and the reactions for which they are to be used, is therefore highly desirable. Although beyond the scope of this text, it is also mandatory to identify the specific health risks associated with reagents and take appropriate precautions. The information presented below is intended to give guidance to the reader in developing a sound approach to the preparation, storage and disposal of histochemical reagents.

3.1 Preparation of Reagents

Precise information concerning both composition and method of preparation are essential. Typical components of histochemical reagents include solvents, dyes, chromogenic substances, enzymes, and buffer salts. For all of these purity is a key factor.

3.1.1 Purity

The degree of purity of a reagent is usually described using one of the following terms:

- *Technical quality:* normal commercial quality, venale, crudum
- *Purified:* gereinigt, practicum
- *Pure:* purum, rein, reagent quality, and pharmacopoeia terms such as Ph. Eur.; Ph. Helv.
- *Very pure:* purissimum, reinst.
- *Analytical grade:* analytical reagent (AR), pro analysi (p.a.), zur Analyse

"Technical quality" and "purified" should not be used in analytical work, because reagents labelled "purified" frequently contain remarkable quantities of contaminants. The choice between "pure" and "very pure" depends upon what the presence of possible impurities may signify for the use.

H. Lyon (Ed.)
Theory and Strategy in Histochemistry

Water quality. The purity of water from different sources is shown in Table 3.1. A measure for the quality of water may be obtained by measuring the conductivity of the water. This is indicated in $M\Omega^{-1}cm^{-1} = \mu Scm^{-1}$, where $M\Omega$ = megaohm; μS = microSiemens (SI unit system (Système International d'Unités)).

Table 3.1. The purity of different qualities of water.

Water quality	Ions	Organic compounds	Gases	Conductivity ($\mu S\,cm^{-1}$) at 25°C
Tap water	Ca^{2+}, Mg^{2+}, HCO_3^- etc.	always some, but variable	especially CO_2	300
Softened water	especially Na^+, some Ca^{2+}, Mg^{2+}	variable	$N_2 + O_2$	300
Demineralized water	0	variable	CO_2, O_2, N_2	0.2–1
Distilled water, technical quality	some Ca^{2+}	nearly 0	CO_2, (O_2, N_2)	1–6
Glass redistilled water	0	0	CO_2, (O_2, N_2)	0.5–1

For analytical work, only demineralized or glass redistilled demineralized water should be used. The advantage of the latter is, providing it is used less than 24 hours following redistillation, that possible traces of ion exchanger and bacterial growth are avoided. Fresh water from these sources is always required for enzyme determinations.

3.1.2 Labelling of Reagents

This should always follow international and local recommendations. Examples are BSC (Biological Stain Commission, Sect.3.3.10), ECCLS (European Committee for Clinical Laboratory Standards), and DIN (Deutsche Institut für Normung). In addition, it is expedient if the following is noted on the label:
1. The use of the reagent
2. Date of purchase or preparation
3. Special requirements regarding storage (e.g. refrigerator)
4. Date and method for discarding.

3.1.3 Storage Qualities of Reagents

Most reagents contain unstable components that are susceptible to particular environmental influences. Important examples include the potential for oxidation by atmospheric oxygen, uptake of carbon dioxide by alkaline solutions leading to the formation of carbonates, and 'spontaneous' cleavage of macromolecules. Exposure to light, particularly ultraviolet light, (a potent catalyst for many chemical processes), and temperature (a major determinant of reaction rate) must always be taken into account.

3.1.4 Storage of Reagents

As far as possible, storage conditions should be optimized for each reagent. In general this means using tightly closed containers, protection from light, and location in a cool environment. Freezing of reagents may give rise to unexpected changes including precipitates and should therefore be undertaken with caution. The container must of course be indifferent towards its contents. Both glass and plastic are well suited for the majority of reagents and the latter is being used with increasing frequency. It should, however, be noted that plastic does not give the same protection against light as dark glass, and that certain substances may be adsorbed.

3.1.5 Discarding of Reagents

When reagents are discarded the regulations for protection of the environment should be strictly adhered to. Choice of method depends on the concentration of the substance, the rate with which it is degraded in nature, and its toxic effects. Three practical approaches are widely used:

1. Sink-disposal using plenty of water to flush through the waste-pipe
2. Detoxification followed by sink-disposal
3. Collection and transportation for incineration

Individual laboratories should establish regulations for the collection and disposal of refuse in accordance with local regulations.

3.2 Solvents

A solvent is a homogeneous mixture. Although solutions may be solid, liquid, or gaseous, in this book, in accordance with general use, the term solution will be used for liquid, homogeneous mixtures.

A solution can be prepared by dissolving a gas, liquid or solid substance in a liquid, which is called the solvent. Clearly, both the potential for the formation of a solution and its properties when constituted are dependant on the physicochemical properties of the substances involved.

Intermolecular forces are of central importance. It is a precondition for the formation of solutions that stronger bonds can be formed between the solvent and the solute than internally between the molecules of the separate compounds. A solvent must also be liquid at room temperature.

Only certain polar and non-polar molecular compounds are liquid at room temperature. Ionic compounds, atomic lattice compounds, and metals, with the exception of mercury, are all solids under ambient conditions. In consequence the following types of solvents are available: *polar* (e.g. water), *non-polar* (e.g. hydrocarbons), and solvents containing both polar and non-polar groups with approximately equal influence, termed *amphiphilic,* e.g. sodium laurylsulphate $C_{12}H_{25}SO_4Na$. The following rules may be applied to solvent-solute combinations:

- *Ionic compounds* form strong intermolecular bonds but many will dissolve in water and other very polar molecular compounds, as the ions can form stronger bonds with the polar molecules
- *Atomic lattice compounds* (e.g. SiO_2) cannot be dissolved in either polar or non-polar molecular solvents due to the strength of their inter-atomic bonds
- *Polar molecular compounds* may be dissolved in other polar compounds
- *Non-polar molecular compounds* can only be mixed with other non-polar compounds

From the practical point of view therefore, polar solvents have more widely applicable dissolving properties than non-polar solvents since the latter can only dissolve other non-polar molecular compounds. The suitability of a polar solvent can be assessed on the basis of the dipole moment (μ) of the compound, its ability to form hydrogen bonds, its dielectric constant (ϵ), and its ability to form chemical bonds with the dissolved compound.

Dipole Moment. This is an expression of the degree of polarization in the molecule. The following equation applies: $\mu = e \times l$, where μ = dipole moment; e = charge; l = distance between charges in the dipole. Examples of dipole moments are given in Table 3.2.

Table 3.2. Examples of dipole moments in solvents.

Solvent	Dipole moment $10^{-30} \times C \times m$[a]	D[b]
Water	6.3	1.9
Ethanol	5.7	1.7
Chloroform	4.0	1.2

[a] C × m = Coulomb metre (S.I. unit system); [b] D = Debye; 1 D = 3.33×10^{-30} C × m.

The greater the dipole moment the more strongly the compound will bind to other permanent or induced dipoles. The dipole moment is, however, not the only factor that determines solubility, the nature of any solvent-solute bonds formed is also important. Both *hydrogen bonds* and complex bonds are relevant in this regard. Compounds containing $-OH$ $-SH$ $-NH_2$ $>NH$ $-COOH$ and $-SO_3H$ are all good at forming hydrogen bonds with each other and with water.

The ability of a polar solvent to dissolve other polar molecular compounds can thus be assessed from the size of the dipole moment and the ability to form hydrogen bonds, while its ability to dissolve ionic compounds may be assessed from its dielectric constant and ability to bind to the compound.

Dielectric Constant. The dielectric constant (ϵ) of a compound is an expression for the ability of the compound to reduce the attraction force between two opposite charges compared to the attraction force in vacuum. The following applies:

$$F = \frac{e^2}{\epsilon r^2}$$

ϵ = dielectric constant; r = distance between charges; e = charge; F = attraction force .

Examples of dielectric constants are:

water	80
ethanol	23
chloroform	5

This means that the attraction force between a sodium ion and a chloride ion in aqueous solution is 1/80 of what it would be in a vacuum, while in ethanol an approximately four-fold greater attraction would be present.

A high dielectric constant is, however, not enough to make the liquid a good solvent for ionic compounds. The molecules of the solvent must also be able to associate with the solute ions, i.e. solvation. Water is an excellent solvent for salts as it has both a high dielectric constant and can form complex bonds to the ions.

The solubility of a compound may increase if an acid/base reaction takes place between solute and solvent. The base pyridine is a good solvent for acidic substances while concentrated sulphuric acid can be used for dissolving organic substances with the characteristics of a base.

Table 3.3 presents physical constants for frequently used solvents.

Table 3.3. Constants for the solvents water, ethanol and xylene.

Solvent	Melting point (°C)	Boiling point (°C)	Dielectric constant	Dipole moment (D)
Water	0	100	80	1.9
Ethanol	– 117	78.5	24	1.7
Xylene			< 2	~ 0
ortho-	– 25	144		
meta-	– 46	139		
para-	– 13	138		

3.2.1 Water

The Structure of Water. Water forms polar, non-linear molecules which are bound together by hydrogen bonds in both solid (ice) and liquid forms.

Ice is composed of a lattice in which each water molecule is bound by hydrogen bonds to four other water molecules. On melting, a decrease in volume takes place and a dynamic equilibrium is established in which each water molecule is bound on average to 3.4 other water molecules by hydrogen bonds. In liquid form "lumps" of water similar to ice crystals are formed, but as the water molecules are not tightly locked in this structure they may form hydrogen bonds to other substances. This makes water an excellent solvent.

Ionization Product of Water. The following equilibrium, which is strongly displaced to the left, applies:

$$H_2O + H_2O \rightleftarrows H_3O^+ + OH^-$$

As H_2O is a constant the law of mass action leads to

$$H_3O^+ \times OH^- = K_w$$

At 25°C $K_w = 10^{-14}$ (mol/l)^2

Water as a Solvent for Polar Substances. Water is an excellent solvent for polar substances with hydrophilic groups due to its large dipole moment and its ability to form hydrogen bonds. The solubility of a polar substance in water can be assessed from the ratio between hydrophobic groups (hydrocarbons) and hydrophilic groups. The smaller this ratio is, the greater the predicted solubility in water.

Water as a Solvent for Ionic Compounds. Water is an excellent solvent for ionic compounds (salts) due to its high dielectric constant and its ability to bind to ions. When a salt is dissolved in water ions are formed. Their activity may be expressed as the *ionic strength* of the solution using

$$I = \frac{1}{2}\Sigma C_i Z_i^2$$

where I = ionic strength; C_i = concentration of ions in mol/l; Z_i = charge of the ion.

The ionic strength for a 0.1 mol/l $MgCl_2$ solution is thus:

$$\frac{1}{2}(0.1 \times 2^2 + 0.2 \times 1^2) = 0.3$$

as $[Mg^{2+}]$ = 0.1 mol/l and charge = 2; $[Cl^-]$ = 0.2 mol/l and charge = 1.

Dependent on which salts are dissolved in water it is possible to prepare:

Buffers	(e.g. NaH_2PO_4 + Na_2HPO_4)
Acids	(e.g. $FeCl_3$)
Bases	(e.g. CH_3COONa)
Neutral solutions	(e.g. NaCl)
Oxidizing agents	(e.g. $KMnO_4$)
Reducing agents	(e.g. $FeCl_2$)
Complex compounds	(e.g. $AgNO_3$ and ammonia water)

Water as a Solvent for Acids. Water is, due to its basic character, a good solvent for acids. The following reaction takes place: $A(1) + H_2O \rightleftarrows B(1) + H_3O^+$, where $A(1)$ = acid and $B(1)$ = corresponding base. The concentration of H_3O^+ in the aqueous solution is expressed by pH which is < 7.

Water as a Solvent for Bases. Water is also a good solvent for bases with the following reaction taking place:

$$B(1) + H_2O \rightleftarrows A(1) + OH^-$$

The concentration of OH^- is expressed using $[H^+]$ as $[H^+][OH^-] = K_w$. In a solution of a base pH is > 7.

Water as a Solvent for Macromolecules. Polypeptides (proteins) polysaccharides, and nucleic acids are generally called macromolecules when their molecular weight exceeds 10 kDa.

The properties of aqueous solutions of macromolecules differ from those of low-molecular weight substances in several important respects. Solubility is highly dependant on side chains which can be acid or basic, and hydrophilic or hydrophobic, as well as the tertiary and quaternary structure. In general, elongated molecules are less soluble in water than globular molecules. In broad terms, if all other factors are equivalent, solubility is inversely related to molecular length. Proteins, polysaccharides, and nucleic acids are called macromolecules as their molecular weight is greater than 10 kDa. When macromolecules are dissolved in water the solutions formed are termed *colloid.* Nucleic acids, proteins, and heteropolysaccharides, all of which are polyelectrolytes, form colloidal electrolytes. Unlike true solutions, colloid solutions do not affect the boiling point or freezing point. They do, however, give rise to an osmotic pressure and, depending on the ability of the molecule to bind water, viscous solutions.

Although the solubility of hydrophilic macromolecules is not influenced by low salt concentrations, high concentrations cause precipitation. Ions with greater ability to bind water have greater effect. The precipitating property of ions decreases in the following sequences:

Li^+ Na^+ K^+

Mg^{2+} Ca^{2+} Ba^{2+}

$SO_4{}^{2-}$ CH_3COO^- Cl^- Br^- I^- NO_3^-

Water as a Solvent for Hydrophobic Substances. While such substances are not soluble in water on their own, they may form emulsions composed of micelles after the addition of a detergent. A detergent is a substance which contains both a hydrophobic radical and a strong hydrophilic group. It may thus bind both to water and to the hydrophobic substance and thereby bring the latter into solution as an oil/water emulsion.

3.2.2 Ethanol is a polar solvent (cf. Table 3.2). It is very versatile because it contains both a hydrophobic group and a hydrophilic group which can form hydrogen bonds. Ethanol is a translucent, colourless, volatile liquid which is easily inflammable. It should be kept in a closed container protected against light. The commercial product is 96% v/v alcohol and absolute alcohol ≈99% v/v. Ethanol is freely miscible with water and is an excellent solvent for the majority of organic and many inorganic compounds, but not for salts. The majority of lipids are very sparingly soluble in ethanol.

3.2.3 Xylene is a purely non-polar or hydrophobic solvent and can be used for dissolving hydrophobic substances such as paraffin. The commercial product is a mixture of o-, m-, and p-dimethylbenzene. It is a translucent, colourless, volatile liquid. The vapours are noxious, and xylene may be absorbed through the skin. Inhalation and contact with xylene must be avoided. Xylene burns with a very sooty flame and can form explosive mixtures with air. It should be kept in well-closed bottles protected against light. Xylene is fully miscible with ethanol, ether, and chloroform.

3.3 Dyes

3.3.1 Definitions

- *Chromogen* (Abrahart, 1977, p.6). Any substance which absorbs electromagnetic radiation in the visible part of the spectrum (400–750 nm)
- *Dye* (in the histochemical sense). Chromogen of aromatic or heteroaromatic nature which is soluble in water or polar solvents and which can bind to other substances
- *Lysochrome*. A chromogen of aromatic or heteroaromatic nature which is insoluble in water, but soluble in organic solvents and has affinity to hydrophobic compounds.

3.3.2 Colour

The pattern of absorption of electromagnetic radiation is characteristic for a particular chromogen. The substance will assume a colour corresponding to the visible light that passes through, the so-called *complementary colour*. In Fig. 3.1 the relationship between the absorbed colour and the observed colour is given.

The absorption spectrum of a dye may be determined using a spectrophotometer. This is a curve showing the absorption at different wavelengths and is characteristic for a particular dye. As an example, the absorption spectrum of Pyronin Y is shown in Fig. 3.2. The absorption shows a maximum at about 550 nm (λ_{max}).

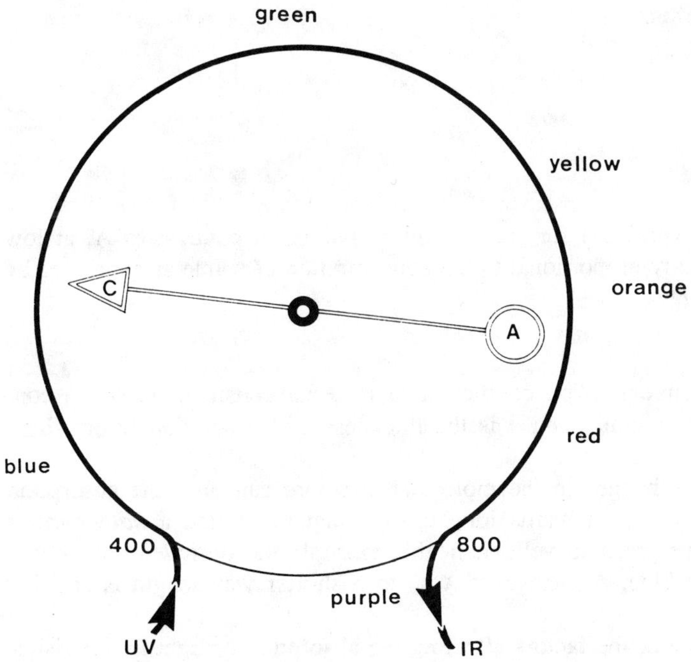

Fig. 3.1. Complementary colour circle showing relationship between wavelength of absorbed (A) and observed (C) colours. Ultraviolet (UV) and infrared (IR) regions.

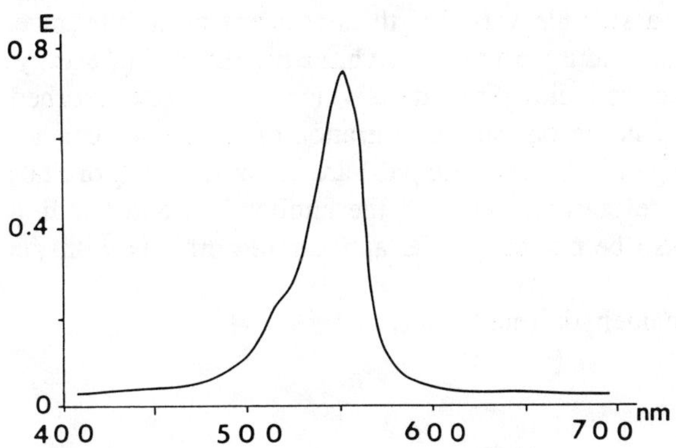

Fig. 3.2. Absorbance curve for Pyronin Y. Absorbance (E).

When plotting absorption curves it is usual to register absorbance (= extinction) instead of absorption.

If I_0 = intensity of incident light and

I = intensity of transmitted light

the *absorption* is defined as

$$A = \frac{I_0 - I}{I}$$

and the *absorbance* as

$$E = \log \frac{I_0}{I}$$

E is thus also an expression for the absorption but is, in contrast to A, at low concentrations, directly proportional to the concentration of solute as expressed by the *Beer-Lambert law:*

$$E = \epsilon \times C \times d$$

where ϵ is the molar extinction coefficient (a material constant), C is the concentration of solute in mol/l, and d is the thickness of the solution in cm (light path).

Relatively small changes in the molecular structure can alter the absorption spectrum. A shift of λ_{max} towards longer wavelength is called a *bathochrome effect*, e.g. from blue-green to yellow. In this example the observed colour will change from red to blue. A change of λ_{max} to a shorter wavelength is called a *hypsochrome effect.*

The modern view of the factors affecting the absorption of light in the visible and ultraviolet spectrum is based on the concept of molecular orbitals. According to this theory, electrons taking part in the formation of a bond form a bonding and an antibonding orbital with different levels of energy. In their ground state, electrons will be found in the bonding orbital which has the lowest level of energy. If electromagnetic radiation of a suitable wavelength is incident on a substance, electrons can be excited to an antibonding orbital which has a higher level of energy. The energy involved comes from the incident radiation which is therefore absorbed to a greater or lesser degree depending on the concentration of the substance.

Electrons which do not take part in bonds (lone pair) are in non-bonding orbitals with an energy level intermediate between those of the antibonding and bonding orbitals. These electrons can also be excited to the antibonding orbitals with the higher levels of energy.

In a substance such as formaldehyde (methanal),

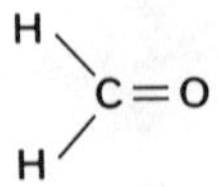

where there are three σ-bonds, 1 π-bond, and two lone pairs, the energy levels will, according to the molecular orbital theory be σ-bonding, π-bonding, n-non-bonding, σ-antibonding, and π-antibonding.

The relationship between the energy levels of the bonds is shown in Fig. 3.3.

Since greater transitions in energy level require more energy input, it can be seen that substances containing only σ-bonds need light of higher energy (shorter wave-

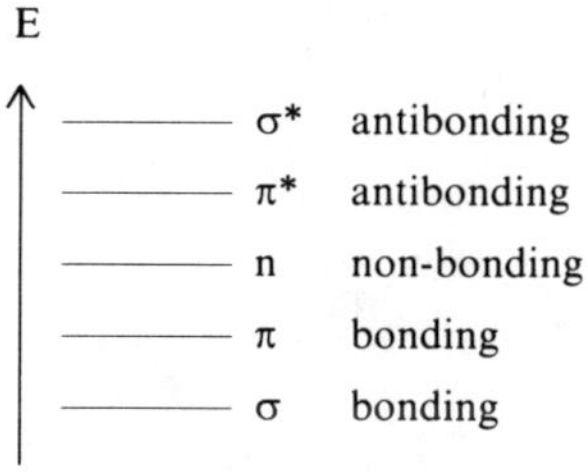

Fig. 3.3. Energy levels of different orbitals. Relative energy (E).

length) to excite the electrons, while substances containing double bonds (π-bonds) and/or heteroatoms can be excited by light of lower energy (longer wavelength).

If a substance contains several π-bonds/lone pairs conjugated with each other, the energy difference between π-bonding and π-antibonding molecular species is reduced, and absorption occurs at a higher wavelength.

From benzene it is known that in cyclic structures the presence of a suitable number of double bonds conjugated to each other gives the molecule its aromatic character. The electrons involved are considered not to be localized in individual π-bonds but delocalized over the whole molecule. A similar delocalization can arise in *planar*, open systems containing conjugated double bonds and/or lone pairs conjugated to the double bonds even if the substance does not have aromatic characteristics.

In larger aromatic and heteroaromatic systems as well as in non-aromatic substances containing many conjugated double bonds, (e.g. carotenoids), the energy difference between the bonding and antibonding molecular orbitals is large enough to allow absorption in the visible spectrum.

Alkanes will thus absorb light with a wavelength of approximately 150 nm, while alkenes (>C=C<) will absorb at around 190 nm. For benzene, where there is a delocalized cyclic π orbital over all six carbon atoms, absorption is seen at 203.5 and 254 nm.

In nitrobenzene

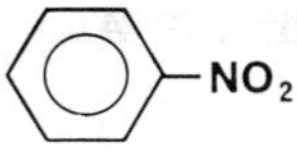

where the size of the delocalized cyclic π orbital is increased compared with benzene absorption occurs at 268.5 nm.

In a dye such as Crystal Violet the size of the delocalized cyclic π orbital has become so great that the absorption has moved into the visible part of the spectrum at 590 nm.

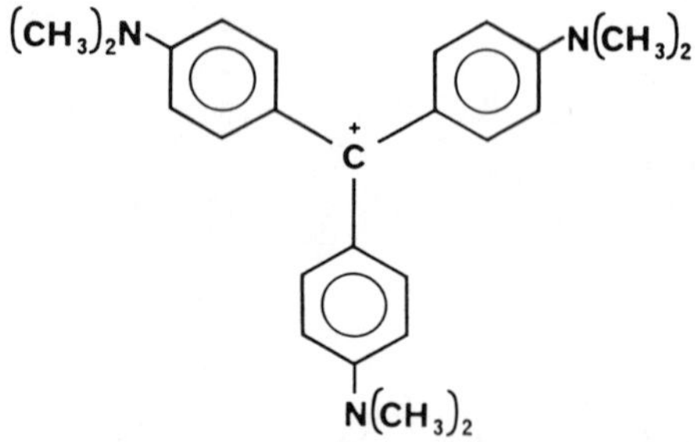

Chromophore groups. Groups that are able to absorb light in the ultraviolet or visible part of the spectrum are called chromophore groups or chromophores. Examples of such groups are:

–N=N–	azo
–N=O	nitroso
$-NO_2$	nitro
>C=C<	alkene
>C=O	oxo

As seen in the previous examples it is necessary to have several chromophores conjugated with each other before a compound will absorb light in the visible spectrum.

Auxochrome groups. These are groups that are not able to absorb light in the ultraviolet or visible spectrum by themselves; however, when they are bound to a chromophore system they can change the light absorption towards longer wavelength and at the same time impart a higher intensity to the absorption. Auxochrome groups contain a lone pair of electrons that, in conjugation with a conjugated system, are able to increase the size of the delocalized cyclic π orbital. Examples are:

–OH	hydroxyl
–SH	thiol
$-NH_2$	amino
–Cl, –Br, –I	halogens

Substitutions in auxochrome groups often produce additional alterations in colour. Substituents of this kind are called *modifying groups* (modifiers). Alkylation will cause a bathochrome shift, acylation a hypsochrome shift.

3.3.3 pH-effect, Anionic and Cationic Dyes

A slight change in pH can frequently radically change the colour of a substance (cf. indicators). This is due to a change in the delocalized π orbital. For example, the protonation of an amino group (pH $<$ 9) removes the lone pair from the delocalization. Simultaneously, the substance acquires a positive charge and with it the facility for binding to negatively charged tissue components through ionic bonds. A positively charged substance of this kind is called a *cationic dye.*

Negatively charged chromogens, *anionic dyes*, can be formed with the help of groups such as OH, SO_3H, and COOH. The OH-group (bound to an aromatic system) is normally ionized at pH > 8–9, with the exception of substances like Picric Acid where the presence of nitro groups gives the OH-group a pK_A of around 1. The negative charge also influences the delocalized electron cloud, so that the ionized compound is a different colour. For example p-nitrophenol is colourless while p-nitrophenolate is yellow. The sulphonic acid group (SO_3H), which is frequently found in anionic dyes, is ionized over most of the pH-scale. Carboxylic acid groups (COOH) are ionized over a similar pH-range but have not found such extensive application in anionic dyes. The group is, however, very important for solubility (cf. Eosins in Fig. 3.4). Eosin precipitates when the carboxyl group is not ionized (pH < 2).

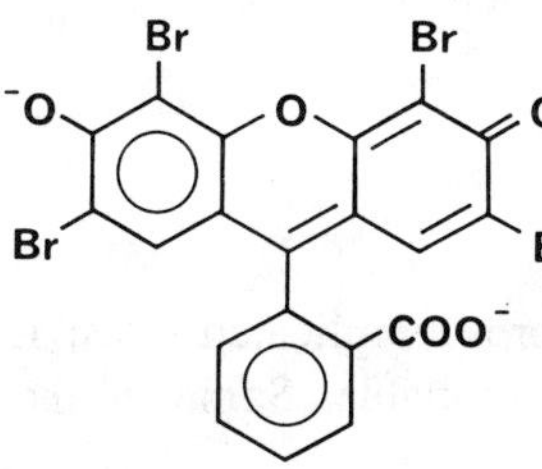

Eosin, C.I. 45380

Solubility in H_2O:
44.2% w/v
Solubility in ethanol:
2.18% w/v

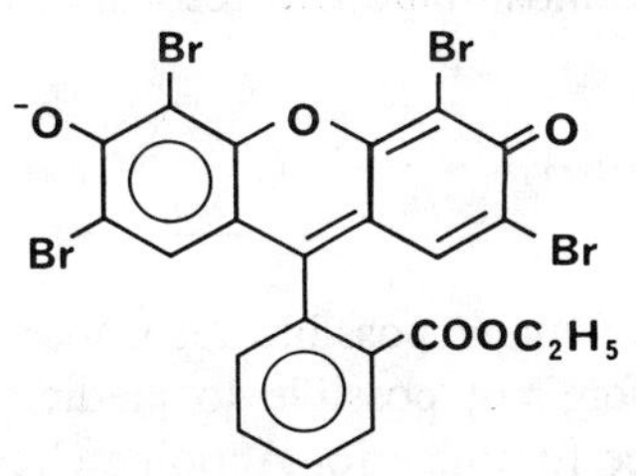

Ethyl Eosin, C.I. 45386

Solubility in H_2O:
0.03% w/v
Solubility in ehtanol:
1.13% w/v

Fig. 3.4. Formulae and solubilities of Eosins.

3.3.4 Resonance Formulae

When writing structural formulae for compounds containing a delocalized cyclic π orbital it is well known that classical symbols for bond valency give an erroneous view of the structure. For benzene the classical Kekulé-structures are:

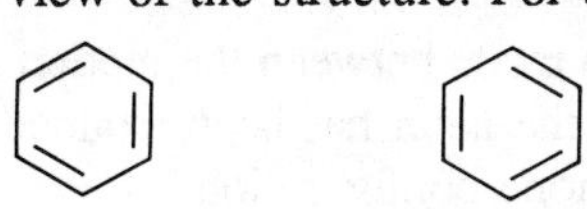

with modern notation:

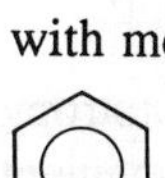

where the ring designates the delocalized cyclic π orbital (the aromatic system).

It is, however, frequently useful to be able to write the structures using the classic formulae as these give a better idea of where possible charges may be found and how large the delocalization is.

Formulae of this kind are called resonance formulae, and the true structure is to be found somewhere between the different possible resonance formulae for the substance. (The electron distribution in the substance corresponds to a superposition of the electron distributions of the possible resonance formulae). This is shown by connecting different resonance formulae for the same substance with a double arrow.

For aniline the following resonance formulae can be written:

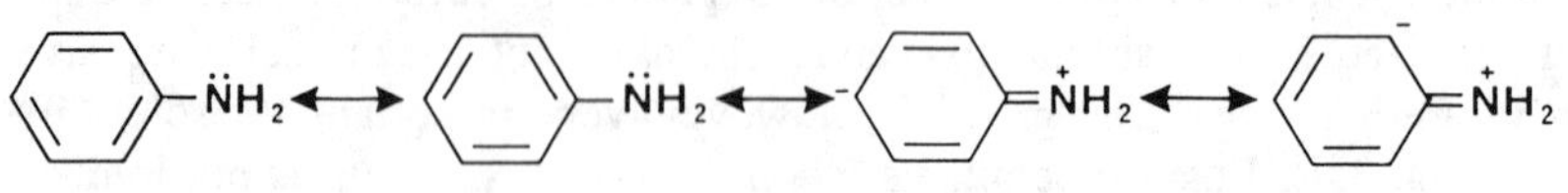

while the most important resonance formulae for the anilinium ion are:

There are many possible resonance formulae for the more complicated dyes; it is therefore not possible to predict which are the most probable. Some of the resonance formulae for Acridine Orange can be written:

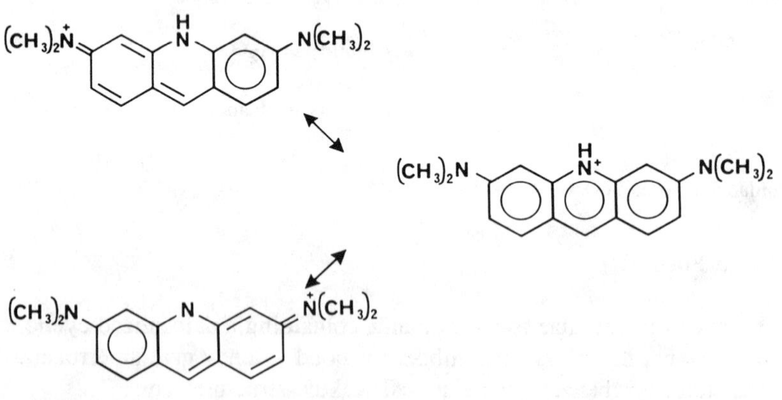

Note that in the rest of this book a compromise has been made between the modern ring formulae and the Kekulé notation. Where relevant, the latter has been retained to show the presence of chromophores such as the quinone configuration.

3.3.5 Phototropy. A colour change ("fading") caused by light is called phototropy (Abrahart, 1977, p.10). This change may in some cases be due to a rearrangement of the dye molecule under the influence of light. The azo dye shown below can exist in a *cis* and a *trans* form:

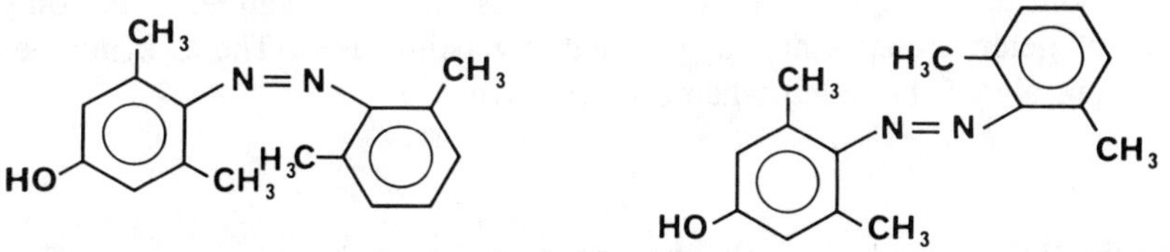

The *trans* form is coloured because the molecule is planar and electron delocalization can take place over the whole conjugated system including the lone pair in the OH group. In the *cis* form the two benzene rings cannot lie in the same plane as the size of the ortho-methyl groups hinders this conformation (steric hindrance). The *cis* compound thus appears colourless. Exposure to light causes a change from the *trans* to *cis* form. On standing in the dark the *cis* compound reverts to the more stable coloured *trans* compound.

3.3.6 The Appearance of the Absorption Spectrum as a Criterion for Purity

Theory concerning the pattern and intensity of absorption in the visible and ultraviolet spectrum for a selected compound is based on the molecular orbital theory. Certain "rules of selection" exist that indicate whether an electron transition is permitted or not (forbidden). These depend on the "symmetry" of the energy levels involved in the transition. Forbidden transitions will often be seen as absorption peaks with low intensity.

When examining the purity of a dye sample it is common to record an absorption spectrum of the substance and compare this with data for the pure dye. Identity does, however, not constitute an absolute criterion of purity.

If the spectrum of the substance under investigation contains more absorption maxima than the pure dye, clearly the substance is impure. If an identical spectrum to that given for the pure dye is found, another (similar) substance with the same absorption maxima may be present. Moreover, there may be absorptions that are not observed with the spectrometer used, or there may be several closely spaced absorptions that "melt" together to one and thus give a "false" impression of purity. To obtain an absolute determination of purity thin layer chromatography (TLC) or high performance liquid chromatography (HPLC) analyses should be made.

3.3.7 Dye Nomenclature

In the Colour Index (Society of Dyers and Colourists, 1971) the dyes are designated by a 5-digit number, C.I. number and a specially constructed name.

The trivial names in more general use are given in Conn's Biological Stains (Lillie, 1977). It should be noted, however, that there are numerous synonyms, i.e. names which may be the same for several dyes. Biological Stains thus distinguishes between e.g. C.I. 42585 Methyl Green and C.I. 42590 Ethyl Green. Ethyl Green is frequently sold as Methyl Green, and in this case the C.I. number is the only certain way of determining which dye is actually being used. The C.I. number should therefore always be noted when ordering dyes.

3.3.8 Classification of Dyes, Dye Precursors, and Pigments

Dyes can be classified according to their chemical structure or according to the purpose for which they are used. No single method is completely satisfactory as the same chromophore system may be present in dyes with very different applications. As a general guide, the presence or lack of solubilizing groups, proton accepting groups, and long aliphatic chains are all key elements in determining the characteristics of a dye and its suitability for a certain application.

Baker (1966, p.89) uses a simplified classification according to the chromophore groups in the dyes:

I. Nitro dyes
II. Azo dyes
III. Quinone (quinonoid) dyes

In Table 3.4 a classification of dyes, dye precursors, and pigments is given. This follows the classification used in the second part of the Colour Index 3rd edition (Society of Dyers and Colourists, 1971). Of the 21 groups described in the Colour Index only those that have specific relevance to histochemistry have been included in the table. The most important difference from Baker's classification is that quinone dyes are subdivided into 5 groups, namely, arylmethane, xanthene, acridine, quinoneimine, and anthraquinone dyes.

Table 3.4. Dyes, dye precursors, and pigments classified according to chemical structure.

C.I. group no.	Class	Example	Formula (the presumed active component)
2.	Nitro	Picric Acid C.I. 10305	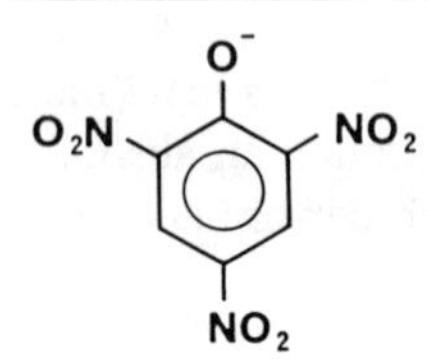
3.	Azo		
	A. Monoazo	Orange G C.I. 16230	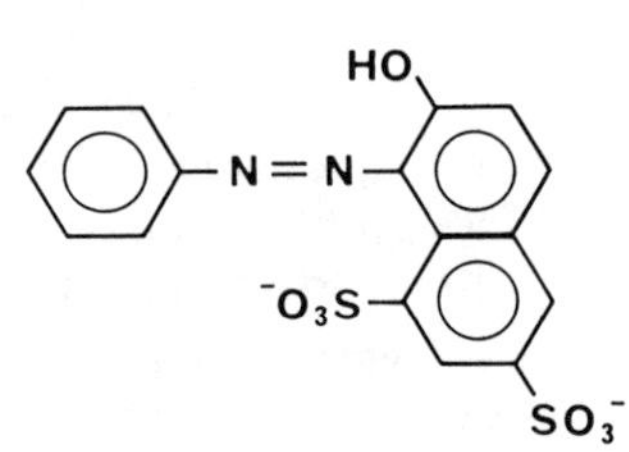
	B. Disazo	Congo Red C.I. 22130	
	C. Trisazo	Chlorazol Black E C.I. 30235	
	D. Tetrakisazo	Sirius Red F3B C.I. 35780	

Table 3.4. (*Continued*).

C.I. group no.	Class	Example	Formula (the presumed active component)
4.	Diazonium salts	Fast Garnet GBC· C.I. 37210	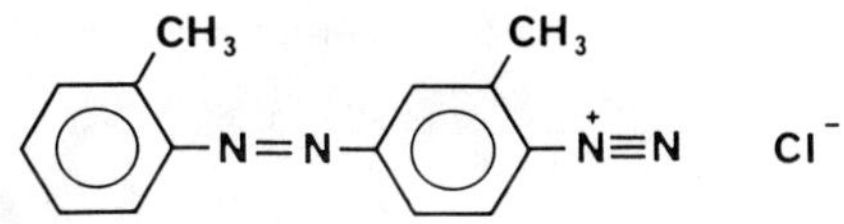
5.	Azo coupling components	Naphthol AS-BI C.I. 37566	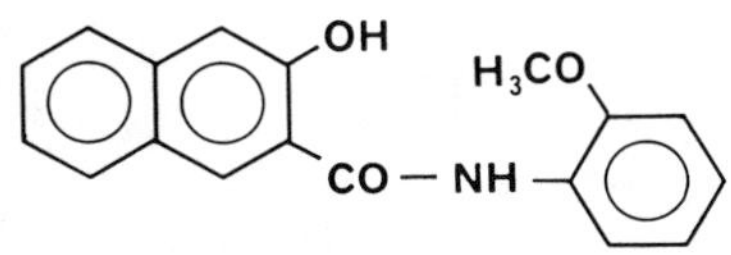
6.	Tetrazolium salts	Nitro blue tetrazolium No C.I. no.	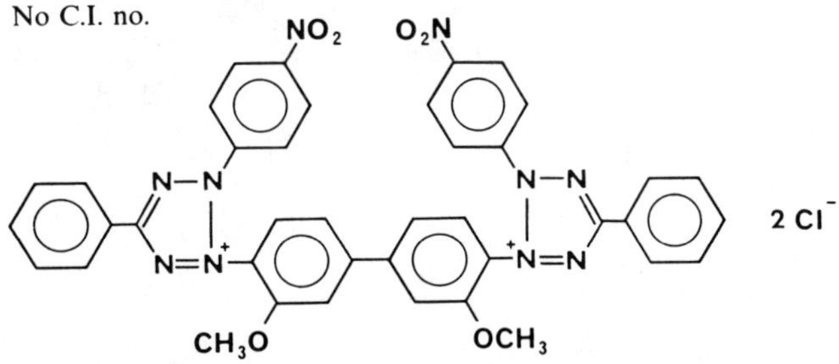
8.	Arylmethane A. Diphenylmethane	Auramine O C.I. 41000	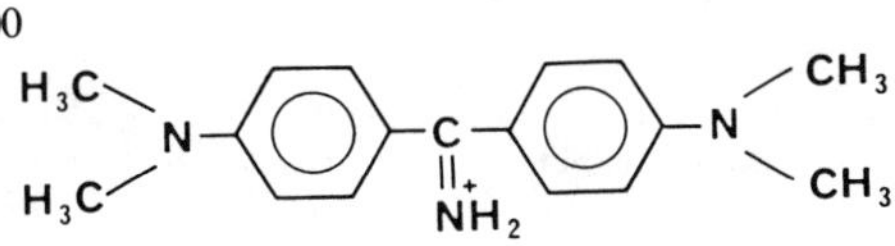
	B. Diaminotriphenylmethane	Fast Green FCF C.I. 42053	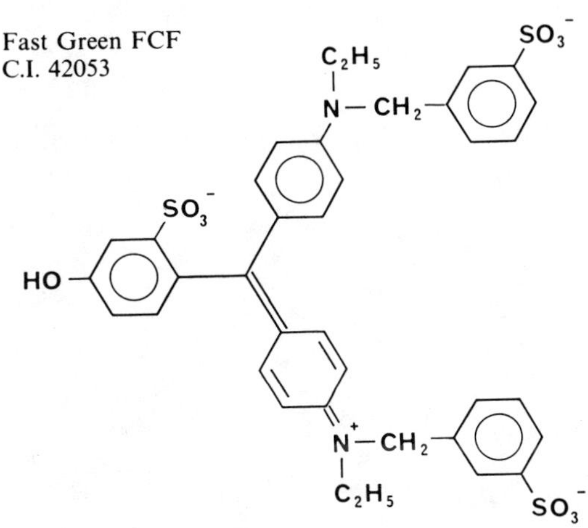

Table 3.4. (*Continued*).

C.I. group no.	Class	Example	Formula (the presumed active component)
	C. Triaminotriphenylmethane	Pararosanilin C.I. 42500	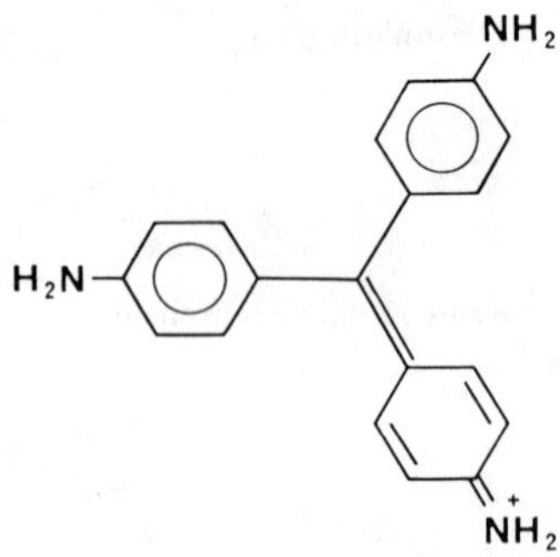
	D. Hydroxytriphenylmethane	Chrome Violet CG C.I. 43810	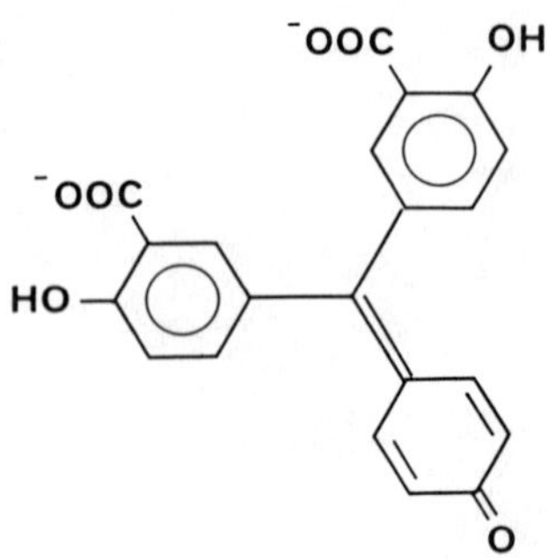
	E. Mono- and dinaphthyl-phenylmethane derivatives	Victoria Blue 4R C.I. 42563	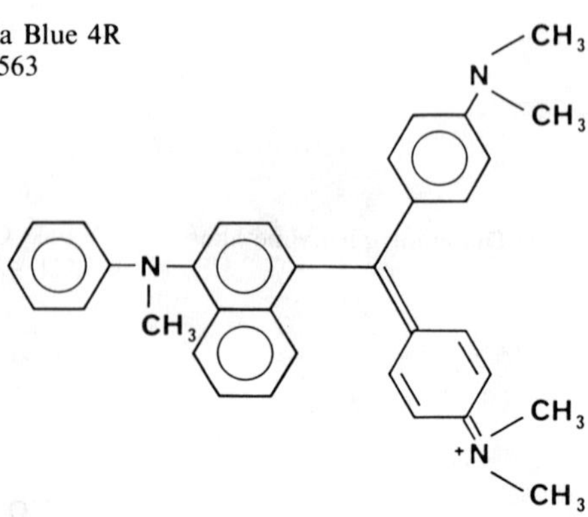

Table 3.4. (*Continued*).

C.I. group no.	Class	Example	Formula (the presumed active component)
9.	Xanthene		
	A. Aminoxanthene		
	a. Pyronin	Pyronin Y C.I. 45005	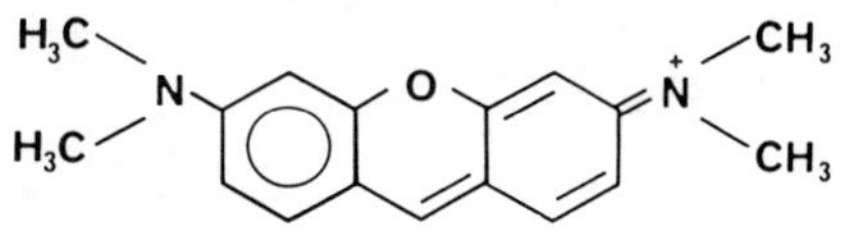
	b. Succinein	Rhodamine S C.I. 45050	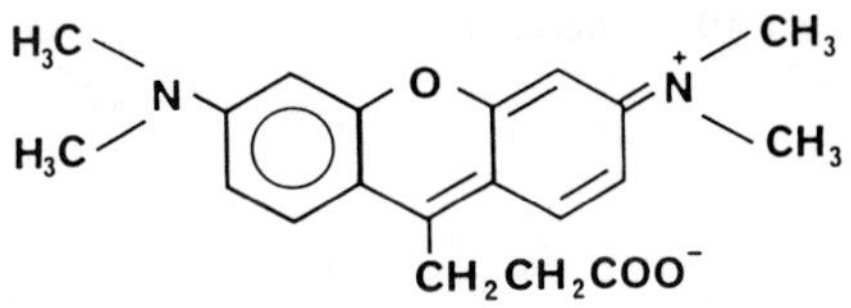
	c. Rosamine	Sulphorhodamine B C.I. 45100	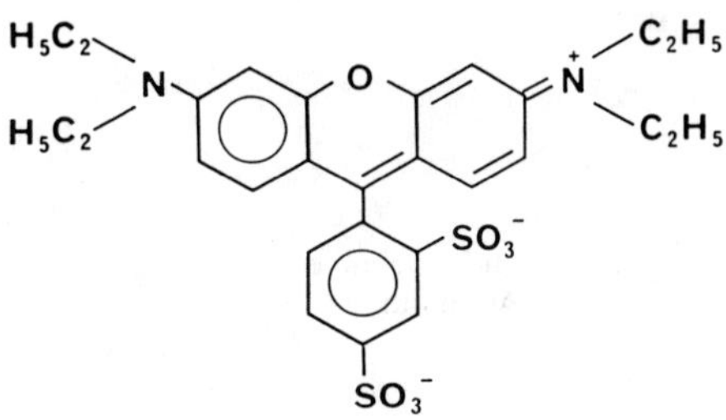
	d. Rhodamine	Rhodamine B C.I. 45170	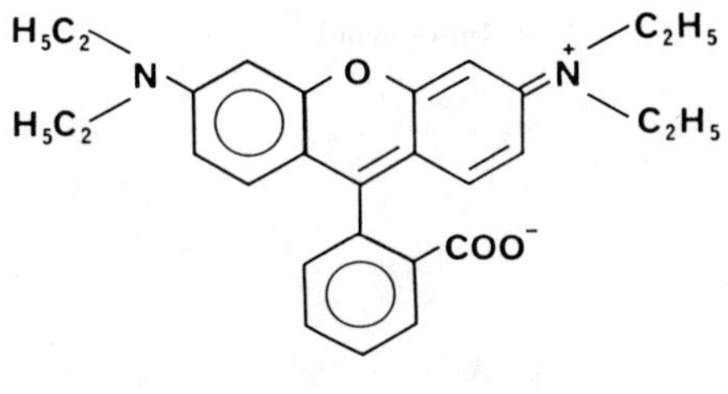

Table 3.4. (*Continued*).

C.I. group no.	Class	Example	Formula (the presumed active component)
	B. Hydroxyxanthene	Eosin C.I. 45380	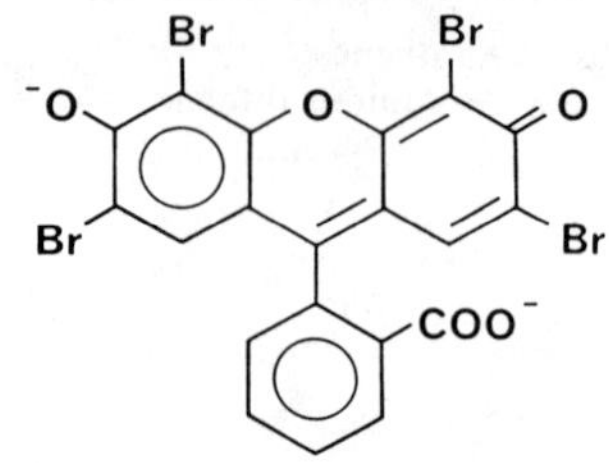
10.	Acridine	Acridine Orange C.I. 46005	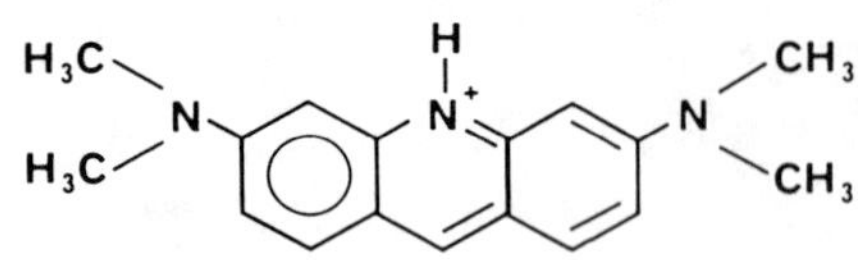
13.	Thiazole	Thioflavine TCN C.I. 49005	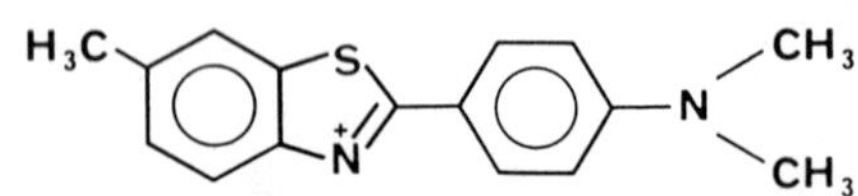
14.	Quinoneimine A. Indamine	Toluylene Blue C.I. 49410	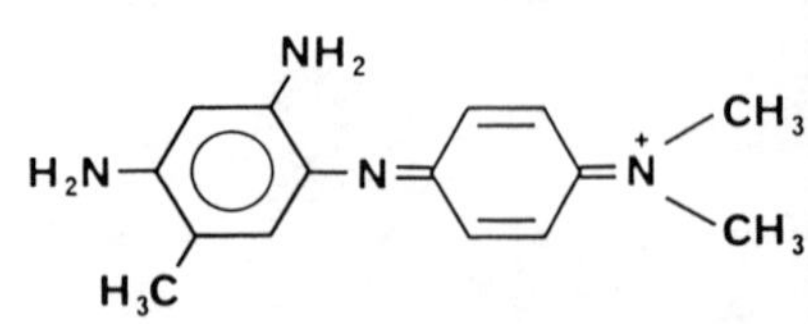
	B. Indophenol	Indo-oxine No. C.I. no.	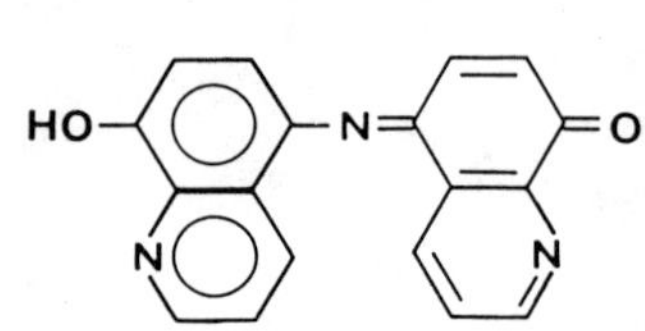
	C. Azin a. Eurhodine	Neutral Red C.I. 50040	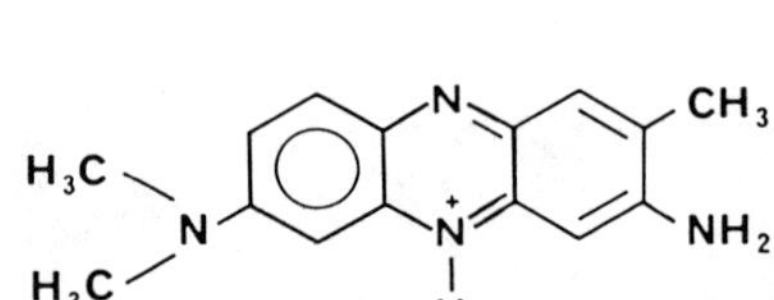

Table 3.4. (*Continued*).

C.I. group no.	Class	Example	Formula (the presumed active component)
	b. Rosinduline	Azocarmine B C.I. 50090	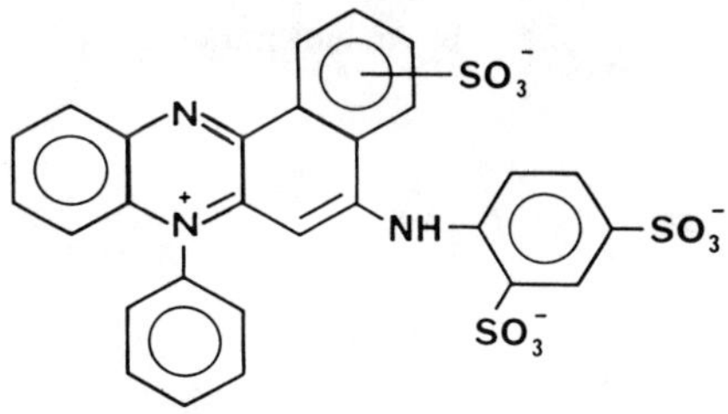
	c. Safranin	Safranin O C.I. 50240	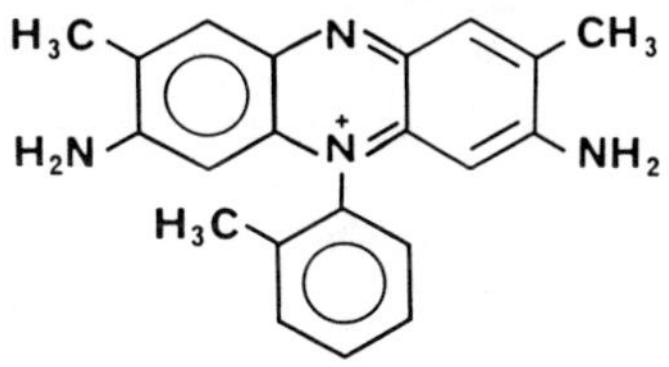
	D. Oxazin	Gallocyanin C.I. 51030	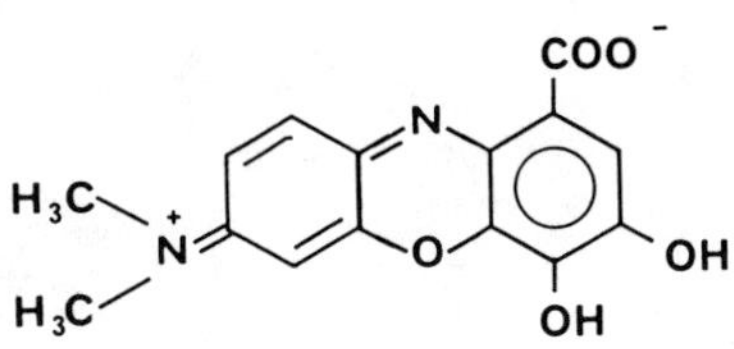
	E. Thiazin	Toluidine Blue O C.I. 52040	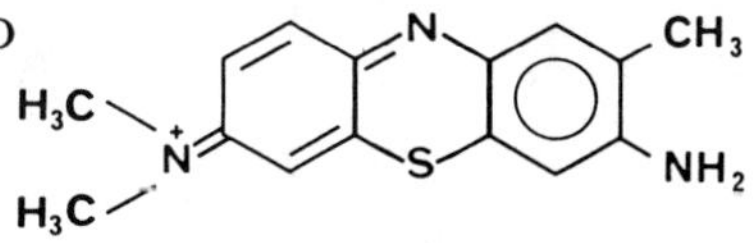
16.	Anthraquinone A. Hydroxyanthraquinone	Alizarin Red S C.I. 58005	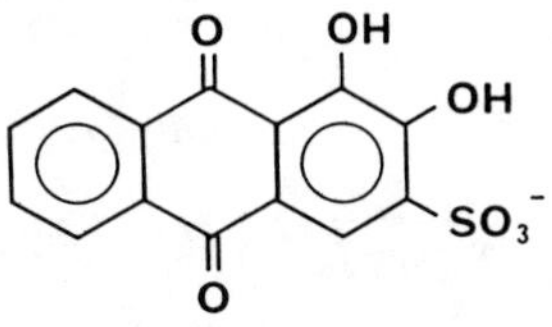

Table 3.4. (*Continued*).

C.I. group no.	Class	Example	Formula (the presumed active component)
	B. Aminoanthraquinone	Kernechtrot C.I. 60760	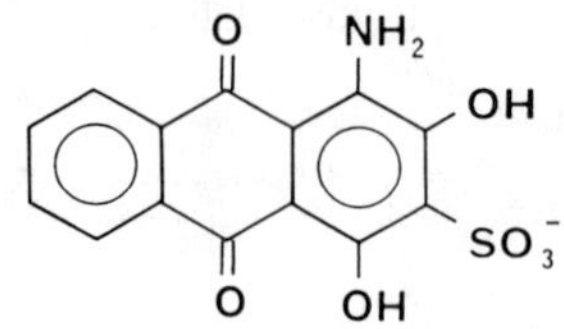
17.	Indigo	Indigo C.I. 73000	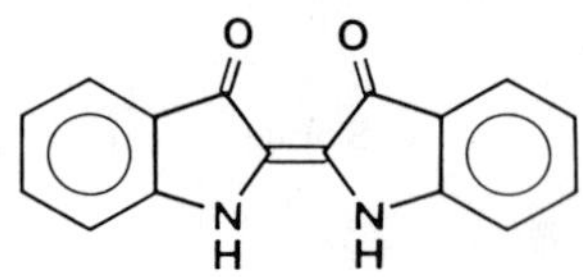
18.	Phthalocyanin	Alcian Blue 8GX C.I. 74240	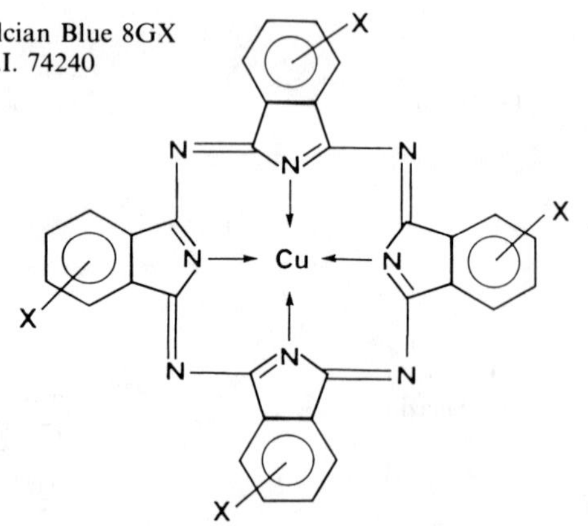 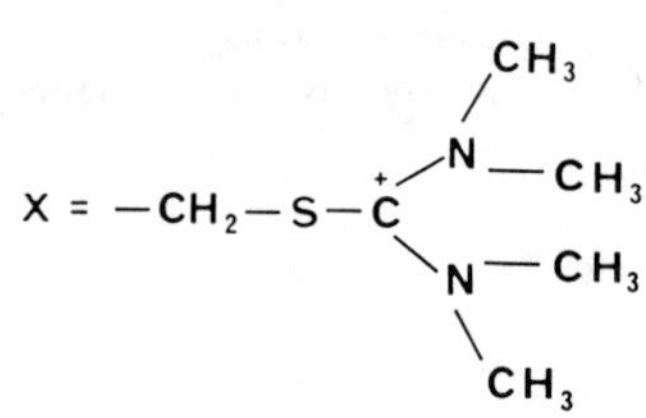

Table 3.4. (*Continued*).

C.I. group no.	Class	Example	Formula (the presumed active component)
19.	Natural	Haematoxylin C.I. 75290	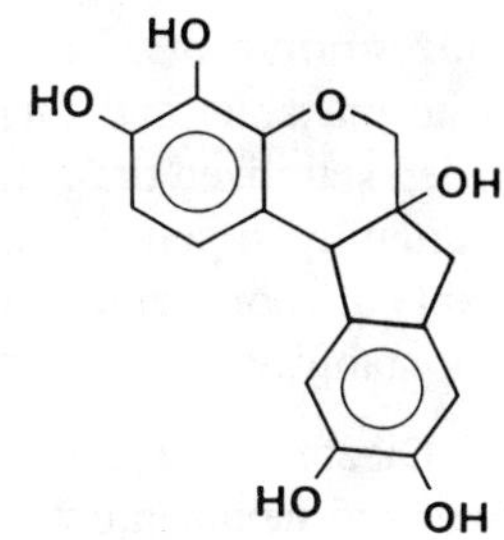
20.	Different chromogenic reagents and metal pigments	Prussian Blue C.I. 77510	$\overset{III}{Fe}_4[\overset{II}{Fe}(CN)_6]_3$

Reactive Dyes. This new group should be added to the list. According to Rys and Zollinger (1972), a reactive dye is defined as a coloured substance containing a group that is able to form a covalent bond between a carbon atom in the dye ion or molecule and an oxygen, nitrogen, or sulphur atom in a hydroxyl, an amino, or a thiol group respectively in the substrate. The first reactive dyes which appeared in 1956 were the procions which contain the reactive group 2-amino-4,6-dichlorotriazine with two labile chlorine atoms:

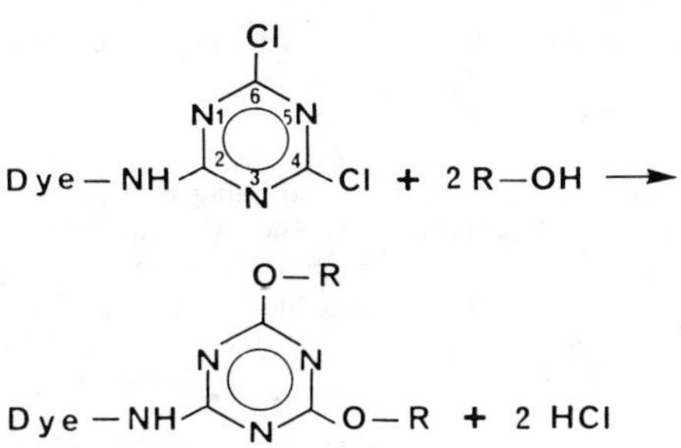

An example is:

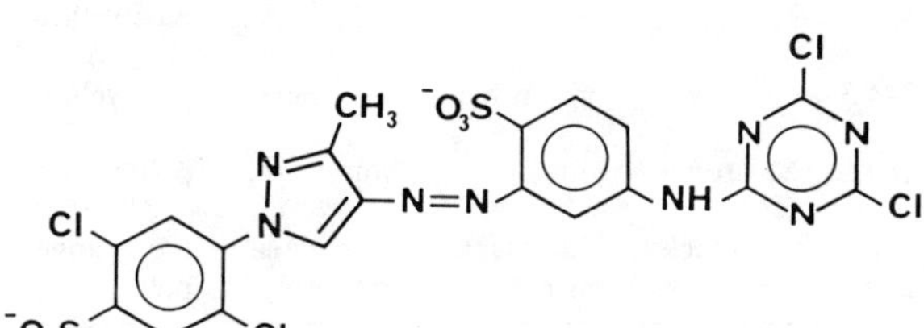

Procion Brilliant Yellow M-6G, C.I. 18971, that has been used for the demonstration of newly formed dentine. Note, that the dye can also be classified as a monoazo dye.

Diazonium Salts. These are salts of the type $R-\overset{+}{N}\equiv N, X^-$. The bases are usually quite unstable, and this is often the case also with the simple halogen salts. So-called stabilized diazonium salts are double salts with zinc chloride, borotrifluoride, sodium hydrogen sulphate, or naphthalene-1,5-disulphonic acid. They are relatively stable at room temperature if kept dry. In solution however, especially at alkaline pH, stability is very limited.

Diazonium salts only contain one $-\overset{+}{N}\equiv N$ group, while tetrazonium salts contain two, and hexazonium salts three. Despite this, the term diazonium salts is widely used to denote the whole group.

Many of the salts are colourless, but when the chromophore azo group $-N{=}N-$ is formed on coupling to tissue bound aromatic amines or phenols or with enzymatically liberated naphthols and naphthylamines colour arises. Many diazotizable amines are already dyes, e.g. Safranin, Pararosanilin, Fast Garnet GBC base, and Fast Black K base. In these cases the new azo dye is usually far more intense and is stable towards treatment with acid. Choice of a suitable diazonium salt for a particular histochemical purpose can be very difficult when confronted with the jungle of currently available commercial products. Some examples of frequently used diazonium salts are given in Table 3.5.

Table 3.5. The colour of some coupling products with different diazonium salts (modified after Lillie and Fullmer, 1976, p. 248).

Name	C.I. no.	Mol. weight (amine)	Entero-chromaffin	Colour on coupling with β-naphthol	naphthol AS	protein
Safranin, diazotized immediately before use	50240	350.8	blue/black	blue	dark blue	red
Fast Red GG, p-nitro-aniline	37035	138.1	red	red	red	yellow
Fast Red B, p-nitro-o-anisidine	37125	168.2	red	red	red	pale yellow
Sulphanilic acid, diazotized immediately before use		173.2	red	red	red	pale yellow
Fast Garnet GBC, o-aminoazotoluene	37210	225.3	red	red	red	pale yellow
Fast Blue B, di-o-anisidine	37235	244.3	red brown	brown black	dark blue	pale yellow
Benzidine, diazotized immediately before use		184.2	red brown	blue	blue	pale yellow
Fast Black K	37190	302.3	blue black	blue black	blue black	red orange
Fast Red Violet LB		260.7	pale yellow	purple	purple	pale yellow

Table 3.6. Properties of different stabilized diazonium salts (modified after Lillie and Fullmer, 1976, p. 148).

Name	Coupling rate	Stability of solution	Specially suited for demonstrating
Fast Red GG	very fast	good	enterochromaffin, bile pigment
Fast Red B	very fast	excellent	enterochromaffin
Fast Garnet GBC	very fast	moderate	enterochromaffin, esterases
Fast Blue B	very slow	good	protein in coupled tetrazonium
Fast Black K	slow	poor, but 3 h at pH 8	protein (direct reddish orange)
Fast Red Violet LB	fast	excellent	phosphatases

Table 3.6 indicates the coupling rate, the stability in solution, and the preferred application for the stabilized salts listed in Table 3.5.

Azo Coupling Components. This group of reagents are not dyes in themselves but, on enzymatic hydrolysis, they produce phenols, naphthols, or aromatic amines. These are transformed into insoluble azo dyes and deposited as precipitates at the site of the enzyme activity by coupling to diazonium salts either immediately (simultaneous coupling) or later (post-coupling).

The azo compound arising from the azo coupling reagent on coupling with a diazonium salt should fulfil the following criteria:

1. Insoluble in water, acid, and base and preferably also in ethanol and xylene
2. Form a finely granular precipitate
3. Give a precise localization
4. Show high substantivity to proteins (Sect.1.2.3)

Early azo coupling reagents did not meet these requirements. The azo compounds of α-naphthol are moderately soluble in water, have only slight substantivity, and form coarse precipitates, while those of β-naphthol are only slightly soluble in water, have a finely granular precipitate, and moderate substantivity, but are easily dissolved in an alkaline solvent. The azo derivatives of 6-bromo-2-naphthol and 6-benzoyl-2-naphthol are insoluble in water, have a finely granular precipitate, and moderate substantivity, but are destroyed in alkali.

In Table 3.7 a survey of modern azo coupling reagents is given. The azo compounds formed with these are all insoluble in water, form very finely granular precipitates, show pronounced substantivity, and give a precise and sharp localization.

Tetrazolium Salts. These are chiefly used in histochemistry in the demonstration of oxidoreductases (Sect.25.1.2). The following factors are of importance in the choice of a tetrazolium salt. It should:

1. be pure, stable in light, easily soluble, and colourless in solution
2. easily be reduced under histochemical conditions
3. have a suitable redox potential

Table 3.7. Azo coupling reagents and their use as substrates for the demonstration of enzymes (modified after Lillie, 1977, p. 217).

Name	Azo coupling reagents: C.I. No.	Mol. weight (amine)	Formula	Compounds	Enzymes
Naphthol AS	37505	263.3	Formula 3.43	phosphate acetate sulphate chloroacetate ($-CO-CH_2Cl$) phenylpropionate ($-CO-CH_2CH_2C_6H_5$)	alkaline and acid phosphatases esterase arylsulphatase protease of chymotrypsin like type protease of chymotrypsin like type
	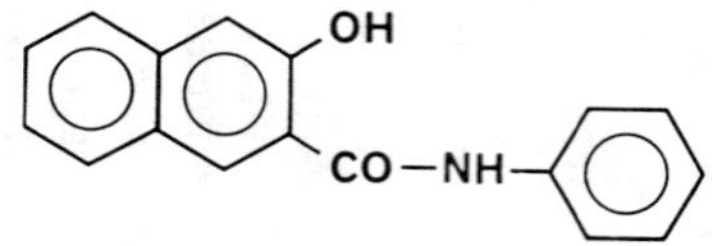				
Naphthol AS-E	37510	297.7	Formula 3.44	phosphate	alkaline and acid phosphatases
	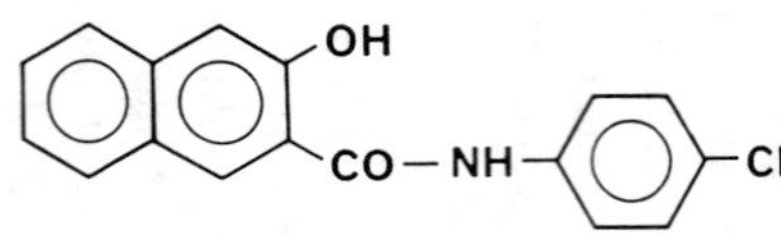				
Naphthol AS-AN	37516	308.3	Formula 3.45	phosphate	alkaline and acid phosphatases
	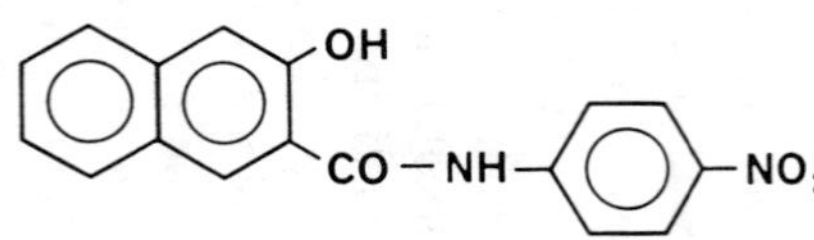				
Naphthol AS-TR	37525	311.8	Formula 3.46	phosphate	alkaline and acid phosphatases
	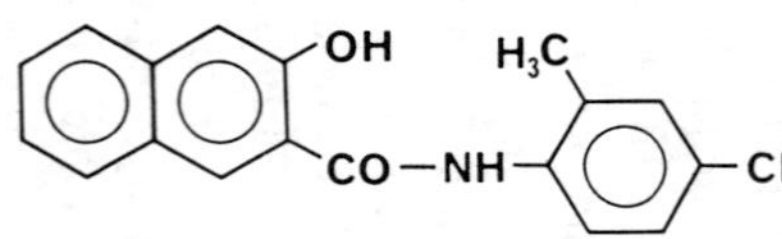				

Table 3.7. (*Continued*).

Name	Azo coupling reagents C.I. No.	Mol. weight (amine)	Formula	Compounds	Enzymes
Naphthol AS-MX	37527	291.4	Formula 3.47	phosphate	alkaline and acid phosphatases
	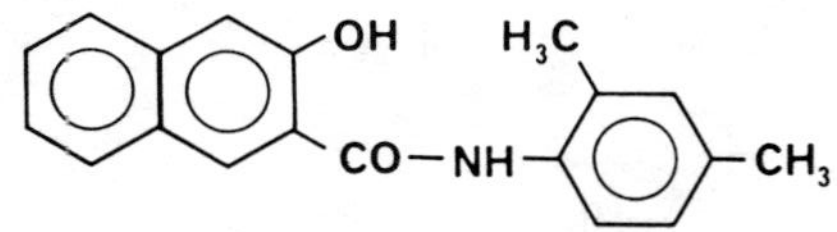				
Naphthol AS-CL	37531	327.8	Formula 3.48	phosphate	alkaline and acid phosphatases
	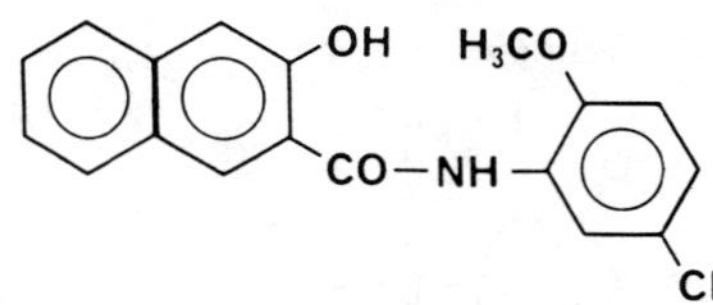				
Naphthol AS-BA	37532	372.2	Formula 3.49	phosphate	alkaline phosphatase
	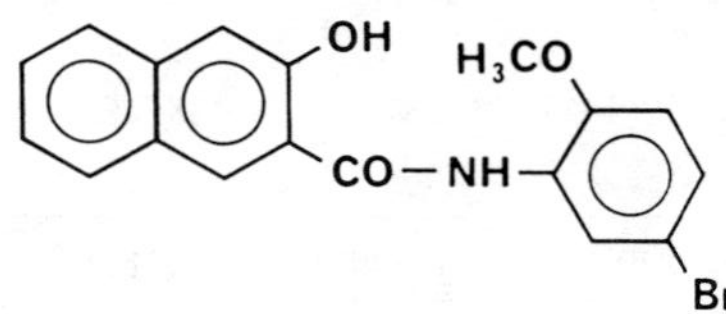				

Table 3.7. *(Continued).*

Azo coupling reagents					
Name	C.I. No.	Mol. weight (amine)	Formula	Compounds	Enzymes
Naphthol AS-LC	37555	357.8	Formula 3.50	phosphate acetate β-D-glucuronide N-acetyl-β-glucosaminide	acid phosphatase esterase β-D-glucuronidase N-acetyl-β-glucosaminidase
			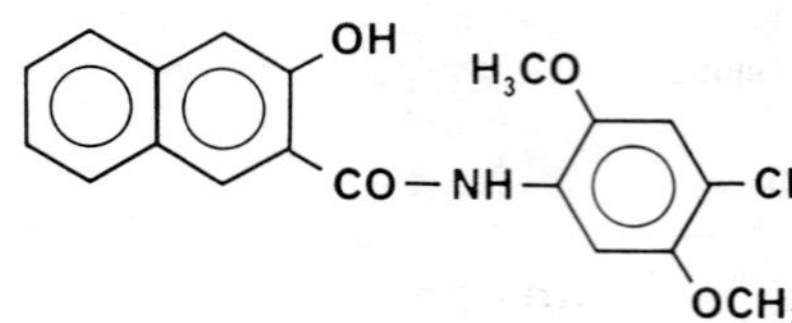		
Naphthol AS-BI	37566	293.3	Formula 3.51	phosphate β-D-glucuronide	acid phosphatase β-D-glucuronidase
			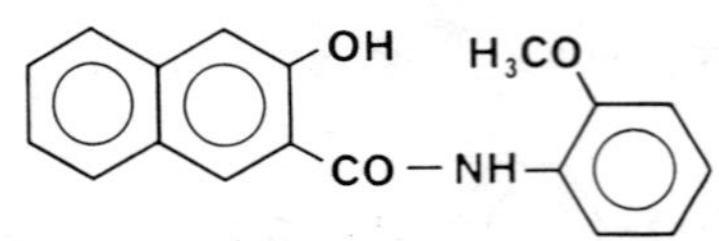		
2-aminoanthroquinone amide of 2-hydroxy-3-naphthoic acid		393.4	Formula 3.52	phosphate	alkaline and acid phosphatases
			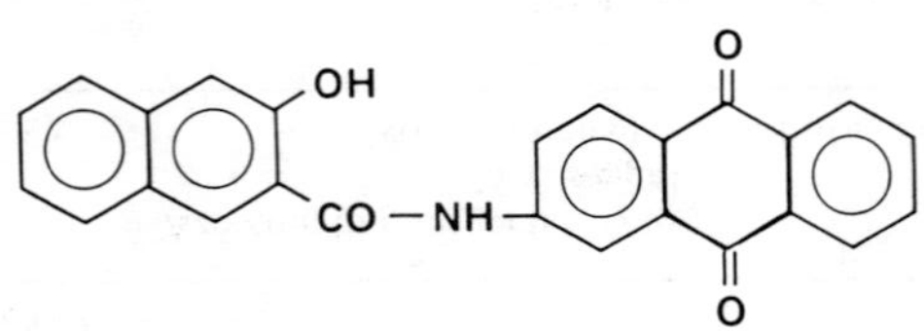		

Aminoazobenzene amide of 2-hydroxy-3-naphthoic acid	367.4	Formula 3.53	phosphate	alkaline and acid phosphatases

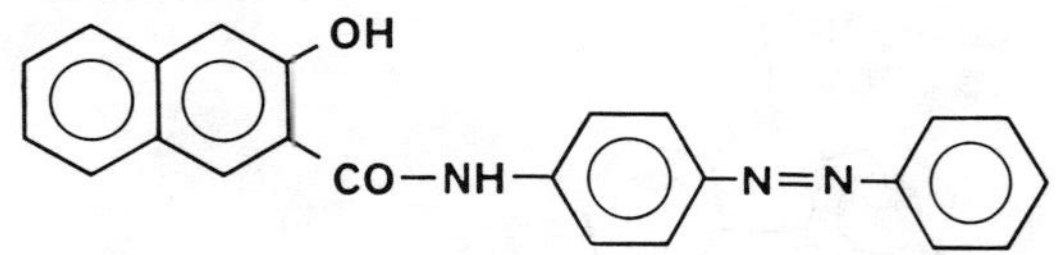

5,6,7,8-β-tetralol-carboxylic acid β-naphthylamide	317.4	Formula 3.54	phosphate	alkaline and acid phosphatases

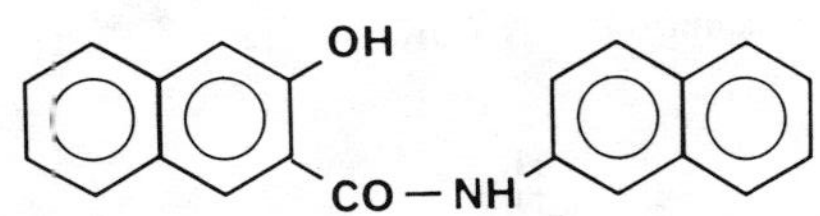

β-naphthylamine	142.2	Formula 3.55	alanyl derivative leucyl derivative	aminopeptidase aminopeptidase

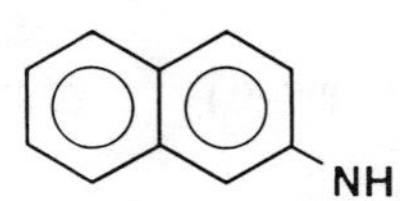

3-aminocarbazole	182.2	Formula 3.56	alanyl derivative leucyl derivative	aminopeptidase aminopeptidase
	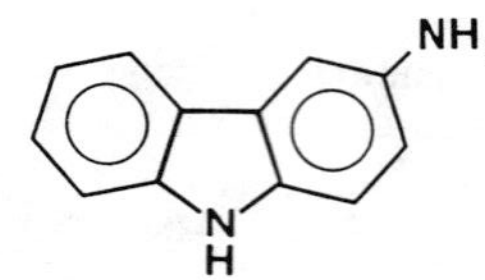			
3-amino-9-ethyl-carbazole	210.3	Formula 3.57	alanyl derivative leucyl derivative	aminopeptidase aminopeptidase
	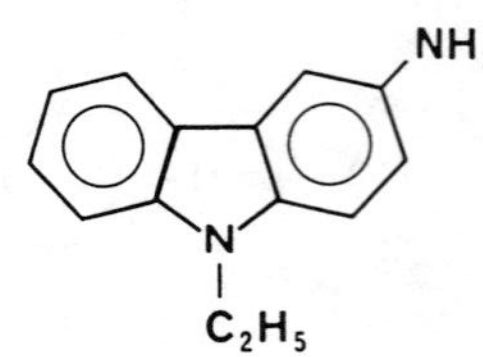			

4. not inhibit the enzyme
5. easily penetrate membranes.

The purity of tetrazolium salts can be easily tested by thin layer chromatography (e.g. with tertiary butanol-water-glacial acetic acid as the solvent. The plate is developed with alkaline ascorbate or ammonium sulphide. Other requirements for tetrazolium salts are enlarged upon in Sect.25.1.2 in conjunction with a consideration of the formazans.

3.3.9 Schiff's Reagent

Schiff's reagent is used for the specific demonstration of aldehyde groups. The reaction with ketone groups may effectively be ignored as this occurs much more slowly (10–20 h against 10–20 min). Uses are listed in Table 3.8.

Basic Fuchsin is used to prepare the traditional Schiff's reagent. Alternatively, Pararosanilin, New Fuchsin, or possibly Rosanilin are now frequently preferred. While it can be bubbled directly through the dye solution, SO_2 is usually prepared from bisulphite, metabisulphite, dithionite or thionyl chloride, which are added as salts, alone or in combination. Coloured impurities are then removed from the primary decolourized product using activated carbon. It is important that this step should not be performed before treatment with sulphite. The finished reagent should be quite clear without colour and should smell of SO_2. pH should be between 1.2 and 2.5.

The precise chemical composition of Schiff's reagent is under debate. For many years with Pararosanilin as the starting material, it was believed that Schiff's reagent

Table 3.8. Histochemical reactions that are completed by the demonstration of aldehyde groups with Schiff's reagent.

Pretreatment	Demonstrated groups*	Demonstrated substance	Reference
None	aldehyde	lysinal aldehyde in young elastin	
Acid hydrolysis	purine–N–C_1–deoxyribose glycoside	DNA	9.9 (Feulgen's nucleal)
Mercuric chloride	acetal phosphatide	plasmalogens	19.6.4 (plasmal)
Ultraviolet light	alkene	unsaturated lipids	9.1.4; 19.6.8 (UV-Schiff)
Oxygen	alkene	unsaturated lipids	9.1.3; 19.6.4 (O_2-Schiff)
Peracid oxidation	alkene	unsaturated lipids	9.1.2 (peracid oxidation-Schiff)
Oxidation	1,2-glycol, α-amino-alcohol	glycoproteins, neutral proteoglycans	9.2.1 (PAS)
Oxidative deamination	lysyl	protein	9.5.1 (oxidative deamination Schiff)

* Schiff's reagent demonstrates all aldehyde groups present, including surplus fixative (see 9.2.1).

was a N,N-disulphinic acid derivative of Pararosanilin leucosulphonic acid (II) (Wieland and Scheuing, 1921).

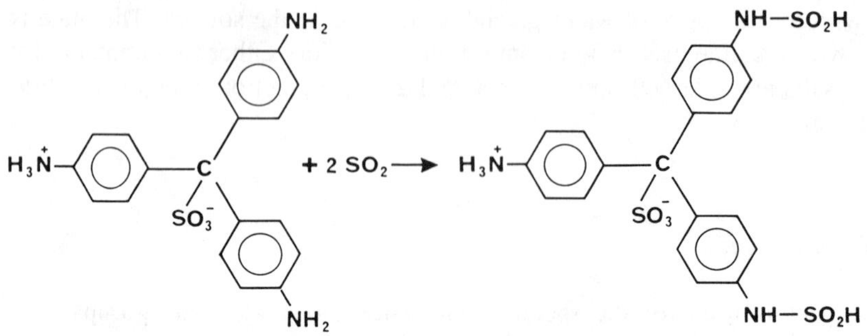

I II

More recent investigations, however, suggest that the reagent is really the leucosulphonic acid (I) itself (Gill and Jotz, 1976). While refrigeration is unnecessary, Schiff's reagent should be kept in a dark bottle. The activity of the reagent can easily be tested by mixing a few drops with one drop of formalin. The magenta red colour should appear immediately. A number of other dyes and fluorochromes can be used to prepare Schiff-type reagents after treatment with SO_2. In contrast to the original Schiff's reagent, these are, however, often coloured. Those prepared from Thionin (blue) and Acriflavine (fluorescent) are amongst the most frequently used.

3.3.10 The Purity of Dyes

Commercial dye products show varying degrees of purity. In the U.S.A. a voluntary arrangement has been made so that manufacturers of dyes can send their products for investigation to the Biological Stain Commission (BSC). BSC has placed certain physical and chemical demands on the dyes and also tests them in different standardized staining procedures. If a dye batch fulfils the demands a certificate is issued stating that the batch contains the dye and that this has a stated degree of purity. A copy of this certificate accompanies the individual samples of this batch when sold by the producer. The physical and chemical tests performed by BSC include a spectrophotometric determination of the absorption maximum of the dye and an evaluation of the symmetry of the absorption peak (see Fig. 3.2).

As stated above, only thin layer chromatography (TLC) or high performance liquid chromatography (HPLC) may be accepted as absolute determinations of purity. In addition, the BSC carries out an assay of the amount of pure dye in the sample. This is performed by spectrophotometry, by precipitation, or in the majority of cases by a redox titration with titanous chloride, $TiCl_3$.

3.4 Enzymes as Reagents

The chief use of enzymes is for the hydrolytic extraction of specific molecules in tissue sections. Many of the same considerations as those discussed in relation to the demonstration of enzyme activity (see Chaps.23–25) also apply in this context.

3.4.1 Factors that Influence the Enzymatic Hydrolysis

Factors influencing enzymatic hydrolysis include concentration of the enzyme, incubation time, temperature, pH, content of activators and inhibitors in the incubation medium, and finally, stability of the enzyme under these conditions. The best results are generally achieved by dissolving the enzyme in glass-redistilled, demineralized water; pH will then be around neutral, frequently slightly lower than 7. The pH-optimum for the enzyme is thus not sought, as any addition of an electrolyte to the solution can change the activity of the enzyme considerably. Likewise, an incubation temperature of 37°C is generally used. The incubation time is, however, varied empirically, though one should be aware that lengthening the incubation time does not automatically lead to an increased extraction of the substrate (the substance to be extracted), if the enzyme is rapidly inactivated. Furthermore, one should be aware that mercury, chromium, and osmium in the tissue (from the fixative) frequently act as enzyme inhibitors. The material used for the container of the enzyme solution can also be critical. Certain glass and plastic types should be avoided as they can release substances which act as enzyme inhibitors.

3.4.2 Factors that Influence the Result of the Staining Reaction

The result of the enzymatic extraction is assessed by comparing the staining results before and after the enzyme extraction. The result depends partly on the fixation method and partly on the choice of staining reaction.

As with other procedures, fixation should, as far as possible, preserve both the concentration and the localization of the substrate prior to enzyme extraction. The fixative should not limit access of the enzyme to the substrate or inhibit the enzyme itself. Finally, the fixative should preserve the substrate in such a manner that the enzymatic reaction products will be able to diffuse freely out of the tissue.

The method for visualization of the substrate before and after the enzyme extraction should have a relatively high selectivity for the substrate in question. Furthermore, the relationship between intensity of the colour arising and the amount of substrate present in the tissue section should be as close as possible. Any form of differentiation of the final product must therefore be avoided, even if this means that an aqueous mountant must be used.

3.4.3 Purity of Enzymes as Reagents

Enzyme preparations of different degrees of purity and activity, priced accordingly, are generally available. It is appropriate to evaluate what level of activity is required, what impurities are present, and what influence these might have on the result. For example relatively impure preparations of RNase contain proteases, chondroitinases, sulphatases etc. It is also important to consider the starting material for production of the enzyme. For example the ground substance of cartilage is quickly degraded by RNase preparations from malt, while RNase preparations from pancreas do not have this effect.

Part 2
Review of Techniques According to Mechanism

4 General Theory for Tissue Staining

A.P. Andersen, P. Prentø, H. Lyon

Although the specific characteristics of the tissue and the choice of staining procedure are the central issue in demonstrating histological features, the pretreatment and processing following staining also make important contributions to the final result. The reactivity of all relevant tissue components must be taken into consideration for each step of the process.

Dye Affinity. The affinity of a dye for a tissue element is a measure of the ability of the dye to bind to a given tissue as opposed to staying in solution. In this context, affinity takes into account the influences of *all* aspects of the staining system (tissue section-dye-solvent-additives). An excellent review on the problems relating to dye affinity is given by Horobin (1982a) in chapter 4 of his book. The key factors involved in determining dye tissue affinity are reviewed in Table 4.1.

4.1 Tissue Sections

Tissue sections predominantly consist of solid material. The retained solids may show very substantial variations in stainability dependent on their content of reactive groups and the accessibility of these. Some reactive groups may be charged (Chap.6); e.g.:

$-SO_4^-$	acid heteropolysaccharides
$=PO_4^-$	nucleic acids
$-COO^-$	acid heteropolysaccharides and proteins
$-O^-$	proteins
$-S^-$	proteins
$-\overset{+}{N}H_3$ and derivatives	proteins

Reactive groups that can be demonstrated by virtue of their metal-binding properties are discussed in Chaps.7 and 8 while Chap.9 deals with so-called reactive or chromogenic reactions (staining reactions involving covalent bonds). Direct non-covalent dye-tissue interactions will be considered here. Lipid droplets are stained

H. Lyon (Ed.)
Theory and Strategy in Histochemistry

Table 4.1. Interacting factors of importance for tissue staining.

Component	Properties
Tissue section (see 4.1)	
Solid material	Different composition Difference in: a) reactive groups, including charge b) distribution of reactive groups c) concentration of reactive groups d) permeability
Liquid material	Hydrophobic lipid droplets
Surfaces	Different reactivity and adsorption characteristics (physical and chemical)
Dye (see 4.2)	Frequently large, organic molecules usually with one or more charges. Characterized by: a) reactive groups b) charge c) colour d) molecular weight e) steric form
Solvent (see 4.3)	
Liquid	Hydrophobic or hydrophilic. Characterized by: a) dipole moment b) dielectric constant c) surface tension
Additives (see 4.4)	
Salts	Influence: a) ionic strength b) pH
Organic substances	Influence: a) ionization b) surface tension c) storage qualities

by hydrophobic dye molecules (Sect.19.6.1) which exhibit higher solubility in the fluid lipid droplet than in the dye solvent.

Surfaces on and in the tissue can bind dyes by adsorption. If the binding to the surface alone is due to van der Waal bonds (Sect.4.5.4) this is called *physical adsorption*. Several layers of molecules can be adsorbed to the surface.

In contrast, where direct non-covalent dye-surface bonds replace the weaker dye-solvent interactions in solution, only a single layer of molecules is adsorbed. This form of adsorption increases with decreasing solubility. Non-electrolytes are adsorbed more strongly than electrolytes, with the exception of large, organic ions, such as dye molecules.

Adsorption depends also on temperature, concentration, the nature of the surface, surface tension, and reactive groups. The adsorption will be dependent on the structure of the tissue components, the fixation, and the embedding procedure.

4.2 Dye

The structural features of a dye that are important in determining its tissue staining properties are considered below.

4.2.1 The Electrical Charge of the Dye

The following ionized groups may be present:

$-SO_3^-$ $-COO^-$ $-O^-$	anionic dyes
$-\overset{+}{N}H_3$ and derived	cationic dyes

The degree of ionization will depend on pH and the dielectric constant of the solvent.

4.2.2 Covalent Reactive Groups

Examples of covalent reactive groups are the chloromercuric groups in Mercury Orange (Sect.9.3.2) and the chlorotriazine group in procion dyes (Sect.3.3.8).

4.2.3 Groups with Potential for Forming Metal Complexes

These groups include carboxyl, hydroxyl, and nitroso groups. See further Sect.7.1.

4.2.4 The Polar Character of the Dye

Easily polarizable dyes with a large dipole moment can form intermolecular (van der Waal) bonds to the tissue (Sect.4.5.4). Particularly pronounced polarizability and large dipole moments are seen in dyes which consist of several aromatic subunits, as is the case for polyazo dyes. Asymmetry of the molecule also contributes towards increasing the dipole moment.

4.2.5 The Hydrophobic-Hydrophilic Balance of the Dye

Decisive for the binding of the dye to the tissue is the ratio between its hydrophobic and hydrophilic groups. The greater the hydrophobic or aromatic part of the dye is, the greater will be the tendency for a dye dissolved in water to bind to hydrophobic areas in the tissue.

4.2.6 Groups with Potential for Forming Hydrogen Bonds

Phenol, amino, azo, and carboxyl groups can all form hydrogen bonds with analogous groups in the tissue. In aqueous solution the hydrogen bonds with water molecules will normally be dominant unless other forms of binding between dye and tissue are taking place at the same time. The binding is usually stabilized by a hydrophilic-hydrophobic interaction between dye, solvent, and tissue (Sect.4.5.5).

4.3 Solvent

As mentioned in Sect.3.2, it is important to distinguish between polar and non-polar solvents. A polar solvent such as water favours hydrophobic interaction between dye molecules, between dye molecules and tissue groups, and between tissue groups mutually. Non-polar solvents such as xylene have the opposite effect, while ethanol has intermediate properties in this regard.

4.4 Additives

Additives to the staining solution can be acids, bases, inorganic salts, organic solvents, or organic compounds such as urea or dimethyl sulphoxide (DMSO).

4.4.1 Acids and Bases

Acids and bases shift pH and consequently charge-based attractions or repulsions with dye molecules. Changes in pH will also affect the amount of water bound by the tissue.

4.4.2 Inorganic Salts Including Buffer Salts

Some inorganic ions such as Na^+, Mg^{2+}, and Cl^- can influence the dye-tissue affinity for anionic and cationic dyes by competing for tissue binding sites (Critical electrolyte concentration, Sect.6.1.2). Other ions including Al^{3+} and Cr^{3+} are hydrated and can form complexes with dyes and/or tissue ligands. These salts have also a pH-effect.

4.4.3 Organic Additives

Some organic additives act as detergents and consequently reduce the surface tension of the solution. Others may alter dye activity by inducing micelle formation.

4.5 Bonds Formed between Dye and Tissue

The different kinds of bonds which can be formed between dye and tissue are enumerated below. In addition, approximate sizes of bonding energy are given in kJ/mol.

	Anhydrous crystals ($\epsilon = 1$)	In water ($\epsilon = 80$)
1. Ionic attraction	800 (max.)	10
2. Covalent bond	400	400
3. Complex bond	400	400
4. Intermolecular bonds	4–40	4–40

4.5.1 Ionic Bond

In polar solvents, and especially in water, ionic forces are much weaker than 800 kJ/mol as expressed by the equation

$$F = \frac{q_1 q_2}{\epsilon \times r^2}$$

where F is the attraction force between two ions with charges q_1 and q_2 separated from each other by the distance r. ϵ is the dielectric constant (Sect.3.2).

For water ϵ is 80 which reduces the attraction from about 800 (for two univalent ions) to about 10, or of the same order of magnitude as hydrogen bonds and ion-dipole forces (Sect.4.5.4). The latter two are of course the main intermolecular forces in water. Between certain di- or trivalent metal ions and water molecules the ion-dipole forces are especially strong and may result in the formation of complex bonds where in fact the interatomic distance is comparable to that in a salt crystal (Sect.7.1).

The ratio of the radii of the cation and the anion determines how many anions surround a given cation (the coordination number). The closer the ions can approach each other the stronger the attraction. It is evident that charged groups in macromolecules, or in dye ions and macromolecules, may interact even though they are not directly opposed. For instance, this is the case in the cooperative bonding which takes place between DNA and histone, and probably also in the bonding between tissue components and dye ions. To avoid the precise term, “ionic bond”,the terms "salt bridge" or simply "electrostatic attraction" are frequently adopted. The possibility for salt bridge formation will thus be determined by:

1. The charge of the dye ion
2. The charge of the reactive groups
3. The dielectric constant, ionic strength, and pH of the solvent.

In general, the hydrated macromolecules, including the majority of proteins, do not show any greater tendency to form ionic bonds with each other than with the small organic or inorganic ions normally present in the cells and tissues.

4.5.2 Covalent Bond

Organic compounds are held together by covalent bonds.

Covalent bonds are formed between non-metals and consist in two atoms sharing one or more electron pairs, whereby both atoms generally fulfil the octet rule. A condition for formation of covalent bonds between dye and tissue is that both contain functional groups with affinity to each other, e.g.

1. Aldehyde group and Schiff's reagent (Sect.9.7.1)
2. Coupling of diazonium ion to aromatic amines or phenols (Sect.9.4.1).

4.5.3 Complex Bond

This bond may be considered as a transition between an ionic bond and a covalent bond. The compound consists of a central atom to which are bound two or more ligands, depending on the coordination number.

The central atom is usually a metal ion with a small ionic radius and high density of charge which exerts a great force of attraction. The ligands are ions or molecules with a free pair of electrons which are drawn towards the central atom.

The ligands are bound directly to the central atom by the joint possession of electrons. The complex compound formed often deviates in properties (colour, solubility, ionizability, reactivity) from its constituent substances. Examples of complex bonds are:

ferric hexaquoion	$[Fe(H_2O)_6]^{3+}$
aluminium hexaquoion	$[Al(H_2O)_6]^{3+}$
silver diammine ion	$[Ag(NH_3)_2]^+$

4.5.4 Intermolecular Forces

In molecules electrons are shared by participating atoms in a manner consistent with the octet rule. The possibilities for covalent bonding are thus used up, but due to existing or induced dipoles, weak bonds may form *between* intact molecules. These are termed intermolecular forces or *van der Waal forces*.

Intermolecular forces (van der Waal forces) are divided into:

a. London forces = dispersive forces (frequently the expression van der Waal forces is used only for the London forces)
b. Dipole bonds which are subdivided into:
   1. Dipole-dipole bonds between permanent dipoles, including the hydrogen bond
   2. Dipole-dipole bonds between permanent and induced dipoles.

London Force or Dispersive Force. This is the weakest kind of bond between molecules (bonding energy about 4 kJ/mol). While there is a factor of r^{-2} in ionic bonds, this factor is r^{-6}, for London forces, where r is the distance between the atoms involved. The bond arises between symmetrical, neutral molecules without any dipole moment. In these molecules, due to the asymmetric movement of the electrons around the nucleus, there will be short-lived induced dipoles that give rise to attractive forces. Under these circumstances involved molecules will be so close to each other that repellent forces between the electrons will be balanced by the attractive forces. Dispersive forces, when they are not reinforced by permanent dipoles (see below), will thus constitute the only van der Waal attraction between the molecules. Due to the extreme dependence on distance (r^{-6}), dispersive bonds are very much affected by temperature.

Dipole Bonds. Dipole bonds (bonding energy as a rule between 10 and 20 kJ/mol) can arise between two molecules with dipole moments or between one molecule with a dipole moment and a second in which a dipole moment can be induced, e.g. a π-electron system. The force of attraction F can be expressed in the same manner as for ionic bonds, though the dependence on distance varies according to whether it is an ion-dipole bond or a dipole-dipole bond.

$$\text{F ion-ion} = \frac{q_1 q_2}{r^2}$$

$$\text{F ion-dipole} = \frac{q_1 \mu_2}{r^3}$$

$$\text{F dipole-dipole} = \frac{\mu_1 \mu_2}{r^4}$$

q = ionic charge; μ = dipole moment; r = distance between charges; F = force of attraction.

A particularly strong dipole bond is the *hydrogen bond* (bonding energy about 40 kJ/mol). This occurs in molecules where a hydrogen atom is covalently bound to a strong electronegative atom, e.g. oxygen, nitrogen, or sulphur.

Atomic groups that can participate in hydrogen bonds include:

$$-\mathrm{OH}, \quad -\mathrm{NH_2}, \quad =\mathrm{NH}, \quad =\mathrm{N}-, \quad \text{and} \ {>}\mathrm{C{=}O}$$

In addition to their importance for solubility in water, hydrogen bonds also play a part in the mutual association of dye molecules and in dye-tissue binding. Both of these latter associations are normally accompanied by hydrophobic stabilization (Sect.4.5.5).

4.5.5 Hydrophobic Interaction

In aqueous solution molecules containing both hydrophilic and hydrophobic groups tend to orientate with the hydrophilic groups facing away. This pushes the hydrophobic groups closely together allowing dispersive bonds to form. The water molecules dispersively bound to the hydrophobic groups are released by this process to bind freely to other water molecules thus forming the maximum number of stronger hydrogen bonds.

This sequence of events is termed *hydrophobic stabilization* and is critical for the formation and stability of cell membranes. It is also very important in determining the structure of macromolecules, particularly DNA and proteins.

In a hydrophilic environment the molecules orientate themselves with their hydrophilic groups facing outwards leaving a hydrophobic core from which water is excluded. This favours the formation of hydrogen bonds. For example the relatively hydrophobic bases in DNA tend to form the maximum number of hydrogen bonds with each other thus stabilizing the double helix structure. Peptide-NH and peptide-CO in proteins similarly form the maximum number of hydrogen bonds with each other, >NH .... OC<, contributing to the stabilization of the secondary structure. These macromolecules can only change their conformation significantly if a large number of hydrogen bonds are broken at the same time (cooperatively). This requires considerable energy and is counteracted by any hydrophobic effects which tend to keep hydrogen bonding groups close together.

Like the bases in nucleic acids, dyes consist of planar ring systems which carry different atomic groups. Some of these groups are ionized and are therefore soluble in water, while others are able to form hydrogen bonds or weaker dipole bonds. In consequence there is considerable potential for hydrophobic interaction in aqueous staining systems (e.g. aggregate formation of metachromatic dyes, Sect.6.1.1). Hydrophobic interaction is very important in the staining of macromolecules.

A special case is the binding of metachromatic dyes to negatively charged carbohydrates. Here the ordered aggregation of the dye is favoured by the neutralization of charges that takes place on binding. In most forms of protein-dye associations, hydrophobic interactions hinder the formation of ordered dye aggregates. Moreover, the distances between ionized groups in protein molecules are frequently too large to allow much dye aggregation. Thus metachromatic staining of proteins is rare.

5 Blocking and Deblocking Reactions

M.R. Barer, H. Lyon

Positive Histochemical Reaction. This is a reaction producing a reaction product that can be seen using a microscope.

Blocking Reaction. A blocking reaction chemically modifies a specific set of reactive groups in a tissue in such a way that, without rendering them visible, the product cannot react further in subsequent processing steps.

Deblocking Reaction. A blocking reaction can frequently be reversed by a deblocking reaction. Deblocking can also be used to reveal demonstrable groups that are naturally concealed.

Blocking reactions may affect several different kinds of reactive group. It should be appreciated that a reaction used to block or deblock under one set of circumstances may often be used in another situation as a step in a positive histochemical reaction. An overview including some of the most important reactions of this kind now follows.

5.1 Classification of Blocking Reactions According to Mechanism

The majority of blocking and deblocking reactions belong to one of the following 6 groups: Oxidation, reduction, formation of acid derivatives, hydrolysis of acid derivatives, deamination, and addition processes.

5.1.1 Oxidation

The use of oxidants can be subdivided into the use of *weak* oxidants (e.g. iodine, bromine, hydrogen peroxide, or a ferric salt), *medium* strength oxidants (e.g. performic acid and persulphate), and *strong* oxidants (e.g. permanganate and chromium trioxide) (Table 5.1).

H. Lyon (Ed.)
Theory and Strategy in Histochemistry

Table 5.1. Histochemical application of oxidants.

Oxidant	Tissue bound group	Compound formed	Use	Reference
weak	thiol	disulphide	demonstration	8.3.1
			blocking	9.3.2
	sulphite blocked aldehyde	aldehyde	deblocking	
medium	alkene	glycol		
		aldehyde	demonstration	9.1; 9.2
			blocking	9.1
	tryptophanyl	o-amino-benzoic acid	blocking	9.4.5
	cyanide or hydroxylamine blocked aldehyde	aldehyde	deblocking	
strong	alkene	glycol		
		aldehyde	demonstration	8.3.2
		carboxyl		
	alkylated carboxyl	carboxyl	deblocking	
	thiol	sulphonic acid	demonstration	

5.1.2 Reduction

Reduction is used for several very different histochemical purposes (Table 5.2). Examples of useful reducing agents include formaldehyde, hydroquinone and sodium or potassium borohydride. Of these, the last two are by far the strongest and are essential where reduction of oxo groups is required.

Table 5.2. Histochemical application of reductants.

Tissue bound group	Compound formed	Use	Reference
aldehyde	primary alcohol	blocking	9.7.2; 22.2.1
ketone	secondary alcohol	blocking	9.7.2
disulphide	thiol	demonstration	9.3.3; 21.3
		deblocking	9.3; 21.3

5.1.3 Formation of Acid Derivatives

As shown in Table 5.3, the formation of acid derivatives (ester, amide) comprises *acylation*, which is performed with an acid anhydride or an acid halogenide, and *alkylation*, which is performed with an alcohol (usually methanol).

5.1.4 Hydrolysis of Acid Derivatives

This is a *saponification* reaction (Table 5.4) with an aqueous or alcoholic solution of a base, usually KOH or NH_3.

Table 5.3. Histochemical use of acid derivatives (ester, amide).

Method	Tissue bound group	Compound formed	Use	Reference
acylation	alcohol	ester	blocking	9.2.1
	phenol	ester	blocking	9.4.1; 9.4.3
	thiol	thioester	blocking	9.3
	primary and secondary amine	acid amide	blocking	9.5.1
alkylation	carboxylate	ester	blocking	9.8.3
	sulphate	ester or alcohol	blocking	9.8.3

Table 5.4. Histochemical uses of the saponification reaction.

Tissue bound group	Compound formed	Use	Reference
acylated alcohol	alcohol	deblocking	9.2.1
acylated phenol	phenol	deblocking	9.4.1; 9.4.3
thioester	thiol	deblocking	9.2.1; 9.3
acylated primary/ secondary amine	amine	deblocking	9.5
carboxylic ester	carboxyl	deblocking	9.8.3
sulphonic ester	sulphonate	deblocking	9.8.3
sulphate ester	sulphate*	deblocking	9.8.3

* Applies only to so-called mild methylation (see 9.8.3).

Table 5.5. Histochemical uses of deamination.

Method	Tissue bound group	Compound formed	Use	Reference
nitrosation	primary amine	alcohol	blocking	9.4.3; 9.5
	secondary amine	nitrosamine	blocking	9.5
oxidative deamination	primary and secondary amines	aldehyde	demonstration	9.5
			blocking	9.5

5.1.5 Deamination

Deamination may be classified as shown in Table 5.5 into *nitrosation* which is performed with nitrous acid, i.e. a nitrite dissolved in acetic acid and *oxidative deamination*, which can be performed with e.g. chloramine T or ninhydrin.

5.1.6 Addition Reactions

Reactions of addition sometimes followed by condensation are used for blocking aldehyde groups (Table 5.6). Cyanide blockade can be reversed by treatment with periodic acid (Table 5.1). After addition of hydrogen sulphite the blockade can be reversed by short treatment with a weak oxidant (Table 5.1) with a weak acid or base, or by longer treatment with Schiff's reagent (9.7.2). Hydroxylamine blockade may be reversed with an oxidant of medium strength (Table 5.1).

Table 5.6. Histochemical use of reactions of addition and possible condensation.

Tissue bound group	Reaction	Compound formed	Use	Reference
aldehyde	cyanide	hydroxynitrile	blocking	9.7.2
aldehyde	hydrogen sulphite	α-hydroxy-sulphonic acid	blocking	9.7.2
aldehyde	arylamine	Schiff base	blocking	9.7.2
aldehyde	hydroxylamine	oxime	blocking	9.7.2
aldehyde	hydrazine	hydrazone	blocking	9.7.2
aldehyde	dimedone	condensation product	blocking	9.7.2

6 Staining of Macromolecules on the Basis of Charge

P. Prentø, M.R. Barer, H. Lyon, E. Hasselager

Chemical Groups that Can Be Ionized. Table 6.1 lists the different chemical groups in tissue that can be ionized detailing name, formula (the ionized form underlined), occurrence, and approximate pK_A.

Table 6.1. Chemical groups which can be ionized. The ionized form is in italics.

Group		Occurrence	pK_A
primary amine	$R\text{-}\overset{+}{N}H_3 \rightleftarrows R\text{-}NH_2 + H^+$	proteins	9–11
secondary amine	$R\text{-}\overset{+}{N}H_2\text{-}R \rightleftarrows R\text{-}NH\text{-}R + H^+$	proteins	8–10
sulphate	$R\text{-}OSO_2OH \rightarrow R\text{-}OSO_2O^- + H^+$	proteoglycans, sulphomucins	< 0
sulphonate	$R\text{-}SO_2OH \rightarrow R\text{-}SO_2O^- + H^+$	induced corresponding to –SH in proteins	< 0
phosphate diester	Formula 6.0	nucleic acids, phospholipids	2
carboxyl	$R\text{-}COOH \rightleftarrows R\text{-}COO^- + H^+$	proteins, sialomucins, proteoglycans	3–5

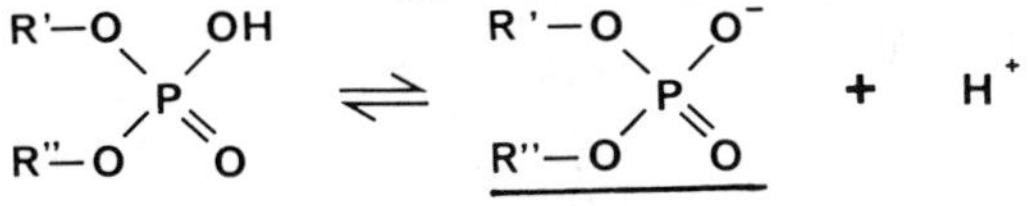

The demonstration of chemical groups that can be ionized naturally falls into two:

- Demonstration of tissue cations
- Demonstration of tissue anions.

In 1877, according to Lison (1960), p.273, Ehrlich introduced the terms *basophil, acidophil,* and *neutrophil* tissue components depending on their differential staining by the so-called basic or acid dyes or a mixture thereof. Basic dyes are now termed *cationic dyes,* and acid dyes are called *anionic dyes* (Puchtler *et al.*, 1985).

6.1 Tissue Groups that Bind Cationic Dyes

Table 6.2 lists the different groups and substances that may be negatively charged. pK_As are also given.

Two preconditions must be met in order to stain these groups:

H. Lyon (Ed.)
Theory and Strategy in Histochemistry

Table 6.2. pK_A-values for different acid groups.

Group	Occurrence	pK_A at 25°C
carboxyl (asparagyl/glutamyl)	proteins	3–5
carboxyl	sialomucin	3–5
carboxyl	proteoglycan	3–5
phosphate	nucleic acids	1–2
sulphate	glycoproteins and proteoglycans	< 0

a. The groups must be ionized when the staining is performed. At a pH equal to the pK_A of the group 50% will be ionized. At $pH = pK_A - 1$ approximately 10%, and at $pH = pK_A - 2$ approximately 1% will be ionized. At $pH = pK_A + 2$ and above practically 100% will be ionized (cf. Fig. 6.6).
b. The total sum of the electric charges on the tissue component must give a net surplus of negative charges. For example a protein at a pH slightly lower than its isoelectric pH will contain negative charges, but the net (external) charge will be positive.

A simplified survey of the electronegative groups, their principal occurrences in tissue sections, and their basophilia is given in Table 6.3.

Table 6.3. The basophilia of different substances (modified from Lison (1960), p. 275).

Group/substance	Visible basophilia
Carboxyl	basophilia at pH 4 and above
proteins	
sialomucins	
proteoglycans	
Ketocarboxylic acid	acid resistant basophilia at pH 2.5–3.0
oxidized lipids	
chromolipids	
Phosphate	basophilia at pH 1.5–2.0 if not protein blocked
nucleic acids	
Sulphate	basophilia specific at pH 1.5 and below
glycoproteins	
proteoglycans	

The Histological Definition of Basophilia. The term basophilia should, according to Lison (1960), p.277, only be used for the selective binding of cationic dyes to groups in staining performed under well-defined conditions. The requirements include:

a. That the staining is performed as a progressive method
b. That the staining is performed using only one kind of cationic dye. If two cationic dyes are used together or in sequence the two dyes will compete for the negative groups. Many of the factors involved in such competition remain undefined.

Only very few tissue components are stainable by cationic dyes at practically all pH-values. They possess an *absolute basophilia*. Examples are the granules in mast cells (heparan sulphate $pK_A < 0$). All other tissue components exhibit a *relative basophilia* (phosphate pK_A 1–2, carboxyl pK_A 3–5), i.e. their stainability with cationic dyes depends upon the staining conditions. These include:

a. pH of the solution
b. Type and concentration of the dye
c. Staining time and temperature
d. Buffer type and concentration
e. Kind of treatment following staining.

Post-treatment is particularly problematic. After the dye bath it is usual to wash briefly in either water or in a buffer at the same pH as the staining solution. If the wash is short the staining result is only slightly affected. If ethanol is used for dehydration, pronounced extraction of cationic dyes frequently takes place. Use of 2-propanol, tertiary butanol (2-methyl-2-propanol), or acetone may reduce this problem (See also Sects.16.2 and 16.3).

What Histochemical Information Does Basophilia Give? As seen in Table 6.3, the lowest pH at which distinct basophilia is retained gives an indication of the chemical substances responsible. Table 6.4 lists additional reactions.

Table 6.4. Additional reactions for the identification of basophilic tissue components

Tissue components	Additional reactions	Section
Nucleic acids	Feulgen's nucleal reaction	9.9
	Chromiun-gallocyanin	7.3
	Methyl Green-Pyronin	6.1.5
Proteoglycans,	PAS	9.2.1
non-sulphated	metachromasia	6.1.1
sulphated	Alcian Blue	6.1.2
Glycoproteins,	Aldehyde Fuchsin	6.1.3
sialomucins		
sulphomucins		
Neutral carbohydrates		
Lipofuscin	acid resistant basophilia	18.4.1
Acid protein	see text	
Uric acid and derivatives	see text	

Acid Protein. In those cases where the additional reactions for nucleic acids, acid carbohydrates, and lipofuscin are negative, a weak basophilia can be due to an acid protein. This can be confirmed with a masked basophilia or metachromasia reaction (Sect.21.4.2).

Uric Acid or Urate Deposits. These are seen in humans with gout (tophi) and also frequently in invertebrates. They are strongly basophilic and can only be identified by negative reactions for all other substances.

6.1.1 Metachromasia

Metachromasia is said to have occurred when, on binding to tissue-ions a dye shows a shift of its absorption maximum and hence of the complementary colour (cf. Fig. 3.1). The spectral shift is due to a change in the resonance of the dye on binding to tissue charges and is therefore most pronounced for small cationic dyes where the charge is a part of the resonance systems of the dye. The non-shifted spectrum for a dye is its *orthochromatic* spectrum, while after the shift the spectrum is called *metachromatic*. The shift of the absorption maximum in metachromasia is normally towards shorter wavelength, hence Safranin O shifts from red towards yellow and the thiazin dyes such as Toluidine Blue and Azure A shift from blue to reddish violet-purple, see Fig. 6.1. The molar extinction coefficient for the new absorption maximum is always less than the extinction coefficient for the non-metachromatic maximum.

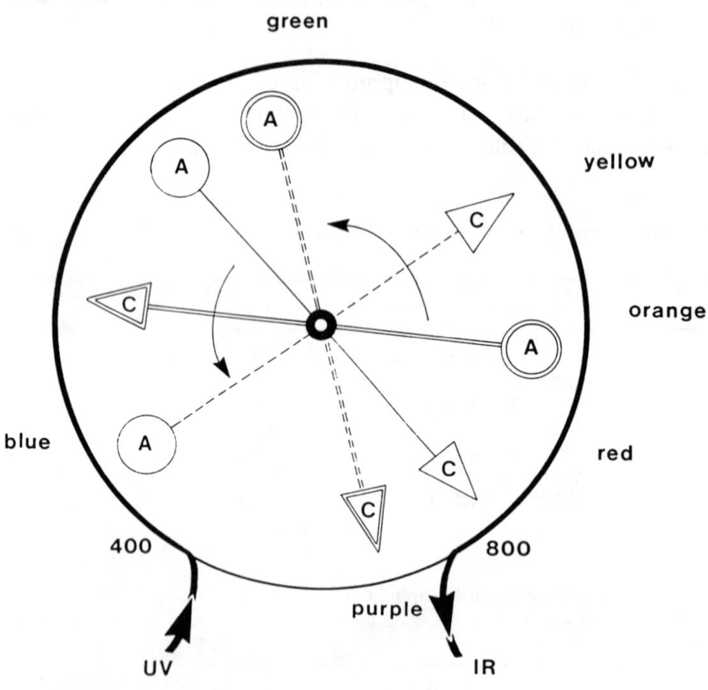

Fig. 6.1. Complementary colour circle showing relationship between absorbed (A) and observed (C) colours. Ultraviolet (UV) and infrared (IR) regions.
The metachromatic spectral shift is shown for two dyes:
Safranin O
——— orthochromatic
- - - - metachromatic
⟶ metachromatic absorption shift
Toluidine Blue
═══ orthochromatic
===== metachromatic
⟶ metachromatic absorption shift

Polyanions are the principle tissue components showing metachromatic binding of cationic dyes. The distance between anionic groups is believed to be critical (Schubert and Hamerman, 1956; Booig, 1958). If the intercharge distance is greater than approximately 0.5 nm the dye is bound orthochromatically; if the intercharge distance is 0.5 nm or less, the dye may be bound metachromatically. More pronounced effects are associated with smaller intercharge distances. The number of closely placed anionic groups (i.e. the charge density) is of similar importance. For mono-ester phosphate ($R{-}OPO_3^{2-}$) a minimum of six closely placed groups is necessary (Pal and Biswas, 1971).

Mechanism. Dyes that show metachromasia form reversible aggregates in water, even at rather low dye concentrations. The two prerequisites for aggregate formation are that the dye molecule should be planar and that it should possess minimal charge. Dyes that have several positive charges, such as Alcian Blue and Methyl Green, do not show visible metachromasia.

A solution of a metachromatic dye will usually have a pronounced asymmetric absorption maximum as an expression of the presence of both monomers and dimers of the dye. In *very* dilute, aqueous solution only the absorption maximum of the monomer, *α-peak*, will be distinct, while the dimer may manifest itself at a somewhat shorter wavelength as a shoulder. In a more concentrated solution the dimer will lead to the presence of a distinct *β-peak*, while the α-band is weakened. In highly concentrated solution a higher degree of aggregate formation can cause a *γ-peak* the breadth of which is an expression of the presence of several different sizes of aggregates; the larger the aggregates, the more pronounced the absorption shift towards shorter wavelength (see Fig. 6.2). Metachromatic dyes thus do not follow the Beer-Lambert law.

Metachromasia in Tissue. This is caused by aggregate formation on bound dye molecules. Binding to the polyanion and the resultant saturation of charges favours the establishment of weak bonds (van der Waal bonds, hydrophobic bonds) between dye molecules. This enables their electron resonance systems to become stabilized. Absorption maximum of dye aggregates is thereby shifted towards shorter wavelength (higher energy) (γ-peak). This is observed as a simultaneous metachromatic shift in observed colour and a fall in the molar extinction coefficient.

The ability to induce metachromasia increases in the sequence:

$R{-}COO^-$	$R_2O_2PO_2^-$	$R{-}OSO_3^-$	$R{-}OPO_3H^-$
carboxylate	diester phosphate	sulphate	monoester phosphate

Hence the degree of ionization and the charge per anionic group both appear important. The most prominent polyanions in tissue are negatively charged carbohydrates, in particular the sulphated polysaccharides. These show a pronounced metachromatic effect. Nucleic acids can sometimes show metachromatic binding due to their high content of diester phosphate. This is most often observed with fresh frozen sections. The metachromasia is, however, usually weak due to steric factors and in-

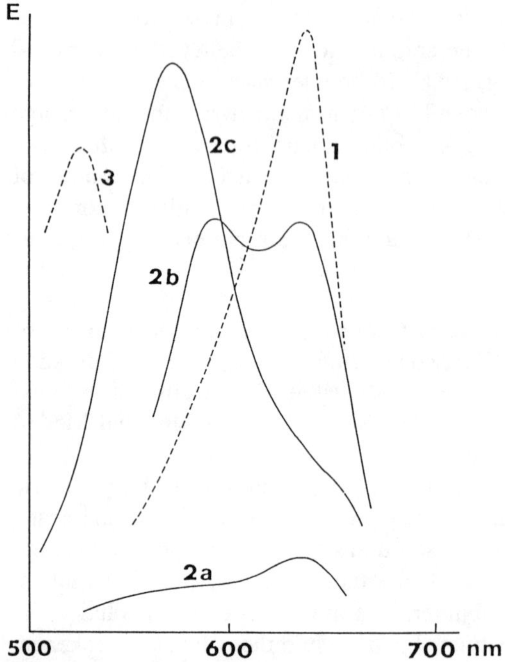

Fig. 6.2. Absorption spectra for Toluidine Blue under different conditions. Extinction (E).

Toluidine Blue in n-butanol		(1)
Toluidine Blue in water		
2.7×10^{-5} mol/l		(2a)
2.0×10^{-4}	"	(2b)
3.0×10^{-3}	"	(2c)
around 10^{-2}	"	(3) (heparin added)

terference from the nucleic acid bases as well as the interaction between phosphate groups and protein-amino groups (particularly histones). The planar dyes (Toluidine Blue, Acridine Orange), which may exhibit metachromasia, tend to intercalate between the heteroaromatic bases when staining non-denatured DNA.

As illustrated above, metachromasia is very sensitive to interference. Alcohols, detergents, relatively low salt concentrations, as well as compounds such as urea and dimethyl urea, are all able to impair or abolish metachromatic binding between dyes and polyanions. With salts the impairment of metachromasia is probably due to a competition between metal ions and dye anions for binding sites on the polyanion. All the other interfering compounds reduce the surface tension of water and thereby diminish the hydrophobic interaction between dye and water.

Metachromasia is thus the result of an ordered aggregation of dye molecules that brings the hydrophobic boundary between dye and water down to a minimum (see Fig. 6.3).

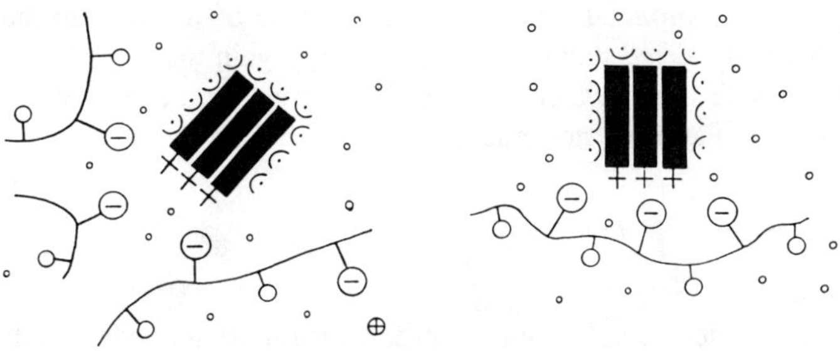

Fig. 6.3. A model of metachromatic bonding between cationic dye and polyanions

o	water molecule
·)	hydrophobic boundary towards water
o–	hydrophobic residue
⊕	cation
⊖–	anionic site
+▬	cationic dye

Proteins do not normally bind dyes metachromatically. This is probably due to the relative stability of the association between dye molecules and hydrophobic groups in proteins compared to the stability of mutual interactions between dye molecules. In addition, the interaction between positive and negative groups within the protein leads to reduced availability of anionic groups. Proteins with many asparagyl or glutamyl residues in sequence or near sequence have an elongated α-helix configuration, a high degree of hydration, and are able to bind cationic dyes metachromatically.

A converse effect, involving metachromatic binding of anionic dyes can be achieved with polyamino-proteins, especially histones, stained under suitable conditions (e.g. Biebrich Scarlet, C.I. 26905). The majority of anionic dyes are, however, unable to show metachromatic bonding.

Practical Applications of Metachromasia. Metachromatic binding to a tissue section provides good evidence for the presence of polyanions. It is more reliable than the binding of Alcian Blue for example. Lack of metachromasia, however, does not imply that polyanions are absent as the reaction is very sensitive to interference (see above).

To optimize the visible effect, the stained section must be mounted in water as the surface tension between water and dye molecules contributes towards stabilizing the aggregates.

Heavily sulphated polyanions have, however, such a powerful electrostatic field that metachromatic bonding is largely preserved, even in dehydrated sections. Therefore, *alcohol resistant metachromasia* indicates the presence of sulphated carbohydrates, especially sulphated proteoglycans.

To demonstrate carboxylated carbohydrates, particularly sialomucins, which show *alcohol labile metachromasia* when stained at pHs above 4, sections must be examined without dehydrating.

Mounting in a hydrophilic mountant does *not* prevent loss of metachromasia and should be avoided. An *air dried* section will, however, regain metachromasia of sialomucins etc. if a few drops of water are added. Examination of the section immediately after staining is therefore not essential.

6.1.2 Alcian Blue

The Alcian Blue dyes are water soluble copper phthalocyanin compounds which were introduced into the textile industry in 1944. Their special advantage was that, after the cloth had been stained, they could be converted to insoluble pigments (Monastral Fast Blue in the case of Alcian Blue 8G) by treatment with alkali.

The formula shown applies to Alcian Blue 8G (previously 8GX) from I.C.I. (Scott, 1973a). The product contained up to 40–50% boric acid as a stabilizer and comparable amounts of sodium chloride and sodium sulphate. These compounds can be removed by gradually diluting a 2–5% aqueous Alcian Blue solution with five times its volume of acetone. Pure dye is precipitated (Scott, 1972a; alternatively McAuliffe, 1983) by this procedure. In some Alcian Blue products other dyes have been added in varying quantities.

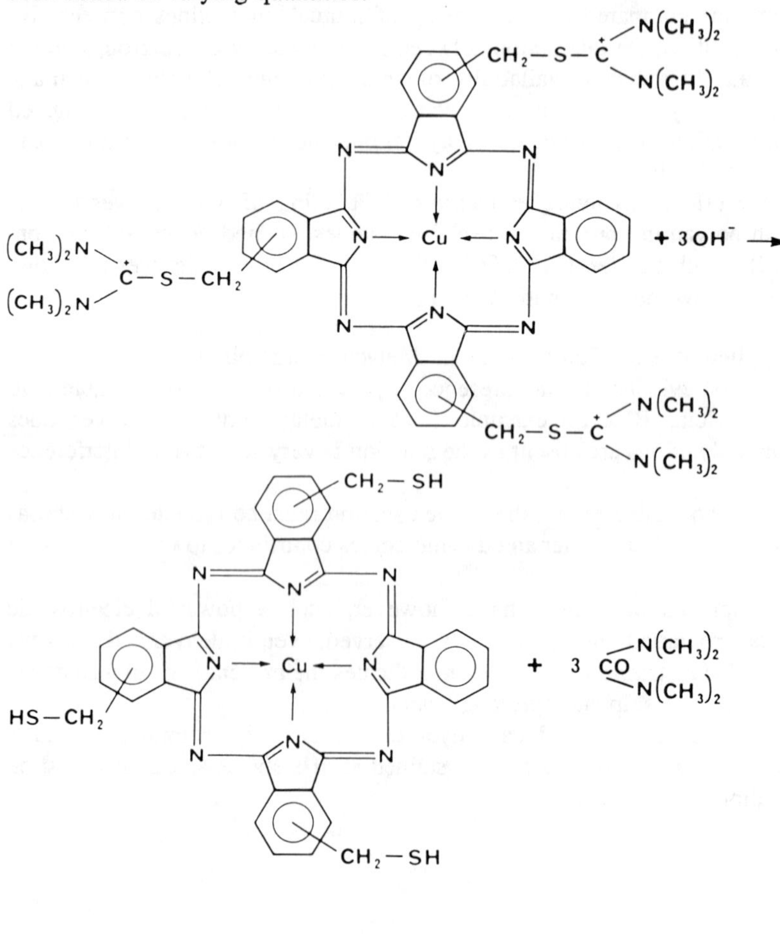

For histochemical applications Alcian Blue samples must fulfil the following criteria (Scott and Mowry, 1970):

a. Solubility in water to a concentration of at least 5% w/v
b. A 1% w/v solution in 0.025 mol/l acetate buffer pH 5.7 with $MgCl_2$ added at 2.0 mol/l should not precipitate on preparation and only very slightly after standing for 24 hours at room temperature
c. The spectrum of an aqueous solution should have a λ_{max} of 623 nm with a shoulder at 674 nm. A solution in ethanol has λ_{max} at 674 nm and a shoulder at 608 nm (Scott, 1970)
d. In paraffin sections (e.g. formalin fixed large intestine or trachea) stained with Alcian Blue at pH 2.5 the nuclei and cytoplasm should remain unstained.

Schenk (1981) gives the requirements of the Biological Stain Commission for certifying Alcian Blue. Lake (1980) discusses the possibilities for acquiring Alcian Blue products of reasonable quality and purity.

Alcian Blue is used for the differential demonstration of acid proteoglycans and acid glycoproteins. Selectivity is controlled by varying either the pH or the inorganic electrolyte content of the staining solution.

Alcian Blue pH-Methods. 1% w/v solutions at pH 0.5–1.0 and pH 2.5–3.0 are generally used.

Mechanism. In contrast to the majority of other dyes (Sects.4.2 and 4.5), dye binding is purely electrostatic.

In aqueous solution the Alcian Blue ions form approximately spherical dimers with 6–8 charges. These are electrostatically isolated against further formation of aggregates. The large ionic diameter and the high charge limit the ability of the dye to stain protein-rich complexes such as chromatin and ribosomes.

Selectivity. Both acid proteoglycans (polycarboxylates, polycarboxysulphates, and polysulphates) and acid glycoproteins (sialomucins and sulphomucins) are stained at pH 3, while only the sulphate containing carbohydrates are stained at pH 1. Nuclei and cytoplasm usually remain unstained at pH 3.

Sulphate Containing Carbohydrates. These are often protein-bound and this may delay or completely inhibit the binding of Alcian Blue. The problem is particularly pronounced with heparin in mast cells and chondroitin sulphate in cartilage. Deposits of calcium salts constitute a further impediment to staining in the latter instance. The polyanion-protein complex is so dense that it impedes penetration by Alcian Blue. The high density is due to the charge saturation of the sulphate groups which reduces the electrostatic field and results in decreased binding of water. Reduction of both Alcian Blue staining and metachromasia by protein can be abolished by treatment with protease, deamination, or by adding 0.1–0.3 mol/l $MgCl_2$ to the dye solution (see Alcian Blue CEC). Sulphomucins almost or completely free of protein can show a similar increase in staining after the addition of salt and lowering of pH. In this instance, these conditions reduce the mutual repulsion between Alcian Blue ions and allow them to bind to polyanions at closer intervals (i.e. increase total dye binding). The protein blocking effect is most important when staining at pHs under 3, when carboxyl groups will be extensively deionized.

After dehydration, carbohydrates containing carboxylic groups are far more intensely stained with Alcian Blue than with most other cationic dyes. This is because the high charge and high molecular weight of Alcian Blue renders the dye only slightly soluble in ethanol and hence non-extractable. In addition, the carboxylic group containing carbohydrates (e.g. many epithelial mucins) are only slightly bound to basic proteins and are therefore easily accessible.

Alcian Blue can stain carboxylic groups down to approximately pH 2. Staining at pH 0.5–1 is therefore a reliable indication for the presence of sulphate groups. The section is washed after staining with a suitable acid buffer solution or – alternatively – the non-bound Alcian Blue can be removed by applying a piece of filter paper before washing with water. Washing directly in water will result in Alcian Blue being removed so slowly from the tissue that considerable staining of carboxyl proteoglycans can take place as the carboxyl groups become ionized when the buffer is removed.

Sensitivity. This is high. The Alcian Blue methods can demonstrate considerably lower concentrations of acid carbohydrates than the metachromasia methods. A further increased sensitivity can be achieved by treating the sections with diluted alkali after staining. Alcian Blue will, as previously mentioned, be transformed into an insoluble pigment that is called Monastral Fast Blue. This process releases negative charges on the polyanion so that they can react with more Alcian Blue (Scott, 1972b).

Blocking. Blocking of carboxylic and sulphate groups can be achieved by alkylation (see Sect.9.8.3).

Alcian Blue CEC Method. Controlled amounts of a salt, usually $MgCl_2$, are added to a solution of 0.05% w/v Alcian Blue in 0.025 mol/l acetate buffer pH 5.8 (Scott and Dorling, 1965). In the absence of $MgCl_2$ there will be a rather diffuse blue staining of the tissue as well as staining of acid carbohydrates. The addition of small amounts of $MgCl_2$ (0.1–0.2 mol/l) ensures that proteins, sialomucins, and polycarboxyl proteoglycans are not stained. Larger amounts of $MgCl_2$ cause a progressive decrease in the staining of sulphated heteroglycans. The more sulphate a compound contains the higher electrolyte concentration required to abolish staining. The effect is shown in Fig. 6.4 where it is also apparent that the concentration interval of $MgCl_2$ over which dye binding shifts from maximum to zero for a given polyanion, is quite sharp. This concentration interval gives rise to the term, *critical electrolyte concentration (CEC)*.

Mechanism. CEC has been investigated in experiments like the one shown in Fig. 6.4 and similar experiments with "models", i.e. "spots" of different polyanions applied to filter paper both with Alcian Blue, other cationic dyes, and the simple organic cation, cetyl pyridinium ion from cetyl pyridinium chloride (1-hexadecyl pyridinium chloride)

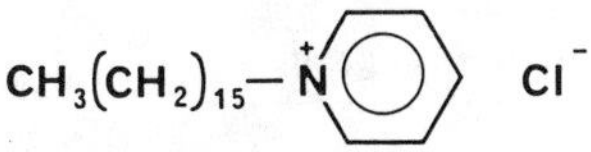

Scott and Dorling (1965) and Scott (1973b) have on this basis put forth a hypothesis for the mechanism on staining with stain solutions with added electrolytes. A polyanion will be precipitated from an aqueous solution by the addition of larger organic cations, e.g. alkaloids, detergents, dyes, and antibiotics. The reaction is an exchange of ions, and by mathematical treatment of the law of mass action they reach the conclusion that the complete removal of bound dye and subsequent precipitation of the polyanion takes place inside a very narrow interval of concentration, or more correctly ionic strength, of the metal salt. This is in agreement with the results of the experiments, but many of the chemical, physical, and mathematical suppositions are doubtful (Horobin and Goldstein, 1974).

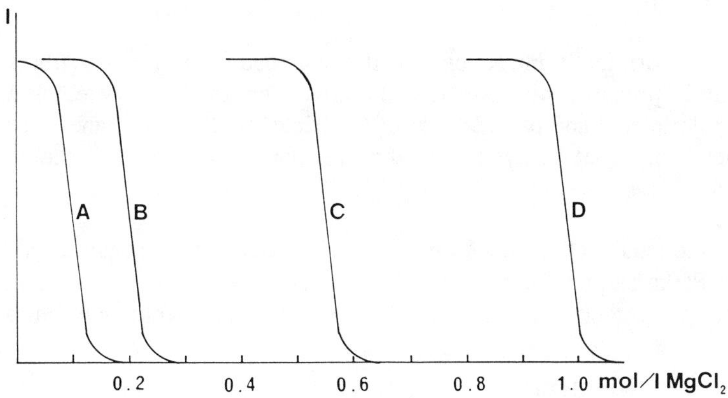

Fig. 6.4. Alcian Blue binding to different polyanions in tissue sections as a function of $MgCl_2$ concentration. The curves illustrate the CEC phenomenon.

Staining intensity (I), Hyaluronate (A), DNA or RNA (not protein masked) (B), Chondroitin sulphate (C), Heparin (D).

Experiments have shown that the CEC principally depends on the type of charged group in the polyanion, so that:

$$CEC(\text{-}COO^-) < CEC(=O_2PO_{2^-}) < CEC(\text{-}OSO_{3^-})$$

For many dyes other than Alcian Blue contributions from other kinds of bonds, e.g. polar or hydrophobic bonds, will cause deviations from the simple CEC-pattern. If the CEC is used for comparing dyes it is worth remembering that the dye powder should only contain the dye salt, and that the concentration of the dye ions should be low (10^{-3}–10^{-4} mol/l). Table 6.5 gives a simplified survey of some CEC-values.

Table 6.5. CEC-values (mol/l $MgCl_2$) for the binding of cetyl pyridinium chloride and different cationic dyes to selected polyanions on filter paper (modified from Scott, 1967).

	Cetyl pyridinium chloride	Alcian blue	Pyronin Y	Methyl green
Hyaluronate	0.1	0.1	0.01	0.05
RNA	0.2	0.1	2.0	0.4
DNA	0.2	0.1	1.0	1.5
Chondroitin sulphate	0.5	0.5	0.05	0.05
Heparin	0.8	0.7	0.05	0.05

Note that CEC-values determined by filter paper investigations should be increased by about 50% when used on tissue sections. Of the dyes shown in Table 6.5, only Alcian Blue behaves like cetyl pyridinium chloride and stains in a pattern consistent with the CEC concept.

6.1.3 Aldehyde Fuchsin

Aldehyde Fuchsin was originally introduced for the demonstration of elastic fibres (Gomori,1950a). It is prepared by dissolving Pararosanilin in 70% v/v ethanol and adding hydrochloric acid and paraldehyde (2,4,6-trimethyl-1,3,5-trioxane). The solution is "ripened" for about 5 days by standing at room temperature. It should be used within a few days thereafter.

Mechanism. The acid catalyzes the decomposition of paraldehyde to acetaldehyde which reacts with Pararosanilin forming Aldehyde Fuchsin which is insoluble in water and does not show metachromasia. The precise formula is unknown, but a possibility is:

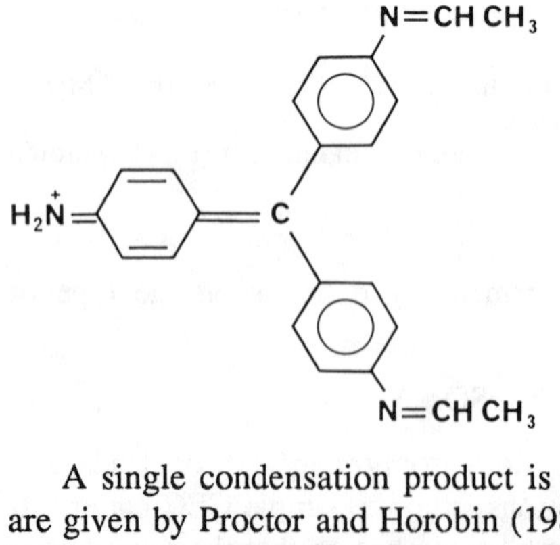

A single condensation product is also a possibility. Several other suggestions are given by Proctor and Horobin (1983). A precise formulation of the mechanism is therefore not possible, but the staining pattern at pH 1–2 (see Table 22.1) closely resembles that obtained by a combination of Alcian Blue and Orcein. Aldehyde Fuchsin probably forms electrostatic and van der Waal bonds to elastin (Sect.21.5)

while the staining of pancreatic B-cell granules may be explained on the basis of covalent bonding (Cole and Nettleton, 1988).

Selectivity. Aldehyde Fuchsin stains nuclei and RNA-rich cytoplasm only slightly or not at all. At pH 1–2 an intense staining of sulphate containing proteoglycans and sulphomucins is obtained. Elastin is also stained strongly. In addition, Aldehyde Fuchsin stains neurosecretory substance, several kinds of APUD cells, and hepatitis B antigen (HB_sAg). For these purposes the staining is preceded by an oxidation of the tissue section, usually with permanganate. While this is necessary for neurosecretory substance, routine use of oxidation is discouraged by Mowry (1978) who recommends a longer staining time (up to 4 hours) instead. It has now been demonstrated (Mowry and Kent, 1988) that true Aldehyde Fuchsin (Aldehyde Pararosanilin) is able to stain purified insulin, oxidized or not. Good results depend on the quality of fixation, for which Bouin's picric acid fixative can be recommended. On formaldehyde fixed material greatly improved results can be achieved by an after-fixation of the *non*-deparaffinized sections in Bouin's fluid (Lyon and Prentø, 1980).

Blocking. This is as for Alcian Blue.

6.1.4 Victoria Blue B

Staining with the cationic dye, Victoria Blue (C.I. 44045) probably involves a substantial degree of non-ionic binding. An analysis of nuclear staining from alcoholic solution has been made by Schulte *et al.* (1988).

6.1.5 Methyl Green-Pyronin

The two cationic dyes Methyl Green (C.I. 42585) and Pyronin Y (C.I. 45005) are used together in the same staining solution. At pH 2.5–4.5 a differential staining of DNA with Methyl Green and RNA with Pyronin is seen.

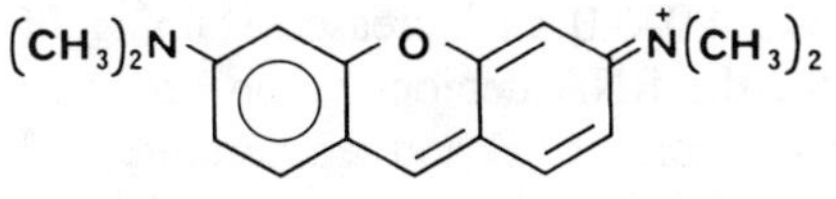

Pyronin Y

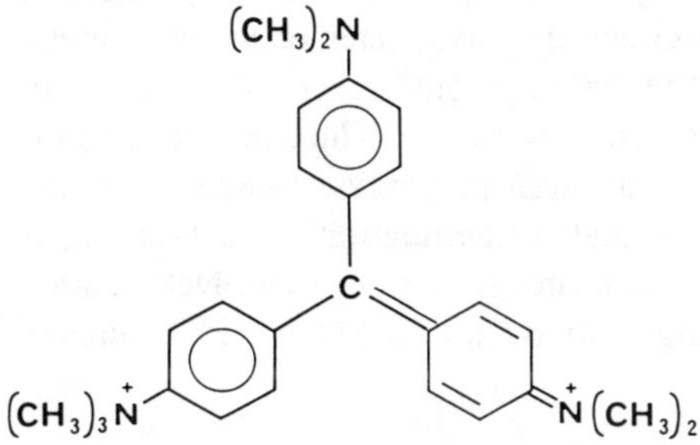

Methyl Green

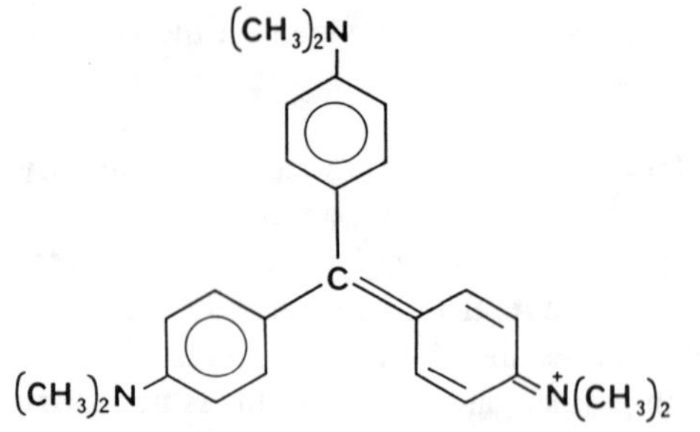

Crystal Violet

Mechanism. Binding of the two dyes to nucleic acids is not purely electrostatic, hydrophobic interactions also play a key role. Methyl Green is intercalated in the DNA double helix by hydrophobic bonding between the aromatic ring of the dye and the purine and pyrimidine bases of DNA. It is likely that the two charged groups of the dye form salt linkages with a phosphate group from each DNA strand (Scott, 1967). When Methyl Green is bound to RNA hydrophobic bonding to the nucleotide bases may take place but the predominantly non-helical structure precludes intercalation. Pyronin Y and other planar dyes with a single charge bind efficiently to both single and double stranded nucleic acids. On the other hand, Methyl Green, which is non-planar and carries two charges, binds much less efficiently to single stranded nucleic acids. Although binding to double stranded nucleic acids is more efficient, it does not reach the same level as Pyronin. In consequence, it is possible to determine a ratio between the two dyes where native DNA is stained by a high relative concentration of Methyl Green while all other nucleic acids are stained with Pyronin. A suitable molar ratio is 2.6 (Høyer *et al.*, 1986).

Selectivity. It is worth noting that the method employs two cationic dyes simultaneously. The normal rules for basophilia are invalid in these circumstances. The dominance of Pyronin staining at pH < 2 is probably due to denaturation of the DNA double-helix. Between pH 2 and 4, Pyronin staining is, apart from RNA, only found in highly sulphated structures such as cartilage matrix and mast cell granules. This rarely gives rise to interpretive problems as these structures give rise to an orange metachromasia. At pH above 4.0–5.0 an increased staining of protein occurs and the value of the method for the RNA demonstration becomes very dubious. If the method is carried out at the correct pH and temperature and consideration is given to the purity of the dyes, Methyl Green is fairly selective for DNA. Commercial samples of Methyl Green are, to greater or lesser degrees, always contaminated with Crystal Violet. The staining solution should therefore be purified by repeated extraction with chloroform before use. The same result may be obtained by filtering an aqueous solution through polyamide (Andersen *et al.*, 1986). Another problem with the method is that differentiation, or extraction, is very considerable when the standard dehydration procedures are used. Dehydration must therefore be performed either by drying in an oven at 37°C or with 1-butanol (Høyer *et al.*, 1986).

6.1.6 The Histological Application of Cationic Dyes

Cationic dyes are particularly useful for the demonstration of nuclei in eukaryotic tissues. They are also widely used for the demonstration of bacteria and other microorganisms.

The Mechanism and Selectivity of Nuclear Staining. The pH of the staining solution should normally be between 3 and 4 in order to ensure complete ionization of DNA. RNA in nucleoli and cytoplasm is also stained under these conditions, as are mast cell granules, some goblet cells, cartilage matrix, and connective tissue matrix (to a limited extent). These latter components, however, usually show some degree of metachromasia (Sect.6.1.1). Blocking of phosphate groups in DNA, due to tightly associated oppositely-charged nucleoprotein amino groups, often prevents adequate staining of nuclei under low pH conditions. Considerably more intense nuclear staining may be achieved under these conditions after deamination (Sect.9.5.1). Using a higher pH, staining intensity is increased, partly due to the carboxylic groups in proteoglycans, glycoproteins, and proteins becoming ionized, and partly due to a diminution of the blocking effect. Extraction of cationic dyes during dehydration can be avoided by using acetone, tertiary butanol, or 1-butanol instead of ethanol or by placing the slides in a thermostat oven at 37°C with a desiccant such as silica gel.

Mechanism and Selectivity of Bacterial Staining. Virtually all bacteria may be demonstrated using cationic dyes. Limitations on particular techniques may be related to differences in the cell wall structure found in different bacterial genera and the degree of contrast and resolution afforded by the technique. Cell wall composition (cf. Sect.2.3) appears to affect both penetration and retention of stain. In this section discussion will be limited to the two most widely used techniques, the Gram and Ziehl-Neelsen methods.

Gram-Staining. This staining method was named after the Danish physician, Christian Gram, who described a method for the selective demonstration of bacteria in tissue sections in 1884. The technique involved a differentiation step that left some bacteria stained so that they could easily be seen on a background of decolourized nuclei and other cell components.

The Gram method consists of the following steps:

1. *Staining with Crystal Violet*, a cationic dye that binds to negatively charged components of bacterial cells
2. *Iodine treatment* leading to complex formation with Crystal Violet
3. *Differentiation:* Using 95% ethanol, the iodine-Crystal Violet complexes are retained by Gram-positive bacteria and extracted from Gram-negative bacteria
4. *Counterstaining* using a second cationic dye (usually giving red or pink staining) restains the Gram-negative organisms and has little or no effect on Gram-positive organisms

The result of the Gram stain both in terms of dye retention and bacterial morphology is of major importance in the detection and identification of bacteria. An understanding of the mechanisms involved and familiarity with the possible range of staining results is essential if misinterpretations are to be avoided.

Mechanism of the Gram Method. The *cationic dye* binds to phosphate groups in bacterial nucleic acid. Any cationic dye that can be precipitated by treatment with iodine, may be used. Generally, Methyl Violet or Crystal Violet are used. Gram-positive and Gram-negative bacteria take up dye in a similar pH dependent fashion. Staining time may be varied from 15 sec to 5 min. One minute is excellent.

Mordanting with Iodine. This makes the stain less soluble in alcohol and water so that it is relatively resistant to extraction during the subsequent processing stages. Lugol's solution (I 1 g, KI 2 g, in 300 ml dist. water), which is widely used, becomes progressively more acidic on contact with air so that the amount of reactive iodine halves over the course of 30 days. An increased iodine content makes the storage period less critical. The dye-iodine complex is formed in a molecular ratio of 1:1 to 1:2 and is identical for Gram-positive and Gram-negative bacteria.

Differentiation. This is achieved because Gram-negative bacteria are destained more quickly than Gram-positive bacteria. 80–100% ethanol is recommended as, at concentrations below 60%, destaining occurs at the same rate in both groups of bacteria. As water dissolves the bound iodine, the specimens must not be rinsed with water for prolonged periods following iodine treatment.

Methanol and acetone destain rapidly. These reagents are widely used in routine bacteriology where staining of tissue sections is rarely performed and where rapid results are required. Considerable experience is essential if over destaining is to be avoided. The higher alcohols act too slowly but 95% ethanol gives a gentle elution time-course allowing variation between 15 sec and 3 min depending on the application.

Counterstaining. Carbol Fuchsin (a phenol containing solution of Basic Fuchsin), Safranin O, and Neutral Red are the most frequently used counterstains. To avoid dissolving the dye-iodine-complex in Gram-positive bacteria staining time should not exceed one min. and washing should not exceed 5 sec.

The Mechanism of Differentiation in the Gram Method. This remains controversial. Lautrop *et al.* (1979) discuss four possible mechanisms:

a. The chemical theory: Gram-positive bacteria contain specific Gram-stainable substances (not necessarily cell wall-associated) that are not found in Gram-negative bacteria
b. As for (a), but exclusively cell wall-associated substances are involved
c. The isoelectric point theory: Gram-positive bacteria have a lower net pI, which gives them a greater capacity for binding cationic dyes than Gram-negative bacteria

d. The permeability theory: The cell walls have different compositions in the two types of bacteria conferring different permeabilities towards alcohol, iodine, and the iodine-dye-precipitate

The permeability theory is the most widely accepted. Gram-positive cells have a much thicker layer of peptidoglycan than Gram-negatives (Sect.2.3.1). Removal of the cell wall (e.g. in spheroplast preparations made by treatment with penicillin) renders "Gram-positive" organisms susceptible to destaining. In addition, peptidoglycan is probably extensively hydrated. Thus removal of water during exposure to organic solvents is likely to compact the cell wall resulting in a reduced permeability to the dye-iodine-complex.

Gram Staining of Histological Sections. Paraffin sections are usually overstained by the standard Gram method. Gram-negative bacteria in particular give variable results. Engbæk *et al.* (1979) have described a variant of Brown and Hopp's method using carbol Fuchsin as the counterstain which overcomes this problem. Garvey *et al.* (1986) report that the inclusion of sodium bicarbonate in the Crystal Violet stain increases the resistance of the Gram-positive organisms to decolourization in the differentiation step which is carried out with Cellosolve (ethylene glycol monoethyl ether), while acetic acid in the Tartrazine counterstain ensures the brilliance of Gram-negative organisms through reduction of background staining.

Ziehl-Neelsen Method. Koch published the first description of tubercle bacilli in 1882, using alkaline Methylene Blue to demonstrate them. Shortly after, Ehrlich (1882) described the acid fast properties of these bacteria. He found that they could *not* be destained using dilute acid after staining with a Fuchsin-aniline oil mixture. Ziehl (1882, 1883) modified the method by using the dye in a solution of carbolic acid (phenol) at a higher temperature. Neelsen (1883) used Basic Fuchsin in a solution of carbolic acid and destained with 25% sulphuric acid. Since then the procedure has generally been called the Ziehl-Neelsen method, although it is really a modification of Ehrlich's method.

Mechanism. Mycobacterial cell walls contain a waxy substance composed of mycolic acids. These are β-hydroxy carboxylic acids with chain lengths of up to 90 carbon atoms. The property of *acid fastness* is related to the carbon chain length of the mycolic acid found in that particular bacterial species. Mycolic acids are found in the group of bacteria known as the *Actinomycetes* and acid fastness is found in increasing degrees in *Corynebacteria* (hardly detectable), *Actinomyces*, *Nocardia*, and *Mycobacteria*.

Staining is achieved with *Basic Fuchsin*, which binds to negatively charged groups in bacteria. The mycolic acids (and other cell wall lipids) present a barrier to dye entry as well as elution and this is partly overcome by adding phenol to a concentrated aqueous solution of the Fuchsin (hence carbol Fuchsin) and partly by combining *heat* and *extended staining time*.

The following combinations are used: 10 min at 70°C; 30 min at 55°C; 2 hours at 37°C; or 4–16 hours at 25–20°C.

Heating. Staining is speeded up by heating, because the lipid content in the bacterial wall becomes more fluid at higher temperature. The phenol in carbol Fuchsin gives an increased lipophilia and aids the passage of the dye through lipids.

Destaining. This is usually performed with hydrochloric acid in ethanol. Mycobacteria here show an "acid fastness" (strictly, acid-alcohol). This is again due to the high lipid content of mycobacteria both in general (20–40% dry weight) and in particular (60% cell wall dry weight).

Some workers (e.g. Harada, 1976) consider that the addition of phenol is only necessary when using water soluble dyes. Lipid soluble agents such as Night Blue and Victoria Blue B can give rise to acid fast staining without phenol. On treatment with bromine, bacteria lose their acid fast properties. This finding supports the view that fluid/non-fluid phase transitions are important. Blocking of carboxylic groups (methylation) and of hydroxyl groups (acetylation, benzoylation) also destroys acid fastness and may therefore be of importance for the actual binding of cationic dyes. Alternatively, the solvents used in the blocking reactions may partially extract cell wall lipids.

Isoniazid inhibits the formation of mycolic acid and is widely used in the treatment of tuberculosis. It is noteworthy that mycobacteria from patients treated with isoniazid lose their acid fastness (Grange, 1983).

Counterstaining. Picric Acid gives good contrast in bacterial smears. For paraffin sections Al-haematein, Methylene Blue, or Janus Green are used.

Other Methods for the Demonstration of Mycobacteria. Some further methods are given in Sect.31.4.1.

6.2 Demonstration of Proteins by Binding of Anionic Dyes

The net charge of a protein depends on the balance between ionized amino groups and ionized carboxyl groups. Whether the groups are ionized or not depends on the immediate environment (e.g. tissue section) and the pK_As of all the ionizable groups (for amino groups see Table 6.6). In general, proteins are positively charged at low pH (deionization of carboxyl groups) and negatively charged at high pH (deionization of amino groups) (see Fig. 6.5).

Table 6.6. pK_A-values for amino groups in proteins.

Group	pK_A at 25°C
Imidazolium (histidyl) in protein	5.6–7.0
α-amino (terminal) in protein	7.6–8.4
ε-amino (lysyl) in protein	9.4–10.6
Guanidinium (arginyl) in protein	11.6–12.6

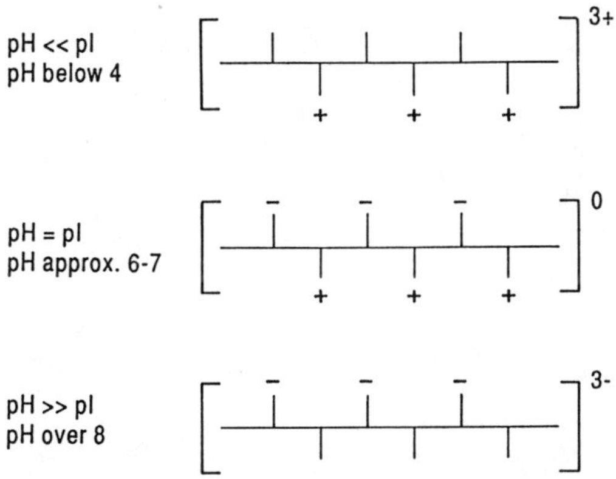

Fig. 6.5 . The influence of pH on the charge of a protein (pI ~ 6-7).
\+ cationic site
\- anionic site

The binding of an anionic dye to proteins by pure ionic bonds depends on a combination of the following factors:

1. pH and the pK_A-values of the different ionizable groups
2. The number of the different ionizable groups in the protein
3. Conformational (steric) considerations
4. The interaction of the protein with other macromolecules
5. Salt concentration

6.2.1 pH and the pK_A-Values of the Ionized Groups

The pK_A-values for different ionizable groups in proteins are shown in Tables 6.1 and 6.6. It is evident that the pK_A-value of an individual group may vary according to position in relation to other groups in the protein (e.g. histidyl residues show values between 5.6 and 7.0). Thus the primary, secondary and tertiary structure of the protein are all relevant. Fig. 6.6 shows the relationship between ionization and pH for protein bound carboxyl groups with pK_A-values of 3, 4 and 5 and for histidyl (pK_A 7), lysyl (pK_A 10) and arginyl (pK_A 12) groups.

6.2.2 Number of Ionizable Groups

The number of ionizable groups clearly depends on the content of amino and carboxyl group bearing amino acids in different polypeptides. Some examples are given in Table 6.7, and the pH dependency of the net charge is shown in Table 6.8.

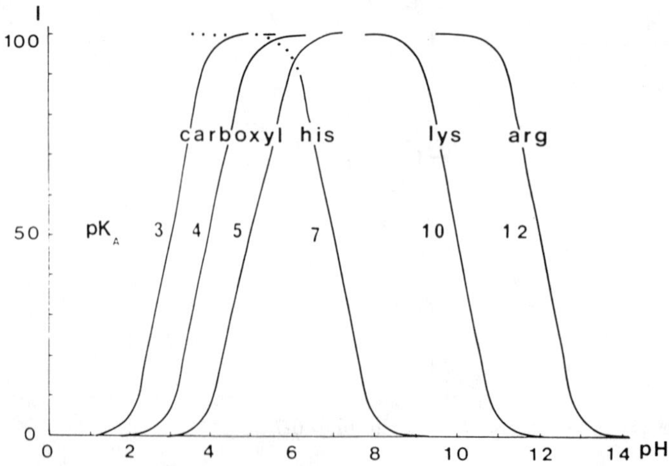

Fig. 6.6. Per cent ionization as a function of pH for protein-bound carboxyl and amino groups (pK_As specified). Degree of ionization in per cent (I), histidyl (his), lysyl (lys), arginyl (arg).

Table 6.7. The content of ionizable groups in some polypeptides.

		Ionizable groups						
		Carboxyl			Histidyl	Amino Lysyl	Arginyl	
pK_A		3	4	5	7	10	12	Total
Polypeptide	A	30	0	2	1	6	1	40
,,	B	12	17	7	8	4	2	50
,,	C	5	5	18	17	6	9	60
,,	D	2	11	11	12	10	4	50
,,	E	1	6	19	6	16	12	60
,,	F	2	3	5	8	7	35	60

The Isoelectric Point of a Protein. Changes in the charge of a protein, as illustrated in Fig. 6.7, often take place over an interval of between 8 and 12 pH-units. Moreover these changes are not necessarily symmetrical around the isoelectric point (pI). This is a natural consequence of the protein containing so many different ionizable groups. The potential for the protein to bind anionic and cationic dyes at different pH-values can be inferred from curves like the one shown in Fig. 6.7. This potential is naturally at a minimum when the pH approaches the pI. Examples of pIs for different proteins are given in Table 6.9.

Table 6.8. Relationship between net charge and pH for the polypeptides listed in Table 6.7.

		Polypeptide					
		A	B	C	D	E	F
pH	pI*	2.4	3.5	6.4	6.3	9.0	12.6
0		8 +	14 +	32 +	26 +	34 +	50 +
1		8 +	14 +	32 +	26 +	34 +	50 +
2		5 +	13 +	31 +	26 +	34 +	50 +
3		7 −	6 +	29 +	24 +	33 +	49 +
4		19 −	7 −	23 +	17 +	28 +	46 +
5		23 −	17 −	13 +	8 +	18 +	43 +
6		24 −	22 −	4 +	2 +	9 +	39 +
7		24 −	26 −	5 −	4 −	5 +	36 +
8		24 −	29 −	11 −	9 −	3 +	33 +
9		26 −	30 −	14 −	11 −	0	31 +
10		28 −	32 −	16 −	15 −	6 −	29 +
11		31 −	34 −	19 −	19 −	13 −	22 +
12		31 −	35 −	23 −	22 −	20 −	8 +
13		32 −	36 −	27 −	24 −	25 −	6 −
14		32 −	36 −	28 −	24 −	26 −	10 −

* pI = isoelectric point.

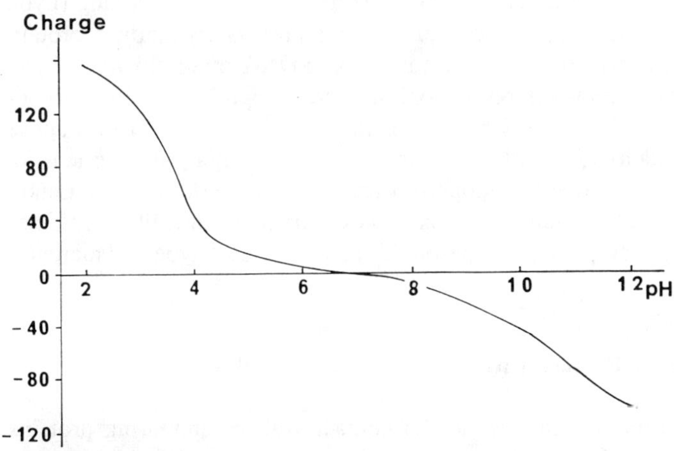

Fig. 6.7. Relationship between pH and net charge for haemoglobin.

Table 6.9. The isoelectric point for different proteins.

Compound	Isoelectric pH
Lysozyme	11.0
Cytochrome C	9.5
Histones	9.0–11.0
Haemoglobins	6.0–7.0
Globulins	6.4–7.2
Myosin	6.2–6.6
Insulin	5.3–5.4
Collagen	4.0–5.0
Pepsin	< 1

6.2.3 Conformational State of Proteins

Isoelectric points are determined on pure preparations of non-denatured proteins. Charges expressed by proteins in tissues are substantially modified by interactions with other tissue components and by tissue processing, particularly fixation. Denaturation (e.g. during ethanol fixation) may reveal ionizable groups that were involved in stable salt bridges in the native protein, thus coagulative fixation may increase dye-binding capacity (Sect.12.2.2). Additive fixatives (Sect.12.2.1) may promote binding of either anionic or cationic dyes depending upon which ionized groups they bind to and whether the binding of fixative is reversible or irreversible. The aldehyde fixatives in particular, strongly promote binding of cationic dyes after prolonged fixation.

6.2.4 Net Effect of Different Fixatives on Basophilia/Acidophilia

Part of the differential effect of fixatives is due to differences in their ability to stabilize various tissue components. For example, although formaldehyde progressively blocks amino groups, it is a slow reacting fixative that leaves a considerable portion of the protein content unfixed and thus able to escape before staining (Lyon and Prentø, 1980). Picric acid is a very powerful, though reversibly binding, protein fixative, but hydrolyzes and extracts RNA quickly and DNA more slowly.

The overall effects of fixatives on acidophilia and basophilia are thus the net result of several factors which may work together (e.g. picric acid) or oppose each other (e.g. formaldehyde). Further examples of fixatives that, like picric acid, promote binding of anionic dyes (acidophilia) are chromium trioxide, potassium dichromate, and trichloroacetic acid, while examples of fixatives that, like formaldehyde, promote binding of cationic dyes (basophilia) are glutaraldehyde and mercuric chloride.

6.2.5 The Interaction of Protein with other Macromolecules

In tissues individual protein molecules are in contact with neighbouring proteins and other macromolecules. Sometimes, true complexes may occur in which macromolecules are very firmly associated through a large number of salt links. This is especially pronounced for nucleohistone complexes where the strongly basic histones cannot be demonstrated at pH 8 due to masking by the phosphate groups. After acid hydrolysis or enzymatic extraction of DNA, an intense reaction may be obtained with an anionic dye (Sect.21.4.2). (see also: masked metachromasia (Sect.21.4.2) and staining of proteoglycans (Sect.6.1.2.))

6.2.6 Concentration of Salt

A high salt concentration may suppress the binding of a dye to an ionized group (see Sect.6.1.2). This is utilized in the demonstration of e.g. elastin and amyloid where the non-electrostatic binding of the dye to protein is emphasized. The normally used concentrations of buffer salts have no appreciable effect on the binding of dye to protein.

6.3 The Mechanism of Staining with Anionic and Cationic Dyes

The principle reason for binding is almost certainly the electrostatic attraction (ionic bond, salt link) between the charged dye and the charged tissue component.

This fundamental principle is, however, heavily complicated by other factors. Some will *reduce* the dye binding – e.g.:

a. Competition with other ions in the staining solution (see critical electrolyte concentration, Sect.6.1.2)
b. Masking of tissue groups by electrostatic interaction (e.g. histone blockade of the phosphate groups of nucleic acids, Sects.6.1.1 and 6.2.5)
c. Steric inhibition (permeability, Sect.6.2.3). The charges in the tissue can be placed in such a manner that the dye ions are unable to come into contact with them. For instance, this can occur if an area surrounding the attractive charge is electrostatically repellent to the dye. Occasionally the concentration of dye matter is so high that this has a direct influence on the penetration of the dye. In both cases the expression "decreased permeability" is sometimes used

Other factors can *increase* dye binding – e.g.:

d. The formation of bonds other than ionic bonds between dye and tissue component. These different categories of bonds are enumerated in 4.5

Under conditions where ionic bonds are unable to assert themselves (e.g. high content of electrolytes) or are weakened (e.g. low pH with cationic dyes) these other weaker bonds predominate. Examples include the selective methods for demonstration of elastic fibres and methods for the demonstration of amyloid (Sects.21.5 and 21.7).

Staining experiments on gelatin with different dyes at pH 4 and pH 9 (Horobin and Bennion, 1973) showed that many cationic dyes stained gelatin to varying degrees at pH 4, just as many anionic dyes stained gelatin at pH 9. In both cases staining must be attributed to the formation of "non-ionic" bonds. Horobin and Bennion also showed that the possibilities for a dye to form non-ionic bonds increases with increasing molecular weight. Since, in their study, higher molecular weight generally meant more aromatic rings, more substituents in the rings, and possibly more heteroatoms, it can be seen that the possibilities for the formation of van der Waal bonds should increase in parallel (Sect.4.5.4).

6.3.1 The Histological Application of Anionic Dyes

In the histological application of anionic dyes, the *aim* is to achieve a histological staining of cytoplasm and intercellular material that clearly contrasts with nuclei stained with a cationic dye or metal complex dye.

Mechanism. This involves binding to positive charges in the tissue, i.e. amino groups in proteins.

Selectivity. This depends upon pH and to some degree on the anionic dye used. An aqueous solution of the dye is frequently used. This may give rise to a rather low staining intensity. By adding a little acetic acid (lowers pH) to the stain, or by dissolving the dye in a buffer at pH 4, intense staining is achieved of red blood cells, eosinophil granules in eosinophils, muscle cells, and the cytoplasm of other cells as well as various connective tissue fibres. Note also that nuclei, particularly nucleoli, are stained by anionic dyes. This is particularly clear if staining with a cationic or metal complex dye is omitted.

6.3.2 Mixtures of Anionic and Cationic Dyes and their Histological Use

In principle, on staining with an anionic and a cationic dye in the same solution, e.g. Azure A-Eosin or a Romanowsky type stain, all structures with ionized groups should be stained. By changing pH, different colour patterns may therefore be obtained. This allows conclusions to be drawn concerning the content of different molecules in the tissue. Such methods are well suited for the identification of different classes of leukocytes, intracellular parasites, and different physiological or pathological conditions where changes arise in the amount of nucleic acids or in their degree of conjugation (e.g. secretory phase in gland cells, acidophil bodies, Mallory bodies). The majority of anionic and cationic dyes precipitate as insoluble complexes on mixing. However, for a few dye combinations, complex formation is less pronounced particularly when the mixture contains organic constituents (alcohols, acetone or, dimethylsulphoxide (DMSO)).

A commonly used combination is Azure A-Eosin which is applied in a buffered solution containing acetone. Staining results are heavily dependent on the pH (Table 6.10) and fixation procedure used (Sect.6.2.4).

Table 6.10. Examples of staining results using Azure A-Eosin.

Tissue component	pH 1	2	3	4	5	6	7	8	9
Mast cell granules	A	A	A	A	A	A	A	A	A
Cartilage ground substance	A	A	A	A	A	A	A	A	A
Ribosomes (e.g. exocrine pancreas)	0	0	A	A	A	A	A	A	A
Chromatin	0	(A)	(A)	A	A	A	A	A	A
Secretion in different gland cells	0, A	0, A	0, A	A	A	A	A	A	A
Sarcoplasm	0	0	(E)	E	E	E (A)	(E) A	A	A
Collagen fibres	0	0	(E)	E	E	E	EA	A	A
Red blood cells	0	0	(E)	E	E	E	E	(E) A	A
Parietal cells	0	0	(E)	E	E	E	EA	A	A
Paneth cell granules	0	0	(E)	E	E	E	E	EA	A
Granules in eosinophils	0	0	(E)	E	E	E	E	E	E

A = stained with Azure A; E = stained with Eosin; 0 = not stained.

The Romanowsky group of dye mixtures are very similar to Azure A-Eosin (including: May-Grünwald, Giemsa, Wright, and Leishman). These mixtures are particularly used for staining blood smears, but are also of value for the examination of other cytological preparations. In addition to Eosin they contain Methylene Blue and/or various oxidized derivatives thereof. The term "polychrome Methylene Blue" is frequently used for oxidatively demethylated Methylene Blue.

The most important dyes which arise on oxidative demethylation of Methylene Blue are Azure B, Azure A, and Thionin (see Fig. 6.8).

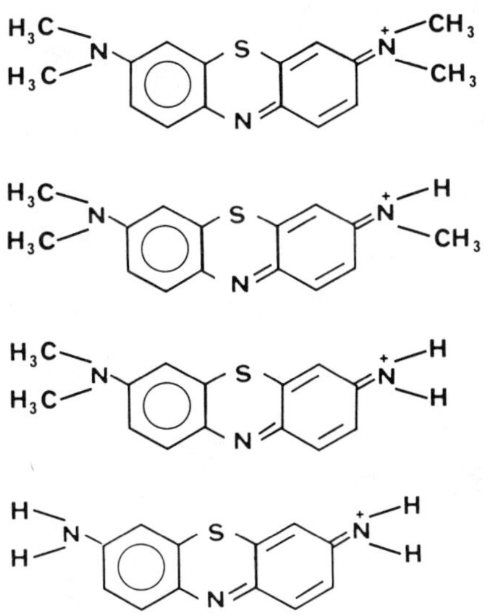

Fig. 6.8. The most important components in polychrome Methylene Blue.

The dyes are frequently sold in a solution which, regardless of the declaration on the label, inevitably comprises a mixture of the different dyes in varying proportions. The dyes are usually dissolved in methanol. It should be noted that the oxidation processes may continue in such preparations, especially if the pH is above 6.6.

The so-called Romanowsky-Giemsa-effect, (i.e. purple staining of nuclei and other structures), cannot be obtained with Eosin alone or with derivatives of Methylene Blue alone, but only when an interaction between Eosin, Azure B, and tissue components is possible. Azure B can to some extent be substituted with Azure A, but not with any of the other thiazin dyes. Eosin (tetrabromofluorescein) cannot be substituted with Eosin B (dibromodinitrofluorescein) but can be substituted with tri- or dibromofluorescein, tetrachlorofluorescein, or with tetraiodofluorescein (Erythrosin B). For a standardized Azure B Eosin procedure see Appendix A.

7 Staining Involving Metal Complex Dyes

P. Prentø, E. Schulte

Metal complex dyes ("Dye Lakes") are frequently used as substitutes for cationic dyes in the staining of chromatin. The relative stability of staining in the face of subsequent processing steps is an important property.

7.1 Complex Compounds

Many metal ions are able to form complexes with water, ammonia, cyanide, and various organic compounds. The resultant compounds are termed complex or co-ordination compounds. They are principally formed by metals with a small ionic radius that exert a particularly strong attraction for negative ions or dipoles such as water or ammonia (O and N respectively form the negative pole). The number of atoms which may bind to the central metal atom is given by the coordination number which is usually twice the charge; for instance, an aluminium ion binds six water molecules forming a so-called hexaqueous ion.

The atoms or atomic groups, which bind to the central metal atom, are called *ligands*. These ligands are characterized by having a free electron pair which is drawn towards the central atom. The bond is therefore intermediate between ionic attraction and a covalent bond.

The binding of water molecules to aluminium in the aluminium hexaqueous ion causes such an extensive displacement of electrons that the electron density of the O-H bond is significantly diminished. The bound water molecules thus have a lower pK_A, and their tendency to release protons is increased.

$$[Al(H_2O)_6]^{3+} \rightleftarrows [Al(H_2O)_5OH]^{2+} + H^+$$

The pH of a 0.025 mol/l aluminium sulphate solution is about 3.

Metal complex dyes (formerly called dye lakes) are complexes between metal ions and so-called ligand dyes. A metal complex dye is formed when two or more complexly bound water molecules are exchanged with the same number of ligand groups from a dye molecule.

$$D \quad + [Al(H_2O)_6]^{3+} \quad \rightleftarrows \quad D \quad Al(H_2O)_4]^{3+} + 2H_2O$$

H. Lyon (Ed.)
Theory and Strategy in Histochemistry

The ring-shaped complex formed in this way is called a chelate. Stable chelates with planar dyes are generally pentagons or hexagons.

The metal ions used most frequently for producing metal complex dyes are Al^{3+}, Fe^{3+}/Fe^{2+}, and Cr^{3+}. The stabilities of the metal complex dyes correspond to a certain extent to the stabilities of the corresponding aqueous ions. While a water molecule is exchanged every minute or so in the Al- and Fe-aqueous ions, the turnover is many hours for the Cr-aqueous ion.

This difference in the rate of ligand turnover is accompanied by corresponding differences in the required staining time for the metal complex dyes. This indicates that tissue binding may be partly coordinate or chelate in nature.

Table 7.1. Dyes which may form histologically useful chelates with aluminium, iron, and/or chromium ions.

Dye	Colour Index No.
Gallein	45445
Coerulein S	45510
Chromogen I	57030
Rhodizonic acid	
Alizarin	58000
Alizarin Red S	58005
Purpurin	58205
Anthrapurpurin	58255
Rubeanic acid	58055
Anthracene Blue SWR	58605
Alizarin Cyanin BBS	58610
Aluminon	43810
Solochrome Cyanin R	43820
Chromoxane Pure Blue BLD	43825
Chromoxane Pure Blue B	43830
Helio Fast Rubin BBL (Nuclear Fast Red)	60760
Alizarin Blue S	67415
Carminic Acid	75470
Gallocyanin	51030
Modern Violet N	51025
Gallamin Blue	51045
Celestin Blue B	51050
Brazilein	75280
Haematein	75290
Morin	75660
Maclurin	75240
Fluorone Black	
Chromoxane Cyanine R	43820

The underlined dyes may replace Haematein/Haematoxylin in some staining methods.

Twenty of the 28 dyes listed in Table 7.1 contain orthodiphenol and/or semi-ortho-quinone, 4 contain orthohydroxy-carboxyl, and the remainder have a variety of different ligand groups (e.g. dithioamide in rubeanic acid). These groups are thought to participate directly in the complex.

The two most important ligand dyes, Haematein and Gallocyanin, both have diphenol and semiquinone groups. The importance of the aromatic nitrogen-carboxyl group in Gallocyanin for chelation is debated.

Dyes with diphenol or semiquinone groups are weakly ionized in aqueous solution at neutral pH. The ionization entails an increase of the dye resonance and a shift of the absorbance towards longer wavelength. The resonance is increased at higher pH, and some of the ligand dyes are used as pH indicators (e.g. Haematein and Alizarin).

The binding of metal aqueous ions to diphenol and semiquinone groups has the same effect as an increase in pH, namely ionization of a phenol group. When forming chelates most ligand dyes therefore exhibit a colour shift and an increase in colour.

This property is exploited in histochemical reactions for tissue bound metals (Sect.17.5.2) where the chelates are insoluble. Some ligand dyes form soluble chelates with certain metal ions. In pure aqueous solution these dyes are mostly weak or very weak dyes. Stain intensity is dramatically increased by chelation with the metal ion which also substantially determines the selectivity of staining. For instance, Chromium-gallocyanin and Chromium-haematein are both selective but slowly acting nucleic acid stains, while chelates of iron and these two dyes are less selective but rapid histological stains for nuclei and other structures.

The following discussion will be limited to metal complex dyes containing Haematein or Gallocyanin.

7.2 Metal-Haematein Complexes

Haematein is a hydroxyquinone dye produced by oxidation of *Haematoxylin.* Haematoxylin is an organic molecule extracted from the heartwood of a small tree, "logwood" (Haematoxylon campecheanum L., Central America). The oxidation of Haematoxylin to Haematein is reversible. In the past, oxidation was achieved (albeit slowly) by exposure to atmospheric oxygen (e.g. Ehrlich's "Haematoxylin"). Iodate, mercuric oxide (pollutant!) or, in Iron-haematein, a ferric salt are now used predominantly. Two equivalents of oxidant per mole of Haematoxylin are used, e.g. $IO_3^- + 3$ Haematoxylin(2H) $\rightarrow$ 3 Haematein $+ I^- + 3H_2O$

As commercial Haematoxylin usually contains some Haematein and several impurities it is generally advantageous to use less than the theoretically required amount of oxidant. This substantially reduces overoxidation which leads to the irreversible formation of *oxyhaematein.* Only complexes with Haematein are usable as histological stains.

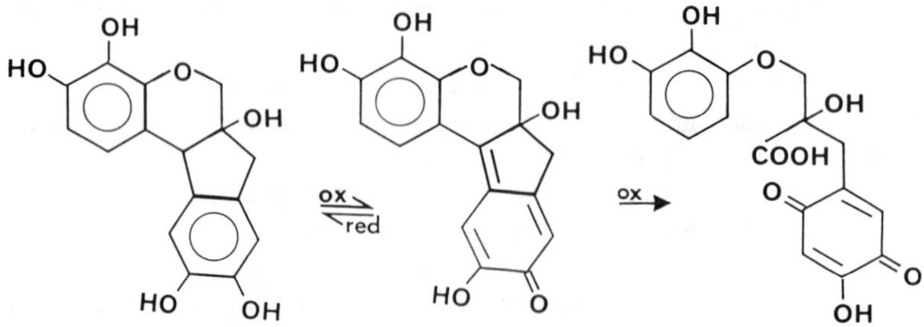

The hydroquinone group in Haematein may form an internal hydrogen bond.

Commercial samples of Haematein should, in general, be treated with suspicion. The complete oxidation of Haematoxylin will give rise to a substantial content of oxyhaematein. Furthermore, Haematein is so sparingly soluble in water and ethanol that the manufacture of usable staining solutions is impossible. (Imes *et al.*, 1969).

Dissociation of a hydroxyl group increases the resonance of the Haematein molecule and hence its light absorbance. The colour of a Haematein solution is therefore much more intense and more blue at high pH (indicator effect). The binding of complex metal ions has the same colour effect as increase in pH.

Haematein is able to bind to many different metal ions; Cr^{3+}, Al^{3+}, and Fe^{2+}/Fe^{3+} are the most important. The Cr-chelates are much more stable than the Al- and Fe-chelates.

The Metal-haematein chelates are positively charged at suitable low pHs. Fe-, Al-, and Cr-chelates are all water-soluble under these conditions and capable of binding to a variety of tissue polyanions such as nucleic acids. Puchtler *et al.* (1986) have reviewed and applied modern chemical concepts to the theory of Metal-haematein formation and staining properties.

7.2.1 Chromium-Haematein

When boiled with chrome alum, both Haematein and Gallocyanin form stable chelates which predominantly stain nucleic acids at low pH. Chromium-haematein (Cr-haematein) is somewhat less selective than Chromium-gallocyanin (Cr-gallocyanin) and the blue colour less bright. The staining procedure and resistance towards subsequent acid treatment is the same as for Cr-gallocyanin.

7.2.2 Aluminium-Haematein

In general aluminium (Al) forms labile chelates, and the staining result using Aluminium-haematein (Al-haematein) is strongly dependent on the molar ratio of aluminium to Haematein (formerly called the mordant ratio), pH, and ionic strength (Sect.3.2.1; Figs. 7.1, 7.2, and 7.3). Furthermore, Al-bound or free Haematein may act as anionic dyes provided pH is not too low (see Figs. 7.1 and 7.3). An Al-Haematein solution with a low Al:Haematein ratio and a high Haematein concentration may therefore stain basic proteins and enterochromaffin granules.

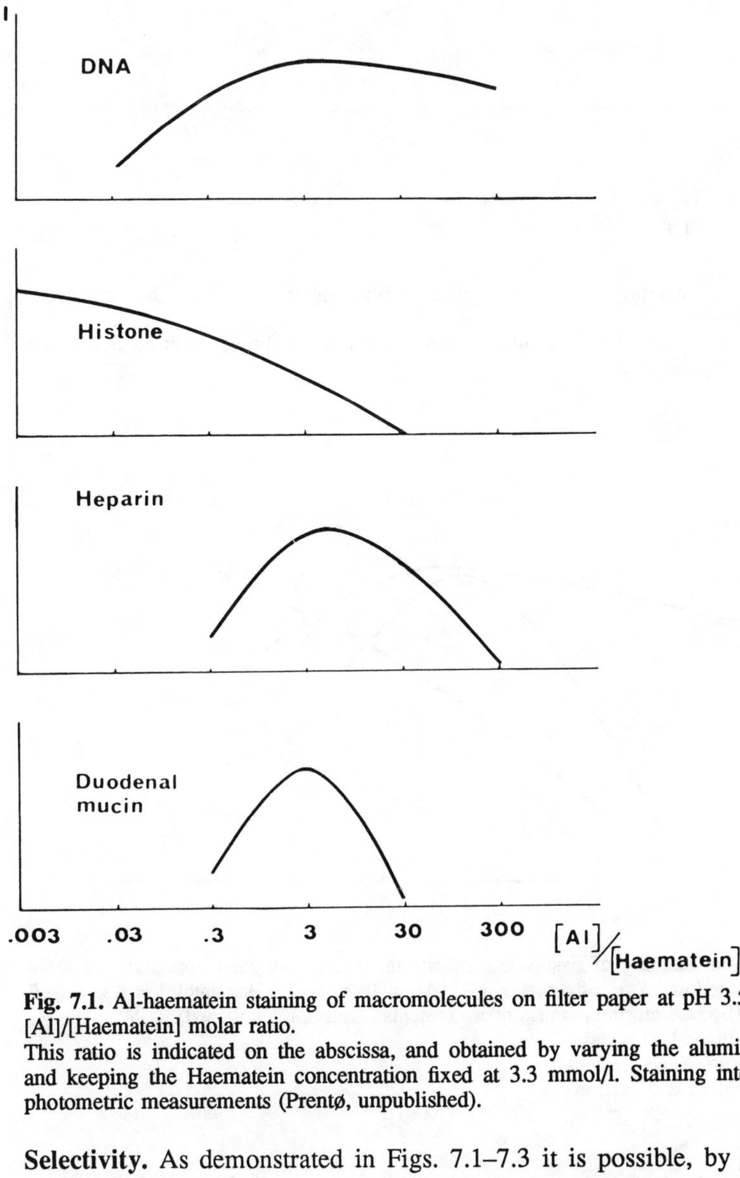

Fig. 7.1. Al-haematein staining of macromolecules on filter paper at pH 3.5: The influence of [Al]/[Haematein] molar ratio.
This ratio is indicated on the abscissa, and obtained by varying the aluminium concentration and keeping the Haematein concentration fixed at 3.3 mmol/l. Staining intensity (I) based on photometric measurements (Prentø, unpublished).

Selectivity. As demonstrated in Figs. 7.1–7.3 it is possible, by judicious choice of Haematein concentration, Al:Haematein ratio, pH, and ionic strength to stain a diverse selection of tissue macromolecules with relative specificity. Clearly a key pre-condition is that the starting concentration of Haematoxylin is known. Unfortunately, this is generally not the case for commercial samples of Haematoxylin so

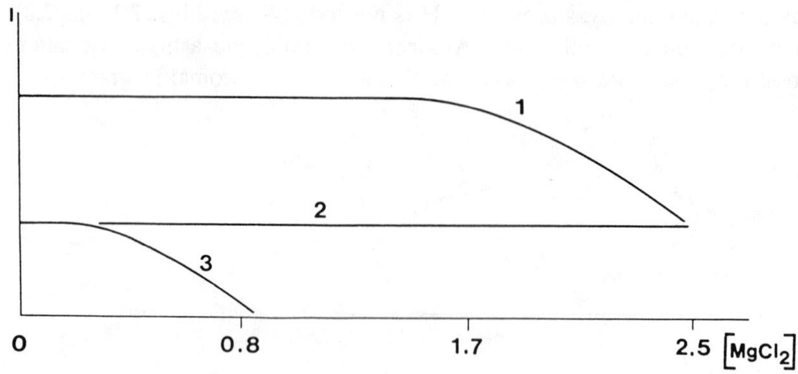

Fig. 7.2. Al-haematein staining of macromolecules on filter paper at pH 3.5: The influence of electrolyte concentration.
[Al] = 10 mmol/l, [Haematein] = 3.3 mmol/l. Staining intensity (I) based on photometric measurements (Prentø, unpublished).
1 DNA
2 histone
3 heparin

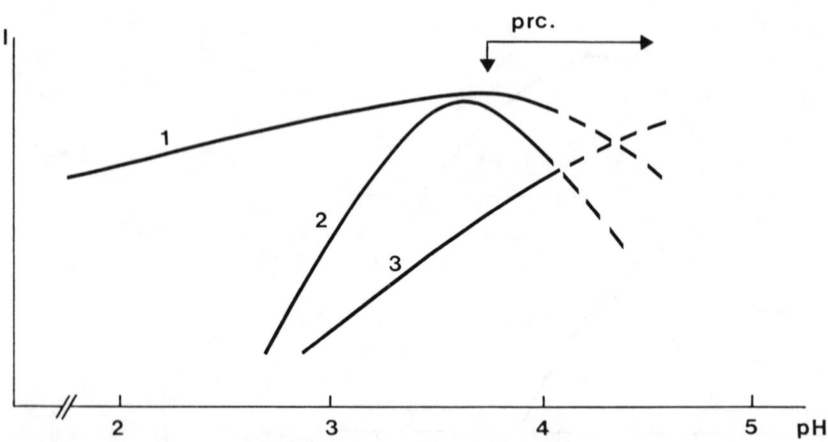

Fig. 7.3. Al-haematein staining of tissue constituents in histone extracted, untreated, or DNA extracted duodenal sections: The influence of pH. [Al] = 10 mmol/l, [Haematein] = 3.3 mmol/l. Staining intensity (I) based on photometric measurements (Prentø, unpublished).
1 nuclei in trypsin-treated sections
2 duodenal mucin (untreated sections)
3 trichloroacetic acid-treated sections
prc. precipitate.

that, in making the desired Al-Haematein solution from a new dye batch, one may have to experiment, especially with the oxidation step.

At pH 3.5 and with an Al:Haematein ratio of about 1 both nucleic acids, basic proteins, and various negatively charged heteroglycans and mucins are stained.

The now obsolete "mucihaematein" method for acid mucins (Mayer, 1891) has an Al:Haematein ratio of about 1.1 and a pH of 3.5–4. The Al-Haematein staining of acid carbohydrates is more reddish violet than the nuclear staining. While reminiscent of metachromasia, this effect is probably not due to dye aggregation but may indicate that the binding is not due to complex formation.

At a pH below 2.5 and an approximate Al:Haematein ratio in excess of 10, relatively selective nucleic acid staining may be obtained. Broadly speaking, a surplus of aluminium salt modifies staining in the same way as the addition of $MgCl_2$ for example (see Critical electrolyte concentration, Sect.6.1.2). In addition to macromolecules, inorganic constituents, especially iron and calcium deposits, may be stained by Al-haematein. This may be due to metal substitution or may involve direct formation of chelates between Haematein and the tissue-bound metal ion (see Sects.17.7.1 and 17.7.4).

Use in Histology. Al-haematein is a very important nuclear stain. Both Haematein and aluminium ions easily penetrate nearly all cell structures, especially at high salt concentration, where charge interaction is diminished. So, in contrast to staining with proper cationic dyes, the staining is relatively independent of electrostatic repulsion or protein-polyanion masking. Regardless of the fixative used, Al-haematein is able to stain the total content of nucleic acids. If the nuclear basic proteins, the histones, and ribosomal basic proteins are preserved, for instance following fixation with Bouin's picric acid/formaldehyde/acetic acid, these may also be stained at the appropriate pH, Haematein concentration, and Al:Haematein ratio. Thus, regardless of tissue pretreatment, adequate nuclear staining may be obtained by use of any of the Al-haematein formulas put forward in Table 7.2. Following acetic acid/ethanol fixatives (Clarke, Carnoy) the nuclear staining is predominantly due to nucleic acids; following prolonged fixation in Bouin's fixative the staining is predominantly due to the basic proteins; and following formalin fixation, and possibly also Lillie's AAF, the staining may, dependent on the fixation procedure and fixation time, reflect nucleic acids, histones, or both.

This independence of tissue pretreatment, that is particularly associated with the non-acidified Al-haemateins, combined with the resistance of the stain to extraction by ethanol are the main reasons for the popularity of Al-haematein in histology.

Table 7.2. The theoretical composition of some Al-Haematein solutions used in histology (assuming pure reagents and total conversion of Haematoxylin to Haematein without overoxidation).

	Mayer	Mallory	Harris	Baker	Delafield
Haematein conc. mol/l	0.003	0.0083	0.0167	0.006	0.021
Aluminium conc. mol/l	0.11	0.11	0.22	0.05	0.132
Al: Haematein ratio	36	13	13	8	6
pH	for all variants 2.9–3.1				
pH after addition of acetic acid up to 2%	for all variants about 2.3				

Differentiation and Blueing. Al-haematein is far less stable towards acid extraction than chromium chelates. Tissue bound Al-haematein can therefore be extracted with acid (e.g. differentiation following overstaining). In contrast, Al-haematein is completely stable towards water and ethanol. This is a considerable advantage if the tissue is to be mounted in a hydrophobic mountant after staining.

Before dehydration, sections are always washed in a weak base, often tap water. This process gives rise to blueing (brownish-red Al-haematein changing to clear blue), an expression of a further ionization of the aluminium bound Haematein corresponding to the formation of aluminium hydroxide by the aluminium-aquoions.

In histological practice Al-haematein is often combined with a red anionic dye, usually Eosin. This combination is exceedingly well suited for general histological studies as the nuclei appear blue against a reddish background showing variations in intensity.

7.2.3 Iron-Haematein

The composition of Iron-haemateins (Fe-haemateins) is very complicated. Fe^{3+} is able to oxidize Haematoxylin to Haematein

$$\text{Haematoxylin} + 2Fe^{3+} \rightarrow \text{Haematein} + 2Fe^{2+}$$

This property is frequently employed in the preparation of Fe-haematein from a ferric salt and Haematoxylin. If the amounts used are appropriate, a pure Ferrous-haematein may be achieved, however, the result will usually be a mixture of ferrous and ferric ions as well as Haematoxylin, Haematein and oxyhaematein. The relative amounts of these components are critical in determining the outcome of staining. While both Ferrous- and Ferric-haematein are able to stain nuclei (Ferric-haematein with low selectivity), the nuclear staining with the latter decreases rapidly with increasing concentrations of the ferric salt. The decrease in nuclear staining may be partly due to surplus ferric ions oxidizing Haematein to oxyhaematein, but competition between free ferric ions and Ferric-haematein for available tissue-binding sites is probably more important. Binding of ferric ions to carboxyl groups, carboxyl-hydroxyl groups, and phosphate groups is considerably stronger than the same interactions involving ferrous ions. Similarly, the Ferric-haematein chelate is far more stable than Al-haematein and Ferrous-haematein chelates.

In agreement with the reaction detailed for ferric ions and Haematoxylin, staining reaches a maximum at a Fe^{3+}:Haematoxylin ratio of about 2, corresponding to complete transformation of Haematoxylin to Haematein and ferric to ferrous ions.

Lillie-Weigert's "Iron-haematoxylin". This Fe-haematein (Lillie and Fullmer, 1976, p.201) is one of the best Fe-haemateins as regards selective staining of nuclei. It is composed of ferric chloride, ferrous sulphate and Haematoxylin in the following amounts:

$FeCl_3$, $6H_2O$	(Fe^{3+})	0.024 mol/l
$FeSO_4$, $7H_2O$	(Fe^{2+})	0.041 mol/l
Haematoxylin		0.008 mol/l

On mixing the reagents together, Haematoxylin is oxidized to Haematein by Fe^{3+}, and the resulting dye solution is composed more or less as follows:

Fe^{3+}	0.008 mol/l
Fe^{2+}	0.057 mol/l
Haematein	0.008 mol/l
Haematoxylin	0

Fe^{2+} therefore accounts for 90% of the total iron and the Fe^{2+}: and Fe^{3+}:Haematein ratios are 7.0 and 1.0 respectively.

Hence the Lillie-Weigert "Iron-haematoxylin" is predominantly a Ferrous-haematein. The selectivity for nuclei and rough endoplasmic reticulum is increased by the addition of HCl (to about 0.01 mol/l), thereby changing pH from around 1.8 to 1.2.

Iron-haematein solutions are unstable. Atmospheric oxygen will oxidize, or reoxidize, ferrous ions to ferric ions leading to an impaired staining result with respect to both selectivity and intensity.

Blueing. Nuclei stained with pure Ferrous-haematein become deep blue after rinsing in alkaline tap water while nuclei stained with Ferric-haematein (and possibly ferric ions) show a less distinct colour change as ferric hydroxide is reddish-brown. Brown to reddish-brown nuclear staining is the norm if the section has been treated with an oxidizing agent such as picric acid after Ferrous-haematein. In a suitable ratio, ferric and ferrous hydroxide are dark brown to black and this feature, combined with the shift in Haematein colour, can give rise to black or nearly black nuclei with certain Iron-haematein stains.

Selectivity and Mechanism. Iron-haematein staining is very pH-dependent. If pH is increased from around 1.2 to 3 (at which point precipitation of Fe-haematein inhibits further reaction) there is a gradual decrease in nuclear staining and a corresponding increase in the staining of proteins. This pH-effect is particularly pronounced if staining time is extended from the usual few minutes to 1–2 hours.

The mechanism of the protein staining is not known. The increasing pH probably leads to the formation of neutral and negatively charged Ferrous-haematein chelates which can bind to amino groups. Impregnation with ferric and/or ferrous hydroxide may possibly also be important, especially if staining is prolonged. Even at pH 1.5 Ferrous-haematein stains strongly basic proteins. For example, Lillie-Weigert's "Iron-haematoxylin" remains an excellent nuclear stain, even after removal of DNA by prior treatment with trichloroacetic acid. This staining becomes very weak if histones have also been extracted. Fe-haematein therefore, like Al-haematein, shows a broad spectrum of nuclear staining with additional enhanced demonstration of nuclear basic proteins. Normally, Fe-haematein does not stain epithelial mucins, probably because the staining pH is too low.

Two-Bath Methods. In the past a two-bath method was commonly used for "Iron-haematoxylin" staining. For this procedure the section is treated with a ferric salt solution (30 min–many hours) followed by washing and 0.5% aqueous, non-acidified Haematoxylin solution for 30 min or longer. Tissue bound ferric ions gradually oxidize Haematoxylin to Haematein and the final result is a mixed Ferrous/Ferric-haematein staining. Alternatively, due to the relatively high pH of the Haematoxylin solution, the overall effect may be that of an impregnation affecting practically all structures in the tissue. The final result is achieved by an extraction, "differentiation", with an acid, an acid oxidizing agent, or a ferric salt solution. Providing fixation is adequate, most cell and tissue structures can be selectively stained using this procedure. Satisfactory results are, however, difficult to achieve and the method has been largely superceded by more satisfactory approaches.

A procedure in which the section is first overstained and then differentiated is called a *regressive* stain (in contrast to a *progressive* stain). Single-bath Fe-haematein and Al-haematein are sometimes used regressively. In this situation the staining solution is not acidified.

7.2.4 Other Metal-Haemateins

Haematein can bind to many other metals besides Cr, Al, and Fe. For example, Copper-haematein (Cu-haematein) and Lead-haematein (Pb-haematein) have sometimes been used in histological staining.

In an equimolar mixture of an aluminium salt and an iron salt, Haematein binds predominantly to iron. This explains why Al-haematein stains tissue iron deposits. For similar reasons zinc and copper may also be stained, while the staining of calcium deposits probably involves a more complex mechanism.

7.3 Chromium-Gallocyanin

Chromium-gallocyanin (Cr-gallocyanin; Gallocyanin-chromalum) was introduced to histochemistry by Becher in 1921. The quantitative aspects of Cr-gallocyanin staining have been investigated in detail by Einarson (1932; 1951) for both cytological and histological material.

Preparation. Gallocyanin (C.I. 51030) is an amphoteric dye of the oxazin group (mol.wt. 336.74 as the chloride salt) which occurs predominantly in the form shown below at pH 1.6.

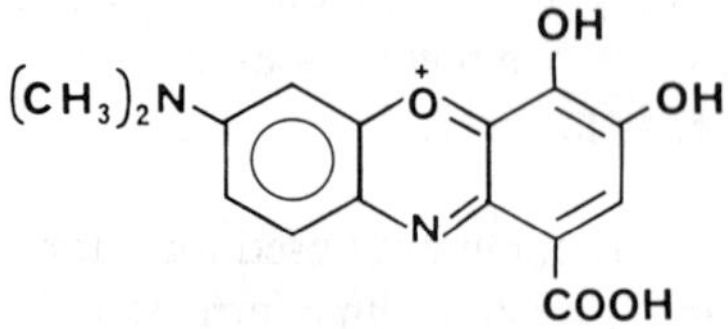

With increasing pH the carboxyl group becomes ionized first followed by one of the phenol groups; the overall charge changes from positive at pH 1.6 to negative at around pH 4.0 (Baker, 1958, p.215).

Boiling with chromalum for between 2 and 20 min yields (1 + 1)- and (2 + 1)-chelates between Gallocyanin and hydrated chromium ions; longer boiling times favour the formation of (2 + 1)-complexes.

The spectral absorption maximum of this Chromium-gallocyanin solution is around 575 nm. Even after boiling for 20 min there is a small proportion of free Gallocyanin which may cause a reddish staining (metachromasia) of sulphate containing polysaccharides or a weak reddish staining of cytoplasmic proteins. Free Gallocyanin can be easily demonstrated by thin layer chromatography (Horobin and Murgatroyd, 1968). Attempts to prepare reagents with no free Gallocyanin result in the formation of histochemically unsuitable Chromium-gallocyanin complexes. We believe that the most selective reagent for staining nucleic acids is prepared as follows (Einarson, 1951; Sandritter *et al.*, 1963; Berube *et al.*, 1966, and Horobin and Murgatroyd, 1968):

Dissolve 150 mg pure Gallocyanin and 15 g $KCr(SO_4)_2$ in 100 ml H_2O, boil for 10 min, allow to cool to room temperature, filter and adjust pH to pH 1.6 with H_2SO_4. Staining time at room temperature is 24 h (16–100 h). The active Chromium-gallocyanin complexes can be isolated by precipitating with ammonia (Berube *et al.*, 1966), flushing the precipitate and redissolving it in 1 mol/l H_2SO_4.

Many different structures have been suggested for the chelates (Einarson, 1951; Sandritter *et al.*, 1963; Horobin and Murgatroyd, 1968). It is highly probable that chromium is attached to the orthodiphenol group in Gallocyanin thus forming a hydroxyquinonimine chelate. Cr-gallocyanin would then have the same configuration as the chromium-chelates of Gallamine Blue (C.I. 51045), Gallo Blue E (C.I. 51040), and Celestin Blue (C.I. 51050) where the only differences from Gallocyanin are that the carboxyl groups have been changed to amides or methyl esters and thus cannot be ionized. However, it is suggested that in the (2 + 1)-complex aminocarboxylate chelation takes place (Marshall and Horobin, 1972). Gallamine Blue and Celestin Blue have been repeatedly tested as substitutes for Cr-gallocyanin (Proescher and Arkush, 1928; Gray *et al.*, 1957).

Binding to Nucleic Acids. Cr-gallocyanin stains DNA and RNA an intense blue colour. As polyphosphoric acid is stained in exactly the same way (Jobst and Sandritter, 1964) the nucleic acid binding probably involves phosphate groups. Cr-gallocyanin behaves therefore, in histochemical terms, like a cationic dye.

The association between Cr-gallocyanin and nucleic acids is highly resistant to both ethanol and acid extraction. This property and the extended time-course of staining (around 100 h to maximum) are both consistent with the view that complex bonds form between Cr^{3+} and nucleic acid-phosphate. In contrast, free Gallocyanin binds rapidly (minutes) to proteins or polysaccharides and does not resist acid or ethanol extraction.

Photometric determinations have indicated a Cr-gallocyanin/DNA-phosphate ratio of 0.86 after 48 h (Sandritter *et al.*, 1963), i.e approaching 1 at maximum staining. Strictly speaking this is only true for cell nuclei with diploid or hyperdiploid DNA content (Jobst and Sandritter, 1964). In haploid cell nuclei (e.g. spermatozoa) chromatin compactness has a marked influence on the quantitative dye/DNA-ratio; in addition, steric factors are probably the main reason for weak staining in condensed nuclei of some diploid cells such as small lymphocytes (Mayall, 1969). Occlusion of binding sites by proteins appears to be relatively unimportant since the long staining time and the low pH lead to extensive histone extraction; however, if proteins are partially removed prior to staining, an increase in stainability can be demonstrated (Polasaky *et al.*, 1964).

Selectivity. If performed correctly, blue Cr-gallocyanin staining is, with the exception of keratohyalin granules, highly selective for nucleic acid phosphate groups and for polyphosphates. The method has, however, been accused of poor selectivity based on the finding that RNase-treatment sometimes (particularly after formaldehyde fixation) fails to remove all the Cr-gallocyanin positive material in the cytoplasm and the nucleolus. It must be remembered that formaldehyde is a protein stabilizing agent (Sect.12.2.1) and may therefore hinder the free access of RNase to ribosomal RNA. These factors should be set against the relative insensitivity of staining to protein blockade (Shires, 1967). In contrast, when acid alcoholic fixatives are used, RNA hydrolysis is normally complete and Cr-gallocyanin does not stain either cytoplasm or nucleoli.

Sensitivity. The molar extinction coefficient for Cr-gallocyanin is high, and the Gallocyanin/phosphate ratio approaches 1:1 (Kiefer,1970). The method is therefore very sensitive provided the staining time is maintained. The reader is cautioned against reducing the staining time by incubating at elevated temperature. Cr-gallocyanin is a complex dye, and staining results from a delicate balance between several variables including temperature. Elevated temperature combined with the standard low pH staining conditions leads to substantial hydrolysis of RNA before it is immobilized by Cr-gallocyanin binding. The long staining time makes Cr-gallocyanin at present unsuitable for application in a routine staining method.

Husain and Watts (1984) have recommended a modified Cr-gallocyanin staining procedure using staining times between 5 and 10 min at pH 1.0. Although their technique has been applied as a quantitative nuclear stain in automated cytology, the theoretical background of the dye-substrate-interaction has not been fully elucidated.

Localization. Substrate diffusion is not a problem as nucleic acids are precipitated at the pH used in Cr-gallocyanin staining (pH 1.6). Localization is therefore dependent only on tissue pretreatment.

Quantitation. The method is well suited for the photometric determination of total nucleic acid content (cf. Sect.28.8.4). RNase treatment allows the direct determination of DNA content, and RNA content may then be calculated by subtraction. The Beer-Lambert law (Sect.3.3.2) is obeyed throughout the visible spectrum (Sandritter *et al.*, 1963) and the acid/alcohol resistance is a clear advantage for both precision and reproducibility.

Considerable fading takes place in stained sections. This is clearly detectable after 3 weeks and stabilizes at about half the original extinction at around 12 weeks. This process occurs regardless of whether the sections are protected from light or not (Vejlsted, 1972). Both microscopy and photometry should therefore be performed within 14 days of staining.

8 Staining Based on Reductants and Oxidants

M.R. Barer, H. Lyon

Redox reactions, apart from those involving enzymes, are widely used in the demonstration of miscellaneous cell and tissue components. Staining depends on the presence of groups capable of reducing or oxidizing staining reagents. Prominent examples include the demonstration of reticulin and neurofibrils using silver.

8.1 Redox Reactions

A simple *redox reaction* can be written as:

$$\text{Red} \rightleftarrows \text{Ox} + e^-$$

where Red is a reducing agent (reductant) and Ox is an oxidizing agent (oxidant). An important precondition is that there should be an electron acceptor substance present (i.e. an oxidant). The general redox reaction is thus:

$$\text{Red}_1 + \text{Ox}_2 \rightleftarrows \text{Ox}_1 + \text{Red}_2$$

where corresponding oxidants and reductants are designated by the same number. A pair of this kind is called a *redox couple*, and is usually referred to by writing Ox_1/Red_1, e.g. Ag^+/Ag. The reductant, Red_1, supplies electrons and is hereby oxidized to Ox_1, while the oxidant, Ox_2, acts as an acceptor of electrons and is reduced to Red_2.

8.2 Classification of Tissue Bound Reductants and Oxidants

Tissue bound reductants and oxidants are surveyed in Table 8.1.

H. Lyon (Ed.)
Theory and Strategy in Histochemistry

Table 8.1. Classification and occurrence of reductants and oxidants in tissue.

Reductant	Occurrence	Oxidant	Occurrence
Thiol	cysteine, free and protein bound	disulphide	cystine, free and protein bound
Phenol	tyrosine, free and protein bound; thyroxine; adrenaline; noradrenaline; dopamine; melanin		
Pyrrole (indole)	tryptophan, free and protein bound; serotonin		
Ascorbic acid		dehydroascorbic acid	
Alkene	unsaturated lipids, including lipofuscin	glycol, aldehyde	
Aldehyde	induced by acid hydrolysis of DNA; by oxidation of 1,2-glycol or 1-hydroxy-2-amino-groups, possibly alkene	carboxyl	

8.3 Methods for the Demonstration of Reductants and Oxidants

There are several very sensitive methods for the demonstration of reductants. These include:

1. The ferric-ferricyanide method
2. Silver salt methods
3. Methods using tetrazolium salts

In contrast, methods for demonstrating oxidants are, in general, not as sensitive. Examples include, nadi (α-naphthol + dimethyl-p-phenylenediamine), leucodichloroindophenol, benzidine, and DOPA. These will not be discussed further in this text.

8.3.1 Ferric-Ferricyanide Method

The ferric-ferricyanide reaction or "Schmorl's reaction" has been described by Golodetz and Unna (1909) and Chévremont and Frederic (1943).

Mechanism. Ferric ions (provided by $FeCl_3$) are reduced to ferrous ions, $Fe^{3+} \rightarrow Fe^{2+}$, and the latter are bound by complex bonds to ferricyanide:

$$K^+ + [Fe(CN)_6]^{3-} + Fe^{2+} \rightarrow K[Fe_2(CN)_6]$$

The complex formed is very insoluble and is called Prussian Blue or Turnbull Blue. The precise composition of the complex is determined by the ratio between ferrous and ferric ions (Sect. 17.7.4). Lillie and Donaldson (1974) have shown that ferricyanide acts as an oxidant in alkaline solutions, but not at pH values under 7. However, ferric chloride easily oxidizes tissue thiol groups, noradrenaline, and serotonin at pH 2.5, the standard pH for the ferric-ferricyanide mixture. These reactions are consistent with the redox potentials for ferric and ferricyanide ions. By changing the molar ratio between ferric chloride and ferricyanide, as well as the reaction time, it is possible to alter the selectivity of the method. As well as thiol groups, melanin, lipofuscin, ascorbic acid, serotonin, adrenaline, and noradrenaline, also give positive reactions; demonstration of adrenaline and noradrenaline, however, depends on prior treatment with a dichromate containing fixative. Formaldehyde fixed material may be used for the demonstration of melanin, lipofuscin, and serotonin.

Selectivity. The method demonstrates all groups that are able to reduce ferric to ferrous ions in acid solution. Schmorl (1928), p.205, described the use of the method for the demonstration of lipofuscin. Naturally, ferrous ions in the tissue will also be demonstrated (Sect.17.7.4).

Sensitivity. This is very high.

Blocking. The reaction of thiol groups in the ferric-ferricyanide method can be blocked by oxidation with weak oxidants (Sects.5.1.1 and 9.3), acylation (Sects. 5.1.3 and 9.2.1), or mercaptide formation (Sect.9.3). Melanin, lipofuscin, and serotonin are not blocked by these procedures but need stronger oxidants such as hydrogen peroxide (Sect.18.4.2).

Localization. This is precise.

Quantitation. This is theoretically possible (cf. Sect.28.8.9).

8.3.2 Silver Salt Methods

These methods involve the reduction of silver ions yielding a fine granular black precipitate. The methods are called argentaffin if a tissue component can reduce a silver salt to metallic silver in the dark and in the absence of a reducing agent. If exposure to light or addition of a reducing agent is required, then the method is called argyrophil. Due to the electron density of silver it is possible to utilize these methods for electron microscopy (Sect.27.2.1).

Argentaffin methods. These methods employ:

a. Simple silver salts in aqueous solution (nearly always silver nitrate) with added acetic acid to a pH of approximately 4
b. Complex silver compounds either as diammine silver $[Ag(NH_3)_2]^+$, pH about 12, (Masson, 1914; Fontana, 1912; or Lillie-Masson (Lillie and Fullmer, 1976), p.277) or as methenamine silver, $[Ag((CH_2)_6N_4)_2]^+$, pH approximately 8, (Gomori, 1946; or Gomori-Burtner-Lillie (Burtner and Lillie, 1949)).

Complex silver compounds allow a suitable concentration of silver ions without precipitation in alkali.

The equilibria

$$[Ag(NH_3)_2]^+ \rightleftarrows Ag^+ + 2NH_3$$

and

$$[Ag((CH_2)_6N_4)_2]^+ \rightleftarrows Ag^+ + 2(CH_2)_6N_4$$

are both substantially displaced to the left so that the actual concentration of silver ions in staining solutions is about 10^{-7} mol/l.

Mechanism. $Red + Ag^+ \rightleftarrows Ox + Ag$. In practice silver nitrate is not easily reduced at pH 4, while at alkaline pH, the complex silver compounds are far more easily reduced as, for example, bases may act as catalysts in the oxidation of aldehyde. Thiol groups reduce neither silver nitrate nor silver complexes. Argentaffin methods can be divided into those that directly demonstrate true argentaffin structures and those that demonstrate induced reducing groups resulting from suitable pretreatment.

Examples of the latter are PASM (periodic acid-silver methenamine) where the aldehyde groups arising on oxidation with periodic acid are demonstrated by reduction of silver methenamine (Sect.9.2.2) and "black Feulgen" where aldehyde groups induced in DNA by acid hydrolysis (Sect.9.9) are demonstrated by reduction of a suitable silver compound. Grocott's method for the demonstration of fungi (1955) includes oxidation with chromium trioxide, silver methenamine, toning, and fixing. The mechanism is probably identical to that proposed for PASM.

Sensitivity. This varies from low for silver nitrate to high for silver complex methods, although the latter are never as sensitive as the ferric-ferricyanide method.

Blocking. The same conditions apply as for the ferric-ferricyanide method.

Localization. This is precise when the reaction takes place in the dark and in the absence of reductants.

Quantitation. This is possible as silver precipitation is a stoichiometric process.

Argyrophil methods. These include methods demonstrating reticulin, neurofibrils, different glial cells and argyrophil cells.

Mechanism and selectivity. The mechanism is not fully elucidated but certain features are common to all methods. There are also a number of similarities to the photographic process from which several of the steps derive their names. The key steps are:

1. Oxidation
2. Reduction
3. Sensitization
4. Silver bath
5. Reduction
6. Toning

7. Fixing

A survey of the key steps in a number of argyrophil methods is given in Table 8.2.

Oxidation. Potassium permanganate is usually the oxidant of choice, but periodic acid and chromium trioxide can be used. The aim is to produce reducing groups in reticulin. Possible groups involved include hydroxylysyl residues in collagen and glycol groups in the carbohydrate matrix surrounding the reticulin fibres. In principle, all tissue homoglycans and glycoproteins can give rise to aldehyde groups and indeed, goblet cell mucin and hepatocyte glycogen granules do become rather black. Alkene groups in lipofuscin can also give rise to aldehyde groups. This seems to occur particularly if the permanganate solution is not acidified. With a permanganate solution pH under 1.5, manganese is reduced from an oxidation state of +7 to +2 instead of +4 (manganese dioxide), and the reducing step can then be omitted.

Reduction. This step is only required following oxidation with permanganate as the aim is to remove the brown manganese dioxide. According to tradition oxalic acid is used. This is oxidized to carbon dioxide and water, while the manganese dioxide is reduced to soluble manganese (II) compounds.

Sensitizing. In this step the tissue components to be demonstrated are made more "sensitive" to the subsequent silver bath. Many sensitizing reagents have been proposed including: salts of uranium, iron, chromium, and silver as well as hydrogen peroxide, iodates, iodine, acetic acid, and ammonia. Mechanisms could involve partial blocking or deblocking of active groups, oxidation, reduction, and pH-induced changes, but the exact effects are unknown. In some methods (e.g. Bodian's neurofibril method), omission of this step makes little difference, while in others (e.g. Gordon and Sweets' reticulin method) it is essential.

The silver bath. Two processes take place here:

a. Reduction of a small number of silver ions to metallic silver. The reduction process takes place on the reducing groups in the tissue in essentially the same way as in the argentaffin methods. However, the density of silver granules (silver nuclei, points of nucleation) is so low that precipitates cannot be seen by light microscopy. The reducing groups are probably aldehydes in the reticulin methods, while it has been suggested that the neurofibril methods involve thiol groups. This suggestion is, however, in conflict with the fact that direct experiments show that thiol groups are unable to reduce silver nitrate or silver complexes. The identity of the reducing groups is not established in the other argyrophil methods. Table 8.1 shows that many possibilities arise. A discussion of formation of silver nuclei and of physical and chemical development is given by LaVelle (1985).
b. A binding of silver ions and possibly charged silver complexes to negative charges in the tissue. This process occurs in the silver bath, is relatively nonspecific, and comprises the formation of ionic bonds between ionized groups in

Table 8.2. Key steps in a number of argyrophil methods.

Steps in the process	Method for the demonstration of: reticulin (Gordon and Sweets, 1936)	oligodendroglia (Del Río-Hortega, 1917)	astrocytes (Ramón y Cajal, 1913; 1916)	nerve cells with processes (Golgi, 1875)	peripheral nerve ends (Bielschowski, 1904; 1909)	neurofibrils and nerve fibres (Bodian, 1936)
1. Oxidation	$KMnO_4$	–	–	–	–	–
2. Reduction	oxalic acid	–	–	–	–	–
3. Sensitization	iron alum	NH_3	–	$K_2Cr_2O_7$	$AgNO_3$	NH_3
4. Silver bath	silver diammine	silver diammine	$HAuCl_4 + HgCl_2$[1]	$AgNO_3$ in H_2O	Silver diammine	silver proteinate
5. Reduction	formaldehyde	formaldehyde	light	light	light	sulphite and hydroquinone
6. Toning	(+)	+	–	–	(+)	+[2]
7. Fixing	+	+	+	–	–	+

Steps in the process	Method for the demonstration of: degenerating axons (Nauta and Gygax, 1951)	argyrophil cells (Grimelius, 1968)	spirochaetes (Faulkner and Lillie, 1945)	nerve fibres (Peters, 1958)
1. Oxidation	–	–	–	–
2. Reduction	–	–	–	–
3. Sensitization	$AgNO_3$	–	–	–
4. Silver bath	silver diammine	$AgNO_3$ in H_2O	$AgNO_3$ in H_2O	silver proteinate
5. Reduction	formaldehyde	sulphite and hydroquinone	hydroquinone	sulphite and hydroquinone
6. Toning	–	–	(+)	+
7. Fixing	+	–	–	+

[1] Gold and mercury instead of silver; [2] followed by renewed oxidation in oxalic acid.
+ = the step is part of the method; (+) = the step is optional; – = the step is not included.

the tissue and silver ions or complexes in solution. The negatively charged tissue groups relevant in this context include, sulphate in carbohydrates, phosphate in nucleic acids, and carboxyl in carbohydrates and proteins. Quantitatively, the carboxyl groups of proteins are the most important when alkaline silver solutions are used. Scopsi and Larsson (1986b) have suggested that the silver proteinate in Bodian's method binds to lysine residues in neurofilaments and also in argyrophil cells such as gastrin (G) cells in the antrum and glucagon (A) cells in pancreas. The possibility that this reaction shares a common chemical background with unspecific immunocytochemical staining is discussed.

Reduction. Reduction or development can be achieved either using light, as in the older methods, or by using reductants (Table 8.2). The process leads to precipitation of silver at sites corresponding to points where silver nuclei arose in the silver bath.

Toning. In this step silver precipitate is replaced by gold. As gold has a higher potential than silver in the electrochemical series the following process occurs on exposure to gold salts:

$$3Ag + Au^{3+} \rightarrow 3Ag^{+} + Au.$$

Clearly, the total amount of metal in the section falls as a consequence of this reaction. Due to the differences in atomic weight the mass of the gold will be approximately 2/3 the mass of the silver. The staining intensity is therefore reduced, and the shift in colour virtually abolishes non-specific background staining. This can be an advantage, but, in the reticulin methods for example, staining of collagen (golden-brown) is so reduced that the methods cannot be used for studies of mature collagen. In some methods, such as Bodian's neurofibril method, the reduction in intensity is so pronounced that a further reduction step, involving gold ions or complexes bound to the negative charges in the tissue, becomes necessary. The gold atoms formed precipitate on the gold nuclei from the preceding gold bath. It should be noted that the process becomes irreversible following toning. Sections should therefore be controlled in the microscope immediately following the primary reduction, as it is possible to remove all the silver using a solution of a ferric salt at this stage.

Toning with palladium (0.05% hexachloropalladate in 4 mol/l HCl) for 10 min has been suggested as a much less expensive substitute for gold (Krikelis and Smith, 1988).

Fixing. This is the final step in the majority of argyrophil and some of the argentaffin methods. The section is treated with sodium thiosulphate (hypo) which removes un-reduced silver or gold by complex formation.

$$Ag^{+} + 2S_2O_3^{2-} \rightarrow [Ag(S_4O_6)]^{3-}$$

If this step is omitted then exposure to light may produce new precipitates by reduction of bound metal ions, and previously stained structures (e.g. spirochaetes) may become unstained.

8.3.3 Tetrazolium Salts

The composition, properties, and range of applications for these compounds is given in Sect.3.3.8. Their use in the demonstration of oxidoreductases is discussed in Sects.25.1 and 25.1.2.

8.3.4 Autometallography

In a series of papers Danscher and coworkers (see for instance Danscher, 1984) have presented the concept of *autometallography*. As described above metallic silver catalyzes the reduction of silver ions to metallic silver in the presence of a reducing agent. A similar catalytic activity on the reduction of silver ions is displayed by metallic gold, metal sulphides, and metal selenides. In autometallography the tissue sections are covered by a photographic emulsion (silver bromide), dried, and exposed to a chemical developer (reducing agent). The developer penetrates the emulsion picking up silver ions released from the crystals and continues into the tissue that now contains both Ag^{+} and reducing molecules. If catalysts (silver, gold, metal sulphides, or metal selenides) are present in the tissue they become "physically developed", meaning that they become surrounded by shells of metallic silver. This reaction continues exponentially until terminated by an acid stop bath (1% acetic acid) before the section is transferred to the fixation bath.

The autometallographic method thus represents a technique for amplifying accumulations of gold, silver, metal sulphides, and metal selenides. The method of Timm, described in Sect.17.7.11, is based on this principle. Amplification of gold deposits (gold-silver methods) are used in both light and electron microscopy for the amplification of colloidal gold as used for instance in immunohistochemistry (Sects.26.2.4 and 27.3).

9 Staining Involving Covalent Bonds

H. Lyon, M.R. Barer, E. Schulte

This chapter covers histochemical methods concerned with the following reactive groups:

1. Alkene (Sect.9.1)
2. Glycol (Sect.9.2)
3. Thiol and disulphide (Sect.9.3)
4. Aromatic and heteroaromatic cyclic ring systems (Sect.9.4)
5. Amine (Sect.9.5)
6. Guanidyl (Sect.9.6)
7. Aldehyde (Sect.9.7)
8. Acid groups (Sect.9.8)
9. Purine-N-C1-deoxyribose glycoside (DNA, Sect.9.9)

9.1 Alkene Groups

Occurrence. These groups are found in unsaturated lipids.

Demonstration. See Table 9.1.

Table 9.1. Survey of methods for demonstrating and blocking –C═C–.

Demonstration method	Section	Blocking
1. Addition of OsO_4	9.1.1; 19.6.5	Br_2 (5.1.1)
2. Peracid oxidation-aldehyde reagent	9.1.2	before peracid: Br_2 after peracid: as aldehyde (9.7.2)
O_2-Schiff	9.1.3; 19.6.4	Br_2
UV-Schiff	9.1.4; 19.6.8	Br_2
3. Dichromate oxidation-acid Haematein	9.1.5; 19.6.7	

H. Lyon (Ed.)
Theory and Strategy in Histochemistry

9.1.1 Addition of Osmium Tetroxide

Osmium tetroxide, OsO_4, adds quantitatively to alkenes (carbon-carbon double bonds, C=C groups) in the ratio 1:1. This reaction is further discussed in Sect.19.6.5.

9.1.2 Peracid Oxidation-Aldehyde Reagent

This method is useful in the investigation of insoluble lipid derivatives (e.g. lipofuscin).

Mechanism. A two step process is involved:

1. Formation of aldehyde groups. Oxidants, such as performic acid, peracetic acid, and potassium permanganate attack carbon-carbon double bonds forming epoxides, glycols, and aldehydes. Permanganate also forms carboxyl groups.

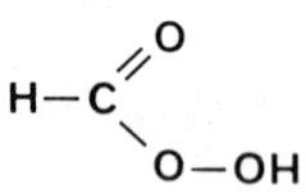

Performic acid

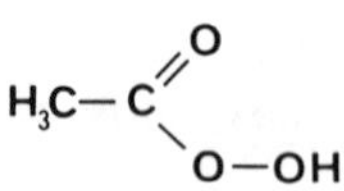

Peracetic acid

   Using performic acid or peracetic acid ninety per cent of the double bonds will be converted to glycol groups. The glycol groups can then be further oxidized to aldehyde groups by periodic acid (Sect.9.2.1)
2. Demonstration of aldehyde groups with Schiff's reagent (see Sect.9.7.1)

Selectivity. This is fairly good but DNA will also react due to acid hydrolysis of the purine-glycoside bonds (Sect.9.9).

Blocking. Blocking reactions include bromination (Sect.5.1.1) carried out before the initial oxidation and aldehyde blocking following the oxidation but before the reaction with Schiff's reagent. The latter can be achieved with aniline, phenylhydrazine, or semicarbazide.

9.1.3 O_2-Schiff

Exposure of unsaturated lipids to air (oxygen) may lead to oxidation of double bonds to aldehyde groups. These can be demonstrated with Schiff's reagent (pseudoplasmal reaction, Sect.19.6.4).

9.1.4 UV-Schiff

This method, where ultraviolet irradiation for 3–4 hours converts alkene to aldehyde groups which subsequently can be demonstrated with Schiff's reagent, is far more specific than the peracid oxidation methods. It is more fully discussed in Sect.19.6.8.

9.1.5 Dichromate Oxidation-Acid Haematein

With dichromate oxidation C=C bonds are broken and complexes form with chromium. These can be demonstrated with Haematein (Sect.19.6.7).

9.2 1,2-Glycol and 1-Amino-2-Hydroxyl Groups

Occurrence. Saccharides, glycolipids, and certain proteins.

Demonstration. This can be achieved by oxidation-aldehyde reagent. Methods of this kind are based on the following reactions:

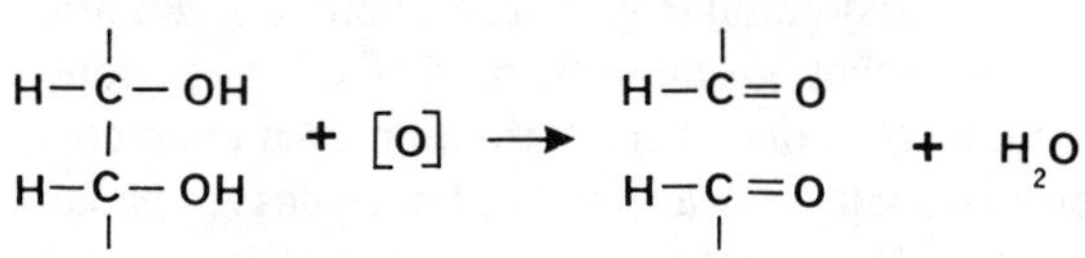

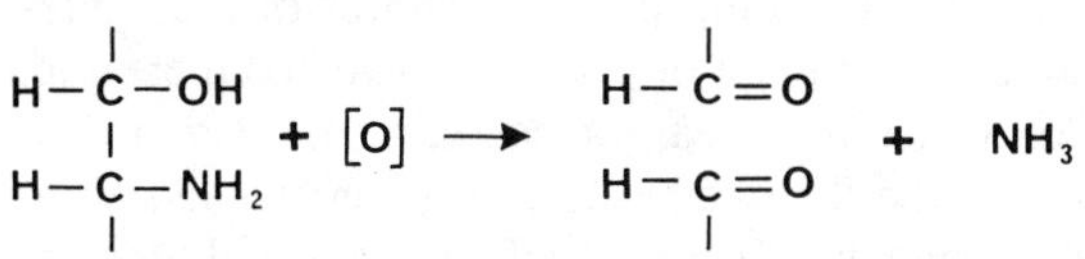

where 1,2-glycol and 1-amino-2-hydroxyl compounds are oxidatively cleaved to two aldehyde groups. If either the hydroxyl or the amino group is acylated the compound does not react. (See, however, Sect.22.2.1).

9.2.1 Periodic Acid-Schiff

The periodic acid-Schiff reaction (PAS-reaction) was described by McManus (1946).

Mechanism. This is a two step reaction:

1. Oxidation with periodic acid to form aldehyde groups. Note, that the carbohydrates still form chains, even though the rings have been opened:

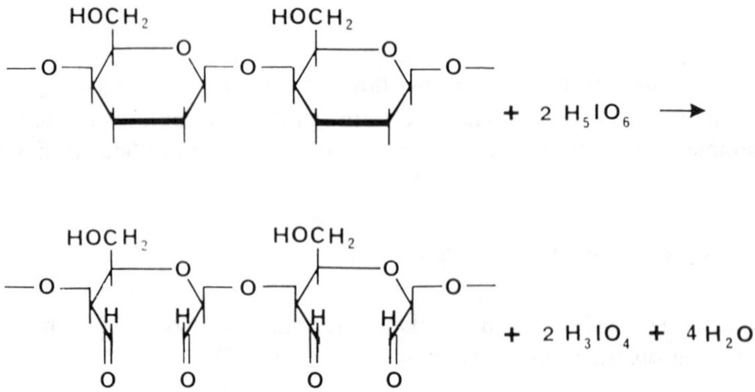

2. Demonstration of the aldehyde groups with Schiff's reagent (Sect.9.7.1)

Selectivity and Results. Using 0.5% w/v (0.022 mol/l) periodic acid (pH 2.1) for 5 min at room temperature the oxidation of 1,2-glycol and 1-amino-2-hydroxyl groups leads exclusively to the formation of aldehyde groups. Alkene groups sometimes react, but this is probably only when partial oxidation to 1,2-glycol groups has already taken place. Higher temperature, increased concentration, or prolonged reaction time with periodic acid can lead to further oxidation to carboxyl groups.

Ovadia and Stoward (1971) investigated the relationship between glycol oxidation by periodic acid after defined times and cellular glycogen content. Even after an oxidation time of 12 days it was found that not more than 24 % of these groups were oxidized. The authors believed this to be due to the formation of anionic complex intermediates formed between periodate ions and the outer glycosyl residues of the precipitated glycogen aggregates. The negative charges formed in this way would prevent free periodate ions from penetrating into the interior of the aggregates. This repulsion would increase as the amount of glycogen oxidized increased. According to the authors this would explain the linear plateau of the oxidation-time curve. This plateau appears after oxidation times of 5–15 min (Gahrton and Yataganas, 1976; Halkær-Kristensen and Ingemann-Hansen, 1979). Ovadia and Stoward (1971) were further able to show that the addition of electrolytes, such as sodium or magnesium chloride, (0.2 mol/l) increases the velocity and extent of oxidation of glycogen in liver sections. They proposed that the cations decrease the repulsion between periodate ions and the negatively-charged carbohydrate-periodate complex. This interpretation fits in well with observations made on the use of critical electrolyte concentration (CEC, cf. Sect.6.1.2).

Preformed aldehyde groups in the tissue and additional reactive aldehyde groups induced by aldehyde, oxidative, or acid fixation can also give rise to a positive PAS-reaction. In addition, several lipids and hydroxylysyl groups in collagen may also react. (Pretreatment of the sections with sodium or potassium borohydride ($NaBH_4$, KBH_4) or with glycine followed by washing in water may eliminate false positive reactions). If the PAS-stained material is not also Sudanophilic, then lipids can be excluded as a source of the PAS-positive reaction; if it is Sudanophilic, then partly oxidized unsaturated lipids (lipofuscin) or glycolipids must be considered. The carbohydrates giving a positive PAS-reaction are the homoglycans, i.e. glycogen (confirmed by extraction with amylase), all glycoproteins and neutral proteoglycans.

Even though they contain glycol groups, acid proteoglycans are PAS negative. This is due to the fact that the C2-C3 glycol groups in glucuronic and iduronic acids are not oxidized by periodic acid. It appears that electrostatic repulsion of periodic acid by the uronic acids may be responsible for this.

Kiviranta *et al.* (1985b) have published a modified PAS method for glucuronic acids. The principle is oxidation with 1% periodic acid at 20°C for 2 hours, blocking with 1% aqueous sodium borohydride (cf. Sect.22.2.1), oxidation with periodic acid for 24 hours at 30°C followed by Schiff's reagent. This method gave selective staining of chondroitin sulphate as shown by control sections subjected to prior digestion with chondroitinase ABC.

Sensitivity. This is high.

Blocking. Hydroxyl groups, amino groups and thiol groups are blocked by *acylation* (Sect.5.1.3), for example:

a. *Acetylation* with acetyl chloride or acetic anhydride with added anhydrous pyridine
b. *Benzoylation.* Benzoyl chloride with added anhydrous pyridine
c. *Tosylation.* Aqueous p-toluenesulphonyl chloride

Primary and secondary amines react forming amides.

$$R_2NH + R{-}COCl \rightarrow R_2N{-}CO{-}R + HCl$$

Hydroxyl (alcohol and phenol groups) and thiol groups react forming esters and thiol esters respectively

$$R{-}OH + R{-}COCl \rightarrow R{-}O{-}CO{-}R + HCl$$
$$R{-}SH + R{-}COCl \rightarrow R{-}S{-}CO{-}R + HCl$$

Deblocking. Deblocking of acylation is achieved by *saponification* with KOH or with a mixture of concentrated ammonia and absolute ethanol (Sect.5.1.4).

9.2.2 Periodic Acid-Methenamine Silver

Mechanism. Periodic acid oxidation is performed as in the PAS procedure. The induced aldehyde groups are demonstrated with silver methenamine (Sect.8.3.2).

Free silver ions in equilibrium with the complex are reduced to metallic silver which is precipitated corresponding to the aldehyde groups.

Selectivity, Sensitivity, and Blocking. These are all as for PAS (Sect.9.2.1).

9.3 Thiol (Mercapto or Sulphhydryl) and Disulphide Groups

Occurrence. Proteins containing the amino acids cysteine and cystine. For example, keratin and certain neurosecretory substances have a high content of these.

Blocking. Blocking of -SH can be performed by

a. *Oxidation* of SH-groups to SS-groups can be achieved with iodine or ferric ions (Sects.5.1.1 and 8.3.1)
b. *Acylation* has been discussed in Sect.9.2.1 and proceeds according to the general equation:

$$R{-}SH + R{-}COX \rightarrow R{-}S{-}CO{-}R + HX$$

where X represents halogen or O–CO–R. Deblocking may be achieved by saponification (Sects.5.1.4 and 9.2.1)
c. *Mercaptide formation* with non-chromogenic mercury compounds, e.g. phenylmercuric chloride and methylmercuric iodide, takes place in the same way as shown below under mechanism
d. *N-ethylmaleimide* reacts specifically with SH groups:

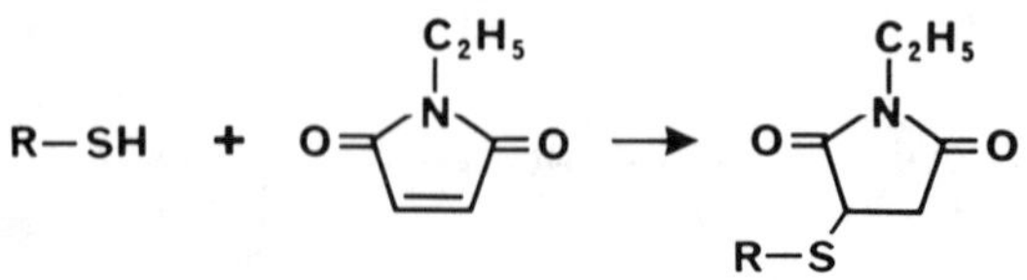

Demonstration. Demonstration of the thiol group can be achieved by metal salt reduction, formation of mercaptides, reaction with dihydroxydinaphthyl disulphide (DDD), and with maleimide derivatives. For the demonstration of disulphide prior reduction to thiol is necessary (Sect.5.1.2).

9.3.1 Metal Salt Reduction Methods

Among these methods the ferric ferricyanide reaction (Sect.8.3.1) is particularly useful.

9.3.2 Mercaptide Formation

The essence of these methods is the formation of a covalent bond between sulphur in a thiol group and mercury in an organic chromogenic mercury compound such as Mercury Orange or Merbromin (dibromohydroxymercurifluorescein).

Mechanism. The reaction with Mercury Orange (p-chloromercuri-phenyl-azo-β-naphthol) is shown here.

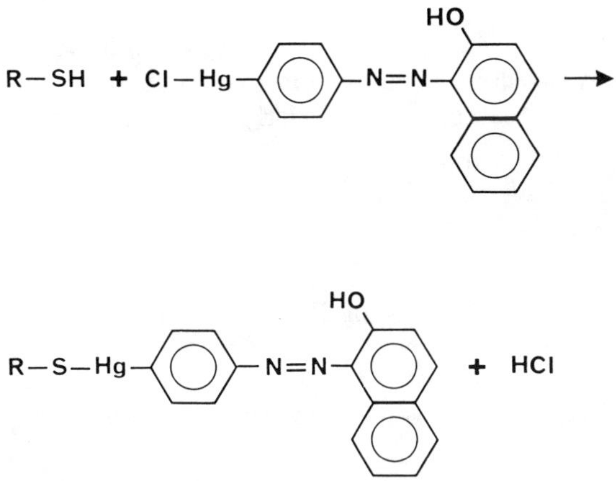

Selectivity. The organomercurial dyes have been claimed to be specific for thiol groups (Barka and Anderson, 1963, p.33; Lillie and Fullmer, 1976, p.238). Horobin and Flemming (1982), however, have clearly demonstrated that Mercury Orange also stains DNA and RNA. Cowden and Curtis (1984) found that both lipids and proteins are stained by Mercury Orange after N-ethylmaleimide blockage of thiol groups and have suggested that this was due to solvent partition. Horobin (1984) concludes that the selectivity of Mercury Orange for thiol groups is questionable.

9.3.3 Dihydroxydinaphthyl Disulphide

The dihydroxydinaphthyl disulphide method (DDD) was described by Barrnett and Seligman (1952).

Mechanism. A two step reaction is involved:

1. Thiol groups in the tissue react with DDD leading to simultaneous scission of the disulphide group and formation of a new disulphide link:

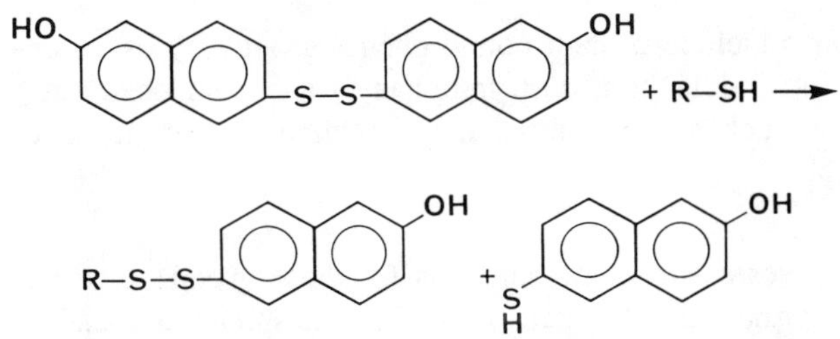

2. The naphthol group bound as a result of this reaction is then demonstrated by coupling with a diazonium salt or a tetrazonium salt (e.g. Fast Blue B, C.I.37235)

Selectivity and Results. The first step in the DDD reaction is specific, however, the second step is only moderately selective as diazonium salts couple with all aromatic amines and phenols. Serotonin can give a particularly strong reaction.

9.3.4 Maleimide Procedures

The basis of these methods is the specific binding of N-ethylmaleimide to thiol groups. The following reagents have been described:
N-(4-aminophenyl)maleimide (APM) (Sippel, 1973)
N-(7-dimethylamino-4-methyl-3-coumarinyl)maleimide (DACM) (Ogawa *et al.*, 1979)
N-(4-(7-dimethylamino-4-methylcoumarinyl)phenyl)maleimide (CPM) (Sippel, 1981)

Mechanism. All these reagents react with thiol groups in tissue sections in a similar manner to N-ethylmaleimide. In the case of APM the bound maleimide derivative is then diazotized and coupled at room temperature (cf. Sect.9.4.3). On binding to thiol the two coumarin compounds (CPM and DACM) become fluorescent (bright blue) and no diazotization reaction is necessary.

Selectivity. These methods appear specific for thiol groups. The reagents are, however, often difficult to obtain and contamination may make purification necessary.

Quantitation. With the coumarin compounds quantitation can be achieved using microspectrofluorimetry.

9.4 Aromatic Groups

Occurrence. Aromatic groups include protein bound phenol groups (tyrosyl), imidazole groups (histidyl), and indole groups (tryptophanyl) and also occur in a number of biogenic amines, such as adrenaline, noradrenaline, dopamine, serotonin, and histamine.

Demonstration. Table 9.2 gives a survey of reactions for these groups. The azo coupling method will be described in detail because of its widespread application.

Table 9.2. Methods for the demonstration of aromatic groups.

Method	Demonstrates	Blocking	Section
Azo coupling	Phenol, imidazole, indole, biogenic amines, bilirubin	acylation, dithionite reduction following coupling	9.4.1
Diazosulphanilic acid – pH 1 Azure A sequence (DAS-AzA)	as for azo coupling		
Millon	phenol	tetranitromethane, acylation	9.4.2
Diazotization coupling	phenol	tetranitromethane	9.4.3
Nitration/diazosulphanilic acid – pH 1 Azure A	imidazole		9.4.4
Postcoupled benzylidene	indole	performic acid, persulphate	9.4.5
Rosindole	indole	performic acid, persulphate	9.4.6

9.4.1 Azo Coupling

The azo coupling reaction for aromatic amines and phenols is used in histochemistry for the localization of tyrosyl, histidyl, and tryptophanyl radicals in proteins. In addition, it can be used to demonstrate bilirubin (Sect.18.4.11) and biogenic amines (Sect.18.4.10) as well as phenols and naphthols arising from different enzymatic reactions in which esters of phenols and naphthols are used as substrates (Sects.24.1.2 and 24.1.3).

Mechanism. Diazonium salts couple to aromatic and heteroaromatic amines and phenols forming an azo group, for example:

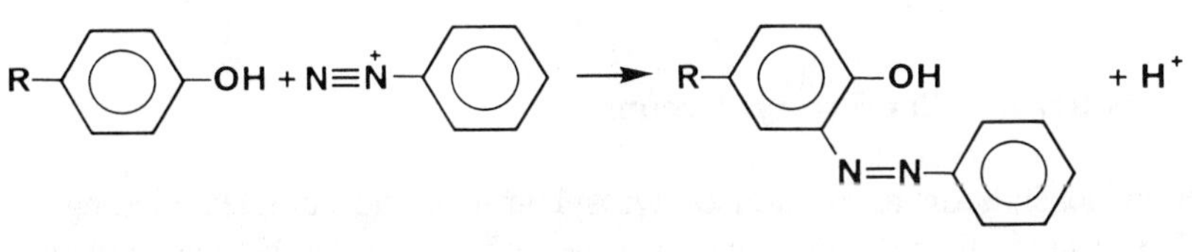

The site of coupling is determined by the coupling reagent, pH, and the substitution pattern of the aromatic ring.

Selectivity. The choice of diazonium salt is determined by the substance to be demonstrated (see Table 3.5). Nucleic acid bound purine and pyrimidine bases do not couple with diazonium salts.

Sensitivity. This is high with modern diazonium salts.

Blocking. This can be achieved with acylation of phenol and amino groups (Sect. 9.2.1). Saponification leads to deblocking (Sect.5.1.4).

Diazosulphanilic Acid-Azure A pH 1 Sequence (DAS-AzA) (Lillie *et al.*, 1973). Diazosulphanilic acid is prepared from sulphanilic acid:

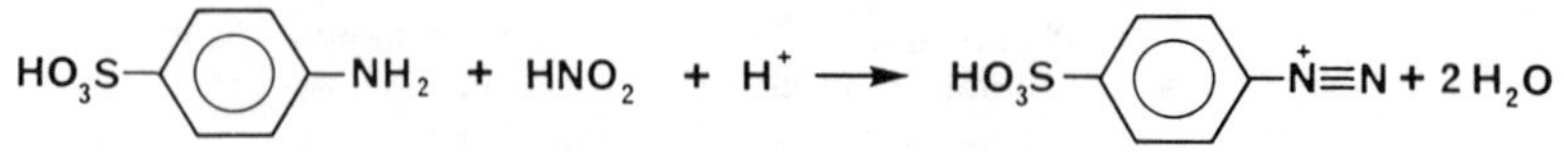

Staining consists of two stages:

1. Coupling to aromatic and heteroaromatic phenols and amines. The reaction with imidazole groups is shown below.

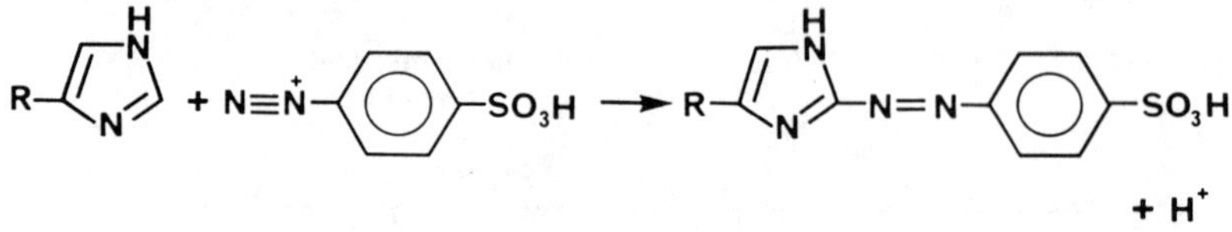

2. Staining with Azure A at pH 1 to demonstrate the sulphonic acid group

Selectivity and results. The coupling reaction is selective for phenol, indole, and imidazole groups. The second step should be controlled by staining a parallel without the coupling step. This will reveal how much Azure A is bound to sulphate and phosphate groups. The method can be made selective for histidine groups by a preceding nitration step (see Sect.9.4.4).

Sensitivity. This is high.

Blocking. This is as described for azo coupling.

9.4.2 Millon's Reaction for Tyrosyl Groups

This reaction depends on nitrosation of tyrosyl groups and complex binding of mercury. It is specific for protein bound tyrosine, but is considerably less sensitive than the coupling reaction described below.

9.4.3 Diazotization Coupling Reaction for Tyrosyl

This reaction was described by Glenner and Lillie (1959).

Mechanism. The phenol group in tyrosine is diazotized and the resultant tissue bound diazonium salt is demonstrated with an aromatic amine or phenol:

1. Formation of diazonium salt in the tissue:

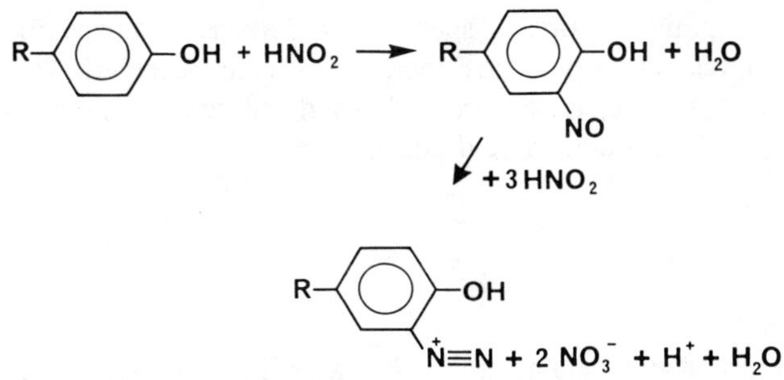

Incubation should be in darkness at around 4°C (refrigerator)

2. Coupling with aromatic amine or phenol. Normally, either S acid (4-amino-5-hydroxy-1-naphthalenesulphonic acid) or H acid (5-amino-4-hydroxy-2,7-naphthalenedisulphonic acid) are used at high pH:

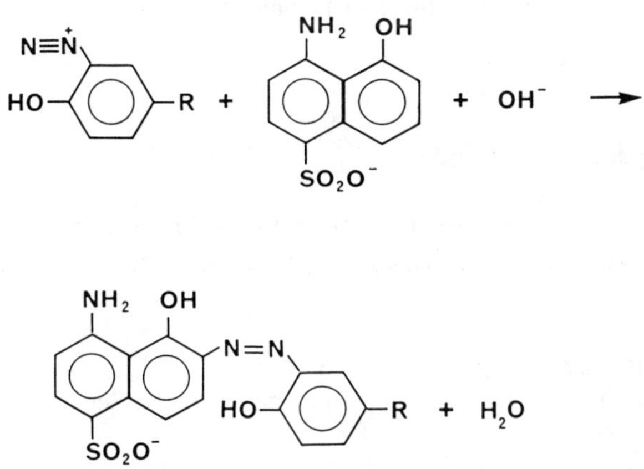

Selectivity. In the diazotization process primary aliphatic amines are converted to alcohols and a deamination takes place (see Sect.9.5.1):

$$R{-}NH_2 + HNO_2 \rightarrow R\text{-}OH + N_2 + H_2O$$

In contrast, primary aromatic amines are converted to diazonium ions:

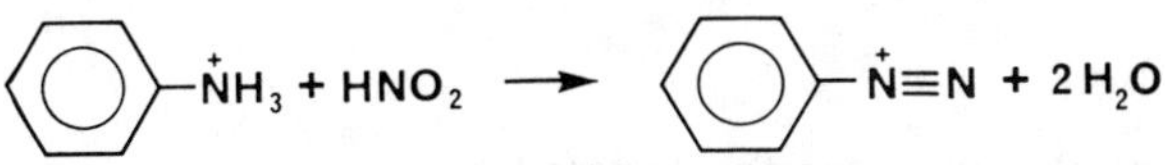

No primary aromatic amines are present in tissue sections. However, secondary amines, both aliphatic and aromatic, do occur. Reaction with nitrous acid leads to formation of nitrosamines:

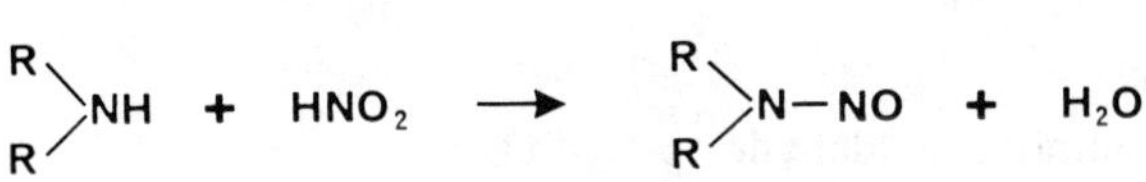

Thus the diazotization coupling reaction is specific for tyrosine. Non-specific staining of some tissue components such as eosinophils has been observed. This staining arises in the coupling step regardless of whether diazotization has been performed or not. It is observed with both S acid and H acid.

Sensitivity. This is high.

Blocking.
a. Acylation of phenol groups (Sect.9.2.1) reversible by saponification (Sect.9.2.1)
b. Treatment with tetranitromethane

9.4.4 Nitration-Diazosulphanilic Acid-Azure A at pH 1

This reaction for histidyl was described by Lillie and Donaldson (1972).

Mechanism.
1. Nitration in
   a. Nitric acid/acetic anhydride/acetic acid or
   b. Tetranitromethane/pyridine/water

   Here tyrosyl and tryptophanyl groups are nitrated so that coupling to diazonium salts is no longer possible at these sites. Histidyl groups are unaffected by this procedure
2. Coupling with diazosulphanilic acid
3. Demonstration of sulphonic acid groups with Azure A at pH 1

   Steps 2 and 3 thus correspond exactly to the reactions described in Sect.9.4.1.

Selectivity. This is high (see Sect.9.4.1). With incomplete nitration a weak tryptophanyl reaction cannot be excluded.

Sensitivity. This is high.

Blocking. Acylation (Sect.9.2.1).

9.4.5 Postcoupled Benzylidene Reaction for the Indole Group

This reaction for indole groups (tryptophanyl) was described by Glenner and Lillie (1957).

Mechanism.
1. Coupling with p-dimethylaminobenzaldehyde (p-DMAB):

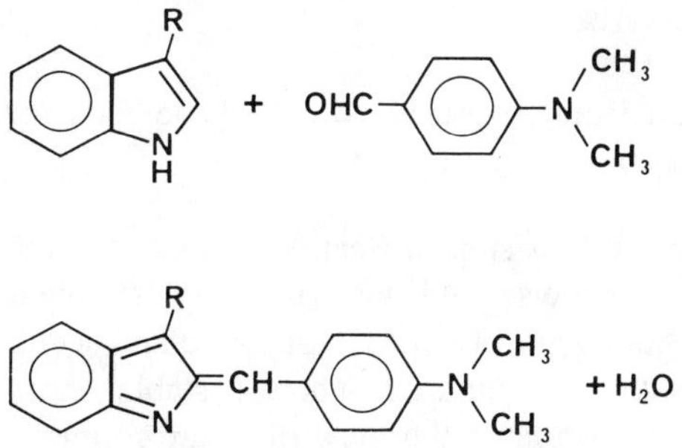

2. Coupling with freshly diazotized S acid gives rise to a blue product:

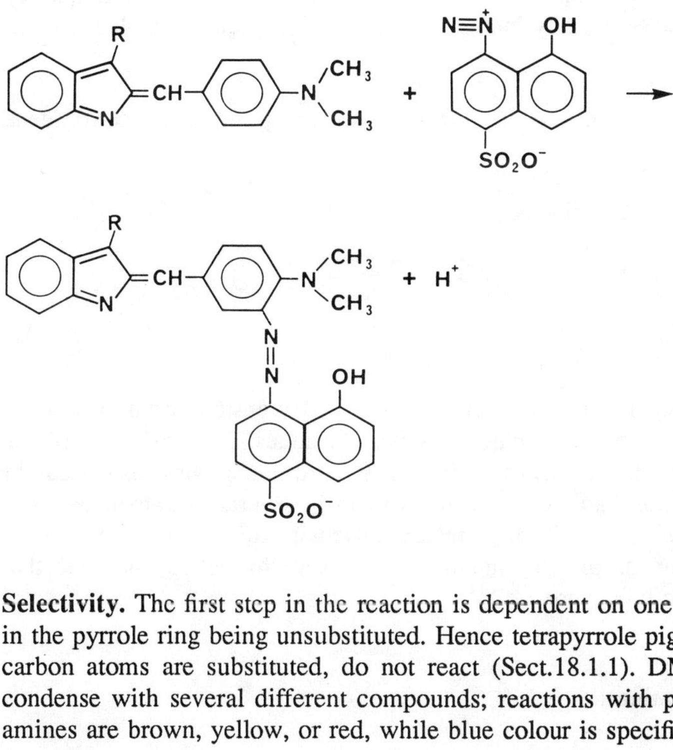

Selectivity. The first step in the reaction is dependent on one of the carbon atoms in the pyrrole ring being unsubstituted. Hence tetrapyrrole pigments, where all the carbon atoms are substituted, do not react (Sect.18.1.1). DMAB will, however, condense with several different compounds; reactions with phenols and aromatic amines are brown, yellow, or red, while blue colour is specific for tryptophanyl.

Blocking. Blocking of tryptophanyl can be achieved with oxidants of medium strength, e.g. performic acid (Sects.5.1.1 and 9.1.2) and persulphate, $S_2O_8^{2-}$. The tryptophanyl group is cleaved from the protein as o-aminobenzoic acid:

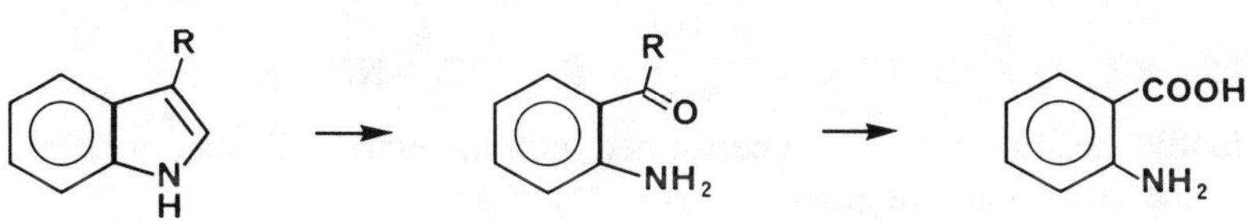

9.4.6 Rosindole Reaction for the Indole Group

Adams (1957) and Glenner (1957) almost simultaneously described methods for the demonstration of tryptophanyl, which differ only in detail.

Mechanism. The first step is identical to the first step in Sect.9.4.5, i.e. coupling with p-dimethylaminobenzaldehyde (DMAB); the second step consists of treatment with sodium nitrite in hydrochloric acid (Adams) or of hydrochloric acid and acetic acid (Glenner). This is not a nitrosation but an oxidation where a stable, more intensely blue dye complex is formed. It is postulated that new rings are formed.

Selectivity. The blue colour is specific for protein bound tryptophan as described under the selectivity of step one in Sect.9.4.5. The tissue morphology is frequently rather badly preserved. especially in Adams' modification, due to the harsh treatment in step two.

Sensitivity. This is high, but not as good as with the postcoupled benzylidene reaction.

Blocking. This is as for Sect.9.4.5.

9.5 Amino Groups

Occurrence. Histochemical reactions can be used to demonstrate primary amines, $R\text{-}NH_2$. These occur in proteins, chiefly as lysyl residues. Demonstration of the arginyl amino group will be discussed in Sect.9.6. Amino groups also occur in phospholipids (e.g. phosphatidyl-ethanolamine) and in several carbohydrates. The degree to which amino groups in these macromolecules are involved is not clear.

The demonstration of primary amino groups is performed by an oxidative deamination-aldehyde procedure.

9.5.1 Oxidative Deamination-Schiff

Mechanism.

1. Oxidative deamination. The amino group is removed, possibly forming ammonia, and the adjacent carbon atom is oxidized to an aldehyde group:

$$R{-}CH_2NH_2 \xrightarrow{\text{ox}} R{-}CH{=}NH \xrightarrow{H_2O} R{-}CHO + NH_3$$

The oxidative deamination can be performed with ninhydrin, alloxan, or chloramine-T, the latter being optimal.

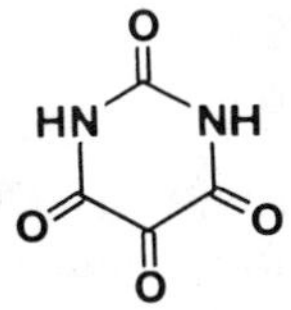

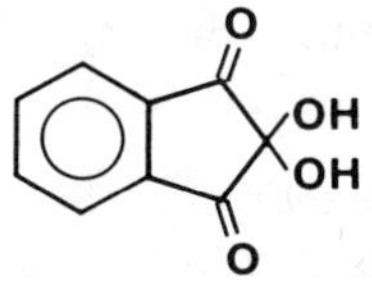

Alloxan Ninhydrin

Chloramine-T (Burstone, 1959) is the sodium salt of N-chloro-p-toluenesulphonamide. It is used at pH 7.5 at 37°C for 6 hours.

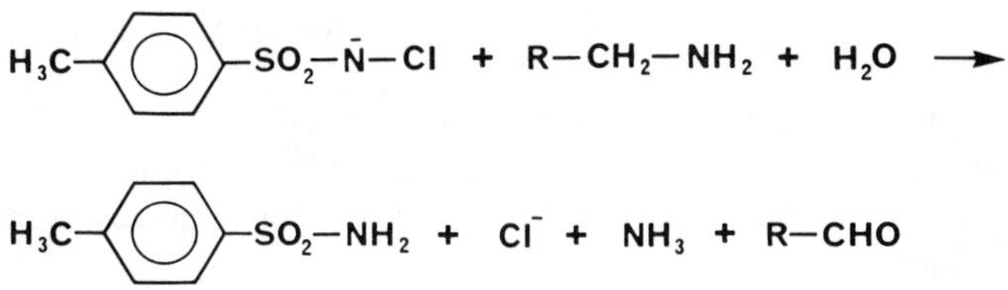

2. Demonstration of aldehyde with aldehyde reagent: Schiff's reagent (Sect.9.7) is normally used here

Ninhydrin Reaction. The Ninhydrin reaction is used in biochemistry and clinical chemistry as a general reaction for demonstrating proteins. At 100°C amino groups react with ninhydrin to form a bluish violet colour. The method has a low sensitivity and the dye formed is diffusible. In contrast, in the *histochemical reaction* (Yasuma and Itchikawa, 1953), aldehyde groups are introduced into proteins, and these are the basis for the subsequent demonstration.

Selectivity. Free amino acids are not present in paraffin sections. The only reactive group corresponding to that shown in step 1 (above) is the ϵ-amino group in lysine. The reaction with terminal amino groups, which has been emphasized in the literature, is in doubt. Secondary aliphatic amines should, in principle, react like primary amines, while tertiary and aromatic amines do not react. See Sects.9.2.1 and 9.7.1. regarding the selectivity of the second step 2 (demonstration of aldehyde).

Sensitivity. The sensitivity of the chloramine-T/Schiff procedure is high. The ninhydrin and alloxan methods are considerably less sensitive.

Blocking. Blocking of amino groups can take place by:

1. *Deamination* which can be achieved by:
   a. *Nitrosation* that is performed with nitrous acid which in practice means a nitrite dissolved in acid. The processes with primary and secondary amines have been shown in Sect.9.4.3. Tertiary amines do not react. Lillie (1958) has shown that solutions of $NaNO_2$ in acetic acid are more effective than solutions in hydrochloric acid. In 1970 Lillie *et al.* were able to show that the necessary time for deamination with van Slyke's reagent increased with increasing fixation times in formaldehyde containing fixatives:

fixation time	(hours)	2	4	24	72
deamination time	(hours)	4	6	12	36

b. *Oxidative deamination.* See "mechanism" (Sect.9.5.1). Clearly this method cannot be used with the chloramine-T/Schiff and similar methods

c. *Hypochlorite*, ClO^-, reacts very effectively with amino groups (Sect.5.1.5). The process probably proceeds thus:

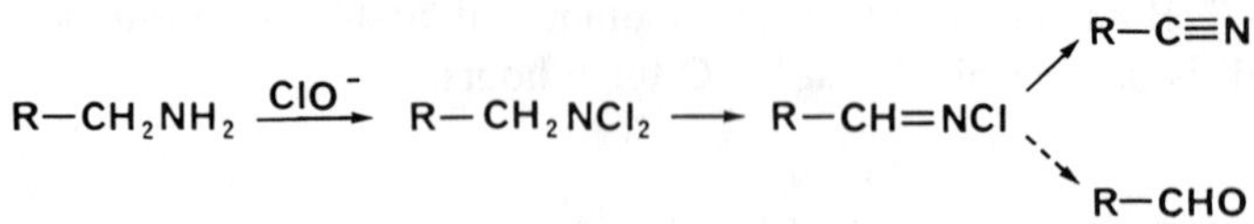

Very few aldehyde groups are formed with this method

2. *Acylation* has been discussed in Sect.9.2.1. Primary and secondary amines react forming amides. Tertiary amines do not react. Deblocking after acylation is achieved by saponification (Sect.9.2.1)
3. *Azomethine condensation* or the formation of a Schiff's base involves the condensation of primary amines with aldehydes:

$$R{-}NH_2 + R'{-}CHO \rightarrow RN{=}CH{-}R' + H_2O$$

9.6 Guanidyl Groups

Occurrence. The guanidyl group is found in arginine (arginyl) which is present in the majority of proteins:

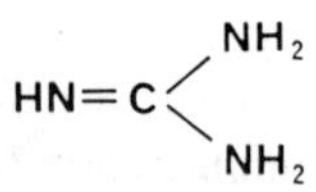

Guanidine

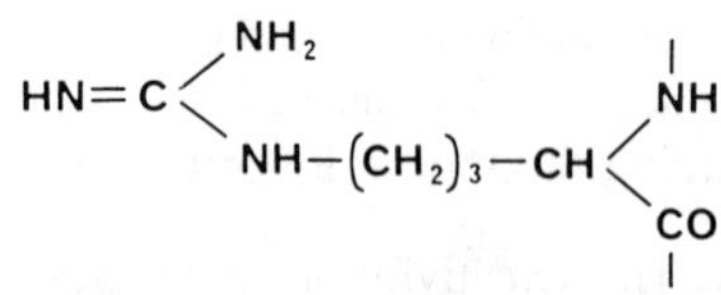

Protein bound arginine

Particularly large amounts are found in certain histones, protamines, and lysozyme.

Demonstration. Methods for the demonstration of the guanidyl group are based on the classical Sakaguchi reaction in which a reddish colour develops on protein-bound arginine after treatment with an alkaline solution of 1-naphthol and sodium hypochlorite.

1-Naphthol is oxidized by NaClO to 1,2- and 1,4-naphthoquinones.

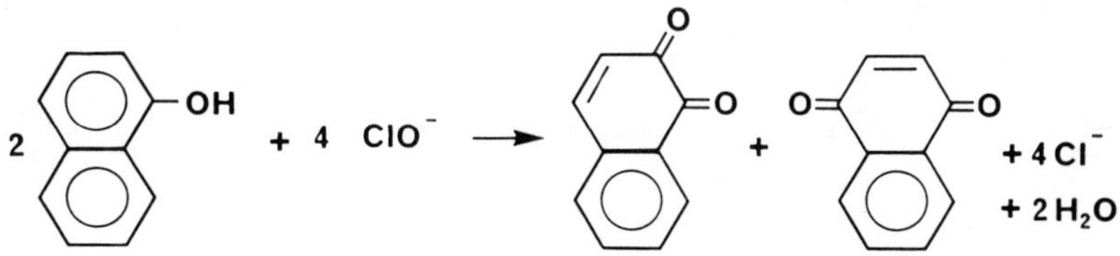

Subsequently the 1,2-naphthoquinone cyclizes with the guanidyl group.

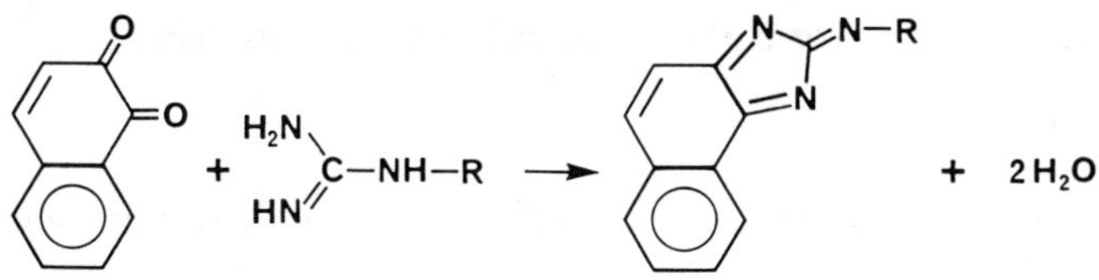

The resultant condensation product is unstable and rapidly breaks down to a soluble product:

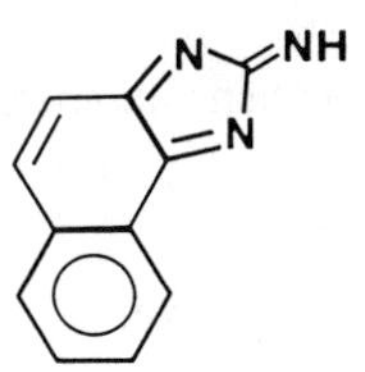

The method thus poses several practical problems,notably, the solubility of the final product in organic solvents, the weak colour, and a pronounced tendency to fade.

These problems have, however, mostly been solved by the advent of the naphthoquinone-sulphonate, and 9,10-phenanthrenequinone methods.

9.6.1 1,2-Naphthoquinone-4-sulphonate Method

The 1,2-napthoquinone-4-sulphonate method (NQS) was described by Lillie *et al.* (1971).

Mechanism. The section is treated with a freshly prepared solution of NQS in an alkaline $BaCl_2$ solution (pH about 13). NQS cyclizes with the guanidyl residue in an analogous fashion to the classical Sakaguchi reaction forming the following tissue condensation product:

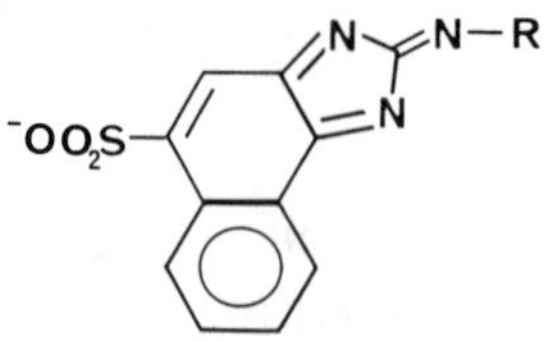

Even if cleavage of the cyclic product takes place between N and R, the added barium ions ensure that the product does not diffuse as the barium compound is insoluble.

Selectivity. The method is specific for the unsubstituted guanidyl residue.

Sensitivity. This is high.

Blocking. Specific blocking can be achieved with alkaline solutions of the following compounds:

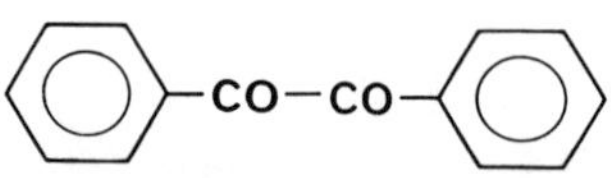

Benzil

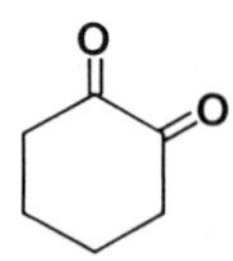

1,2-cyclohexanedione

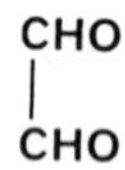

glyoxal

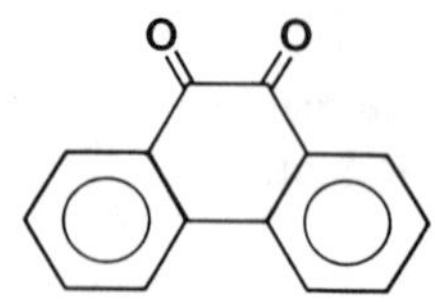

9,10-phenanthrenequinone

These act in the same way as described for NQS above.

Hypochlorite. Hypochlorite pretreatment renders the NQS reaction negative.

Nitrosation. This is inappropriate for blocking arginyl. There is no demonstrable effect after 6 hours and only around 50% blocking after 24 hours.

9.6.2 9,10-Phenanthrenequinone Reaction

The 9,10-phenanthrenequinone reaction (PQ) for guanidyl groups was published by Magun and Kelly (1969).

Mechanism. The mechanism is as described in Sect.9.6.1:

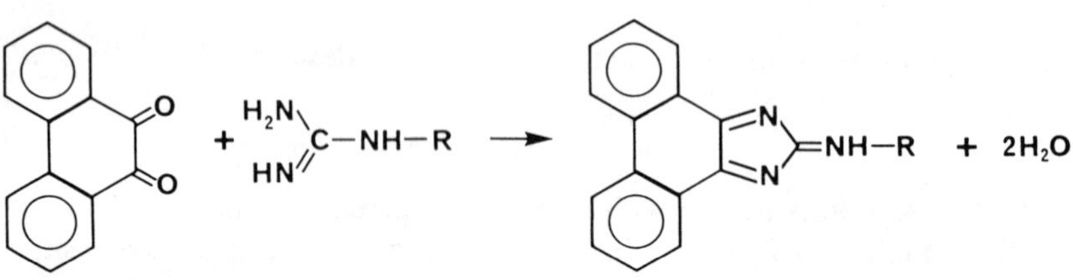

The reaction product is fluorescent.

Selectivity. The reaction is specific for guanidyl groups. The cyclic product gradually breaks down and the preparations can only be considered stable for 1 hour.

9.7 Aldehyde Groups

The carbonyl group comprises aldehydes and ketones.

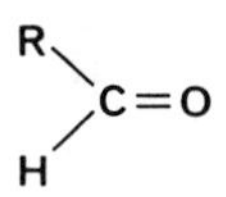

Aldehyde

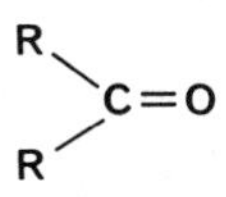

Ketone

Occurrence. Aldehyde groups are only present in very small amounts in tissue. In young vertebrates lysinal aldehyde is found in elastic tissue arising from the oxidation of lysine during the synthesis of elastin. The great importance of the aldehyde group is that, unlike the ketone group, it can be induced in several different macromolecules and lipids by a number of pretreatments (see Table 9.3. below).

Table 9.3. Induction of aldehyde groups.

Pretreatment	Group	Compound	Section
Peracid oxidation	alkene	lipids	9.1.2
Ultraviolet light	alkene	lipids	9.1.4
Oxidation	1,2-glycol- and 1-amino-2-hydroxyl	carbohydrates, glycolipids, and collagen	9.2.1
Oxidative deamination	primary amine	proteins	9.5.1
Hydrolysis	deoxyribosyl	DNA	9.9

9.7.1 Demonstration

This may be performed with Schiff's reagent or certain metal salts.

Schiff Reagent. Schiff's reagent (see Sect.3.3.9) has been of such importance that Lison (1960), p.159, has aptly called it the chameleon of histochemistry. According to Kasten (1960), it reacts with aldehyde groups through its sulphinic acid groups (carbonyl reaction) forming covalent bonds:

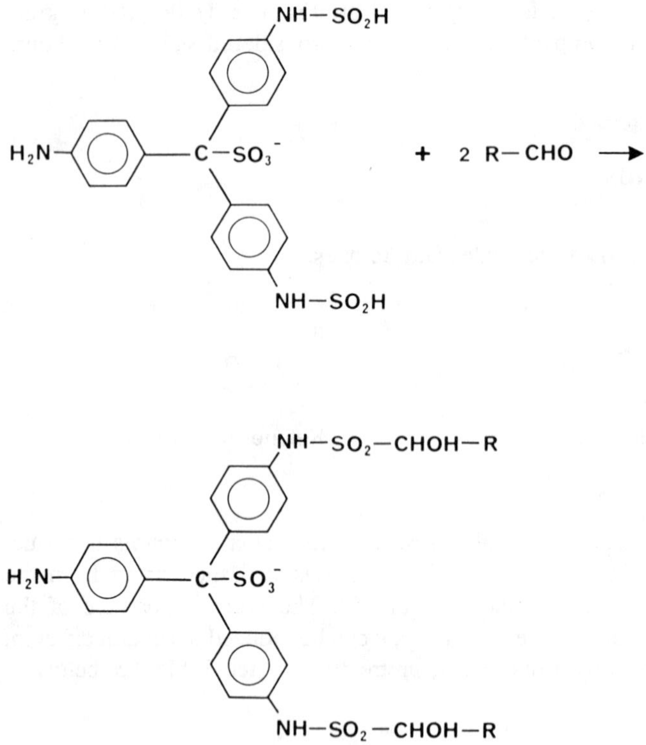

Gill and Jotz (1976) propose an alternative theory:

1. Excess sulphur dioxide in Schiff's reagent reacts with the tissue aldehyde forming an alkyl sulphonic acid

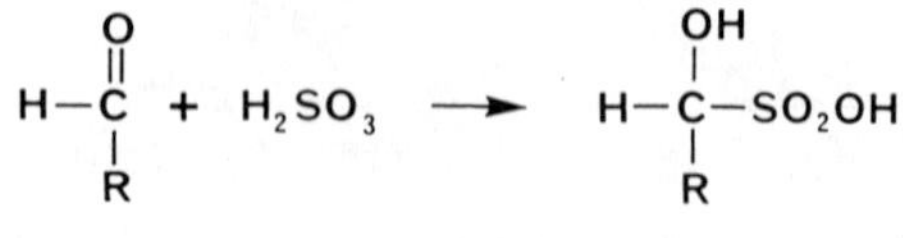

2. The sulphonated carbon atom is bound to an amino group in Pararosanilin leucosulphonic acid:

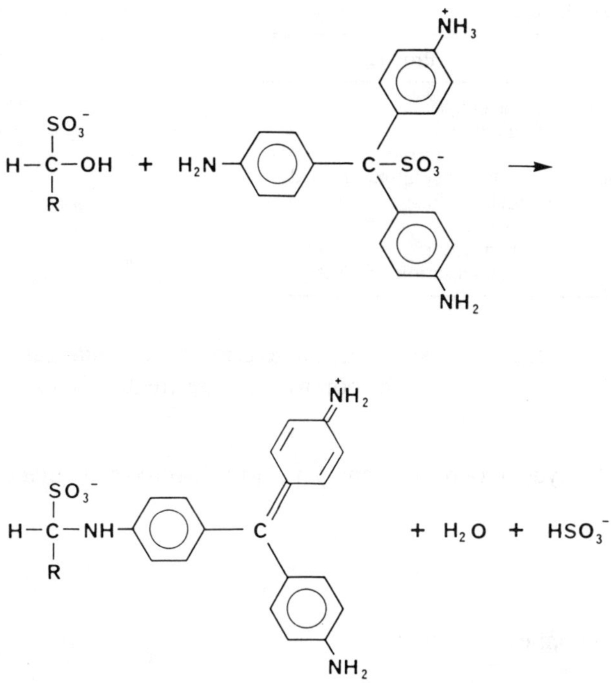

The sulphonic acid group must be removed at this point and this can be achieved by increasing pH to around neutrality. Unbound Schiff's reagent/Basic Fuchsin is normally removed by washing in running tap water. This restores the paraquinone structure of the molecule. In critical work this can be avoided by interposing a "sulphite rinse" comprising a solution of potassium metabisulphite or an approximation to Schiff's reagent without the Pararosanilin. This allows removal of all the non-reacted Schiff's reagent from the section without any possibility of this forming Pararosanilin. Alternatively the section can be washed in 0.01 mol/l hydrochloric acid.

Metal salts. Ferric ions are reduced by aldehyde groups (see ferric ferricyanide reaction Sect.8.3.1), while silver ions are reduced to metallic silver (Sect.8.3.2).

9.7.2 Blocking

Blocking of aldehyde groups can be performed by the methods listed in Sect.5.1.6 and Table 9.4.

a. *Oxidants*. Even with weak oxidants, aldehyde groups are easily oxidized to carboxyl groups. Hypochlorite or bromine can be used for this purpose. In contrast, ketones can only be oxidized using strong oxidants

Table 9.4. Blocking reactions for aldehyde groups.

Process	Product
a. Oxidation	carboxyl
b. Reduction	alcohol
c. Addition of cyanide	hydroxynitrile
d. Addition of hydrogen sulphite	α-hydroxysulphonic acid
e. Addition of arylamines	Schiff's base
f. Addition of hydroxylamine	oxime
g. Addition of hydrazine	hydrazone
h. Addition of dimedone	condensation product

b. *Reductants.* Aldehydes and ketones are easily reduced to primary and secondary alcohols respectively. Sodium and potassium borohydride are frequently used for this purpose

Addition reactions. The aldehyde group is strongly polarized, (dipole moment 2.5–2.7 Debye, cf. Sect.3.2).

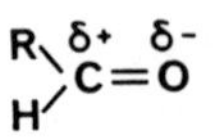

This allows addition of nucleophiles such as:

c. *Cyanide*

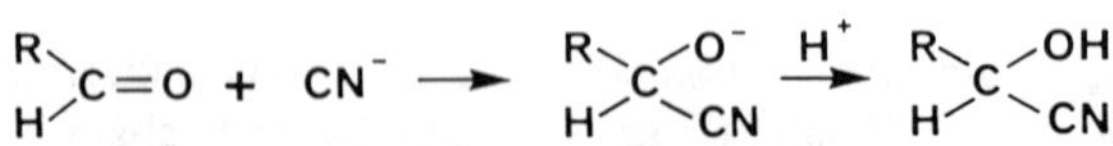

The net result is addition of hydrogen cyanide and the formation of a hydroxynitrile. This blockade can be broken using periodic acid.

d. *Hydrogen sulphite* (bisulphite) reacts with aldehydes to form α-hydroxysulphonic acids:

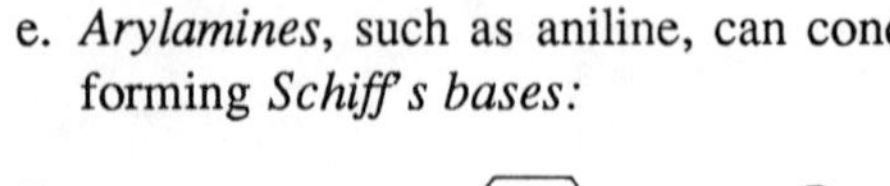

The hydrogen sulphite compound is easily broken (deblocked) with weak acid or alkali, this also occurs after extended (2–3 hours) treatment with Schiff's reagent.

e. *Arylamines*, such as aniline, can condense with tissue bound aldehyde groups forming *Schiff's bases:*

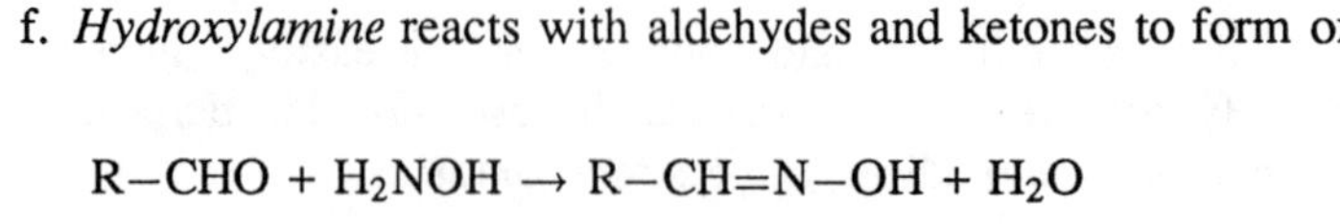

f. *Hydroxylamine* reacts with aldehydes and ketones to form oximes:

$$R{-}CHO + H_2NOH \rightarrow R{-}CH{=}N{-}OH + H_2O$$

g. *Hydrazines* are added to carbonyl groups forming hydrazones, e.g. phenylhydrazine:

h. *Dimedone* or *5,5-dimethylcyclohexane-1,3-dione* condenses with tissue aldehydes:

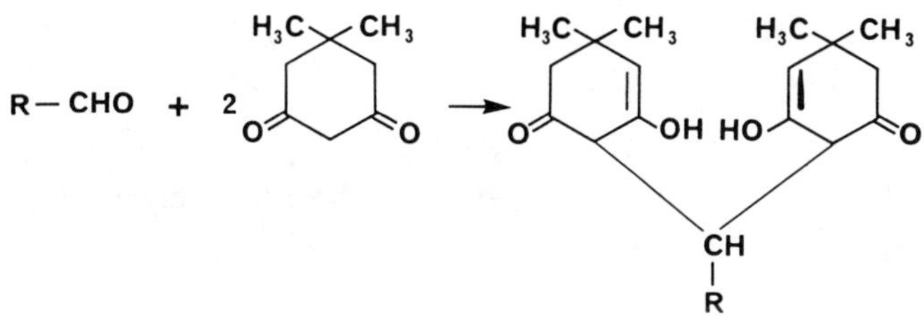

Steric conditions influence condensation and it is therefore sometimes not possible to achieve complete blocking.

These blocking reactions serve two purposes:

a. Selective blocking of aldehydes (e.g. blocking of free aldehydes prior to the plasmal reaction (Sect.19.6.4))
b. Confirmation that the tissue groups responsible for a positive Schiff-reaction are aldehydes. Aniline is the best selective blocking reagent

9.8 Acid Groups

This section includes a diverse series of reactive groups for which the single common characteristic is a tendency for protons to dissociate leaving a negative charge. Since their demonstration depends on this charge it is logical to consider them together. The groups comprise sulphate, sulphonate, phosphate, and carboxylate.

9.8.1 Occurrence

The sulphate group, $R{-}OSO_2O^-$, occurs in a number of proteoglycans (e.g. chondroitin sulphates and heparin) and in sulphomucins. The sulphonate group, $R{-}SO_2O^-$, is not found naturally in tissue but can be introduced into proteins by oxidation of thiol groups and possibly also of disulphide groups.

The phosphate group is found as phosphate diester in nucleic acids and phospholipids and as phosphate monoester in phosphoproteins, respectively:

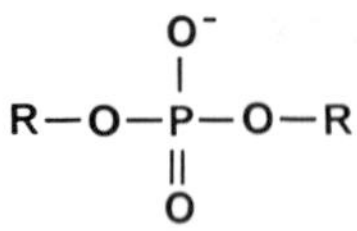 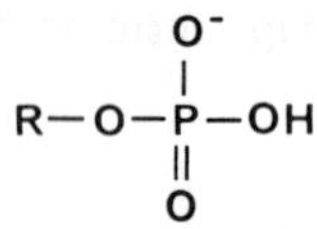

The carboxylate group, R–COO$^-$, occurs as a side group in proteins (asparagyl and glutamyl) and in small amounts corresponding to the C-terminal carboxyl. It also occurs in sialic acid (gangliosides and sialomucins), uronic acid (proteoglycans), and in free fatty acids (e.g. lipofuscin, cf. Sects.18.1.2 and 18.4.1).

9.8.2 Demonstration

The demonstration of acid groups is performed using cationic dyes (Sect.6.1) and metal complex dyes (Chap.7). These dyes are bound to the negatively charged groups, and therefore depend on ionization.

9.8.3 Blocking

Blocking of carboxylate, sulphonate, and sulphate groups can be achieved by alkylation (Sect.5.1.3). This is nearly always performed with methanol (i.e. methylation) which for the carboxyl group leads to ester formation:

$$\mathrm{R{-}COOH + R{-}OH \rightleftarrows R{-}CO{-}O{-}R + H_2O}$$

The process is reversible and is, in principle, identical to that described under acylation (Sect.9.2.1). The reaction is catalyzed by protons. Displacement to the right is favoured by a water-free environment and large excess of reagent. The equilibrium is shifted to the left by treatment with alkali in a saponification reaction (cf. Sect.9.2.1).

Similar reactions with methanol might be expected with phosphate, sulphonate, and sulphate groups. This impression is reinforced in several of the newer textbooks (e.g. Lillie and Fullmer, 1976, p.318 and Bancroft and Stevens, 1990, p.204). However, phosphate groups in DNA are not blocked by methylation (Prentø, 1980). The situation with sulphate groups is more complex: it is believed that *"mild methylation"* (i.e. 0.1 mol/l HCl in absolute methanol for up to 1 hour at 60°C or not more than 4 hours at 37°C) results in a real, saponification reversible, methylation. Treatment for longer periods of time (*"drastic methylation"*) results in cleavage of the sulphate group:

$$\mathrm{R'{-}O{-}SO_2OH + R''{-}OH \rightarrow R'{-}OH + R''{-}O{-}SO_2{-}OH}$$

The process is thus an alcoholysis or an "interesterification" and tissue-bound sulphate groups cannot, therefore, be regenerated by saponification. The results of "drastic methylation" are not in dispute while the effects of "mild methylation" are somewhat more doubtful.

9.9 Purine-N-C1-Deoxyribose Glycoside Bond

Occurrence. This bond is solely found in DNA.

Demonstration. This is achieved with *Feulgen's nucleal reaction* (Feulgen and Rossenbeck, 1924).

Mechanism. The reaction comprises two steps:
1. Acid hydrolysis
2. Aldehyde demonstration

Acid hydrolysis of the purine-N-C1-deoxyribose glycoside bond. The DNA with purine bases removed is sometimes designated "apurinic acid (*apa*)" and contains potential aldehyde groups.

Many different reagents have been proposed for the acid hydrolysis. Nowadays, in practice, only two are used:

a. 1 mol/l hydrochloric acid at 60°C ("hot hydrolysis")
b. 5 mol/l hydrochloric acid at room temperature ("cold hydrolysis")

The kinetics of hydrolysis, including the formation of "apa", can be visualized when Feulgen staining intensity is plotted against time of hydrolysis. Irrespective of the technique used, the graphs of acid hydrolysis may be subdivided into four main phases (Kjellstrand and Lamm, 1976; Kjellstrand, 1977):

- *Phase A)* Hydrolysis of the purine-N-C1-deoxyribose glycoside bonds associated with a continuous increase in the number of aldehyde groups in "apa". Staining intensity, therefore, also increases in relation to the length of phase A
- *Phase B)* The peak phase during which hydrolysis leading to maximal staining intensity is achieved
- *Phase C)* The plateau phase – staining intensity remains constant over a certain period of time
- *Phase D)* Decrease of staining intensity

Phases A–D are the result of two histochemical reactions:

a. Formation of aldehyde groups
b. Fragmentation of DNA and extraction of DNA-fragments

In phase A formation of aldehyde groups predominates, and the loss of DNA-fragments is negligible. The result is an increase in the number of stainable aldehyde groups. The peak of the hydrolysis curve (phase B) with maximum staining intensity will be reached *either* when all available aldehyde groups have been generated *or* when the rate of DNA-fragment extraction (= decrease of staining intensity) matches the rate of aldehyde group formation. In phase C there is an equilibrium between the two reactions, and the net number of aldehyde groups remains constant; in phase D DNA-extraction predominates over aldehyde formation.

The four phases can only be clearly discriminated with cold hydrolysis. With hot hydrolysis, the peak and plateau phases are very short.

The hydrolysis time yielding the maximum staining intensity with the aldehyde reagent is referred to as the *optimum hydrolysis time*. This is a compromise between the time necessary for extensive or complete hydrolysis of purine-N-glycoside bonds and phosphate ester bonds and the subsequent extraction of DNA-fragments. At 60°C, hydrolysis of the phosphate ester bonds results in rapid extraction of the DNA-fragments, and the margins for the optimum hydrolysis time are very short. *It is therefore difficult to achieve optimum hydrolysis at 60°C.* Table 9.5 gives optimum times of hydrolysis using HCl at 60°C for tissues fixed in a number of different ways.

Table 9.5. Optimum hydrolysis times for the Feulgen reaction using 1 mol/l HCl at 60°C.

Fixative	Optimum hydrolysis time (min)
Aqueous neutral formaldehyde solutions	5
Absolute ethanol	5
Acetic acid ethanol fixatives	7

Notes:

a. The required time for hydrolysis of the glycoside bonds is highly dependent on the choice of fixative (Böhm, 1968; Deitch *et al.*, 1968)

b. The use of 5 mol/l HCl at room temperature yields maximum staining intensity after 15–20 min hydrolysis virtually independent of fixative. The phosphate ester bonds in DNA appear less susceptible to hydrolysis and extraction is therefore reduced. If hydrolysis is extended for 12 h, a minimal decrease in staining occurs and this is really only clearly discernible after 18 h. *Hydrolysis at room temperature is therefore strongly recommended*

The required time for hydrolysis of the glycoside bonds is also influenced by the compactness of the nuclear chromatin. Highly condensed chromatin is less sensitive to acid hydrolysis than decondensed chromatin (Böhm *et al.*, 1968; Kiefer *et al.*, 1973; Duijndam and van Duijn, 1975). Consequently, different cell types show different hydrolysis curves. Photometric comparison of Feulgen staining intensities between different cell types is therefore invalid unless staining has been performed after the optimum hydrolysis time for each cell type to be compared.

Aldehyde demonstration. After hydrolysis, the purine bases dissociate, and free C1-hydroxyl groups appear on the deoxyribose molecules. These deoxyribose molecules occur in both furanose and aldehyde forms:

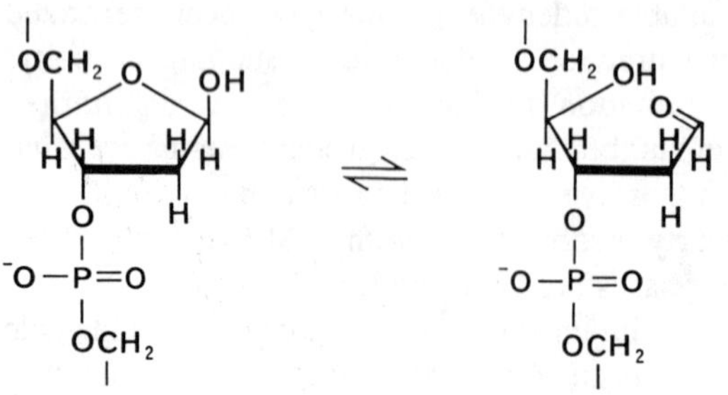

The equilibrium is strongly displaced to the left, but as the reaction with the aldehyde reagent is irreversible (Sect.9.7.1), all C1-atoms will eventually react.

Selectivity. The glycoside and phosphate ester bonds in DNA and RNA are hydrolyzed at different rates as shown below:

	Purine-N-C1-pentose glycoside	Pyrimidine-N-C1-pentose glycoside	Phosphate ester
DNA	++	0	+
RNA	+	0	++

++ rapid hydrolysis
+ slow hydrolysis
0 no hydrolysis

This means that, under appropriate hydrolysis conditions, in DNA only purine-N-C1-deoxyribose glycoside bonds are broken, while RNA is decomposed to nucleotides. Under routine conditions of hydrolysis RNA-fragments are extracted from cytoplasm and nucleoli.

Sensitivity. Although the method is very sensitive it does not demonstrate mitochondrial DNA.

Localization. As phosphate ester bonds in DNA remain intact and Schiff's reagent binds covalently to the induced aldehyde groups in "apurinic acid", there are no problems with diffusion of reaction products. Feulgen's nucleal reaction therefore gives quite precise localization of DNA provided that the nucleic acids are precipitated during fixation and remain insoluble during the subsequent tissue processing.

Quantitation. The Feulgen nucleal reaction is a popular method for quantitative determination of DNA content as the reaction may be assessed microspectrophotometrically (see Sect.28.8.4). Control sections (no acid hydrolysis) are essential.

Part 3
Tissue Processing

10 Tissue Processing: I. Overview

H. Lyon

Certain aspects of living cells may be studied directly using the light microscope. However, for most work where chemical and structural information is sought, it is necessary to "stabilize" the cells or the tissue before the examination can take place. This can be achieved either by freezing the tissue or by using chemical treatments that act upon or react with tissue components, notably protein. A detailed consideration of the former approach is given in *tissue processing II, freezing* (Chap.11) while further discussion of chemical treatments can be found in *tissue processing III, fixation, general aspects* (Chap.12) and *IV, fixation, applied* (Chap.13).

After the stabilization/immobilization process the tissue must be dehydrated and the water substituted with a suitable solid (embedding) substance. These processes are discussed in *tissue processing V, embedding* (Chap.14). The special considerations relevant to hard tissue (i.e. bones and teeth) require a dedicated section (*tissue processing VI, hard tissues*, Chap.15). Finally, the treatment of sections after staining is discussed in *tissue processing VII, posttreatment* (Chap.16).

In every histochemical analysis of tissue the following points must be considered:

- What is to be demonstrated?
- Which histochemical method is best suited?
- What demands do these factors place on the pretreatment of the tissue?
- Is a freezing procedure or a chemical fixation preferable?
- In the latter case, which is the optimum fixative and how should the fixation take place?

Sectioning of Unfixed Non-Frozen Tissue. It is currently possible to cut non-fixed, non-frozen tissue and several commercially available instruments have been developed for this purpose. A popular instrument is Smith & Farquhar's *Tissue Sectioner*, "Tissue Chopper", (Sorvall, Inc.) which, after the tissue block has been embedded in agar, allows sections with a thickness of between 25 and 250 μm to be cut. An alternative instrument is the *Vibratome* (Oxford Instruments) on which it is possible to cut relatively thin sections (4–6 μm). Sections cut with these instruments are often well suited for performing more specialized histochemical reactions such as enzyme reactions and immunohistochemical reactions. After the reaction, the sections can be embedded in epoxy resin and sectioned for ultrastructural studies.

H. Lyon (Ed.)
Theory and Strategy in Histochemistry

In Figs. 10.1–5 below, the options for preparing biological material for microscopic examination are presented in schematic form.

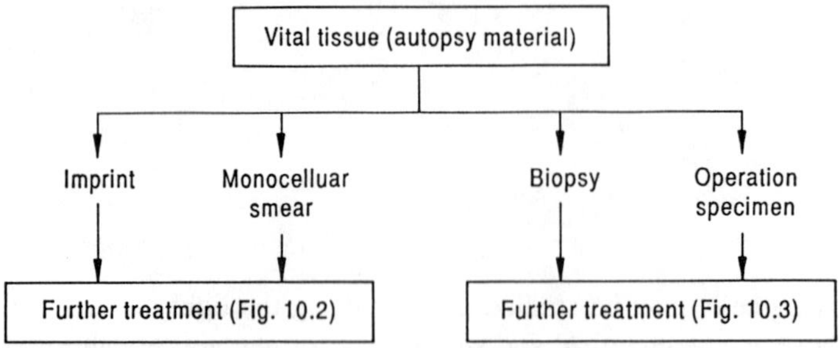

Fig. 10.1. Initial pretreatment of material for microscopic investigation.

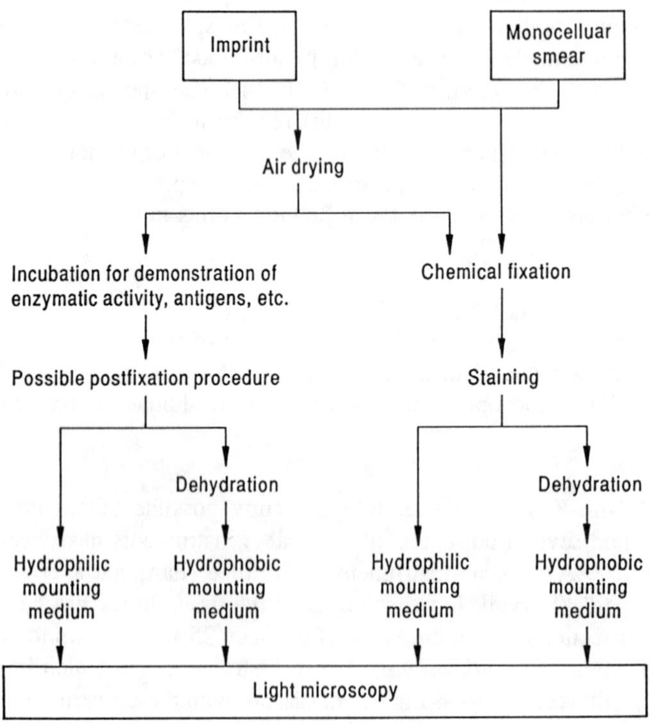

Fig. 10.2. Further treatment of imprints and monocellular smears.

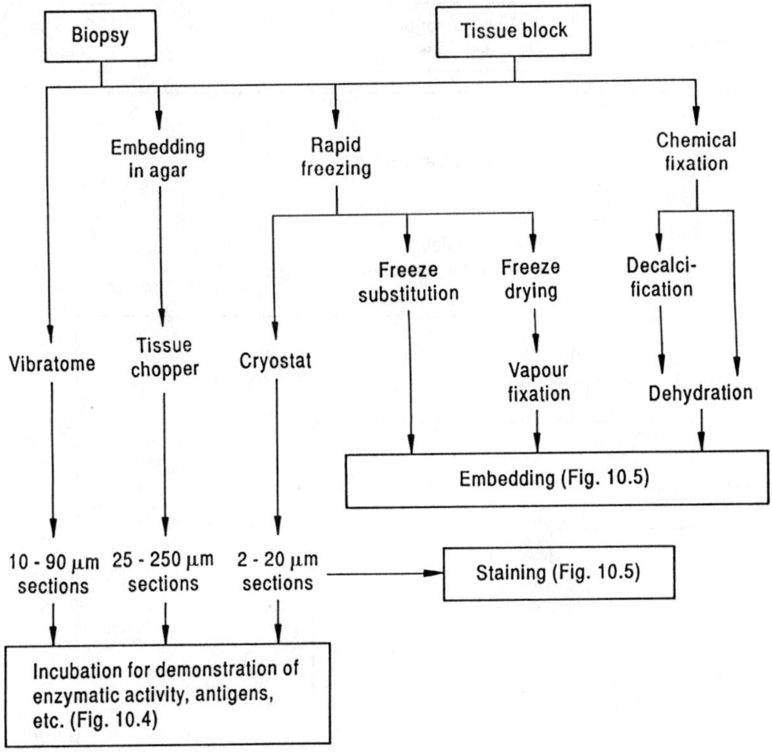

Fig. 10.3. Further treatment of biopsies and tissue blocks.

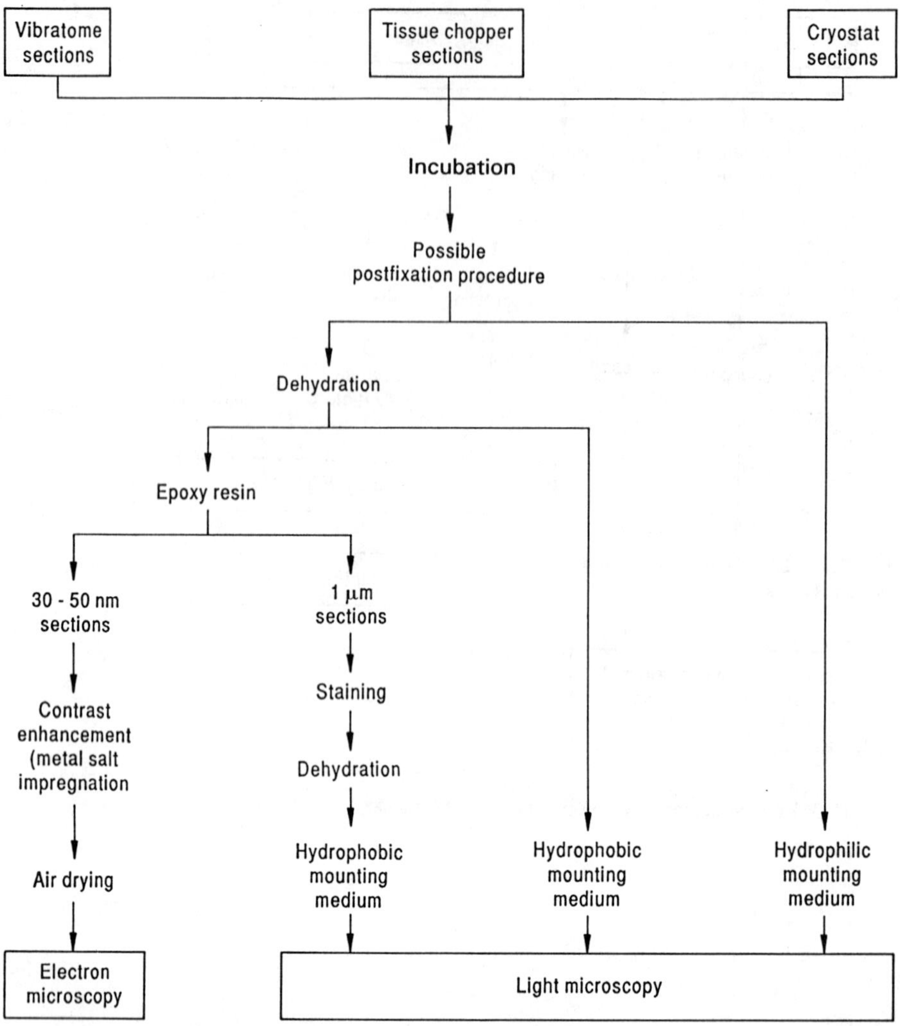

Fig. 10.4. Further treatment of Vibratome, Tissue Chopper, and cryostat sections.

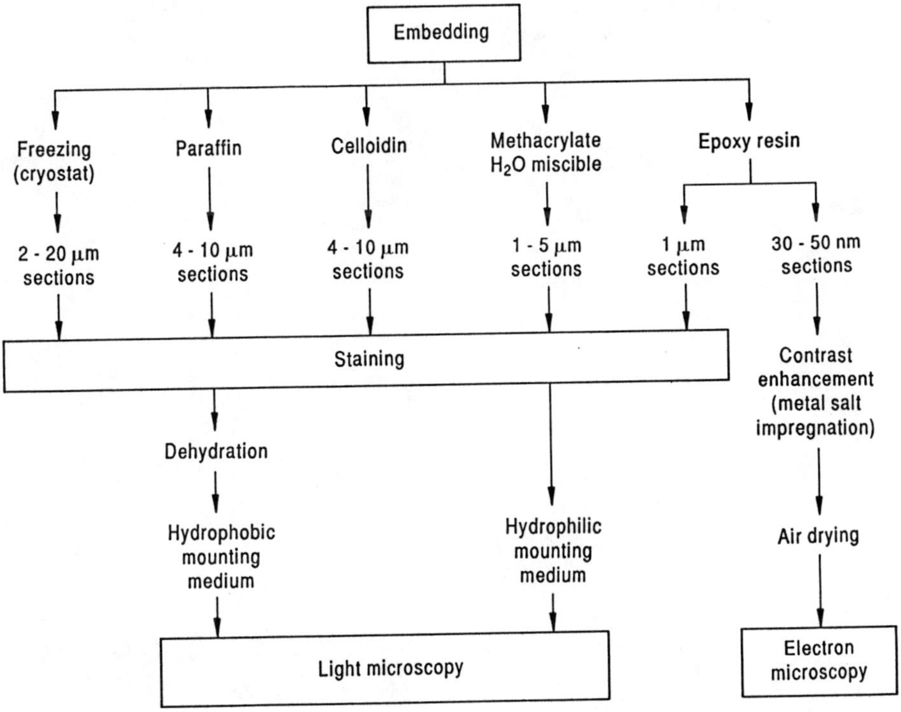

Fig. 10.5. Further treatment of cryostat sections and of material embedded in paraffin, celloidin, methacrylate, and epoxy resin.

11 Tissue Processing: II. Freezing

M. Møller, P.E. Høyer

11.1 Purpose of Freezing Tissue

The purpose of freezing tissue immediately after sampling is to *immobilize* easily diffusible compounds.

In practice freezing is used:

1. To immobilize and preserve activity in enzyme histochemical investigations. (frequently)
2. To preserve antibodies and antigenic groups for immunohistochemical investigations. (sometimes)
3. To investigate highly diffusible compounds such as inorganic cations and anions. (usually)
4. In most lipid investigations to avoid lipid solvents
5. For rapid diagnostic work during surgical operations

Freezing of tissue is sometimes preceded or followed by chemical fixation.

11.2 The Mechanism of Freezing

11.2.1 Freezing of Water

Water is converted into crystalline ice by cooling to 0°C at a pressure of 1 atmosphere (101 kPa). On further cooling, the temperature at which the structure of ice changes from a dynamic to a static state is attained; this is called *the point of recrystallization*. The point of recrystallization is believed to be about −130°C for pure water.

In tissue this point is probably considerably higher, but cannot be determined unambiguously. Values of about −80°C to −90°C are regarded as reasonable approximations. At temperatures above the recrystallization point, smaller crystals tend to rearrange into larger crystals due to exchange of water molecules. In consequence, the faster the cooling from 0°C to the point of recrystallization, the smaller the size of the ice crystals. Ideally, the freezing process should produce crystals that are too small to be resolved by electron microscopy (vitrification, di-

H. Lyon (Ed.)
Theory and Strategy in Histochemistry

ameter less than 1 nm). To achieve this, cooling would have to occur at a rate of 10^5 °C per second. Since this is not possible in practice, the technical objective is to keep the inevitable ice crystals as small as possible.

The so-called "cryoprotective" agents provide some assistance in this regard by raising the point of recrystallization. These agents fall into two groups: low molecular weight compounds such as glycerol, glucose, and dimethylsulphoxide (DMSO), which penetrate into cells, and high molecular weight compounds such as dextran and polyvinylpyrrolidone which do not.

11.2.2 Freezing of Tissue

The extra- and intra-cellular fluids both have an osmotic activity equivalent to a 0.9% w/v solution of NaCl in water (300 mOsm). As tissue is cooled (Franks, 1978), extracellular fluid starts to freeze at slightly below −10°C. Initially, salt-free ice crystals are formed so that the remaining salt solution becomes more concentrated and highly osmotically active. This preferential freezing of pure water continues down to −21°C; only then does freezing of the salt solution take place.

The increased extracellular tonicity during freezing draws water out of cells and may lead to pronounced shrinkage. This problem can be minimized by rapid freezing. The dynamics of intracellular freezing, which starts later, are more complex and less well understood. In general, however, the same considerations regarding crystal size and speed of freezing apply. The fastest freezing times that have been achieved with biological tissue are between 5×10^3 and 10^4 °C per second, a factor 10 below the theoretically desirable rate. Moreover, these results have only been achieved with exceedingly small pieces of tissue (less than 0.1 mm^3). It is therefore absolutely vital that small pieces of tissue should be used. In practice this means that at least one dimension (the thickness) is held to a minimum.

Good results are facilitated by making the contact between tissue and the heat exchange surface as intimate as possible. This may be achieved by using two different approaches either alone or in combination:

1. Placing the tissue on a polished metal surface (copper, aluminium, or silver) with good heat conducting properties
2. Placing the tissue in a cold liquid medium. The medium must be carefully chosen so that it does not boil on contact with the tissue and form a mantle of gas around the "warm" block. For instance, if a tissue block is placed directly into liquid N_2, freezing takes place slowly in spite of the low temperature, as gaseous nitrogen insulates the tissue (The domestic phenomenon of a drop of water surviving on a hot plate due to a cushion of steam is analogous.)

Table 11.1 reviews the practical options for freezing tissue. Direct freezing followed by sectioning on a cryostat is the most frequently used approach. The following procedure is recommended, partly because it is easy to perform and partly because it has given reliable and reproducible results in numerous quantitative cytochemical investigations also at the ultrastructural level (von Bülow and Høyer, 1983). The arrangement shown in Fig. 11.1 is recommended.

Table 11.1. Range and effectiveness of techniques for freezing of tissue.

	Methods	Temperature in cooling medium (°C)	Efficiency
Tissue in cooled liquid	1. Organic liquid surrounded by liquid gas[a]	– 196	Fastest method; feasible method for electron microscopic investigations
	2. Organic liquid surrounded by carbon dioxide ice in organic liquid[b]	– 79	Suitable for most purposes in light microscopy
	3. Carbon dioxide expansion cooler	– 70	Good for routine work
Tissue in contact with cooled metal	4. Thermo-electric cooling	– 40	Poor
	5. Metal block cooled with a cooling coil[c]	– 30	Poor

[a] Suitable organic liquids for tissue immersion are pre-cooled, often in liquid nitrogen, and include isopentane and commercial mixtures of butane and propane in liquid form.
[b] n-hexane or isopentane pre-cooled in carbon dioxide ice mixed with ethanol, methanol, or acetone, are suitable here.
[c] Bench- and cryostat-mounted models are available.

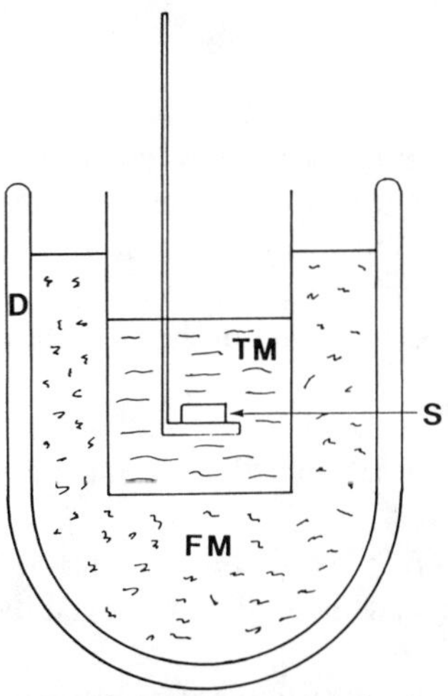

Fig. 11.1. Arrangement for rapid freezing of tissue.
S tissue specimen on metal holder
TM cold transport medium (must not boil when the tissue is immersed)
FM freezing mixture (ethanol + CO_2-ice: −79°C; liquid N_2: −195°C)
D Dewar flask

A beaker or preferably a metal container is half filled with n-hexane (free of aromatic hydrocarbons, boiling point 67–70° C). The container is placed in a glass vessel which is nearly filled with a mixture of crushed solid carbon dioxide and absolute ethanol. The sides and the bottom of the outer vessel are surrounded with a thick layer of insulating material, and the temperature in the n-hexane stabilizes to between −70°C and −75°C after a few minutes. Taking care to avoid touching the sides of the inner vessel, a small tissue block (up to 5 × 5 × 3 mm) is rapidly immersed in the n-hexane. After about one minute, using pre-cooled forceps (−70°C), the block is removed quickly but carefully, and the tissue can now either be mounted on a metal tissue holder and placed directly in a cryostat chamber or kept in a freezer wrapped in aluminium foil at −70°C or lower.

There has been no detectable loss of enzyme activity in any of the systems examined so far if the tissue is used within three days. Several of the other methods listed (e.g. isopentane precooled with nitrogen) can give equally good results.

These procedures have, however, hitherto received minimal use in quantitative cytochemistry. It is worth noting that, providing the freezing process occurs from all surfaces of the block, surprisingly good morphology is achieved by freezing with a carbon dioxide expansion cooler even when relatively large tissue blocks are used (e.g. in qualitative diagnostic work).

11.3 The Further Preparation Following Freezing

11.3.1 The Cryostat Microtome

This is a microtome housed in its own low temperature (usually between −16°C and −30°C) cabinet. Very thin sections (4–6 μm) for use in histochemical and cytochemical reactions can be made using a cryostat. In spite of freezing, structural integrity is well preserved, at least at the light microscopic level.

Cryostat design itself is beyond the scope of this text but some of the following points should be stressed:

1. The temperature of the cabinet should usually be between −10°C and −30°C, most often −17°C to −20°C. The temperature is varied in order to adjust the hardness of the ice to the tissue hardness
2. It should be possible to cool the knife to around −70°C so that heat generated during cutting is effectively removed from the tissue. This may be achieved by fixing a tray containing solid CO_2 to the knife
3. The knife should have a correctly sharpened cutting edge without curvatures, and it should be set at the correct angle with respect to the tissue
4. The anti-roll plate should be very carefully adjusted with respect to its height above, angle to, and distance from the knife edge
5. The cryostat should be motorized so that constant cutting speed and thereby section thickness are ensured. It is worth noting that, even if the micrometer screw is set for a particular μm-value, a higher speed will give thinner sections, and vice versa (see Table 28.2). This variation also depends on the nature of the tissue

6. The section should be transferred to slide (or cover slip) using the correct technique

If the section is transferred to the slide with a cold brush or by placing the slide in direct contact with the section, there is a considerable risk of mechanical injury to the tissue. Many workers attempt to avoid this problem by cooling the slide to cryostat temperature. Unfortunately, the supercooled water in the section forms ice crystals as soon as room temperature is reached. Both of these disadvantages can be avoided in the following manner: A slide held at room temperature, or even heated to 37°C is moved in a parallel plane towards the section.When the slide is around 2 mm from the section, the temperature gradient between the two (50–100°C) causes the water to evaporate from the section and condense on the knife almost instantaneously. At the same time the section, assisted by a kind of jet-effect, adheres to the slide. The rapidly dehydrated section can now be brought to room temperature without risk of ice crystal formation as there is no longer sufficient free water for this to occur.

11.3.2 Freeze-Drying

Freeze-drying is a simple dehydration technique which can be used when aqueous fixatives are to be avoided. It is often used when retention of water soluble molecules is desired. The technique utilizes the sublimation properties of ice. The process may be accelerated by applying a vacuum and the consequent need for very low pressures make freeze-drying apparatuses expensive.

11.3.3 Freeze-Substitution

This is another specialized form of dehydration-fixation technique which can be used when retention of small water soluble molecules is desired. In freeze-substitution the tissue is frozen and placed in a substitution medium, e.g. ethylene glycol + glutaraldehyde, at −50°C. The ice in the tissue block is substituted by the ethylene glycol and tissue fixation occurs simultaneously.

12 Tissue Processing: III. Fixation, General Aspects

P.E. Høyer, H. Lyon, M. Møller, P. Prentø, B. van Deurs, E. Hasselager, A.P. Andersen

12.1 Definition

Fixation is used in this text to denote a chemical treatment leading to *immobilization and stabilization* of tissue components. Displacement and extraction of native components should be minimal and, ideally, those properties which characterize the tissue *in vivo* should be preserved. Certain methods of investigation such as electron microscopy, enzyme histochemistry, and immunohistochemistry are very exacting in their fixative requirements and therefore demand a detailed understanding of fixation.

12.2 Classification of Fixatives

Since proteins constitute the major macromolecular component of cells it is not surprising that fixation of tissue depends predominantly on fixation of proteins. This can be accomplished with either *coagulant* or *cross-linking* fixatives. Coagulant fixatives irreversibly denature proteins, whereas cross-linking fixatives produce either minor or no denaturation.

Biochemically, *denaturation* implies complex changes in secondary and tertiary protein structure extensively disrupting the compact native conformation. These changes may be reversible but are usually permanent. Denaturation does not inevitably lead to stabilization or "fixation"; hence, carbamide, SDS (sodium dodecylsulphate), and acetic acid are recognized as denaturants, but they are not fixatives. Conversely, irreversible denaturation is not identical to fixation; hence, acetic acid, dilute sodium hydroxide, and ultraviolet irradiation can cause irreversible denaturation without causing fixation.

Denaturation often leads to *precipitation* of proteins but precipitation does not necessarily imply irreversible denaturation or fixation. For example, concentrated solutions of ammonium sulphate and other neutral salts cause precipitation of protein but the precipitate can be reconstituted with minimal loss of biological activity. This is also often the case with ice-cold acetone or ethanol precipitation.

H. Lyon (Ed.)
Theory and Strategy in Histochemistry

An irreversible denaturation associated with fixation occurs when proteins are partly precipitated and partly cross-linked, and locked in an open, non-native conformation.

12.2.1 Cross-Linking Fixatives

These compounds (e.g. glutaraldehyde, formaldehyde, crotonaldehyde, and carbodiimide) effectively stabilize the native conformation of proteins by forming cross-links both between polypeptides within individual proteins and between neighbouring proteins (Fig. 12.1). The reaction is reversible in varying degrees depending on the reagent used and the net result is that the entire protein phase in the specimen forms a gel (Fig. 12.1). This comprises the cross-linked network of proteins with a liquid phase, usually water, bound to the network by dipole-dipole and hydrogen bonds.

The fixative molecules themselves participate in the cross-linking, so that the process is *additive* (Fig. 12.1). Additive fixation is the only means by which fixation can be achieved without denaturation since *direct* cross-links involving reactive amino acid residues would require extensive conformational changes.

The principle advantage of cross-linking fixatives is that they maintain proteins in native or near-native conformation. Subsequent dehydration does not, therefore, cause coagulation and this is very important in electron microscopy. The conformational stability also makes a substantial range of enzyme histochemistry possible, provided that the fixative has not affected the catalytic (active) site.

12.2.2 Coagulant Fixatives

These compounds (e.g. ethanol, picric acid, chromium trioxide, mercuric chloride, platinic chloride) exert their fixative effects by denaturing protein irreversibly (Fig. 12.1). Conformational changes open out the protein tertiary structure and expose both reactive and hydrophobic groups (Fig. 12.1). This results in further forms of stabilization involving groups from different proteins reacting with each other to form larger, disordered aggregates (Fig. 12.1).

In principle, coagulation can be achieved in three ways:

1. Increase in thermal motion leading to disruption of both tertiary and secondary structure. This is achieved by heating
2. Increase in the electrostatic repulsion between different areas of the protein (e.g. by blocking water-soluble groups). Treatment with strong acid or additive metal salts such as mercury compounds will do this. Picric acid acts both as an acid and as an additive
3. Removal of water from the protein and its surroundings by dehydration (e.g. with ethanol)

In the last of these, *dehydrating fixation*, the aqueous phase is extracted and substituted with the solvent. This leads to exposure of previously concealed groups

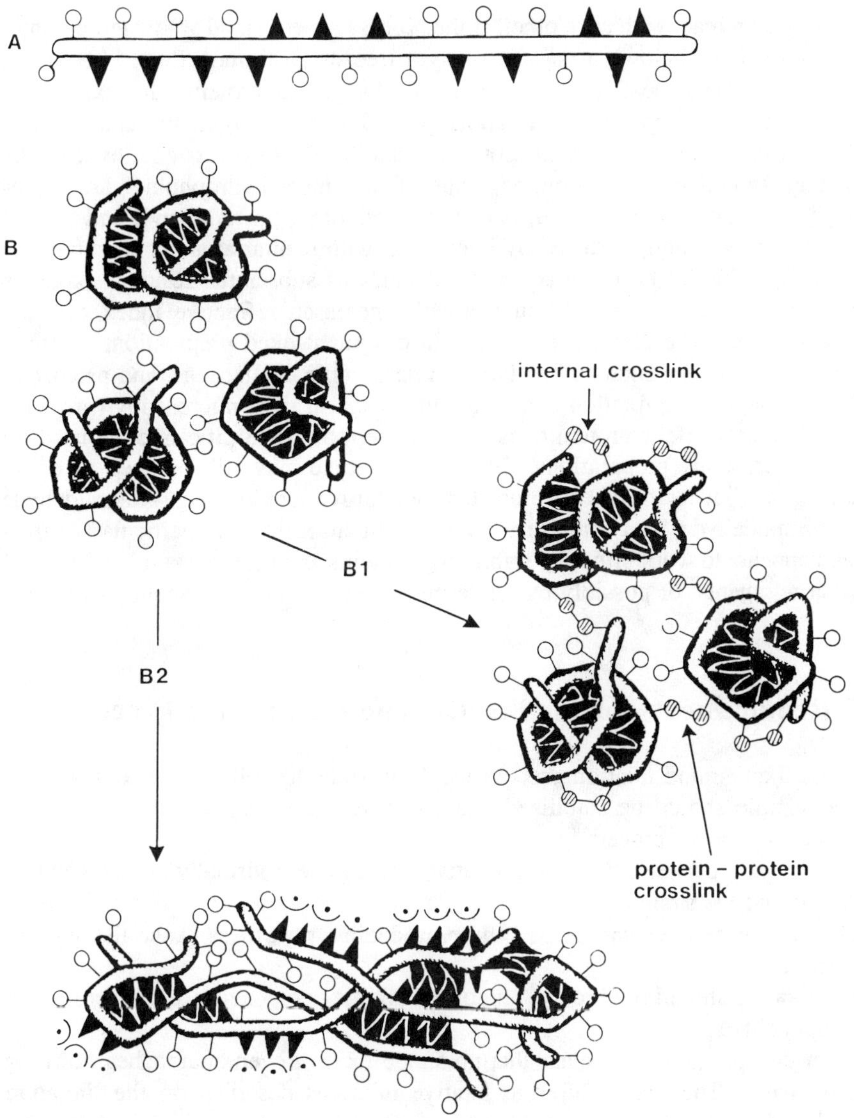

Fig. 12.1. Simplified view of the effects of cross-linking and coagulant fixatives on protein conformation.
A Unfolded protein
B Native protein conformation
B1 Cross-linking
B2 Coagulation

hydrophilic (polar) amino acid residues; (hatched if crosslinked)
hydrophobic (non-polar) amino acid residues
hydrophobic boundary

which may then react with each other. If the process is performed at low temperature (around −40°C or below), it is effectively a freeze-substitution (Sect.11.3.3) and does not necessarily lead to conformational changes in proteins. In contrast, at room temperature extensive denaturation (Sect.12.2) and coagulation take place.

The native tertiary structure of proteins is stabilized by hydrophobic/hydrophilic interaction. In coagulant fixation, exposure of the "inner hydrophobic" areas (notably glycyl, alanyl, valyl, leucyl, isoleucyl, methionyl, phenylalanyl, and tyrosyl residues) is only slightly reduced by interaction with similar areas in neighbouring polypeptides (Fig. 12.1). In an aqueous solvent, this substantial residual exposure of hydrophobic residues results in a greatly increased refractive index between tissue and water. The effect can be seen in coagulant-fixed preparations of small transparent organisms such as fish larvae, microscopic crustaceans, and protozoa.

While coagulating fixation can be an advantage in light microscopy (provided the resulting network of protein does not become too coarse), it cannot be used for electron microscopy preparations. Ethanol or acetone may still be used as dehydrating agents but they must be preceded by additive fixation (e.g. glutaraldehyde (+/− formaldehyde) and osmium tetroxide). Glutaraldehyde is particularly effective and appears to work either by stabilizing proteins against the coagulant effect of dehydrating agents, or possibly by rendering the effects of these agents reversible.

12.3 Requirements for the Fixative and the Fixation Process

The ideal fixative and fixation process would include the following features:

a. The sample should be stabilized against unwanted changes due to the subsequent preparative procedure
b. The reaction between tissue and fixative should occur virtually simultaneously throughout the sample
c. The fixation process should be fully reproducible regardless of the tissue composition
d. The fixative should not induce "abnormal" tissue components or extract normal components

From the foregoing it is clear that no single fixative even approaches fulfilling these demands. The many different fixative mixtures described in the literature must, therefore, be regarded as an attempt to balance the desired and the unfavourable effects of the individual components. Clearly, careful consideration of what is to be demonstrated is essential prior to fixation.

When using any fixative one should have knowledge of its:

1. Chemistry (Sect.12.3.1)
2. Method of application (Sect.12.3.2)
3. Reactions with different tissue components (Sect.13.1)
4. Safety and disposal problems (Sect.12.3.3)
5. Removal from the tissue (Sect.12.3.4)
6. Effects on tissue storage properties following fixation (Sect.12.3.5)

Knowledge of several additional factors is also desirable if the results of fixation are to be fully appreciated:

7. Optimum concentration (Sect.12.3.6)
8. Optimum pH (Sect.12.3.7)
9. Optimum osmolarity (Sect.12.3.8)
10. Ionic strength (Sect.12.3.9)
11. Optimum temperature (Sect.12.3.10)
12. Optimum time of fixation (Sect.12.3.11)

In practice, for the majority of fixatives, very little is known about these additional factors.

12.3.1 The Chemistry of Fixatives

Primary Fixatives. The reagents mentioned in Sects.12.2.1 and 12.2.2 are known as primary fixatives. With the exception of substances that can be used as gases (e.g. formaldehyde), fixatives are nearly always used in a defined solution comprising salts and other compounds influencing pH, osmolarity, and electrode potential (see below). The form of commercial products (e.g. degree of polymerization of glutaraldehyde) and the presence of impurities and stabilizers (e.g. formic acid and methanol in formalin) should be ascertained as far as possible and purification considered. Finally, the form and extent of storage should be defined.

Fixative Mixtures. Table 12.1 surveys primary fixatives that can be used alone or in combination with other reagents as fixative mixtures.

Table 12.1. Primary fixatives. Classification and application.

	Form of application				
Primary fixative	gas	liquid	solution	alone	in mixture
Cross-linking					
formaldehyde	+		+	+	+
glutaraldehyde	+		+	+	+
glyoxylic acid	+		+	+	
osmium tetroxide			+	+	+
potassium dichromate			+		+
Coagulant					
ethanol		+			+
picric acid			+		+
chromium trioxide			+		+
mercuric chloride			+		+
platinic chloride			+		+
trichloroacetic acid			+		+

Table 12.2 gives the composition of some widely used mixtures of fixatives. A number of these (e.g. Zenker's and osmium tetroxide containing reagents) must be prepared immediately before use as they decompose fairly rapidly.

Table 12.2. Examples of composite fixatives.

		Primary fixatives				Non-fixing compounds			
Name	pH	Ethanol	Picric acid	Mercuric chloride	Formaldehyde	Potassium dichromate	Sodium sulphate	Acetic acid	Water
Bouin	1.5		+		+			+	+
Clarke	—	+						+	
Helly	3.6			+	+	+	+		+
Lillie's AAF	1.5	+			+			+	+
Zenker	2.3			+		+	+	+	+

Carnoy's fixative should also be mentioned; it is essentially Clarke's reagent with chloroform added.

The miscibility of some primary fixatives is shown in Table 12.3. Amongst the incompatibilities, reduction of chromium trioxide, osmium tetroxide, and potassium dichromate by ethanol and formaldehyde and, inactivation of potassium dichromate by chromium trioxide, deserve specific mention. Potassium dichromate only acts as a cross-linking fixative if the pH is above 3.8. Below pH 3.4 it behaves like chromium trioxide.

Table 12.3. The miscibility of different fixatives.

	CH_3COOH	$K_2Cr_2O_7$	OsO_4	$OHC(CH_2)_3CHO$	$HCHO$	CrO_3	$HgCl_2$
C_2H_5OH	+	0	0	+	+	0	+
$HgCl_2$	+	+	+	+	+	+	
CrO_3	+	0	+	0	0		
$HCHO$	+	0	0	+			
$OHC(CH_2)_3CHO$	+	0	0				
OsO_4	+	+					
$K_2Cr_2O_7$	0						

+ : can be mixed; 0: should not be mixed.

12.3.2 Application of the Fixative

Fixatives can be applied either by immersion or by perfusion.

Immersion Fixation. For fixation by immersion the sample should be placed in at least 20 times its own volume of fixative. If the sample is small enough and the interval between removal and immersion is minimal, the results are usually excellent. For most pathology departments this will be the only practical possibility.

In Situ Fixation. This is a modification that is valuable in experimental work and may be considered in pathology. The fixative is added drop by drop directly to the area of interest in an anaesthetized animal. This approach has many advantages from a physiological viewpoint but it remains limited by the size of the piece of tissue to be fixed.

Perfusion Fixation. In perfusion fixation the fixative is passed under pressure through the circulatory system or other hollow cavities for a period of 10–15 min (Fig. 12.2). Individual organs or even an entire animal may be treated this way. Many of the problems inherent to the immersion approach are avoided but new ones arise.

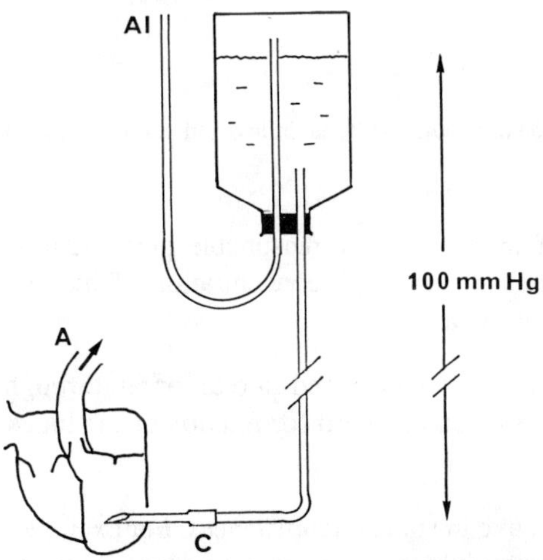

Fig. 12.2. Perfusion fixation of an intact animal. Inflow is through a hypodermic needle (C) placed in the left chamber of the heart. Aorta (A). The fixative bottle is raised to give a suitable hydrostatic pressure. Air intake (AI).

Perfusion fixation usually implies a *vascular perfusion* through the heart. The technique is popular, particularly in electron microscopy laboratories working with smaller animals (e.g. mice and rats). In principle, the whole animal is "fixed" although there is a pronounced tendency for the forelegs and head to become "well" fixed.

The major advantage of vascular perfusion is that the fixative is delivered to all tissues at roughly the same time. Fixation takes place quickly because the diffusion path is reduced to half the distance between two capillaries. The denser the capillary network for a particular organ is, the shorter the diffusion path and the greater the potential flow of fixative and consequently the more rapid the fixation.

If only one particular organ is to be fixed then "*local perfusion*" is always preferable. For example, the liver may be accessed via the portal vein and the kidneys may be fixed by retrograde perfusion from the abdominal aorta below the point where the renal artery branches off. Infusion pumps (e.g. Braun Melsungen) are preferable to the bottle system for such local "miniperfusions" (Fig. 12.3.).

Examples of other hollow cavities that may be used include the oesophagus, intestine, renal tubules (using micropipettes), and the ventricular system in the brain.

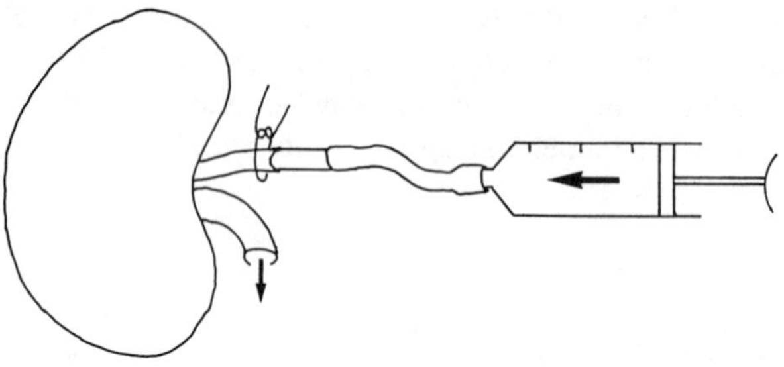

Fig. 12.3. Arrangement for local perfusion fixation of a kidney (arterial inflow and venous outflow)

In many of these instances the tissue of interest is in a reasonable physiological condition to start with and the rapid exposure to a high concentration of fixative effectively constitutes an *in situ* immersion fixation.

Infusion Fixation. In infusion fixation the fixative is forced into an organ through an inlet which is then clamped. This approach can be used for fixation of the lungs or the urinary bladder.

Fixation with Gas. This is a useful alternative in some circumstances. For example, the respiratory system can be fixed *in situ* by inhalation of gaseous glutaraldehyde (this also applies to laboratory staff!). Smears and cell cultures can also be fixed with gaseous fixative in small, closed containers.

12.3.3 Safety and Waste Disposal

Aldehydes. These pose many health risks. *Acrolein* is particularly toxic and should not be used unless it offers very clear advantages. *Formaldehyde* and *glutaraldehyde* should always be kept in tightly sealed containers and all processing, including cutting of blocks from fixed specimens, should take place in well ventilated work areas using suitable gloves where direct handling is required. After use, aldehydes should be collected as organic chemical waste for destruction.

Heavy Metals. *Chromium, mercury,* and *osmium* are all heavy metals constituting potential health hazards and therefore require careful handling. Osmium, which is used as osmium tetroxide, is strongly corrosive and its vapours are noxious to eyes and other mucous membranes. Waste is collected as for other inorganic materials.

Protein Denaturing Compounds. These, including *acetic acid, picric acid,* and *ethanol,* are not as hazardous as the compounds mentioned above. Concentrated acetic acid is corrosive, while picric acid and picrates are explosive in their anhydrous forms, so careful storage is essential. Waste is treated as solvent waste for the alcohols, and as organic chemical waste for the acids.

12.3.4 Removing Excess Fixative

In general, this does not pose any serious problems (Table 12.4). The majority of primary fixatives can be removed by thorough washing in water. An exception is mercury chloride, which leaves 1–3 μm diameter black precipitates of metallic mercury in the tissue. These may be removed by treatment with an alcoholic solution of iodine: $Hg + I_2 \rightarrow Hg^{2+} + 2I^-$. Excess iodine will itself stain the tissue, however, this is easily removed by washing with 70% v/v ethanol or a sodium thiosulphate solution (much quicker): $2S_2O_3^{2-} + I_2 \rightarrow S_4O_6^{2-} + 2I^-$. Glutaraldehyde can only be partly removed by washing in water or aqueous buffer. The remaining free aldehyde groups can, however, be blocked with glycine or reduced with borohydride (cf. Sect.9.7.2).

Table 12.4. Survey of the possibilities for removing excess fixative.

Fixative	Wash with: water	ethanol	xylene
Formaldehyde	+	+	–
Glutaraldehyde	(+)	–	–
Chromium trioxide	+	–	–
Potassium dichromate	+	–	–
Mercuric chloride	–	–	–
Osmium tetroxide	+	– *	–
Ethanol	+	–	+
Acetic acid	+	+	–
Picric acid	+	+	–

+ complete removal; (+) partial removal; – ineffective; *50% v/v ethanol will remove osmium tetroxide.

12.3.5 Storage After Fixation

Storage of fixed tissue allows preparation of supplementary tissue blocks and possibly the use of alternative embedding techniques (e.g. frozen sections and lipid staining). As far as possible, the tissue should be maintained in the same state as when the original sections were cut.

In general, fixatives are not suitable for storing tissue, and several of these reagents should be washed out with water after fixation (Sect.12.3.4). Bouin's fluid contains picric acid which destroys tissue after prolonged periods and also makes subsequent paraffin embedding and cutting of sections more difficult. It should therefore be washed out with several changes of 70% v/v ethanol.

70% v/v ethanol is also frequently used as a storage medium, an application that conveniently complements its use for the first step in the normal dehydration procedure. Prolonged storage (weeks) may lead to inconvenient tissue hardening. This effect is particularly pronounced at higher concentrations of ethanol.

In neuropathology some workers use 4% w/v formaldehyde as a medium for extended storage (years). The main draw-back is that staining properties gradually alter. For example, lipid stainability using lysochromes is slowly reduced due to progressive oxidation. The decline in stainability can be minimized by using neutral buffered formaldehyde.

The volume of storage fluid should be just sufficient to protect the tissue against desiccation and microbial degradation. Considerable space can therefore be saved by packing the fixed tissue and a small volume of liquid in heat-sealed plastic bags. Vapour-tightness of the wrapping material and the seams can be tested by simple weight control of a test specimen.

12.3.6 Optimum Concentration

The solvent in which the fixative is dissolved is frequently referred to as the *vehicle*. This is nearly always water containing buffers, electrolytes, and other compounds (Sects.12.3.7, 12.3.8, and 12.3.9). The optimum fixative concentration for a given application gives a balance between rapid fixation and induction of artifacts. With the exception of dehydrating fixatives, this is not usually achieved with concentrated fixatives as the need for control of pH, osmotic effect, etc. generally lead to the optimum concentration being much lower.

In practice, the optimum fixative concentration is usually defined in terms of a concentration interval within which the desired feature may be reliably demonstrated. Thus the highest permissible concentration is often determined by the need to retain enzyme activity (Chaps.23–25). Glutaraldehyde, which gives good results at between 0.5 and 5% w/v, is additionally limited by its tendency to polymerize at higher concentrations (Polymerization leads to a reduced availability of the active monomer).

12.3.7 pH

Fixatives are generally used at between pH 1.5 and pH 7.5. Acid fixatives (Table 12.2) dissociate iron from haemosiderin (Sect.13.7) and hydrolyze nucleic acids, so that purines are dissociated and phosphate-ester bonds are cleaved (cf. Feulgen reaction, Sect.9.9). The resultant extraction may, however, be avoided and good preservation maintained if a nucleic acid precipitating compound, such as ethanol or chromium, is added (Sect.13.3). When fixation is necessary prior to the histochemical demonstration of enzyme activity, it is important to know the pH optimum as many enzymes are particularly susceptible to denaturation at this pH.

For light microscopic examination of cells, an aldehyde fixative at pH 7.2–7.4 is normally preferred. Similarly, ultrastructural morphology is best preserved when fixative pH is close to 7.4.

12.3.8 Osmolarity

Although the primary fixative has some influence on osmolarity, the composition of the vehicle is usually far more important in this regard (Table 12.5).

If a mammalian cell is subjected to a hypertonic environment, it will shrink, and conversely, in a hypotonic environment it will swell. To avoid these effects, an isotonic vehicle (about 300 mOsm) is generally used. Alteration of the fixative content, therefore, makes the solution more or less hypertonic.

Table 12.5. The osmotic activity of different fixatives and buffers.

Fixative	Buffer	Total osmotic pressure (mOsm)
1% FA		~ 300
1% GA		~ 100
	0.1 mol/l phosphate	240
	0.1 mol/l cacodylate	190–200
1% GA	0.1 mol/l phosphate	340
2% FA + 2.5% GA	0.1 mol/l cacodylate	1050

FA = formaldehyde; GA = glutaraldehyde.

As some fixatives chemically bind to tissue (e.g. aldehydes), the osmolarity of these reagents decreases during fixation. This is one reason why apparently hypertonic fixatives can give satisfactory results. Hence, a 3% glutaraldehyde solution in a buffered vehicle can easily have an isolated osmolarity of 600 mOsm but, after contact with tissue, has a practical osmolarity of only 300 mOsm.

12.3.9 Addition of other Compounds to the Fixative

Non-fixing compounds may be added to the fixative to control osmolarity (salts, sucrose), stabilize membranes and reduce the extraction of lipids (Ca^{2+}), or to give the vehicle a high colloid osmotic pressure (dextran, polyvinyl pyrrolidone). This last approach appears to limit changes in volume that tend to occur between endothelium and epithelium with perfusion fixation in loosely built tissues such as kidney, pancreas, and intestine (Bohman and Maunsbach, 1970). The high colloid osmotic pressure probably counteracts the high hydrostatic pressure caused by the perfusion (Fig. 12.4). This is analogous to the normal transport of material across the capillary wall (Fig. 12.5): About half way through a capillary the hydrostatic pressure is equal to the colloid osmotic pressure. In fixing tissue by vascular perfusion without the addition of colloid, the hydrostatic pressure is, in principle, sustained without the counterbalancing influence of colloid osmotic pressure. Fixation, therefore, tends to produce an increased extracellular space (Fig. 12.4) just as oedema occurs in conditions where blood protein levels are reduced (e.g. proteinuria).

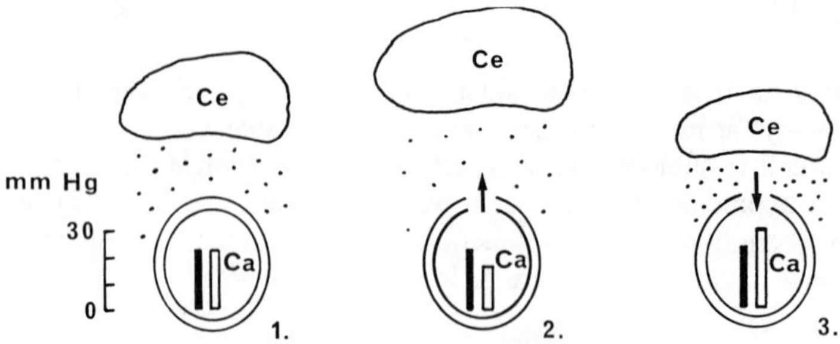

Fig. 12.4. The spatial relationship between capillary (Ca), ground substance (dots), and a cell (Ce) under normal physiological conditions (1.), following perfusion with fixative with low colloid osmotic pressure (2.), or high colloid osmotic pressure (3.). Hydrostatic pressure (▌), colloid osmotic pressure (||). Net transport of fluid through the wall of the vessel (⟶)

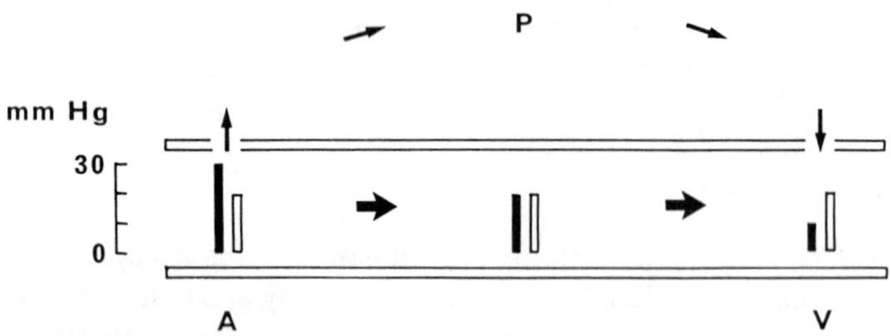

Fig. 12.5. Hydrostatic pressure (▌) and colloid osmotic pressure (||) along a capillary, resulting in a net outflow through the wall of the capillary at the arterial end (A) and a net inflow at the venous end (V). Paracapillary circulation (P).

12.3.10 Temperature

Both penetration of the fixative and the fixation process itself (Sect.12.3.11) are temperature dependant. This is particularly useful in perfusion fixation where the diffusion path is already short. If high flow rates can be ensured, room temperature or even body temperature can be used with advantage.

Autolytic processes place an upper limit on optimum temperatures for immersion fixation. If the temperature is lowered from room temperature to 0–4°C, the combined reduction in speed of fixation and the autolytic process gives acceptable results. It must be stressed that, unless the specimen is very small, fixation at temperatures above 4°C leads to the tissue only being suitable for a rather coarse light microscopic diagnostic assessment. Unfortunately tissue removed at operations is traditionally, but nonetheless detrimentally, fixed at room temperature.

Fixation should always take place at 0–4°C for histochemical work. It should be noted, however, that the addition of CPC (cetylpyridinium chloride) to a concentration of 0.5% w/v for improved fixation of proteoglycans (Sect.13.4) leads to precipitates in the fixative at temperatures below 0°C.

12.3.11 Duration of Fixation

According to the definition (Sect. 12.1) fixation is a process of immobilization which can either be a cross-linking or a coagulation (Sect. 12.2). The *duration of fixation* for a specific cell or tissue component with a certain fixative under well defined conditions is *the time from the moment at which the whole tissue sample is exposed to the fixative (e.g. by immersion) to the moment when that component has become immobilized.* The duration of fixation thus consists of two phases (Flitney, 1966; Hopwood, 1969; Hayat, 1970, p.31):

1. The first phase (diffusion) comprises the time it takes for the fixative to penetrate to the component in question and achieve a sufficiently high concentration to effect fixation (cf. Fig. 12.6 and Table 12.6)
2. The second phase is the time it takes for the actual fixation (immobilization process) to take place (cf. Fig. 12.6)

The duration of the first phase depends on the *diffusion rate of* the fixative through the tissue and, with certain reservations, obeys Fick's law.

Diffusion describes the net flow of molecules from an area of high concentration to one of low concentration in the absence of any other driving force (Freifelder, 1976, p.346). The diffusion coefficient, K_d, of a molecule can be defined by Fick's law, which states that the number of molecules, dn, passing through an area, A, in time dt is related to the concentration gradient, dc/dt, by the equation:

$$\frac{dn}{dt} = -K_d A \frac{dc}{dt}$$

It can be shown that the magnitude of K_d depends on the frictional coefficient which depends in turn on the size and shape of the molecule (Freifelder, 1976, p.346).

Fick's law does not take into account the "binding" of fixative molecules on their way to the site under consideration. Clearly allowance should be made for this factor when making calculations in relation to additive fixatives (e.g. formaldehyde).

The diffusion time is influenced by:

1. The nature of the fixative, i.e. size and shape of fixative molecules and their inclination to react with each other (polymerization) (van Holde, 1971, p.86)
2. Concentration of the fixative (see Fig. 12.6)
3. Temperature of the fixative (Sect. 12.3.10)
4. Diffusion path, i.e. where the cell or tissue component under consideration is situated in relation to the point of application of the fixative (Sect. 12.3.2 and Fig. 12.6)
5. Type of tissue (see below)
6. Cross-linking or coagulation of the tissue (see below)

It would appear that the kind of tissue should be an important determinant for the rate of fixative diffusion. In practice, very few experiments have addressed this issue. It has, however, been shown for a number of fixatives that the diffusion

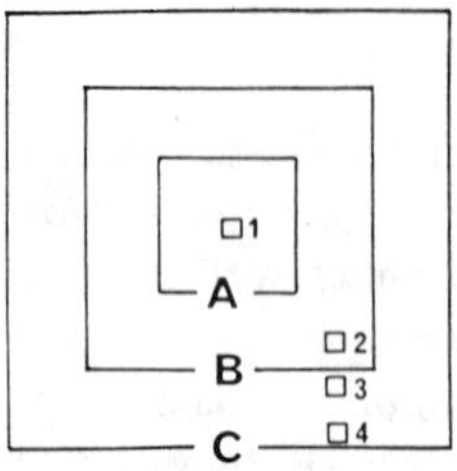

Fig. 12.6. Diagram of tissue block illustrating the time course of fixation by immersion at a given point in time.
A Extent of fixative penetration
B Boundary line indicating zone outside which the fixative has achieved a concentration that allows cross-linking/coagulation
C Boundary line between tissue and pure fixative
1 Location as yet unexposed to fixative
2 Location exposed to suboptimal concentration of fixative
3 Location exposed to optimal fixative concentration (Time insufficient for complete fixation)
4 Location exposed to optimal fixative concentration (Time sufficient for complete fixation).

rate is considerably slower in liver tissue than in a gelatin-egg albumin gel (Baker, 1966, p.24).

Cross-linking of the superficial parts of a piece of tissue do not seem to have any appreciable effect on fixative diffusion whereas coagulation can markedly hinder further diffusion of the fixative into the tissue (Baker, 1966).

Medawar (1941) derived the following equation for the dependence between diffusion path (d) and diffusion time (t) from Fick's law:

$$d = k\sqrt{t}$$

where k is a constant, the size of which depends on the diffusion coefficient, K_d, for the fixative in question. With d measured in mm and t in hours k-values have been measured for a number of fixatives at different concentrations partly for diffusion into gelatin-egg albumin gel and partly into liver tissue. Some k-values measured in this way are given in Table 12.6. Note, that *k denotes the number of mm the fixative penetrates during the first hour*. If, for example $k = 2$, the fixative will have penetrated $2\sqrt{1} = 2$ mm after 1 hour and $2\sqrt{2} = 2.8$ mm after 2 hours.

Finally, it should be mentioned that problems may arise from the vehicle (buffer) diffusing more quickly into the tissue than the fixative.

The length of the second phase (process of immobilization) depends on the *reactivity* of the fixative and its concentration. Knowledge of this is rather limited. However, it is clear that formaldehyde cross-links very slowly in comparison to the dialdehyde, glutaraldehyde.

The influence of temperature and concentration of a fixative in solution can be illustrated by the following observations on the times taken for fixation of the central parts of a 3 mm thick slice of liver tissue:

Table 12.6. Survey of diffusion rate, immobilization reaction, and total rate of fixation for some fixatives.

Fixative	Concentration %	First phase diffusion rate	k°C^x gelatin-egg albumin gel	k°C liver	Second phase immobilization process	Total fixation rate	References
Formaldehyde	4	fast	3.6^{20}	2.0^{4}	slow	moderate	Baker, 1966, p. 43 Flitney, 1966
Glutaraldehyde	3	slow		0.19^{4}	fast	slow	Ericsson and Biberfeld, 1967
	4	slow		0.34^{20}	fast	slow	Hopwood, 1967,
Osmium tetroxide	1	slow	1.0^{20}	0.2^{4}	moderate	slow	Baker, 1966, p. 48
Ethanol	70–90	moderate			fast	slow	Baker, 1966, p. 33
Mercuric chloride	7	moderate	2.2^{20}	0.84^{20}	fast	moderate	Baker, 1966, p. 35
Chromium trioxide	0.5	slow	1.12^{20}		fast	slow	Baker, 1966, p. 37

^{x}k = penetration of fixative in mm during first hour.

4%	w/v	formaldehyde	solution	at	4°C	for	96	hours
4%	w/v	–	–	–	24°C	–	48	–
4%	w/v	–	–	–	35°C	–	24	–
8%	w/v	–	–	–	55°C	–	3	–

It is interesting to note that routine fixation of a 3 mm thick liver slice in 4% w/v solution of formaldehyde at room temperature thus is *not* adequate after 24 hours according to these figures.

12.3.12 Microwaves in Fixation

Microwaves have during the last few years been increasingly applied not only in fixation procedures and other steps of tissue processing but also in staining procedures (Suurmeijer *et al.*, 1990; Kayser and Bubenzer, 1990).

Clearly one of the chief advantages of using microwaves is the much accelerated rate at which the processes take place. In fixation a discrimination should be made between microwave stabilization without the presence of chemical fixatives and microwave-stimulated fixation with fixatives present (Ruijgrok *et al.*, 1990).

It is not the intention in this place or elsewhere in the book to give a detailed survey of the field of microwave application. It should, however, be understood (Horobin and Flemming, 1990) that a thorough understanding of the physics underlying the processes taking place in the microwave oven is a prerequisite for being able to fully utilize this intriguing addition to the equipment in the laboratory.

A recent review of the use of microwaves in tissue processing has been given by Kok and Boon (1990). Descriptions of practical applications of microwaves may be found in Boon and Kok (1987; 1988).

13 Tissue Processing: IV. Applied Fixation

H. Lyon, B. van Deurs, P.E. Høyer, P. Prentø, M. Møller

13.1 Reactions of Fixatives

Although reactions with proteins are of overriding importance in tissue fixation (cf. Sect. 12.2), an appreciation of the reactions involving other macromolecules and lipids is also important. A selection of these are surveyed in Table 13.1.

Table 13.1. The reactions of primary fixatives with nucleohistones, lipids, and carbohydrates.

Fixative	Nucleohistones	Lipids	Carbohydrates
Aldehydes e.g. formaldehyde and glutaraldehyde	nucleic acids not stabilized	triglycerides, cholesterol, and glycosphingosides are preserved without being fixed. Phospholipids are gradually decomposed	glycogen preserved without being fixed
Chromium trioxide	nucleic acids precipitated in an insoluble form	normal fixation time gives very little effect. Prolonged treatment leads to fixation	glycol groups oxidized to aldehydes
Potassium dichromate	nucleic acids not stabilized	fixed after prolonged treatment	no reaction
Mercuric chloride	no stabilization	complex formation with phospholipids	no reaction
Osmium tetroxide	extensive extraction	fixation takes place by reaction with double bonds	no reaction
Ethanol	nucleic acids precipitated	weak solvent effect	glycogen precipitated without being fixed
Acetic acid	nucleic acids precipitated, histones extracted	preserved without being fixed	no reaction
Picric acid	nucleic acids not stabilized, histones precipitated	no reaction	glycogen precipitated without being fixed

Figs. 13.1 and 13.2 show the processes that are thought to occur on exposure to the additive fixatives formaldehyde and mercuric chloride.

According to Horobin (1982), p.23, the effects of formaldehyde (Fig. 13.1) are largely reversed by washing in water. Puchtler and Meloan (1985), however, claim that only loosely bound formaldehyde is removed even after extensive washing (several hours). The residual bound formaldehyde is not dislodged by washing extended over weeks. Before accepting the latter view it should be remembered that the fixation process is time dependent (Sect.12.3.11). The resistance of bound formaldehyde to extraction is probably dependent on the extent of cross-linking (R–NH–CH_2–NH–R in Fig. 13.1). Clearly formaldehyde molecules participating in cross-links are less easily extracted by hydrolysis than those bound by single ester linkages (R–NH–CH_2OH in Fig. 13.1). Cross-links are formed progressively over a period of hours-days-weeks (Sect.12.3.11). The majority of the links formed in the early stages of fixation are not cross-links.

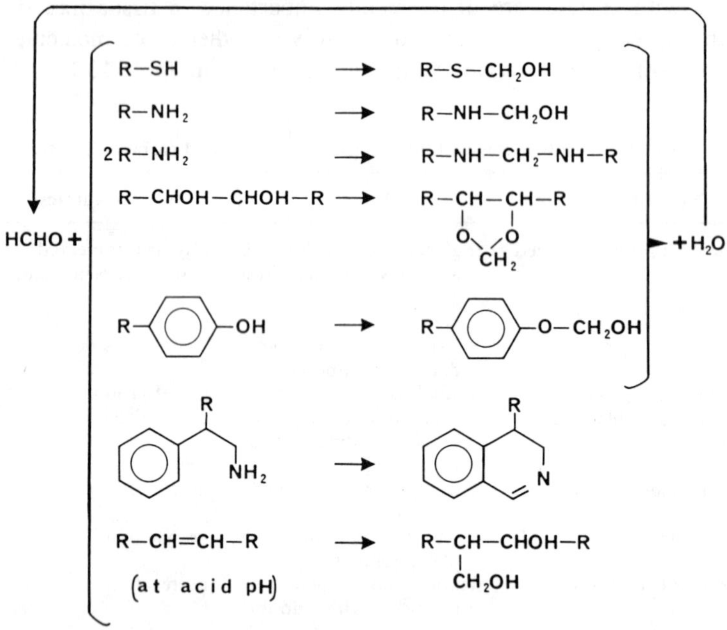

Fig. 13.1. Reactions involving formaldehyde (modified from Horobin, 1982). In the presence of excess water (washing) all the reactions leading to net production of water are fully reversible (cf. Table 12.4)

13.2 Fixation of Proteins

As already discussed, aldehydes fix proteins by forming cross-links between different reactive groups. The amino acid residues that can participate in such bonds are those possessing free amino groups, thiol groups, or aromatic ring systems. Whether an individual group will react depends on steric conditions (i.e. the tertiary and sec-

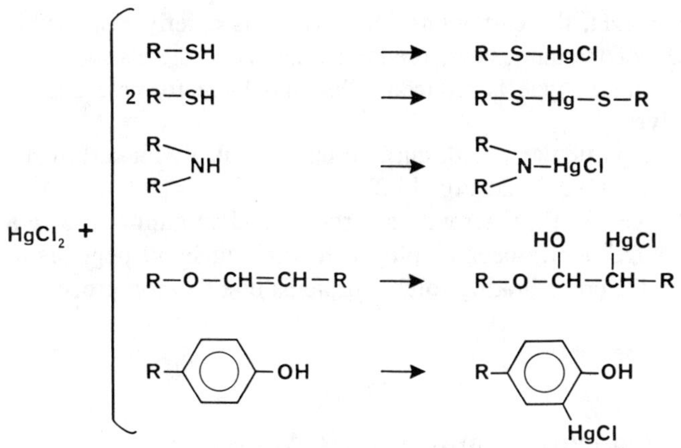

Fig. 13.2. Reactions involving mercuric chloride (modified after Horobin, 1982). All are irreversible (cf. Table 12.4)

ondary structure of the protein). As well as being time and concentration dependent, the protein fixing effect is also sensitive to the relative abilities of different aldehydes to form cross-links under the same conditions. For example, the reaction with formaldehyde is largely reversible and cross-links tend to be hydrolyzed with prolonged washing in water. In contrast, the cross-links formed after fixation with glutaraldehyde are considerably more stable.

In aqueous solutions of formaldehyde, a number of equilibria are established (Jones, 1973):

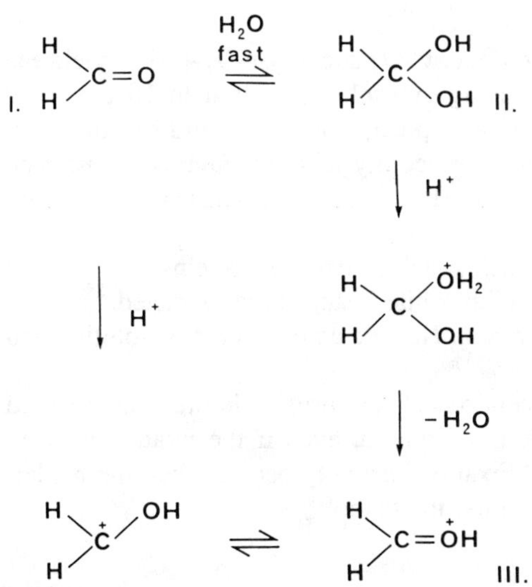

I = formaldehyde; II = methylene glycol; III = carbonium ion

At low pH, III, (CH_2+ OH, the carbonium ion), which is chiefly responsible for the fixative properties of formaldehyde, predominates. Thus the fixing activity in neutral, buffered formaldehyde is reduced. Glutaraldehyde forms a similar methylene glycol derivative.

The mercuric ion reacts particularly with thiol groups, but also with carboxylic and hydroxyl groups (see Sect.9.3.2 and Fig. 13.2).

Carbodiimides, $R-N=C=N-R'$, react with a carboxyl and an amino to give a peptide. This reaction has frequently been employed for linking small peptides to larger proteins but seldom for cross-linking soluble proteins in a fixation procedure (Pearse, 1980, p.107).

13.3 Fixation of Nucleoproteins and Nucleic Acids

Nucleic acids are precipitated at pH 2 (the pK_A of phosphoric acid), i.e. in acid fixatives. Nucleic acids are also precipitated in solvents that are less polar than water (e.g. alcohols and organic solvents including acetone, ether, and chloroform).

No reagents are able to fix nucleic acids in the proper sense of the word. DNA and especially RNA may be extensively extracted during the subsequent tissue processing. Treatment with chromic acid containing solutions yields the best stability against extraction. Chromic acid treatment, however, hinders the effect of RNase and is not widely used. Nucleosides, nucleotides including coenzymes and products of hydrolysis such as di-, tri-, and other oligonucleotides, are all extracted in aqueous solvents and during dehydration, while polynucleotides are retained.

Specific comments concerning:

a. *Primary fixatives*
    - *Solutions containing chromic acid* (i.e. CrO_3 and $K_2Cr_2O_7 + H^+$) yield the best nucleic acid fixation by producing an insoluble precipitate. Histones and acid proteins in the chromatin are also precipitated. It should be noted that in *neutral solutions*, the precipitating activity of *potassium dichromate* on nucleoproteins is *abolished* while the effect on nucleic acids is substantially reduced.
    - *Osmium tetroxide* does not fix nucleic acids or nucleoproteins.
    - *Acetic acid* precipitates nucleic acids while histones are extracted.
    - *Picric acid* precipitates nucleoprotein. Nucleic acids remain in solution and histones are strongly coagulated

b. *Mixtures of fixatives.* With few exceptions all the usual mixtures can be used when DNA and RNA demonstration is required, even if the fixative in itself does not precipitate nucleic acids. "Fixation" actually occurs when the nucleic acids are precipitated during dehydration in ethanol

The Individual Compound Fixatives.

3.6% w/v Formaldehyde Solution. When this fixative is applied for 12–24 hours, it offers a number of advantages for histochemical purposes. The non-histones in chromatin are fixed while the nucleohistones are *not* fixed, but are preserved *in situ* in the chromosomes. As the fixative is a non-electrolyte the bonds between nucleic acids and histones are not broken and there is little or no hydrolysis in the fixative. Some extraction of RNA and DNA does, however, take place in the peripheral areas of the sample.

Clarke's Fixative. Clarke's fixative (ethanol + acetic acid) gives reasonably good fixation of DNA as it has a strong coagulant effect on nucleoproteins and precipitates DNA. Nuclear morphology is well preserved.

Carnoy's Fixative. This fixative corresponds to Clarke's fixative with the addition of chloroform. The chloroform facilitates fixative penetration by dissolving tissue lipids. As with Clarke's fixative the nucleic acids remain sensitive to RNase and DNase treatment. It should be noted that both fixatives give a very coarsely coagulated cytoplasm with destruction of all organelles. The ribosomes together with their content of RNA are trapped within the protein net.

Flemming's Strong Fluid. Flemming's strong fluid (CrO_3, OsO_4, acetic acid) is a chromic acid containing solution (pH 1.4) and is therefore an excellent fixative for nucleoproteins. Nucleic acids tolerate subsequent processing very well and, even after treatment with hydrochloric acid (Feulgen), DNA is not extracted to a significant degree. The cytoplasm coagulates in a finely meshed network and there is no dislocation of the ribosomes. Flemming's strong fluid is well suited for histologic investigations of DNA and RNA but its somewhat drastic effects on other cell components preclude its use for finer histochemical investigations.

Zenker's Fixative. Zenker's fixative ($HgCl_2$, $K_2Cr_2O_7$, acetic acid) effectively precipitates nucleoproteins. The cytoplasm forms a rather delicate network and as RNA becomes relatively resistant to extraction during subsequent processing, the fixative is well suited for investigations concerned with cytoplasmic RNA content.

Lillie's AAF. Lillie's AAF (ethanol, acetic acid, formaldehyde) is an excellent fixative for nucleoproteins. It precipitates nucleic acids but the formaldehyde ensures that coagulation of the cytoplasm is far finer than with Clarke's fixative so that localization of RNA is much improved. RNA is, however, easily lost during the subsequent treatment.

Bouin's Fixative. Bouin's fixative (picric acid, formaldehyde, acetic acid) directly dissolves nucleic acids.

13.4 Fixation of Carbohydrates

This is difficult. To achieve reasonable results at all it is necessary to use different fixatives for the different kinds of carbohydrates.

Glycogen. The glycogens that occur in animal tissues show substantial differences in their degree of polymerization. The lower molecular weight varieties are both more soluble in aqueous media and more difficult to fix. Glycogens formed in the different glycogen storage diseases are generally of low molecular weight (Sect.31.10.8).

In extreme cases, poor fixation of glycogen leads to complete extraction. Slightly improved fixation leads to pronounced artifactual localization with the glycogen polarized according to the diffusion path of the fixative through the tissue. This so-called *streaming artifact*, is especially pronounced for fixatives containing ethanol used at room temperature.

The mechanism in "fixing" of glycogen is unknown but physical retention in a network of fixed protein is highly probable. With glycogens of low molecular weight it can be advantageous to coat the sections of fixed tissue with nitrocellulose prior to staining.

The majority of formaldehyde containing fixatives give poor retention of low molecular weight glycogens (e.g. in muscle cells and placenta) and improved results with higher molecular weight deposits (e.g. in liver). Bouin's fixative (picric acid, formaldehyde, acetic acid) at 4°C gives reasonably good preservation of glycogen but does not avoid the streaming artifact. Good results can be achieved using freeze substitution with Lison's "Gendre fluid" (ethanol, picric acid, formaldehyde, acetic acid) at −73°C. If small tissue blocks (less than 1 mm^3) are used comparable results can be obtained with absolute ethanol at −73°C. Finally, freeze-drying followed by fixation with formaldehyde gas can give good results in some tissues.

Glycoproteins. These macromolecules chiefly consist of protein and can therefore be fixed with all good protein fixatives.

Proteoglycans. Proteoglycans are very highly soluble in water and to some degree also in 70% v/v ethanol. They can be precipitated with absolute ethanol but remain partly soluble in water.

Proteoglycans can be fixed with lead salts (Lillie and Fullmer, 1976, p.48).

The introduction of aqueous formaldehyde solution with added cetylpyridinium chloride (CPC) (Williams and Jackson, 1956), however, represented a considerable advance. CPC binds to the carboxylic and sulphate groups in proteoglycans leading to a precipitation effect which is combined with the protein fixing effect of formaldehyde. 50% v/v ethanol with added 5-amino-acridine chloride (Williams and Jackson, 1956) also fixes proteoglycans by a similar mechanism. CPC containing fixatives should not be stored or used below 0°C (cf. Sect.12.3.10).

It should, however, be noted that after a study of the effect of different fixatives on Alcian Blue staining of glycosaminoglycans, Tuckett and Morriss-Kay (1988) discouraged the use of CPC. Due to its detergent effect CPC causes cell membrane disruption and vesicle formation. At the light microscopic level, stained cell membrane-associated glycosaminoglycans appear dissociated from the cell surface in the extracellular matrix. In contrast, in an electron microscopic study of extracellular matrix using high iron diamine (HID) (Sect.22.2.2), Albedi *et al.* (1988) noted improved preservation of sulphated glycosaminoglycans at the ultrastructural level when CPC was added to the fixative.

13.5 Fixation of Lipids

Sections of fresh frozen tissue should be used for lipid histochemical investigations. The reactions should be performed without delay after cutting as auto-oxidation of the fatty acids rapidly leads to significant changes in the physico-chemical properties of the lipids. Methods relying on oxidation of double bonds (OsO_4, UV-Schiff) as well as the plasmal reaction (and essential controls) require absolutely fresh, unfixed sections. Unfixed sections are also necessary for experiments involving differential extraction of lipids. If same-day processing is not possible then changes in the lipids can be minimized by keeping the sections sealed against oxygen and light in deep freeze.

Several lipid reactions necessitate a preceding stabilization procedure. This should be performed with formol calcium, possibly with added cadmium chloride, and should be as short as possible.

In certain lipid reactions fixation is an integrated part of the method. This is especially true for chromation (Sect.19.6.7) and osmium tetroxide (Sect.19.6.5) methods. Following chromation ($K_2Cr_2O_7/CaCl_2$, pH 3.5–4) the greater part of the lipids become insoluble in organic solvents, and a chromated piece of tissue can then be transferred to paraffin, epoxy resin, or methacrylate without loss of lipid. In a similar way osmium tetroxide ($OsO_4/CaCl_2$, pH 7–7.5) gives general fixation of lipids and permits epoxy resin or methacrylate embedding of tissue if the lipid content is not too high. Osmium fixation renders paraffin infiltration more difficult, but can be used in combination with other fixatives before paraffin embedding. It should be noted that free osmium tetroxide is reduced to osmium black by ethanol.

Under controlled conditions, all the fixatives mentioned in Table 13.2 give excellent histological preservation. In addition, the aldehyde fixatives produce virtually complete dissociation of phospholipoprotein complexes and emphasize the negative charge of the phosphoryl group. They therefore usually increase stainability with Sudan Black, Nile Blue, and Luxol Fast Blue.

Aldehyde fixatives for use in lipid histochemistry should be of the highest purity to avoid undesirable effects. For preference, formaldehyde solutions should be prepared from paraformaldehyde as technical formalin contains both formic acid and methanol which can increase the hydrolysis and extraction of lipids during fixation. Buffered formaldehyde should be used with added calcium or cadmium chloride as these immobilize the phospholipids. Salt concentrations above 0.1 mol/l

Table 13.2. Fixatives suitable for lipid histochemistry.

Fixative	Lipids that can be fixed	Reactive groups	Lipid demonstration
Osmium tetroxide	most *unsaturated lipids*, not cholesterol	alkene	reactive lipids black, oxidative methods for alkene blocked, lipid extraction impossible
Potassium dichromate	all unsaturated lipids, not cholesterol	alkene	oxidative methods for alkene blocked, extraction of reactive lipid impossible
	phospholipids sphingomyelin and lecithin	phosphoryl choline	Nile Blue impossible can be demonstrated with acid Haematein, chromated lipids can, dependent on type, be demonstrated with Oil Red O or Sudan Black B at 60°C
Formaldehyde	especially phosphatidylethanol amine	amino	lipid extraction unreliable
Glutaraldehyde	all N-containing lipids	amino and choline	phospholipid extraction impossible, carbonyl-lipids cannot be demonstrated
Acrolein	all N-containing lipids	amino and choline	phospholipid extraction impossible, carbonyl-lipids cannot be demonstrated, demonstration of glycolipids unreliable

should be avoided as phospholipids become hydrophobic under these conditions and this would interfere with procedures such as the OTAN reaction. The concentration of alkali metal ions should be kept as low as possible as these have an extractant effect on phospholipids (Sect.19.2).

13.6 Fixation of Enzymes

As enzymes are proteins they are affected by all fixatives but, if activity is to be demonstrated, then the active site (Sect.23.1.2) should remain intact. Total preservation using any chemical fixative must be considered impossible and, for all quantitative and many qualitative investigations (particularly those concerned with oxidoreductases) cryostat sections of fresh frozen tissue are essential (Sects.11.1 and 23.2.1). Fixation has, however, been used to avoid diffusion of soluble enzymes (Sects.24.2 and 25.1.1).

Currently formaldehyde is widely used for the fixation of hydrolases. The alternatives include formol calcium and paraformaldehyde in either dilute (0.05 mol/l) phosphate or cacodylate buffer and these should be used at 0–4°C. The choice of buffer depends on the enzyme to be demonstrated (e.g. β-glucuronidase is inhibited by cacodylate buffer). If formol calcium is used, this should be prepared from paraformaldehyde as formalin contains sufficient methanol to denature some enzymes.

The ability of different hydrolases to tolerate fixation is highly variable, so established successful protocols for the pretreatment of individual enzymes should be followed closely. Particular care should be taken not to exceed the recommended fixation time.

The sample should be washed thoroughly after fixation using either buffer or better cold acacia gum/sucrose. This halts fixation and removes loosely bound formaldehyde, thereby reactivating a considerable proportion of the formaldehyde inactivated enzyme.

13.7 Fixation of Inorganic Components

The problems associated with correct localization of inorganic tissue components make the use of fresh frozen cryostat sections preferable (Sects. 11.2.2 and 11.3.1). Freeze dried (Sect. 11.3.2) or freeze substituted (Sect. 11.3.3) material can also be used. If chemically fixed material is to be used then neutral buffered formalin is suitable. Acid fixatives must be avoided as they inevitably lead to excessive extraction and displacement of immobile salts such as Fe^{3+} in haemosiderin (Sects. 17.1 and 17.4).

13.8 Fixation of Pigments

As most pigments are insoluble in water there is little problem with localization. The biogenic amines do, however, constitute an exception as they are easily diffusible. Aqueous formaldehyde is reasonably effective at fixing serotonin in enterochromaffin cells (Sect. 30.2.1), and the amine remains demonstrable after dehydration and paraffin embedding. Adrenaline and noradrenaline are not normally demonstrable in the adrenal medullary chromaffin cells after this procedure. As a rule, one needs to use freeze drying followed by treatment with formaldehyde gas for this purpose (Sect. 30.2.1). With this method the biogenic amines can also be demonstrated in mast cells, ganglion cells, and nerve terminals.

Noradrenaline and, to some extent, adrenaline can be fixed by sequential fixation in neutral buffered glutaraldehyde solution followed by potassium dichromate at pH 4. The chromaffin reaction (Sect. 18.4.5) which consists of an oxidation of biogenic amines to coloured, insoluble products, is usually performed with a solution of potassium dichromate alone or with added potassium chromate. This produces simultaneous fixation and demonstration of biogenic amines in chromaffin and enterochromaffin cells.

14 Tissue Processing: V. Embedding

H. Lyon, B. van Deurs, P.E. Høyer, P. Prentø, M. Møller

14.1 Embedding of Tissue

Embedding of tissue enables it to withstand sectioning. One simple approach is to freeze the material and section it in a cryostat (Sects.11.2.2 and 11.3.1). The formation of ice has an immobilizing effect and also creates an embedding medium. Embedding in its more traditional sense implies the use of gelatin, paraffin, or plastic materials.

Many of the substances used in these procedures pose health hazards. This must be taken into account when planning working practice by giving due attention to both the procedures themselves and the disposal of waste in accordance with legal and regulatory guidelines. The importance of using a fume cupboard for organic solvents cannot be overemphasized. This should be placed in well-ventilated accommodation with careful attention to whether the expected fumes are heavier or lighter than air, and good facilities for the collection of waste in suitable receptacles.

14.2 Dehydration

In order to embed a tissue in a hydrophobic medium its water content must be substituted with a suitable organic solvent. This process is termed dehydration.

Dehydration can be effected by two different principles:

1. *Dilution.* This is frequently achieved by using ethanol. The tissue is successively placed in a series of solutions containing increasing concentrations of the dehydrating agent. The use of an extracting process leads to a variable degree of tissue shrinkage and extraction of other tissue components (differential extraction). The duration of the dehydration procedure should be kept to a minimum in order to reduce these effects. In contrast, the time for each dehydrating step should be sufficiently long to ensure a gradual substitution of water with solvent. Excessively rapid dehydration can lead to violent osmotic changes which may produce pronounced structural disturbances. The optimum dehydration time depends mainly on the size and nature of the tissue sample. Incomplete dehydration is a common cause of tissue shrinkage and hardening

H. Lyon (Ed.)
Theory and Strategy in Histochemistry

in paraffin blocks after sectioning and storage (Sect.14.5). This problem may be avoided by ensuring a sufficient number of changes of undiluted dehydrating agent prior to the clearing process (Sect.14.3)

2. *Chemical binding of water.* This can be achieved with a compound such as 2,2-dimethoxypropane

14.2.1 Notes on the Individual Dehydrating Agents

Ethanol. Mechanism: Water extraction. Pronounced hardening effect, especially with larger pieces of tissue. Shrinkage of the tissue is unavoidable even with optimum fixation. The shrinkage occurs at concentrations between 50 and 90% v/v.

Methanol. As for ethanol.

2-Propanol. Mechanism: Water extraction. Considerably less hardening effect than ethanol.

Acetone. Mechanism: Water extraction. Less hardening effect than ethanol, but can make the tissue brittle—well suited for small pieces of tissue but high inflammability makes it less suitable for use in automatic tissue processing machines using volumes of one litre.

1,2-Ethanediol and 1,2-Propanediol. Mechanism: Water extraction. Minimal hardening effect and shrinkage.

2-Ethoxyethanol, Ethyl Cellosolve. Mechanism: Dilution effect. Considerably less shrinkage and hardening than ethanol. Differential extraction of tissue components may be a problem.

Propylene Oxide, 1:2-Epoxypropane. Mechanism: Dilution effect. Can be used as a dehydrating agent, but in practice is nearly always used as a clearing agent in connection with tissue processing for electron microscopy (Sect.14.4.2).

1,4-Diethylene Dioxide or Dioxane. Dioxane has been recommended as a dehydrating agent which has the advantage that it is miscible with both water and paraffin. It is, however, rather difficult to work with technically and its use is discouraged due to its pronounced toxicity.

2,2-Dimethoxypropane. Mechanism: Chemical effect. Dimethoxypropane is insoluble in water and therefore reacts primarily with a narrow zone of tissue. Here water reacts with dimethoxypropane forming methanol and acetone. The process is progressive, and a 2–3 mm thick piece of tissue can be completely dehydrated in around 15 min.

14.3 Clearing

The use of a clearing agent is necessary when the dehydrating agent (e.g. ethanol) is not miscible with the embedding medium (e.g. paraffin). A clearing agent should therefore be miscible with both the dehydrating agent and the embedding agent. There are numerous organic liquids with appropriate properties in this context. Many of these have a refractive index corresponding to that of protein and this may lead to the tissue becoming transparent, hence the name clearing.

If ethanol has been used as the dehydrating agent and the embedding agent is to be paraffin, the choice of a suitable clearing agent will depend on the following factors: viscosity, boiling point, vapour pressure, miscibility with ethanol and paraffin, effect on section quality, and toxicity.

The majority of clearing agents are inflammable liquids which require considerable caution, especially when automatic tissue processing machines are used. On transfer to hot paraffin, liquids with low boiling points are usually removed more rapidly than liquids with high boiling points and, where applicable, this factor may be accentuated by vacuum. An important exception is chloroform which has a low boiling point but is removed considerably more slowly than xylene. The viscosity also influences the speed with which the clearing agent penetrates the tissue.

Table 14.1 details important properties of several clearing agents.

Table 14.1. Refractive index, boiling point, and inflammability of clearing agents.

Name	Refractive index	Boiling point (°C)	Inflammability
Xylene	1.50	140	+
Toluene	1.50	111	+
Chloroform	1.45	61	
Benzene	1.50	80	+
Tetrachloromethane	1.46	77	
Petroleum	1.44	196–252	+
Amyl acetate	1.40	142	
Methyl benzoate	1.51	200	
Methyl salicylate	1.54	223	
Cedarwood oil	1.51	260	
Clove oil	1.48	200	

+ = low flash point.

Notes on Individual Clearing Agents. Xylene. Usually quite satisfactory if the thickness of the tissue slices does not exceed 3–4 mm. The immersion time of the tissue in xylene must not be prolonged if the block is to have a reasonable consistency. Xylene is inflammable and toxic.

Toluene. Toluene has similar properties to xylene. Tissue is harmed less by prolonged immersion. Toluene is inflammable and toxic.

Chloroform. This acts more slowly than xylene or toluene, but causes less brittleness. Thicker blocks of tissue (up to 1 cm) can be used provided sufficient time is allowed for both clearing and replacement of chloroform with paraffin. The slightest trace of chloroform remaining after embedding leads to difficulties in sectioning. Chloroform vapours are not inflammable but, on heating, the toxic gas phosgene is liberated.

Benzene. This substance has properties similar to xylene but should be avoided due to its carcinogenic properties.

Tetrachloromethane. This compound has similar properties to chloroform but is considerably more toxic. It also produces phosgene on oxidation in air.

Liquid Paraffin. Good results can be achieved with liquid paraffin as the clearing agent, although the processing time will vary according to the composition of the mixture of hydrocarbons. It is less inflammable than the agents mentioned above and is therefore comparatively less dangerous.

Petroleum. This can be used as a clearing agent if the tissue does not contain too much lipid. Particular advantages are reduced fire hazard and toxicity.

Amyl Acetate. This agent is seldom used, but removes ethanol fairly quickly and is itself removed with reasonable speed by melted paraffin. It has a rather unpleasant smell and is expensive.

Methyl Benzoate. This clears tissue at a moderate rate and causes very little change in the tissue. Due to its penetrating smell it is, however, not used very frequently, but can be of value when making museum preparations.

Methyl Salicylate. Properties as for methyl benzoate.

Cedarwood Oil. This slowly acting agent causes little hardening and tissue shrinkage. The need for longer processing time in the melted paraffin, however, outweighs its otherwise mild effects. Due to its rather low volatility cedarwood oil is less effectively removed by vacuum treatment than the previously mentioned clearing agents. It is expensive.

Clove Oil. This has similar effects to those of cedarwood oil but is even more expensive.

14.4 Embedding Media

An ideal embedding medium would fulfil the following requirements:

1. A suitable melting point relative to the temperatures during embedding and sectioning
2. Uniform infiltration of all tissue components
3. No chemical reactivity with the tissue
4. Sufficient hardness after infiltration to make cutting of sections possible
5. Easy removal after sectioning or, if this is not possible permeability to staining reagents and a refractive index corresponding to that of proteins

14.4.1 Embedding for Light Microscopy

Water. The use of water as an embedding medium is discussed in Sect.11.2.2.

Gelatin. The major advantage of using gelatin as an embedding medium is that dehydration is not necessary. As a result, even if they are unfixed, the majority of lipids, are preserved within the tissue. On the negative side, it should be noted that gelatin is generally not removed after sectioning and is therefore stained by the majority of staining procedures. Moreover, sections thinner than 5 μm are difficult to prepare and do not form ribbons.

Blocks may be embedded in gelatin directly after fixation but the fixative must be washed out first. This is particularly important if, as is the case with aldehydes and osmium tetroxide, gelatin itself is fixed by the reagent used.

Gelatin embedding can be used to advantage when the tissue is very brittle.

Agar. An aqueous solution of agar can be valuable with very brittle tissues or with exudates (e.g. examination of contents of lipid and oil in pulmonary exudates).

Paraffin. Although there are other embedding media that have melting points and other properties that cause less damage to tissue, paraffin is still the most frequently used. This is largely due to the ease with which a large number of tissue blocks can be prepared in a relatively short time (see water soluble resins Sect.14.4.2). In addition, sectioning and subsequently staining pose fewer difficulties than most of the other media amongst which paraffin is virtually the cheapest.

Paraffin is a mixture of hydrocarbons with overall properties that vary considerably. The melting point is in the region of 40–70°C. At any particular temperature below the melting point, paraffin with high melting point will be harder than a corresponding paraffin with lower melting point. In order to prepare ribbons of sections on a rotary microtome it is necessary to choose a paraffin that is reasonably hard at room temperature. In general, a melting point between 54–58°C gives reasonable results.

Prolonged heating to temperatures around 10–20°C above the melting point leads to evaporation of some of the fractions with lower melting points. The paraffin develops a more soap-like consistency and the previously crystalline structure more or less disappears. Deliberate treatment of paraffin in this way has been recommended in order to facilitate production of ribbons. A number of different substances have also been added in order to modify the consistency. The aim has been partly to increase the hardness of the paraffin so that thinner sections could be cut, partly to give the necessary support to the tissue, and partly to improve formation of ribbons. Many compounds have been suggested but mention of beeswax, which should result in better adherence between the sections, will suffice here. Latterly the addition of selected plastics to commercial paraffins in order to increase hardness has become common.

Celloidin. When embedding in celloidin (nitrocellulose) the fixed tissue is dehydrated with ethanol and then gradually infiltrated with nitrocellulose dissolved in a mixture of ether and ethanol. This infiltration takes a long time (days to months) depending on the size of the tissue block. The nitrocellulose is transformed into a dense gel by treatment with chloroform in which it is insoluble. The method, which is rarely used today, can be applied to very large pieces of tissue without appreciable shrinkage. Celloidin embedded material is usually cut in thick sections (25 μm).

14.4.2 Embedding for Electron Microscopy

This places special demands on the embedding medium as very thin sections (20–60 nm) are necessary in order to benefit from the resolving power of the electron microscope. Such thin sections are made possible partly by reducing the section area and partly by embedding in considerably harder materials than those used for light microscopy. The embedding media used are plastics which may be water soluble or water insoluble.

Plastic. Plastic is a *polymer*, which, in the present context, is also termed a *resin.* The majority of the water insoluble media fall into three main groups:

1. Acrylic
2. Polyester
3. Epoxy

while the water soluble media are derivatives of epoxy or methacrylate. The tissue is infiltrated with a monomer which is transformed into the plastic (polymer) on the addition of a hardener and an accelerator. The monomer should ideally have a low molecular weight, low viscosity, be stable, and polymerize uniformly (preferably at room temperature) with a minimum of shrinkage.

The resultant polymer must be transparent, have a very small grain size, and adhere well to tissue so that the latter is not torn out during sectioning. Further, the polymer should, as far as possible, be indifferent to the heating effect of the electron beam. The monomer should not influence or alter the tissue chemically and, with regard to possible histochemical reactions, it is advantageous for the polymer to be water soluble. Last, but not least, the embedding material should not be toxic. Unfortunately none of the embedding materials for electron microscopy meet all of these demands and none of the classical media used for light microscopy (see Sect.14.4.1) are suitable.

Methacrylate. Methacrylate (butylmethacrylate) was the first compound successfully used to embed material for electron microscopy. The monomer has a low molecular weight and viscosity and rapidly infiltrates the fixed, dehydrated tissue at room temperature. The polymer is easy to section and the sections are stainable so that ultrathin and 1 μm sections can be used for light microscopy. Unfortunately, considerable damage to the tissue frequently occurs during polymerization. Furthermore, polymerized methacrylate is very sensitive to the electron beam, and a heat induced depolymerization takes place in the electron microscope which leads to very pronounced swelling.

Polyester Resins. These are fairly good embedding media for electron microscopy but they are relatively viscous before polymerization. Acetone is preferred to ethanol for dehydration due to the low hydrophilicity of the medium.

Epoxy Resins. The epoxy resins (Epon 812 and others) comprise a large group of very frequently used embedding media with a wide range of properties (molecular weight, viscosity, hydrophilicity, toxicity). They are all di-epoxides, i.e. they have two epoxy rings.

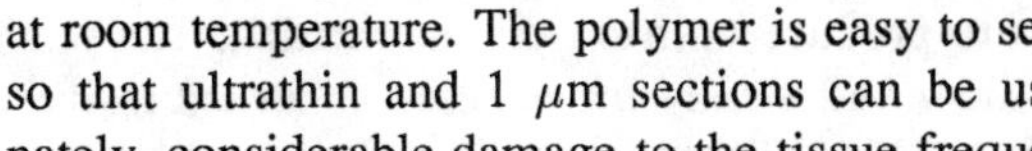

The polymerization occurs in the presence of an ample amount of a hardener (e.g. an organic acid anhydride, amide, or amine) and an "accelerator" (frequently a strong organic base, e.g. a tertiary amine such as dimethylaminoethanol (DMAE) or tri(dimethylaminomethyl)phenol (DMP-30)). By changing the ratio between the different hardening agents it is possible to vary the hardness of the block. Good sectioning properties are obtained by individually matching the hardness of the embedding medium and the tissue.

Epon 812, which is the most frequently used resin, is sufficiently polar to be soluble in ethanol. In principle, it should therefore be possible to pass directly from ethanol to the resin after infiltrating a mixture of ethanol and resin. It is, however, preferable to use propylene oxide (epoxypropane) as clearing agent first as this allows a shorter infiltration time and other advantages. A typical embedding procedure will thus comprise dehydration in ethanol of increasing strength up to absolute, then a few changes in propylene oxide, followed by propylene oxide: Epon, in the ratio 1:1 and finally infiltration in pure Epon in a thermostat oven at 60°C for 18–24 hours. It should be noted that only very small pieces of tissue are used (Sect.11.2.2).

In comparison with methacrylate which is very sensitive, the polyester resins, and even more so the epoxy resins, are only slightly susceptible to the electron beam. However, a certain loss of mass must still be anticipated (40–50% for methacrylate, approximately 5–20% for epoxy resins) as a result of radiation.

All resins used in electron microscopy lead to a degree of shrinkage during polymerization. On polymerization, methacrylate shrinks the most, the polyesters somewhat less, and the epoxy resins least. The relatively high viscosity of Epon 812 (Table 14.2) frequently leads to problems in achieving a sufficiently homogeneous mixture of the components to produce uniform polymerization. The resin needs to be stirred with relatively slow and steady movements as an excessively vigorous approach can introduce air bubbles. These interfere with embedding and can, possibly by oxidation, inactivate the accelerator.

Table 14.2. Viscosity (mPa s = milli Pascal second; Pa s = $kg\,m^{-1}\,s^{-1}$ at 25°C) for some epoxy resins.

Resin	Viscosity
Araldite	2500
Epon 812 (A:B = 1:1)	1400
Vestopal W	900
Vinylcyclohexene dioxide (VCD; Spurr Plastic)	60
Diglycidyl ether (DGE)	23
1,2,7,8-diepoxyoctane (DEO)	20.5

Several newer epoxy resins with low viscosity have been developed (VCD, DGE, DEO, see Table 14.2). These resins are well suited for the infiltration of larger pieces of tissue and they are also easy to cut on the ultramicrotome.

Strict regulations apply to all work with epoxy materials and their waste products and these should be closely adhered to. The health hazard associated with epoxy resins decreases with polymerization providing the components have been used in well balanced amounts. Allergic reactions can, however, be elicited by hardened material (try to avoid too many shavings during sectioning). The aromatic amines which comprise part of the hardener are particularly toxic.

Water Soluble Resins. Finally, the water soluble resins must be considered. These are used to avoid extraction of certain substances, such as lipids, during dehydration. They are also used when resins with hydrophilic properties which permit histochemical staining, particularly enzymatic reactions, are required. Examples of such water soluble resins include glycolmethacrylate (2-hydroxyethyl methacrylate), the epoxy products Durcupan and Aquon, and the relatively new products JB-4 (Polysciences) and EFL-67 (Serva). It should, however, be noted that at least some of these resins (glycolmethacrylate) are themselves very strong organic solvents.

There is little doubt that the water soluble resins will be much used in the future (plastic sections), as, with better preservation of structure than paraffin, they enable performance of light microscopic staining methods and electron microscopy of ultrathin sections on a single tissue block. This is clearly very attractive to areas such as pathology.

14.5 Storage of Embedded Material

Most embedding media will either crystallize or harden further on storage. The process is accelerated by heat and light, and, in paraffin, by the presence of reagents such as water or clearing solvents. Tissue blocks are therefore often too hard or too brittle if sectioning is delayed. Dry and cool storage in darkness is generally recommended. Paraffin blocks which *have been sectioned* should be sealed by dipping the cut surface in melted paraffin.

15 Tissue Processing: VI. Hard Tissues

H. Lyon

The hard tissues, which include bones and teeth, pose special problems in histological technique due to their content of crystals of mineral salts. Bone may be divided into dense cortical bone and thc fine network which forms cancellous bone.

The organic matrix in bone tissue and in two of the dental tissues, cement and dentine, is predominantly composed of collagen arranged in parallel bands and lamellae. Collagen is absent from enamel, the third hard tissue of teeth, and an incompletely characterized protein is found instead.

Crystals of sparingly soluble mineral salts are deposited on and in the collagen fibres. Based on dry weight, the inorganic component may be estimated to comprise 60–75% in bone and 95% in enamel with around 15% of the mineral salts as calcium carbonate ($CaCO_3$) and 85% as hydroxyapatite crystals ($Ca_{10}(PO_4)_6(OH)_2$). Within the hydroxyapatite lattice a small proportion of calcium atoms are substituted by Mg and Sr as well as by Li, Na and K, while some of the hydroxyl groups are substituted by F and Cl.

Preparation Technique. As ever, the choice of technique depends on the aim of the investigation. A summary of the different possibilities is given in Figs. 15.1-3.

15.1 Specimen Types

Bone specimens can be:

a. Biopsies, frequently from the iliac crest
b. Amputation specimens (whole or part of an extremity)
c. Resection specimens (include part of a bone with attached soft tissues)
d. Autopsy specimens

15.2 Fixation

Neutral buffered formaldehyde is suitable for the majority of procedures.

H. Lyon (Ed.)
Theory and Strategy in Histochemistry

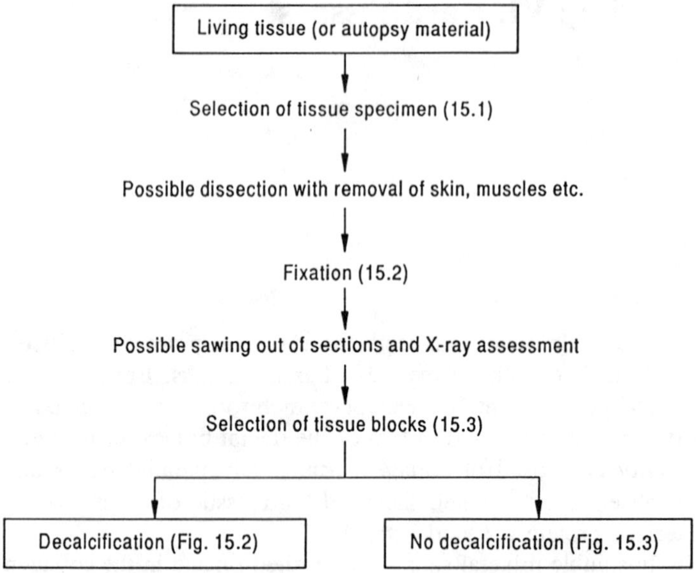

Fig. 15.1. Initial procedure for the preparation of sections of hard tissue for microscopy (references to sections in this chapter)

15.3 Selection of Tissue Blocks for Further Preparation

It is often useful to saw biopsies from the iliac crest into two pieces, decalcify one and prepare ground sections from the other. With large fragments, X-ray imaging is invaluable in defining the nature and size of lesions and in the selection of suitable tissue blocks for further processing.

15.4 Decalcification

Preparation of satisfactory paraffin, celloidin, or frozen sections from hard tissue requires prior removal of the mineral salts. This is achieved by treating with reagents which react with calcium. At low pH the following process takes place:

$$CaCO_3 + 2H^+ \rightarrow Ca^{2+} + H_2O + CO_2$$

Strong acids (e.g. HCl) rapidly lead to coarse artifacts, while weak acids (e.g. formic acid or chromium (III) sulphate) produce far less artifactual change.

The mildest decalcification can be achieved using chelating agents (cf. Sects.7.1 and 17.5.2) such as disodium *e*thylene*d*iamine*te*tr*a*cetate (EDTA). Calcium ions are so strongly bound to EDTA that they cannot participate in other chemical reactions. As the Ca-EDTA complex is water soluble the process acts as a continual removal of Ca^{2+} from the solution.

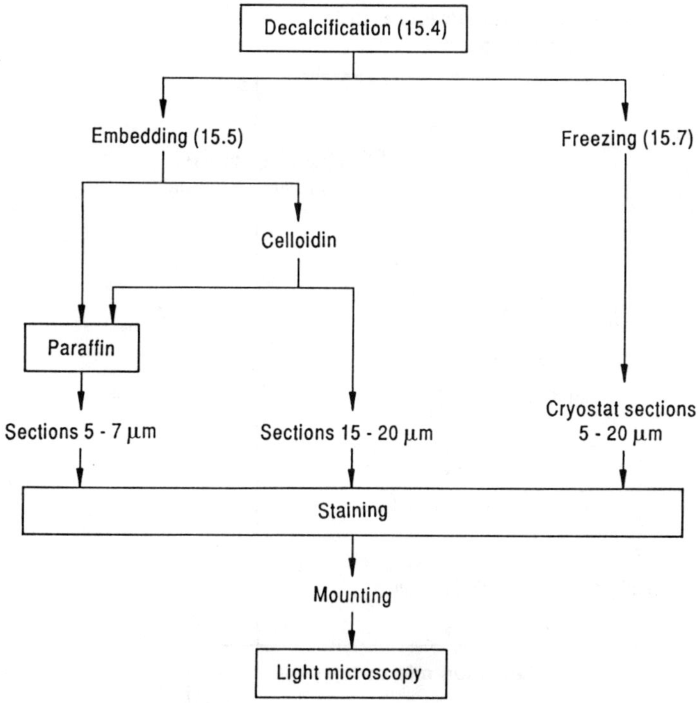

Fig. 15.2. Processing of decalcified hard tissue.

This kind of decalcification can take place at a physiological pH and, in contrast to acid decalcification, it conserves nucleic acids. A survey of the properties of decalcifying agents is given in Table 15.1.

15.4.1 Choice of Decalcifying Agent

The following factors should be considered when choosing a decalcifying agent:

1. The need for rapid processing (e.g. in diagnostic work)
2. The degree of mineralization of the tissue
3. The aim of the investigation and the proposed staining methods

The faster the decalcification (i.e. the stronger the acid) the more decalcification will influence the result of subsequent staining. Extraction of nucleic acids, especially RNA, may be extensive. For example, chromatin is far more weakly stained with metal complex dyes and cationic dyes in tissue decalcified with trichloroacetic acid than in more gently decalcified tissue.

Investigations into the enzymatic activity of the cells in hard tissue are impossible on acid decalcified tissue but they can be performed on EDTA decalcified tissue. Some guidelines for the selection of decalcifying agents are given in Table 15.2.

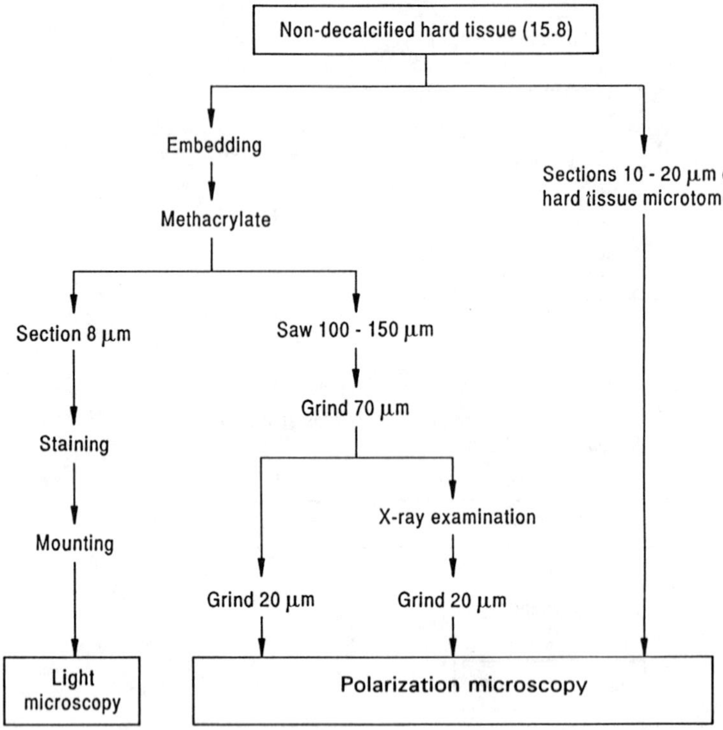

Fig. 15.3. Procedure for the preparation of sections from non-decalcified hard tissue.

Table 15.1. Decalcifying agents, concentration, relative decalcification time, and artifact formation.

Decalcifying agent	Type	Concentration %	Relative decalcification time	Artifacts: CO_2-formation	Artifacts: Precipitates
Nitric acid	strong acid	5–10	fast	+ +	none
Hydrochloric acid	strong acid	5–10	fast	+ +	none
Sulphuric acid	strong acid	5–10	very slow	+ +	$CaSO_4$
Trichloro-acetic acid	strong acid	5–10	fast	+ +	none
Phosphoric acid	moderately strong acid	5–10	slow	+	$Ca_3(PO_4)_2$
Sulphurous acid	weak acid	5–10	moderate	+	$CaSO_4$
Picric acid	weak acid	5–10	slow	+	none
Formic acid	weak acid	5–10	slow	+	none
Acetic acid	weak acid	5–10	very slow	+	swelling
Chromium(III) sulphate	weak acid	0.5	slow	+	none
EDTA	chelating	10–15	very slow	none	none

+ + – substantial, + – slight.

Table 15.2. Selection of decalcifying agents.

Purpose	Decalcifying agent	Time
Fast decalcification of acute biopsies with slight mineralization	nitric acid	2 up to a maximum of 48 h
Diagnostic material where speed is not essential, excepting cortical bone and enamel	formic acid	1–10 days
Cortical bone and teeth, especially enamel	chromium(III) sulphate	5–15 days
Demanding morphologic and histochemical purposes	EDTA	2 days to 8 weeks

The composition of a number of commercial decalcifying agents is unknown, but from the speed of decalcification it would seem probable that an important ingredient in these compounds is a strong acid.

15.4.2 Factors Influencing Speed of Decalcification

The decalcification process is influenced by:

1. Concentration of the decalcifying agent: Generally a higher concentration leads to faster decalcification, but also to more pronounced tissue damage
2. Temperature: Higher temperature gives faster decalcification, but with acids this also leads to an increase in artifacts
3. Amount of fluid: An adequate supply of protons or EDTA ions is essential to the decalcification process so at least 100 ml of reagent is allowed per gram of tissue
4. Change of fluid: If sufficient fluid is allowed (see point 3) daily changes serve no purpose
5. Movement in the decalcifying agent: Manual shaking a couple of times every day gives just as good results as continual stirring

15.4.3 Determination of the End Point for Decalcification

Several methods for determining the end point of the decalcification process are available:

1. Mechanical methods, in which the content of calcium salts is estimated on the basis of the mechanical properties of the specimen (hardness, flexibility), are not recommended. These methods are very unsatisfactory partly because they induce artifacts in the tissue and partly because small, central, undecalcified areas are easily overlooked
2. Chemical methods involve testing for the presence of Ca^{2+} in freshly changed decalcifying fluid. Examples include oxalate and the fluorescent product which arises on the addition of Calcein (Fluorescein bismethylene iminodiacetate)

3. Radiographic (X-ray) methods provide much the quickest and safest means of controlling the course of decalcification. Fully decalcified specimens give a homogeneous shadow when a suitable voltage, current, and exposure time are used

The frequency at which the course of decalcification should be assessed depends on the degree of mineralization and the decalcifying agent. Using nitric acid or hydrochloric acid on a biopsy, a test after 6–18 hours, depending on the expected amount of mineral, is usually sufficient. With a weak acid, such as formic acid, a single daily test is appropriate, while with EDTA the interval should be extended to a week.

15.5 Further Preparation Following Decalcification

Surplus acid must be removed by quickly washing in tap water before any further fixation or dehydration procedures are undertaken. If EDTA has been used the sample should be placed in a formaldehyde solution with added NaCl for 12 hours before transfer to 70% v/v ethanol. This prevents precipitation of EDTA in the tissue by ethanol. Careful dehydration and infiltration with embedding medium are critical in the preparation of blocks from hard tissue.

15.6 Staining of Decalcified Sections

After decalcification with EDTA most staining methods present no problems. This is also true for acid treatment providing this is stopped promptly when decalcification is complete. If, however, acid treatment is continued beyond the optimum time, extraction of nucleic acids occurs (cf. Table 15.3) and modification of methods involving the use of metal complex dyes and cationic dyes becomes necessary. For example, a standard Haematoxylin-Eosin stain yields strong red cytoplasmic staining but no nuclear staining. Table 15.3 surveys staining methods used in the investigation of hard tissue.

15.7 Frozen Sections

Accepting a risk of chipping the knife, firm tissue which feels only slightly gritty can be cut when frozen without preceding decalcification. If frozen sections are to be made from decalcified tissue the sample must first be washed thoroughly in tap water to avoid damaging the microtome and the knife.

Table 15.3. Staining methods and hard tissue.

Staining method	Demonstrates	Comments
Haematoxylin-Eosin	general morphology	see above
van Gieson	connective tissue	not very suitable
Trichrome methods	connective tissue	better than van Gieson, but not particularly good on paraffin sections; better on plastic or celloidin
Toluidine Blue	cartilage	excellent metachromasia reliably indicates cartilage
Alcian Blue-CEC 0.4–0.5 mol/l $MgCl_2$	cartilage	almost specific
Periodic acid Schiff	newly formed bone and cartilage undergoing calcification; glycogen in young osteoblasts	prolonged treatment in acid should be avoided
Oxidation-argyrophilia	reticulin	particularly valuable in primary and secondary bone tumours; N.B. sections have a tendency to detach from the slide during the reaction
Schmorl's Picrothionin	bone canaliculi	celloidin or plastic sections are recommended; often capricious probably due to variations in the quality of Thionin
Polarization microscopy of unstained sections	connective tissue	best method for the investigation of collagen

15.8 Preparation of Non-Decalcified Material

Cancellous bone can be cut, but needs special support. On the other hand cortical bone can only be cut with special hard tissue knives or saws. Cortical bone can also be ground to a suitable thickness.

The procedures involved are:

a. With a very stable microtome, the new hard tissue knives of specially hardened material can be used directly to cut sections of undecalcified hard tissue. It is an advantage to use a motorized ground sledge microtome with a cutting speed of around 2 mm per second
b. Sawing produces irregular sections with a thickness of 100–150 μm. These are ground between two rotating pieces of plate glass with powdered pumice and water to a thickness of 70 μm which is well suited for X-ray structural examinations
c. Grinding can be extended until the minimum achievable thickness of around 20 μm is attained. A lot of tissue is lost with this method

The staining methods listed in Table 15.3 are all applicable with the possible additions of Chromoxane Cyanine R (see below) and the von Kossa method for calcium (Sect. 17.7.1). It should be noted that staining times for methyl methacrylate sections are considerably prolonged due to the slow diffusion in the methacrylate, for example with Haematoxylin-Eosin a 60 + 30 min procedure is required.

Special Staining Methods Suited for Methacrylate Sections. Chromoxane Cyanine R, C.I. 43820 (Solochrome Cyanine) with iron alum is excellent for differentiating between osteoid tissue and mineralized bone.

Von Kossa's method (Sect. 17.7.1) is very useful for the demonstration of calcium salts.

15.8.1 Methyl Methacrylate Embedding

In this procedure the tissue is first infiltrated with methyl methacrylate monomer. This is then polymerized and hardened. The surface of the tissue is exposed by sawing or grinding so as to avoid unnecessary strain on the knife. Finally, sections are prepared with a hard tissue knife or D-knife or by sawing and grinding as described above.

15.8.2 Hydroxyethyl Methacrylate Embedding

The hydrophilic monomer, hydroxyethyl methacrylate, can also be used for undecalcified bone tissue. The advantage is that it is possible to use the "ordinary" staining methods without prolonging staining time or increasing the temperature.

15.9 Other Special Methods for Hard Tissue

Fluorescence marking can be used for investigations into the calcification process. For example, tetracycline, which binds to calcium ions, is taken up *in vivo* into areas of active calcium deposition. Thereafter, either on frozen sections of fresh, non-decalcified material or on ground sections of fresh tissue or material fixed in 70% v/v ethanol, the tetracycline can be visualized in the fluorescence microscope as golden yellow deposits on a blue autofluorescent background (collagen). Administration of two or more doses at defined time intervals enables an estimate of the rate of the calcification process.

16 Tissue Processing: VII. Post Treatment

H. Lyon

16.1 Introduction

For an image to be seen with an unstained section in the light microscope, different tissue components must have different refractive indices from each other and from the mounting medium. Pieces of tissue, which are predominantly protein, have a refractive index of 1.53–1.54 when fixed with formaldehyde. If they are subsequently infiltrated with a colourless, transparent liquid with the same refractive index as the tissue (e.g. methylsalicylate) they become invisible and no image is seen in the microscope. Contrast can be achieved in three different ways:

1. The tissue can be *stained.* The section becomes visible because light of certain wavelengths is absorbed by the dye or dyes bound to tissue components but is relatively unaffected by the mounting medium. A mounting medium with the same refractive index as protein is now an advantage as this gives an optically clear image
2. The tissue can be left unstained, but can be made visible by mounting it in a liquid with a refractive index which differs slightly from that of proteins. The section may then be examined by *phase or differential interference contrast microscopy*
3. A transparent piece of tissue can be made visible by direct microscopy if *it is mounted in a liquid with a refractive index substantially different* from that of protein. This method is only applicable for low power examinations and it should be noted that boundaries between the tissue and mounting medium appear as sharp lines which have no *in vivo* correlates

After staining, histochemical reaction, blocking reaction etc., the section must be made suitable for microscopic examination. For most procedures the requirements are:

1. Removal of surplus reagent(s)
2. Mounting in a medium which does not interfere with the reaction product or with the subsequent microscopic examination

H. Lyon (Ed.)
Theory and Strategy in Histochemistry

16.2 Removal of Surplus of Reagent

The removal of surplus of reagent is usually achieved by washing the section in distilled water. If staining has been performed with anionic dyes at <pH 7 or with cationic dyes at >pH 7, water at neutral pH will alter the charges on ionizable groups involved in dye-binding and produce some dye extraction (cf. Sect.6.1). As distilled or demineralized water take up CO_2 from the atmosphere on storage they are usually weakly acidic (pH 4–5) and extraction (differentiation) of anionic dyes is often less than expected. Optimally, the section should be washed in a buffer solution with the same pH and ionic strength as the staining reagent. Washing should generally be short (e.g. two changes of buffer solution, 5 seconds in each) as complete extraction of dye may occur if washing is unduly prolonged.

If surplus staining reagent cannot be removed without simultaneously giving rise to excessive dye-extraction, the slide can be blotted with filter paper. Direct contact between the section and the filter paper must be carefully avoided.

16.3 Further Treatment of the Section

The next stage depends on the choice of mounting medium (Sect.16.4). If the latter is miscible with water (hydrophilic) further treatment is unnecessary, however, if a hydrophobic medium is to be used then the section must be dehydrated. Sequential exposure to an ascending series of ethanol concentrations is frequently used (e.g. 70% v/v, 95% v/v, and 99% v/v). It should be noted that cationic dyes, especially Thiazin dyes can be extracted by ethanol. Anionic dyes are also extracted in water-ethanol mixtures. As paraffin sections have already shrunk as much as they are able to, sections can be moved directly from water to absolute ethanol. This approach, however, leads to dilution of the ethanol with water from the sections after relatively few slides have been processed. Due to the problems with extraction of dyes in ethanol (Sects.6.1 and 6.1.6), several other compounds have been suggested for use as dehydrating agents. Examples include 2-propanol, 1-butanol, 2-methyl-2-propanol (tert-butyl alcohol), and acetone. Finally, dehydration can also be performed by drying the section in air, preferably at around 40°C in a thermostatically-controlled oven containing silica gel. A vacuum oven can also be used for this purpose.

16.4 Mounting Media

A mounting medium is a substance in which cells or tissue sections are placed before examining them under the microscope. The medium must be transparent and completely or almost completely colourless, and must at the same time be able to fill all spaces between the tissue components. Furthermore, it must not dissolve or corrode any of the tissue components which are to be examined nor should it be a suitable substrate for fungal or bacterial growth.

Solid objects that cannot shrink such as ground sections of bone (Sect. 15.8), can be mounted dry (i.e. with air as the mounting medium). Mounting media can be fluid, mixtures of several fluids, or solutions of one or more solid compounds in a liquid solvent. Volatile mounting media can be used as the edges of the cover slip can be sealed to the slide (e.g. with gelatin). Compound mounting media often use the volatility of the solvent as they become increasingly viscous and finally solidify with prolonged storage.

As stated above, mounting media can be divided into hydrophobic and hydrophilic and both these groups can be subdivided into adhesive (e.g. Eukitt^{R}) and non-adhesive (e.g. methylsalicylate). The advantage of the non-adhesive mounting media is that the final refractive index is known as there is no change in the composition of the mounting medium on evaporation.

The exact refractive index for an adhesive mounting medium cannot be given as it is not possible to control the extent of evaporation. The limits for the refractive index are normally given (i.e values for the fresh mounting medium and after complete solvent removal). It should be appreciated that complete solvent evaporation from a mounted section cannot be expected. Against these problems, adhesive mounting media offer the advantage of secure binding of the cover slip to the slide.

16.4.1 Hydrophilic Mounting Media

Hydrophilic mounting media contain or are miscible with water thereby allowing the sections to be directly transferred from water. Hydrophilic mounting media are used when low refractive indices are needed or when dehydration is to be avoided.

Hydrophilic mounting media are valuable in investigations concerning lipids as these are preserved regardless of whether they are fixed. Lysochromes retain their colour well in these media but some dyes, such as Al- and Fe-haematein, are not quite stable.

Non-Adhesive Hydrophilic Mounting Media. Glycerol is an example of a non-adhesive hydrophilic mounting medium.

Adhesive Hydrophilic Mounting Media. Examples of adhesive hydrophilic mounting media are glycerol-gelatin and polyvinylpyrrolidone containing media.

The effect of glycerol in glycerol-gelatin is that the gelatin does not crack when the water evaporates. Dehydration after mounting is best achieved by placing the mounted slides in a thermostat oven at 60°C overnight. A particular disadvantage of glycerol-gelatin is that it must be heated prior to use.

Polyvinylpyrrolidone (PVP) and several related reagents in contrast do not need to be heated, and the majority of commercial, hydrophilic, adhesive mounting media are drawn from this group.

16.4.2 Hydrophobic Mounting Media

These media are not miscible with water, and the tissue sections must therefore be dehydrated first (Sect. 16.3). After dehydration the section is usually immersed in the solvent that forms the basis for the mounting medium and from here to the medium itself. In practice it is usually possible to achieve good results by transferring the "wet" sections from the final dehydrating step (e.g. absolute ethanol or 1-butanol (Lyon *et al.*, 1983)) directly to the hydrophobic mounting medium.

The adhesive, hydrophobic mounting media "set" as a result of solvent evaporation. The solvent is usually xylene which evaporates sufficiently slowly to allow the cover slip to be placed as desired. The congealing process can be accelerated by placing the mounted slides on a hotplate or in a thermostatically controlled oven at around 40°C. Most hydrophobic mounting media have high refractive indices quite close to that for formaldehyde fixed protein. They therefore usually give more transparent preparations than hydrophilic mounting media.

Non-Adhesive Hydrophobic Mounting Media. These media, e.g. methylsalicylate, are normally only used when it is necessary to know exactly what the refractive index of the mounting medium is.

Adhesive Hydrophobic Mounting Media. The key component of these media is a resin, an amorphous, water-insoluble, organic compound which, due to being composed of molecules of varying length, has an ill-defined melting point.

The natural resins (e.g. Canada balsam) ooze out from the bark of different trees. They are oxidation products of terpenes. Modern plastics that have similar physical properties also bear this name. An example (Baker, 1966, p.131) of a synthetic resin of this kind is polystyrene:

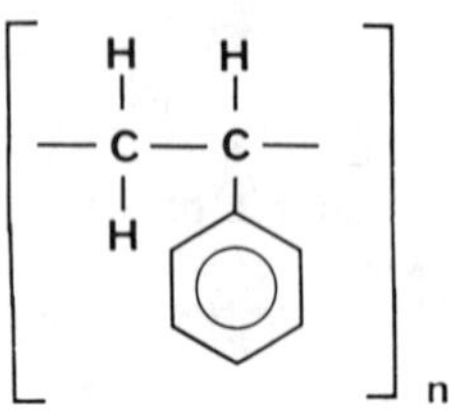

where n = about 800.

It is found, for example, in DPX and is a colourless, solid compound which is easily soluble in xylene. If a tissue section is mounted in a solution of polystyrene in xylene the polymer retracts gradually from the edges of the cover slip as the solvent evaporates. This drawback is reduced or eliminated by the addition of a so-called "plasticizer" to the compound (Baker, 1966, p.132). Typical "plasticizers" are liquids, which are non-volatile and which function as solvents for the plastic, on which they act. It is believed that the "plasticizer" binds the resin molecules together in a three-dimensional network, so that the polymer becomes softer and does not retract in the way it otherwise would when the volatile solvent evaporates. An example of a "plasticizer" is tri-p-tolylphosphate, which is a colourless, non-volatile liquid with high refractive index (about 1.56).

Part 4
The Staining of Chemical Entities

17 Metals and Metal Salts

P. Prentø

17.1 Occurrence

Some metals are normal constituents of living tissues (calcium, magnesium, iron, selenium, zinc, copper, sodium, and potassium); some (copper and selenium) are only needed in small amounts, and larger concentrations are nearly always due to industrial or environmental contamination, while the occurrence of others (lead, mercury, thorium, cadmium, gold, etc.) is always pathological and reflects contamination.

Many invertebrates accumulate relatively large amounts of metals. Earthworms, for instance, accumulate zinc and tunicates accumulate vanadium. It is not known whether this represents contamination (industrial or otherwise) or a functional activity.

From a histochemical point of view, inorganic constituents of tissues may be present in three states:

1. *Diffusible ions*. Precise localization is difficult or impossible. Examples include sodium, potassium, chloride, and bicarbonate. See, however, Sect.28.8.10.
2. *Immobile salts* are generally directly detectable and localization is usually demonstrable without difficulty. Examples include ferric iron in haemosiderin and calcium and magnesium phosphates
3. *Masked, inorganic constituents*. These are not directly reactive but are covalently or complexly bound to organic molecules. Examples include iron in haemoglobin (only detectable after demasking with alkaline 30% H_2O_2) and certain mercury compounds (where Hg is covalently bound to thiol groups).

Before describing the "conventional" histochemical methods for detecting metals, two methods based on very different principles will be outlined

17.2 Micro-Incineration

This method dates back to Raspail (1833). The organic constituents are incinerated by heating to 500–700°C, then the ashed section is examined by polarization microscopy or other techniques, possibly in combination with a chemical or his-

H. Lyon (Ed.)
Theory and Strategy in Histochemistry

tochemical detection method for metal salts. The micro-incineration sublimates or modifies some inorganic substances (e.g. mercury) after which they cannot be detected histochemically. Moreover, many of the detectable substances are delocalized by the incineration or the subsequent histochemical reaction. The latter may sometimes be limited by prior coating with celloidin. In practice only calcium, magnesium, iron, and silicon are detectable with certainty while the method is generally inapplicable to the detection of diffusible ions.

17.3 Electron Microscopical X-Ray Microanalysis

In electron microscopical X-ray microanalysis (EMMA), a 50–200 nm thick section is submitted to a focussed electron beam which strips inner electrons from the atomic nuclei of the tissue constituents. Outer electrons in affected atoms make a quantum jump and fill the empty orbital(s) by emitting energy equivalent to the difference between the outer and the inner shell energy levels as X-rays.

Most elements have several outer electrons at different levels, and each electron is capable of jumping to the vacant shell. Several different quantum jumps are therefore possible for one element and X-rays of different energies and wavelengths are emitted in a spectral pattern characteristic for each different element. In addition, the total amount present within a given area may be inferred from the amplitudes of the emissions. The detection limit is very low, often around 1–0.1 fg (10^{-15}–10^{-16} g), and localization can be within 0.01 μm^2.

In principle all chemical elements with a sufficiently high atomic number can be detected. For most X-ray detection systems the atomic number has to be above 11–12, hence detection of sodium is not possible and magnesium is difficult. An additional problem is that some elements have overlapping emission spectra. For instance an M-line from osmium and the K-lines from phosphorus overlap making it impossible to use osmium tetroxide as a fixative when investigating phosphates. Analysis of metals which form part of the tissue support system (grid, grid support, etc.) is of course also out of the question.

X-ray analysis gives no information regarding the chemical state of the elements. Detection of the element sulphur says nothing about whether sulphur is present as free sulphate, ester-bound sulphate, thiol/disulphide in proteins or even free sulphur. Similarly, the method does not distinguish free and ester-bound phosphate but gives the sum of both.

The most important errors inherent to the method are extraction- and diffusion artifacts.

X-ray microanalysis is a complicated and time-consuming method and the equipment is very expensive. In most cases analysis using the classical histochemical approach is preferred.

17.4 Processing of Tissue

In preparing tissues to be examined for inorganic salts, acid fixatives should be avoided. In some situations, for example the detection of calcium-lipid complexes, even dehydration and embedding in paraffin should be omitted. The best starting point for an investigation of inorganic constituents is rapid freezing at −70°C or below, followed by freeze-substitution, freeze-drying or freeze-sectioning. Freeze-sectioning is very well suited for investigation of immobile metal deposits in tissue. Buffered formalin may sometimes be used before freezing.

17.5 Principles for the Histochemical Detection of Metals

Metal salts are a substantial, and by far the most heterogeneous, component of inorganic tissue constituents. Metal ions may be detected, directly or indirectly, by use of substitution methods, or by chelation with appropriate metallochromes.

17.5.1 Substitution Methods

Substitution is based on inorganic precipitation, and the range of possibilities for satisfactory histochemical procedures using inorganic reactions is limited. One of the ions of the metal salt is substituted by an inorganic ion leading to the formation of a coloured or black insoluble product. There are two forms of substitution. In one case (I) the anion is substituted, in the other case (II) the metal ion itself is substituted (metal substitution).

I	Metal-A tissue	$+ D^{e-} \rightarrow$	Metal-D coloured tissue	$+ A^{e-}$
II	Metal-A tissue	$+ C^{+} \rightarrow$	C-A coloured tissue	+ Metal^{+}

A = anion; D = dye; C = cation

Mcthod I is used in the Prussian Blue reaction for iron (Sect.17.7.4). Method II is of course not in itself a demonstration of the metal, and the result is often difficult to interpret. Nonetheless, some of these methods, such as the silver substitution method of von Kossa which detects certain calcium deposits, are valuable.

17.5.2 Chelating Methods

The potential application of these methods is very extensive as they depend on the selective binding properties of chelating dyes known as *metallochromes*. The properties of these reagents (chelating ability, colour, solubility, etc.) can be modified by pH, solvent, or other complex-forming substances such as cyanide, tartrate etc. (cf. Sect.7.1 for chelation). The properties of some histochemically useful metallochromes are compared in Table 17.1. (See also Feigl, 1958).

17.7.1 Calcium

In tissues, calcium is present both as free ions and precipitated as a constituent of carbonates, phosphates, oxalates, and soaps. The deposits are usually mixed regarding both metal ions (magnesium, zinc, strontium, etc. may be present in varying amounts) and anions. The most important biological calcium salts are carbonate, phosphate, and oxalate.

Calcium carbonate is common in many invertebrate groups in the form of shells, spikes, and various exo- and endoskeletons. In vertebrates it forms otoliths and it is the main constituent of the eggshells of reptiles and birds.

Calcium phosphate occurs as inclusions in bacteria and protozoa, in the invertebrate brachiopods and arthropods, etc., and in vertebrate bone, dentine, and tooth enamel.

Calcium oxalate may be found in the larval cuticle and eggs of some insects and in some soft vertebrate tissues. Its occurrence is more widespread in plants where it is found as cell inclusions in most species of mosses, ferns and seed plants.

Insoluble calcium salts may be detected by metal substitution (i.e. indirectly) or by chelation.

Metal Substitution Methods. Many metal ions (silver, cobalt, iron, copper, etc.) can be exchanged with the calcium present in calcium salt deposits. The best known method is *von Kossa's silver substitution method* (Lillie and Fullmer, 1976, p.539; Pearse, 1985, p.982). First silver is substituted for calcium then the bound silver is reduced to form black, metallic silver; e.g.

$$CaCO_3 + 2Ag^+ \rightarrow Ag_2CO_3 + Ca^{2+}$$

$$Ag_2CO_3 + 2H \rightarrow 2Ag + H_2O + CO_2$$

Even though von Kossa's method demonstrates the presence of insoluble phosphates, carbonates, and possibly oxalates, rather than calcium as such, selectivity is in practice quite good as these anions will nearly always be in the form of calcium salts in vertebrate material. *Calcium oxalate* presents a problem as the silver oxalate intermediate is much more soluble. Silver substitution of calcium oxalate may therefore result in complete dissolution of the oxalate and is in any case a slow process. Effective demonstration of calcium oxalate may be performed by adding hydrogen peroxide to the silver nitrate solution or by prior micro-incineration.

Errors. Uric acid and *urates* reduce silver ions and may be mistaken for calcium deposits. They may, however, be removed by treatment with saturated, aqueous lithium carbonate before substitution. The problems associated with oxalates are discussed above.

Other reducing substances in tissue may sometimes (independent of light or post-reduction) produce a spontaneous reaction (Pearse, 1985, p.982). The presence of *melanin* can mimic a positive von Kossa reaction as can carbon particles. Acid proteoglycans may bind silver ions giving a variable amount of additional staining.

Table 17.1. Chelate colour after chelation of metal ions with some commonly used metallochromes.

Metal	Dithizone	Rubeanic acid	Alizarin Red S	Na-rhodizonate	Bathophenanthroline
Ca^{2+}	(greenish)	—	ORANGE-RED, pH 6–9	red–brown, pH 7	—
Cu^{+}	red-brown	GREEN, pH 4–8	—	—	—
Cu^{2+}	red-brown	GREEN, pH 4–8	—	(yellow)	—
Fe^{2+}	yellowish brown[1]	(yellow)	purple[1], pH 5–8	—	RED
Fe^{3+}	pink[1]	(orange)	purple to black[1]	—	—
Zn^{2+}	RED	(yellow)	—	brownish violet, pH 7	—
Pb^{2+}	red	(yellow)	orange, pH 4–5	PURPLE, pH 2.8 violet, pH 7	—
Hg^{2+}	orange	brown to black	purple, pH 4.5–5.5	brownish red pH 2.8–7	—
Ag^{+}	violet	brown to violet, pH 7–9	—	black, pH 2.8–7	—
Ni^{2+}	brownish red	BLUE-VIOLET, pH 7.8	—	—	—
Al^{3+}	orange	—	SCARLET	(yellow)	—
Sr^{2+}	—	—	ORANGE RED	RED, pH 7	—
Ba^{2+}	brown	—	SCARLET	RED, pH 2.8 black, pH 7	—

CAPITALS: insoluble chelate, colour intense; histochemically useful.
( ): chelate weakly coloured; histochemically unusable.
—: no chelate formed or the chelate soluble.
[1] Demonstrates only some inorganic Fe-salts, but not Fe in ferritin and haemosiderin.

17.6 Detection Limits

The distribution of metals in tissue is very uneven and frequently involves the formation of separate particles. This makes quantitation of metal deposits very difficult. A reasonable estimate can, however, be obtained if the *detection limit* is known for the metal and the particular method to be used (sensitivity). This is rarely the case.

The detection limit for a metal depends on the molar extinction coefficient (ϵ) for the coloured reaction product (e.g. metal-dye chelate or metal pigment such as Prussian Blue) and on the local concentration of the reaction product. With ϵ = 20,000 cm^2/mmol and an area of 1 μm^2 the detection limit is 10^{-13} mmol which, when used for a metal with an atomic weight of 56 (iron), corresponds to a mass of 5×10^{-14}–5×10^{-13} g (50–500 fg).

Table 17.2 presents some limits of detection for light microscopy calculated from spot tests and, in two cases, from combined histochemical and spectrophotometric investigations (Prentø, 1979).

Table 17.2. Limits for the histochemical detection of metals by light microscopy using metallochromes. Calculated from data on spot tests (Feigl, 1958).

Metal	Metallochrome	Limit of detection in spot test (µg) (according to Feigl, 1958)	Microscopic limit of detection pg per μm^2
Ca^{2+}	Glycol bis-(2-hydroxyanil)	0.05	0.2
	Alizarin Red S		0.08[1]
	Na-rhodizonate	1	4
Mg^{2+}	Titan Yellow	1.5	6
Cu^{+}/Cu^{2+}	Dithizone	0.1	0.4[2]
	Rubeanic acid	0.006	0.02
	Na-diethyldithiocarbamate	0.002	0.008
Fe^{2+}	Bathophenanthroline	probably about 0.01	0.04
Fe^{3+}	K-ferrocyanide	0.05	0.2
Zn^{2+}	Dithizone	0.025	0.01[3]
Pb^{2+}	Dithizone	0.04	0.16
	Na-rhodizonate	0.1	0.4
Hg^{2+}	Dithizone	0.25	1
Ag^{+}	Dithizone	0.05	0.2

[1] Experimental value (Prentø, 1979). This level of calcium is easily detectable, the real limit of detection is probably 0.01–0.03 pg.
[2] Only applies to Cu^{+}, as Dithizone does not react with Cu^{2+}.
[3] Experimentally (Prentø, 1979), this level of zinc is easily detectable and the real limit of detection is probably around 0.005 pg.

17.7 Individual Methods for the Detection of Inorganic Constituents

The histochemistry of calcium, magnesium, copper, iron, lead, zinc, mercury, silver, and of carbonate, phosphate, polyphosphate, and oxalate will be discussed below.

Under normal circumstances none of these sources of error give rise to confusion and most problems can be avoided by comparing the von Kossa treated section with an untreated control and with a section where reduction has been omitted. Sometimes intracellular calcium deposits may be difficult to identify with certainty. Alternative methods for calcium demonstration should then be considered. Additional evidence can be obtained by showing that there is no staining when the calcium reaction is preceded by treatment with weak acid or 2–5% EDTA.

Interestingly, pure crystalline calcium carbonate or calcium phosphate does *not* react with the von Kossa method. The amorphous nature of mineral deposits found in biological material (probably due to the close association with organic materials) seems to be necessary for a positive reaction. In very dense deposits only the outer zone is positive due to poor penetration by the silver salt. While the von Kossa method generally performs well on vertebrate material, it is not suitable for invertebrates (Prentø, 1979), probably because their calcium deposits have a different composition from those associated with vertebrates.

Other Cation Substitution Methods. Aluminium ions exchange with carbonate- or phosphate-bound calcium. Hence non-acidified Al-haematein stains calcium deposits deep blue. This basophilia is not specific for calcium but is nonetheless very useful as an indication for performing a more specific calcium method.

Many cationic dyes, e.g. Toluidine Blue, can stain calcium deposits. The mechanism is probably substitution of Ca^{2+} by the dye ion, which binds to the phosphates, carbonates, or other anions present in the calcium deposit.

Localization with these methods is excellent but the selectivity is very poor, and the methods should always be followed by the von Kossa reaction or, particularly with invertebrate material, a chelation method.

Anion Substitution Methods. By treatment with certain acids the anions of calcium deposits may be exchanged with the anions from the acid, leading to the formation of well defined crystals. This method is specific but the crystal formation leads to very coarse localization.

Application of dilute sulphuric acid leads to the formation of gypsum crystals ($CaSO_4$) in and above the section, while the original calcium deposit disappears.

Chelation Methods. In a manner analogous to the formation of metal complex dyes (Cr-gallocyanin, Al- or Fe-haematein), Ca^{2+} is able to bind to certain metallochromes. Alizarin sulphonate (*Alizarin Red S*, C.I. 58005) and glyoxal bis-(2-hydroxyanil) (GBHA) are the most widely used in histochemical work.

Both metallochromes are suitable for the detection of small calcium deposits and several procedures have been described. Highly selective methods include the Alizarin Red S procedure of McGee-Russell (1958) and the GBHA procedure of Kashiwa and Atkinson (1963).

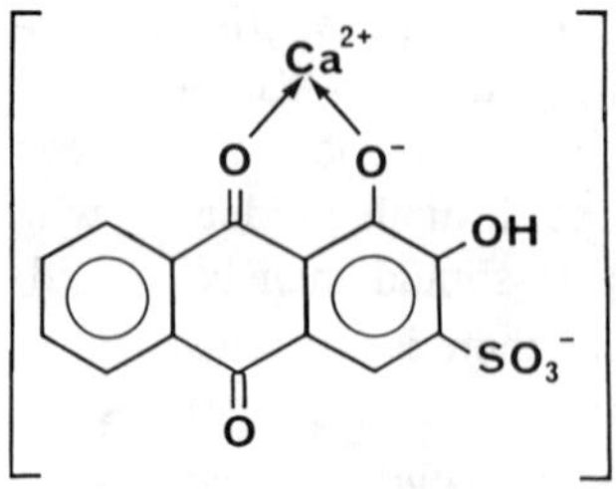

Ca-Alizarin sulphonate chelate

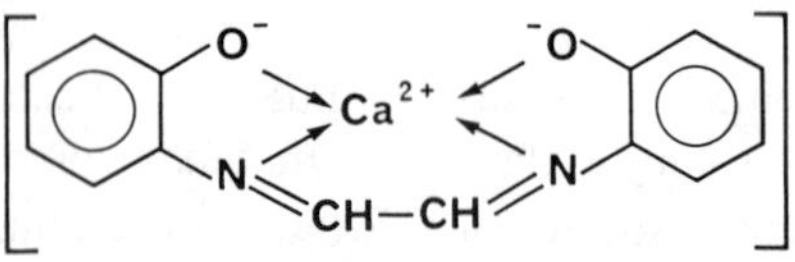

Ca-GBHA chelate

For chelation methods it is essential that all details in the description of the method are strictly adhered to and that the appropriate control reactions for interfering metals are done. For instance, Alizarin Red S may form red chelates with several other metals in addition to calcium (see Table 17.1).

17.7.2 Magnesium

Magnesium can only be detected by chelation methods. However, the ability of the magnesium ion to form chelates is relatively poor. Chelation has to be performed at high pH (12–13), which favours magnesium chelation and precipitates most other metals in the form of non-reactive metal hydroxides. (Above pH 13 even magnesium forms insoluble, non-reactive magnesium hydroxide). At best the reaction is slow and the high alkalinity disrupts the tissue. The best chelating agents for Mg are *Thiazol Yellow G* (Pearse, 1985, p.997), also called "Titan Yellow", C.I. 19540, and *Magneson*, 4-(p-nitrophenylazo)-resorcinol (Pearse, 1985, p.997).

Titan Yellow imparts a flaming red colour to magnesium deposits. The colour disappears after a few hours. Detection limits are shown in Table 17.2. Magneson stains magnesium deposits bright blue and interference from other metals may be prevented by adding potassium cyanide.

17.7.3 Copper

In vertebrates copper is only found as Cu^{2+} and, with the exception of certain pathologic conditions, the levels are below the limits of histochemical detection. In some cases invertebrates contain relatively large amounts of copper, which is present in both the cuprous and the cupric form.

Even if copper is present in detectable amounts a large part of it may be so firmly bound to proteins that it cannot be visualized directly. Demasking may be performed with hydrogen peroxide or better with HCl vapour, but this is by no means 100% effective. Copper is best detected with rubeanic acid (dithiooxamide) or the zinc or sodium salt of diethyldithiocarbamate (DDTC). The so-called *oxidation-catalyst* reactions should generally be avoided. They are sensitive and selective, but are based on the oxidation of o-tolidine or benzidine, both of which are carcinogenic, and require the presence of free cyanide.

Uzman's Rubeanic Acid Method. In this method (Uzman, 1956) rubeanic acid is a very sensitive reagent for both Cu^{2+} and Cu^{+} (see Table 17.2 for detection limits). The chelate is green to greenish black and cannot be mistaken for the iron, cobalt, and lead chelates, which are yellow to yellowish brown, or with the nickel chelate, which is blue.

The rubeanic acid method is the method of choice for vertebrate material.

Diethyldithiocarbamate, Zn- or Na-Salt. This compound (DDTC) was introduced by Waterhouse (1951). DDTC reacts with both Cu^{2+} and Cu^{+}. Detection limits are given in Table 17.2. The chelate is yellowish brown, and this may lead to problems with interpretation if the tissue contains lipofuscin or similar pigments. DDTC also chelates with Fe but this interaction is so insensitive that it is generally of no importance. For both the DDTC- and the rubeanic acid method, the tiny amounts of copper detectable make it essential to keep copper containing materials well away from the tissue and reagents to minimize contamination.

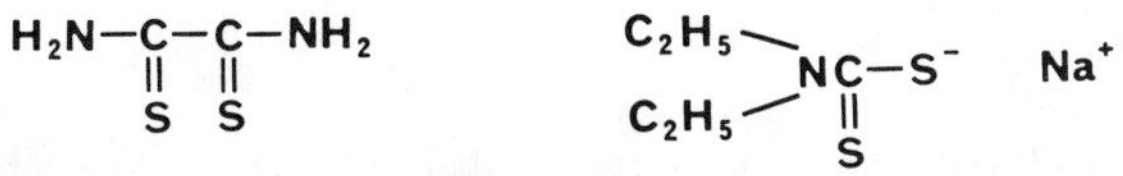

Rubeanic acid   Na-diethyldithiocarbamate

17.7.4 Iron

In tissue, iron does not normally occur as inorganic deposits but is bound to proteins which include cytochromes, haemoglobin, myoglobin, ferritin, and haemosiderin.

Both ferritin and haemosiderin bind ferric iron in a manner that allows direct detection provided the pH is sufficiently low. The properties of tissue iron are summarized in Table 17.3.

Iron deposits are stainable with freshly prepared, non-oxidized Haematoxylin, giving a blue to bluish black colour. The reaction is (cf. Sect.7.2.3)

$$Fe^{3+} + Ht.H_2 \rightarrow Fe^{2+} + Ht.$$

$$Fe^{2+} + Ht. \rightarrow Fe\text{–}Ht.\ \text{chelate}$$

The method is sensitive, but shows poor selectivity.

The most useful methods for the detection of iron are Perls' Prussian Blue reaction and the Bathophenanthroline method (Hukill and Putt, 1962).

Perls' Prussian Blue Reaction. This reaction (Lillie and Fullmer, 1976, p.501; Pearse, 1985, p.974) is an anion substitution method in which the section is exposed to a freshly prepared, acidified solution of potassium ferrocyanide. The acid releases

Table 17.3. Properties of iron in tissue.

	Haem and haemin proteins	Ferritin, haemosiderin	Free iron mineral
Oxidation state of Fe	Fe^{2+}, $Fe^{2+/3+}$, Fe^{3+}	Fe^{3+}	Fe^{3+}, (Fe^{2+})
Masking	total	partial	depends on mineral type
Extractable at pH 7	nil	nil	nil
— — pH 4	nil	partial	partial (depends on mineral type)
— — pH 1	nil	total	total (except for haematite Fe_2O_3)
Can be dissolved with 1% $Na_2S_2O_4$	nil	total	total
Prussian Blue reaction	0	+	+ (haematite needs several hours)
30% H_2O_2 – Prussian Blue reaction[1]	(+)	+	+ (haematite needs several hours)
Turnbull Blue reaction	(+) ??	+ ??	occasionally +
Bathophenanthroline reaction	0	+	occasionally +
Reduction – Bathophenanthroline reaction	0	+	+
Benzidine-H_2O_2 reaction[2]	+	0	0
Leuco Patent Blue-H_2O_2 reaction[3]	+	0	0

[1] Not recommended.
[2] The method demands higher H_2O_2 concentration (about 0.05%) than the peroxidase reactions proper. Benzidine is *carcinogenic* and the method should not be used.
[3] According to Lison-Dunn (Lillie and Fullmer, 1976, p. 447). Patent Blue V, C.I. No. 42051.

ferric ions from the tissue anions (generally the proteins apoferritin and aposiderin). The ferric ions on liberation immediately form a complex with the ferrocyanide anion, for instance

$$K^+ + [Fe(CN)_6]^{4-} + Fe^{3+} \rightarrow K[Fe_2(CN)_6]$$

The resultant highly insoluble complex is called Prussian Blue. The exact composition of the complex depends on the proportion of ferrous to ferric ions. The intense colour of the Prussian Blue complex is associated with the fact that most of the ferrous electrons derived from the ferrocyanide are very loosely bound and in consequence enhance light absorption. In the formed coloured complex it is therefore not possible to attribute the ferrous or ferric state to any individual Fe atom.

There are several modifications of Perls' Prussian Blue reaction. Without doubt, the most valuable is Lison's (Pearse, 1972, p.975), which employs equal parts of 0.25 mol/l HCl and 2% potassium ferrocyanide, giving a pH of 1.5 which is optimum for the reaction. At higher pH not all the ferric ions in ferritin or haemosiderin are released, and at lower pH the rate of ferric ion release is too high in relation to the rate of Prussian Blue formation. If iron is present in large amounts, or loosely bound, it may be profitable to carry out the reaction at a higher pH (e.g. around 2) or, even better, the section may be preincubated in a non-acidified ferrocyanide solution and the HCl added after 5–10 min.

Selectivity and Sensitivity. The formation of Prussian Blue is specific for ferric ions and very small amounts may be detected (Table 17.2). In order to avoid non-specific precipitates, it is necessary to use absolutely clean glassware, demineralized or distilled water, *analytical* grade chemicals, and to avoid exposure to metal objects of any kind. Dust particles (or tobacco ash or smoke) are also significant contamination risks for sections, chemicals, and solutions.

The ferrous iron in the ferrocyanide used for the Prussian Blue reaction is quite rapidly oxidized to ferric iron by atmospheric oxygen. If the reaction time is extended non-specific Prussian Blue formation may occur, even if the reaction has been performed correctly in other respects. The ferrocyanide solution must therefore be *freshly prepared* and the reaction time should not exceed half an hour, counting from the time of preparation. If a prolonged reaction time is required, the section should be transferred to a fresh ferrocyanide solution every half hour.

Accumulations of *asbestos fibres* bind ferric ions in vivo and may therefore be Prussian Blue positive to varying degrees. Similar reactions may also occur with *calcium deposits* or lipofuscin.

Deposits of *exogenous* iron may sometimes contain ferrous ions. If these are not too firmly bound, they may be detected by use of potassium *ferri*cyanide. Although the resulting blue complex is sometimes called *Turnbull Blue*, it is the same as Prussian Blue. Ferrous iron is more readily demonstrated by the bathophenanthroline method, which is also well suited for the general demonstration of iron.

Bathophenanthroline Method. In this method (Hukill and Putt, 1962) Bathophenanthroline (4,7-diphenyl-1,10-phenanthroline) is a chelating agent which forms a red complex with ferrous ions in the pH range 3–7. While both the reagent and the chelate are practically insoluble in water, they are soluble in ethanol and most lipid solvents. Bathophenanthroline itself is, however, soluble in weak acetic acid (pH 3–4), while the iron chelate is not.

Selectivity. The formation of a red chelate is specific for ferrous ions. Copper forms a colourless, and cobalt a light yellow chelate. Addition of 0.25 mol/l thioglycolic acid to the reagent ensures reduction of ferric ions to ferrous ions and allows demonstration of the entire non-haem iron content of the section without having to use acid demasking. In contrast to the Prussian Blue reaction, Bathophenanthroline cannot give rise to false positive reactions. However, contamination is a potential problem due to the sensitivity of the reaction for ferrous ions (see Table 17.2).

The reaction time for Bathophenanthroline should not exceed two hours, as ferrous iron from haemoglobin and other haem proteins gradually chelates with Bathophenanthroline. As the iron chelate is soluble in organic solvents, the stained section is mounted in an aqueous mounting medium.

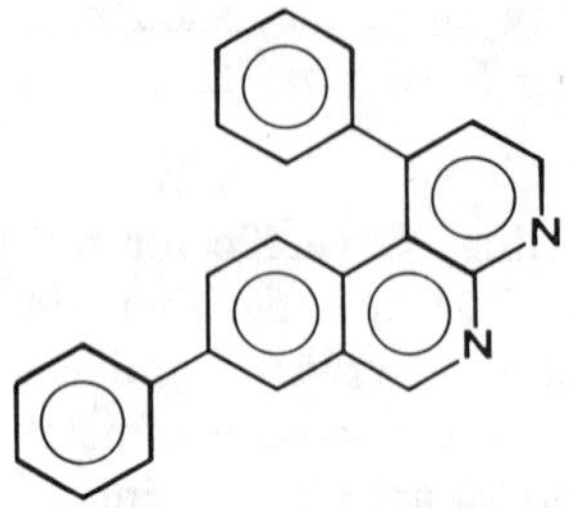

Bathophenanthroline (4,7-diphenyl-1,10-phenanthroline)

17.7.5 Zinc

Zinc is a natural constituent of insulin in most mammals, it is also present in carbonic anhydrase, superoxide dismutase, certain dehydrogenases, carboxypeptidases, and many other enzymes. In mammals zinc is found in particularly large amounts in prostate gland and spermatozoa. Many invertebrates contain inorganic zinc deposits of unknown function.

Dithizone Method. The best method for the detection of zinc ions is chelation with dithizone (diphenylthiocarbazone) (Mager *et al.*, 1953).

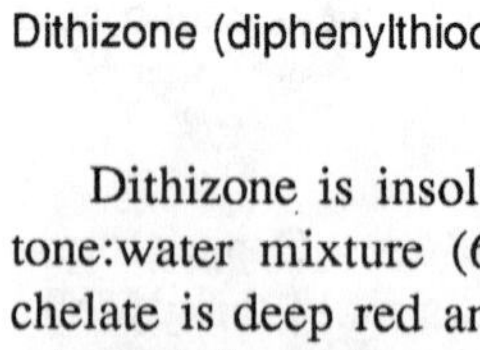

Dithizone (diphenylthiocarbazone)

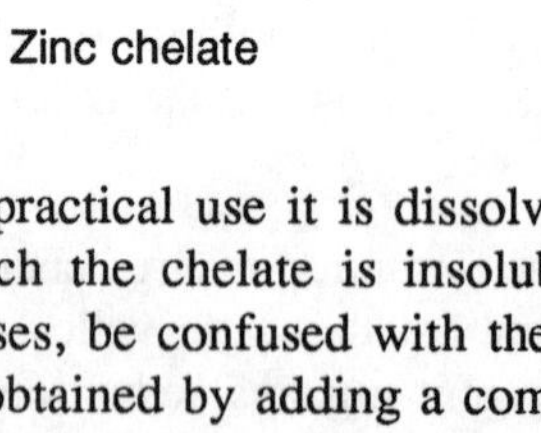

Zinc chelate

Dithizone is insoluble in water, so for practical use it is dissolved in an acetone:water mixture (60% acetone), in which the chelate is insoluble. The zinc chelate is deep red and it may, in some cases, be confused with the lead or cadmium chelate. A specific reaction may be obtained by adding a complex-forming buffer, containing sodium thiosulphate and potassium cyanide, to the dithizone solution. In most cases this is not necessary.

Dithizone has lysochrome properties and fat droplets are coloured slightly pink. If, as is preferable, frozen sections are used, two sections may be used for zinc detection. One section is pretreated with 2.5–5% neutral EDTA or 1 mol/l nitric acid, and the other with absolute acetone for 15 min. A positive reaction in the first section is due to lipid, while a positive reaction in the second section is due to zinc.

Sensitivity. See Table 17.2. The method demonstrates the natural occurrence of zinc in areas such as the pancreatic beta cells or cells with a high content of carbonic anhydrase.

17.7.6 Lead

Lead does not occur naturally in organisms but it may be accumulated as a result of environmental pollution or through an occupational hazard. Organisms that store salts of calcium, magnesium, or zinc are particularly prone to accumulate lead. In tissues lead is most often present in the form of salts. Lead in bone is not removed during decalcification, provided the decalcifying fluid contains more than 10% sulphate. This ensures that lead is precipitated as lead sulphate during the decalcification process but concomitant delocalization is probably significant.

Lead ions may be demonstrated by anion substitution with H_2S (lead sulphide methods), or by chelation with rhodizonate.

Lead Sulphide Methods. In these methods (Lillie and Fullmer, 1976, p.547), treatment with H_2S or an ammonium sulphide solution causes lead salts to form brown insoluble lead sulphide. The methods are very poorly selective (see Demonstration of heavy metals, Sect. 17.7.11). The selective demonstration of lead may be improved by bleaching the section with H_2O_2 (which gives rise to colourless, insoluble lead sulphate), followed by re-exposure to hydrogen sulphide gas or ammonium sulphide solution.

$$\mathrm{Pb{-}X} \xrightarrow{H_2S} \mathrm{PbS} \xrightarrow{H_2O_2} \mathrm{PbSO_4} \xrightarrow{H_2S} \mathrm{PbS}$$

Rhodizonate Method. The rhodizonate method (Lillie and Fullmer, 1976, p.547; Pearse, 1985, p.996) using sodium or potassium rhodizonate is able to form chelates with many different metal ions. At pH 2.5–2.8 (e.g. in dilute acetic acid) red chelates are formed with barium, cadmium, mercury (non-covalently bound), and lead. Pretreatment with dilute sulphuric acid dissolves barium, cadmium, and mercury salts, while lead remains in the tissue as insoluble lead sulphate, and therefore renders the rhodizonate method specific for lead. In contrast to the prolonged decalcification procedures, brief treatment with dilute sulphuric acid or with a sulphate solution does not appreciably affect the localization of lead deposits.

The chelating ability of rhodizonate is very high; even PbS reacts. At higher pH, 6–7, lead rhodizonate changes colour to more or less blue-violet. For *sensitivity*, see Table 17.2.

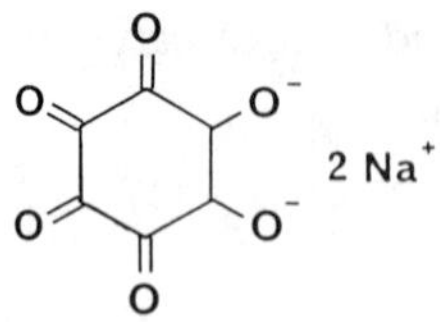

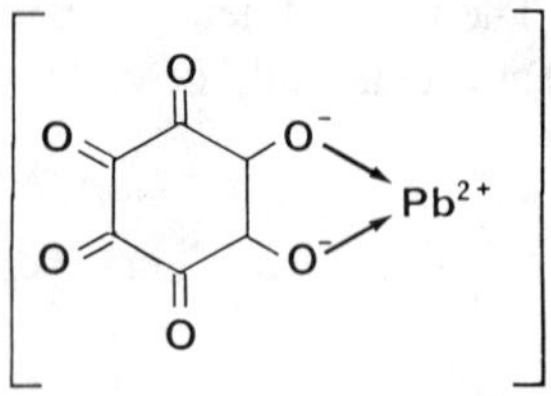

Sodium rhodizonate, yellow   Lead rhodizonate, red

17.7.7 Barium and Strontium

The presence of these metals in organisms is generally considered to be due to contamination. At pH 6–7, where the lead chelate is purple, both barium and strontium give insoluble red chelates with rhodizonate. As the sulphides of barium and strontium are practically colourless in tissue sections and soluble in water, the metals may be differentiated from lead by exposure to hydrogen sulphide gas followed by a rinse in slightly acidified water (0.1% acetic acid). After this procedure lead is still able to react with rhodizonate. Waterhouse (1951) differentiates between barium and strontium on the basis of the properties of their chromates. Pretreatment of barium deposits with potassium chromate leads to the formation of insoluble barium chromate, which cannot be stained by rhodizonate. Posttreatment with chromate decolours barium rhodizonate due to the formation of colourless barium chromate. In contrast, chromate-bound strontium is chelated by rhodizonate and the chelate is not destroyed by chromate.

17.7.8 Cadmium

Cadmium is a very poisonous metal and its presence in the tissues is probably always due to contamination. Unfortunately, the demonstration of cadmium is subject to interference from a variety of competing metal ions. Sumi (1980) and Sumi *et al.* (1982) have described a method with increased selectivity towards zinc and cadmium ions using various *benzothiazolylazonaphthol* (BTA) derivatives. For example, after treatment of the section with BTA-β-naphthol, cadmium, zinc, nickel, and manganese deposits were stained blue, while most other interfering metal ions stained red to purple. Post-differentiation with 1% oxine (8-hydroxyquinoline) in 75% v/v ethanol turned tissue cadmium from blue to red, whereas zinc remained blue. The colour of other interfering metals remained red or their chelates were decolourized by the oxine solution.

17.7.9 Mercury

Mercury in tissue is predominantly associated with protein, probably covalently bound, and therefore difficult to detect. Unfortunately micro-incineration, which removes the protein, simultaneously sublimates mercury. There is no satisfactory histochemical reaction for mercury salts. Mercury reacts with several metallochromes (Table 17.1) and with *sulphide methods* (Sect.17.7.11). In contrast to other metal sulphides, the brown to black mercury sulphide resists oxidation with strong hydrogen peroxide (15% H_2O_2 for 15 min). Mercury sulphide may possibly also be found in the red, less intensely coloured, cinnabar form.

17.7.10 Silver

Silver appears in the tissues as brown to black granules. The brown granules are blackened by treatment with hydrogen sulphide gas or ammonium sulphide solution. The state of tissue-bound silver is uncertain; it may be partly protein-bound and partly in the metallic state.

Silver salts are dissolved by solutions of potassium cyanide, sodium thiosulphate, or dilute sulphuric or nitric acid. *"Metallic"* silver may thus be extracted following pretreatment with an iodine/iodide solution (Weigert's or Lugol's iodine). One per cent potassium ferricyanide in 20% sodium thiosulphate simultaneously bleaches and extracts "metallic" silver. The treatment removes all silver from the section within 30 minutes, leaving melanin granules and carbon particles unchanged.

There is no acceptable histochemical method for the demonstration of silver.

17.7.11 "Heavy" Metals

Heavy metals form insoluble sulphides with hydrogen sulphide gas or ammonium sulphide solution. These sulphides are mostly brown to brownish black. The brown colour may be blackened by substitution of the heavy metal by silver as shown below.

Timm's Silver Sulphide Procedure. See Pearse (1985) p.996.

1. metal-X + $H_2S \rightarrow$ metal-S
2. metal-S + $2Ag^+ \rightarrow Ag_2S$ + free metal ion

After step 1 salts of silver, gold, cadmium, cobalt, copper, iron, mercury, nickel, lead, and zinc are black to brownish black. At this stage silver sulphide is of course black or brownish black, while the sulphides of gold, cobalt, copper, iron, mercury, nickel, and lead generally appear brown, and the sulphides of cadmium and zinc are uncoloured in tissue sections (cf. also Sect.8.3.4).

17.7.12 Anions

The anions present in biological minerals are predominantly carbonate, (ortho-)phosphate, polyphosphate, oxalate, and fatty acids. With the exception of polyphosphate, they mostly occur together with calcium (see Sect.17.7.1). The inorganic anions may be characterized as presented in Table 17.4 (for fatty acids, see Sect. 19.6.3).

Table 17.4. Characterization of anions in biological minerals.

	Toluidine Blue	Acid methods	Molybdate methods	Pb^{2+}-binding at pH 3
Carbonate	0	dissolves during formation of CO_2 bubbles	0	+/0
(Ortho-)phosphate	often blue	dissolves without bubble formation	blue colour, often *very* diffuse	+
Oxalate	0	does not dissolve	0	0
Polyphosphate	red	does not dissolve	0	+

Carbonates and Phosphates. Carbonate and phosphate are best distinguished by using the *acid method.* After the section has been mounted in water or paraffin oil, concentrated acetic acid is infiltrated under the edge of the coverslip and the mineral deposit is observed by phase contrast microscopy or with the condenser slightly defocussed. In Bunting's carbonate method (Pearse, 1985, p.1007) the presence of carbonate is revealed through the formation of barium carbonate. If no gas bubbles form, the mineral anion may be phosphate.

Molybdate Reduction. Molybdate reduction methods are specific for phosphate. Addition of molybdic acid leads to the formation of phosphomolybdate. This in turn is demonstrated by reduction to Molybdic Blue. The localization is, however, often *very* diffuse. The method is generally unsuitable for invertebrate material, possibly because invertebrate phosphate deposits have, for the most part, very little organic matrix to which the phosphomolybdic acid or Molybdic Blue may adsorb.

Cheng's (1956) variation is to be preferred, as it does not use benzidine for the reduction step.

Polyphosphates. Phosphate polymers with more than 7 phosphate units with anhydride linkage are termed polyphosphate and exhibit ethanol-resistant metachromasia with Toluidine Blue and other metachromatic dyes. In contrast to sulphated carbohydrates, polyphosphates bind lead ions at pH 3–4, and may therefore be demonstrated by a lead nitrate-ammonium sulphide sequence (Ebel *et al.*, 1958). The occurrence of polyphosphate granules ("volutin" granules) is doubtful in higher animals and plants, but they are relatively common in bacteria, fungi and protozoa (Sect.2.3).

18 Pigments

H. Lyon

Pigments are defined and classified in Sect.2.1.7. Metals and metal salts are discussed in Chap.17.

18.1 Description of Pigment Groups and Individual Pigments

18.1.1 Haem Group

This group of pigments is derived from the haem molecule which is characterized by the heteroaromatic pyrrole ring:

(also cf. tryptophan and its compounds, e.g. serotonin, Sect.18.1.4). The majority of haem pigments contain four pyrrole rings connected by methene groups and are called tetrapyrrolic pigments. One pigment, Dubin-Johnson pigment, may only comprise two rings and is therefore dipyrrolic. The basic structure in the tetrapyrrolic pigments is the porphin molecule:

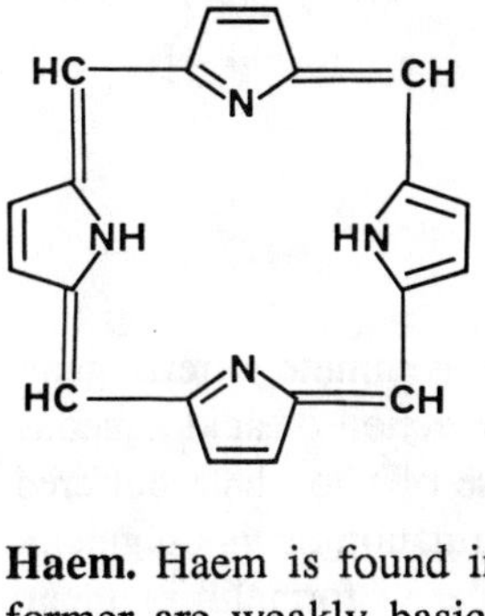

Haem. Haem is found in haemoglobins, myoglobins, and cytochromes. The two former are weakly basic and consist of a colourless protein moiety, globin, and

H. Lyon (Ed.)
Theory and Strategy in Histochemistry

a pigment moiety, haem. Haem comprises a ferrous iron (Fe^{2+}) complexed with porphyrin (a substituted porphin molecule):

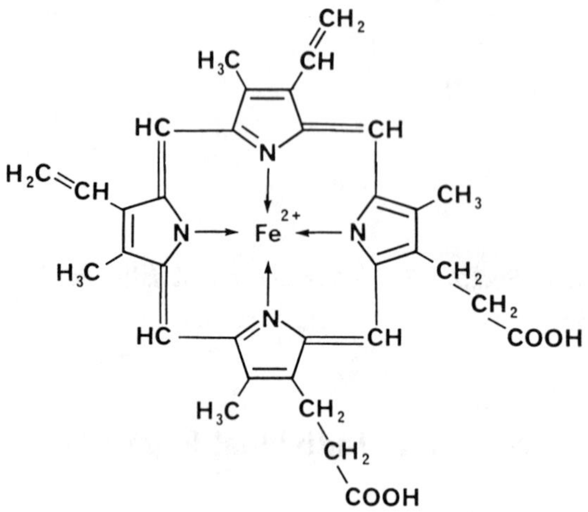

Porphyrins. These are derivatives of porphin which do not contain iron. In addition, the 8 H-atoms are substituted as shown above in haem. Porphyrins may be found in tissues in both congenital and acquired porphyria.

Haemosiderin. This is a golden brown, non-fluorescent pigment which is soluble in acid. It contains iron and consists of ferric ions which are bound, probably by complex bonds, to a glycoprotein which is termed aposiderin. It is formed in macrophages in the bone marrow and spleen during the normal degradation of red cells.

In pathological conditions haemosiderin is also formed in macrophages at other sites including Kupffer cells, around haematomas, and in congestive conditions. The latter are exemplified by the so-called heart failure cells in the lungs. The intestinal macrophages in pseudomelanosis coli and melanosis coli contain pigments which show characteristics for both haemosiderin and lipofuscin and possibly represent mixtures of these in varying proportions.

Acid Haematins. These comprise

1. Formalin pigment
2. Malarial pigment
3. Hydrochloric acid haematin

Acid haematins are dark brown pigments which occur as minute, birefringent, needles or rhomboids. Formalin pigment is formed by the action of acid aqueous solutions of formaldehyde on tissue rich in blood. The use of phosphate buffered formaldehyde solution at pH 7 effectively prevents the formation of this pigment. Malarial pigment, which has an appearance identical to that of formalin pigment, is found in malaria parasites around capillaries in the brain and in phagocytes in spleen, liver, bone marrow, and lymph nodes. An identical pigment is encountered in schistosomiasis.

Hydrochloric Acid Haematin. This is sometimes formed in small haemorrhagic lesions of the surface of the stomach.

Bile Pigments. These are formed as part of the normal degradation of haemoglobin from red cells. The haem part is dissociated from the globin and an oxidative opening of the ring system then leads to the formation of the iron containing product *verdohaemin.* The iron is transferred to apoferritin and the remaining product is *biliverdin.* This is produced mainly in the phagocytes of bone marrow and spleen and is transported via the blood to the liver, where, under normal conditions, it is reduced to *bilirubin* in hepatocytes. Bilirubin then forms a mono- or diglucuronide with one or both of the propionic acid residues to C1-hydroxyl in glucuronic acid. The conjugated bilirubin is secreted into the bile capillaries and passed on to the intestine where further reduction leads to *urobilinogen* (stercobilinogen). This is subsequently oxidized to *urobilin* (stercobilin), the yellowish-brown pigment in stool and urine.

Haematoidin. This is the golden-yellow to orange to reddish-brown, crystalline or amorphous pigment which arises in haemorrhagic infarcts and in old haemorrhagic foci. Some investigators identify this substance with bilirubin, while others claim that it is chemically different.

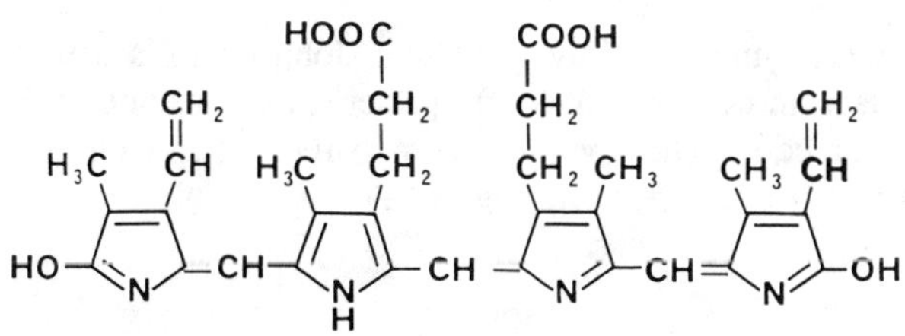

biliverdin

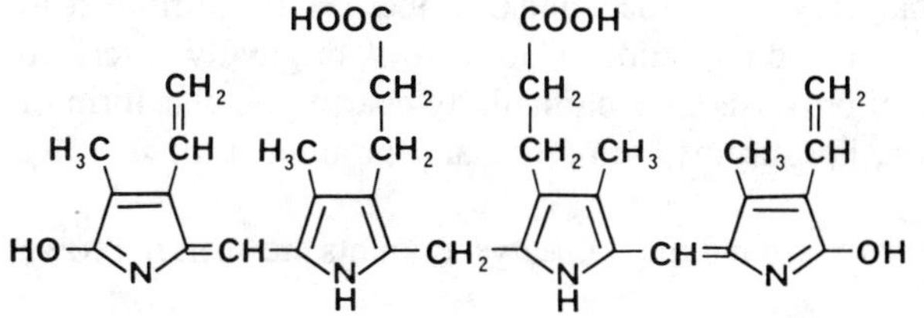

bilirubin

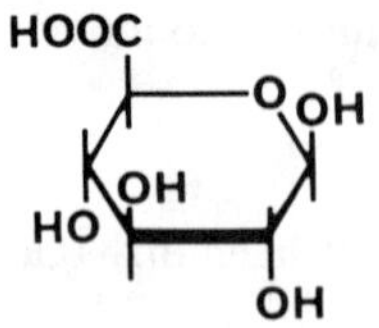

glucuronic acid

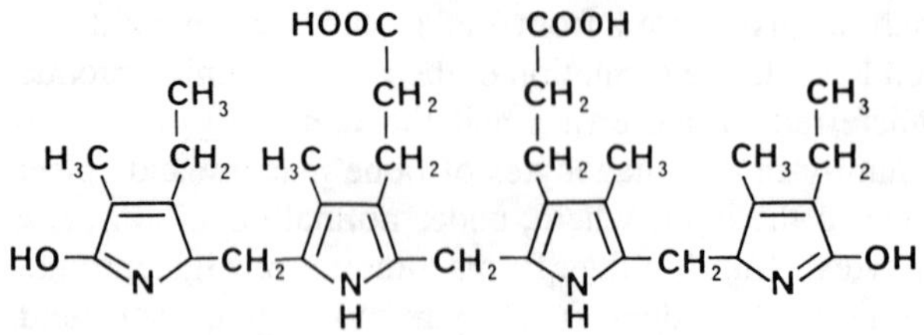

urobilinogen

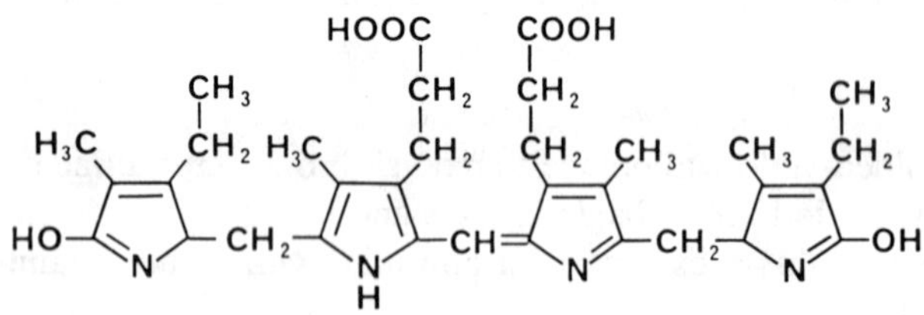

urobilin

Dubin-Johnson Pigment. This is formed in the liver in Dubin-Johnson's disease. It is ochre-yellow to brownish-black in colour and occurs principally in hepatocytes but occasionally also in Kupffer cells. The chemical composition is not clearly established in spite of claims that it is a dipyrrolic pigment.

18.1.2 Lipofuscin

Lipofuscins are yellowish-brown pigments which are derived from the degradation of fatty acids. Their somewhat varying histochemical properties are attributed to varying degrees of polymerization and oxidation. Thus, *ceroid*, originally described in Kupffer cells in experimental cirrhosis, is a particularly coarse granular form of an early lipofuscin. In addition, lipofuscins are particularly prominent in:

- liver cells
- cardiac muscle cells, especially around nuclei. Large amounts are seen in brown atrophy
- zona reticulosa of the adrenal cortex
- testis, especially Leydig cells

- ovary, especially surrounding corpora lutea
- cytoplasm of nerve cells in brain, spinal cord, and ganglia

18.1.3 Melanin

Melanins are complex polymer substances derived from tyrosine. They are found bound to protein in the form of granules. Melanins comprise the black to brown pigment encountered in epidermis, hair follicles, hair, naevi, and cutaneous malignant melanomas. Melanins further comprise the ocular pigments in the iris, ciliary body, choroid and here and there in the meninges. Malignant melanomas from the eye seem to develop from the choroid. The pigments in the substantia nigra, locus coeruleus and nucleus dorsalis nervi vagi in the brain differ histochemically from the above described melanins and are termed *neuromelanins*.

The biosynthesis of melanin in the skin and probably in the eye takes place in melanocytes, where, in the presence of oxygen, the enzyme tyrosinase slowly catalyzes the oxidation of tyrosine to dihydroxyphenylalanine (DOPA) and more rapidly the further transformation of DOPA to DOPA quinone.

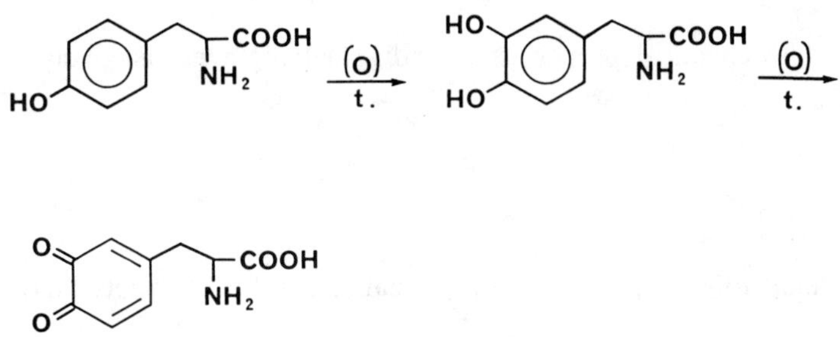

t. = tyrosinase

The subsequent steps occur spontaneously with oxidation and polymerization of several molecules and finally conjugation to protein.

18.1.4 Biogenic Monoamines

These include serotonin (5-hydroxytryptamine, 5-HT) and the catecholamines adrenaline, noradrenaline, and dopamine. They are not pigments but they do react with both formaldehyde and chromate during fixation to form chromogenic compounds which may be investigated histochemically. In some reactions these appear similar to melanin and lipofuscin. All the biogenic monoamines are found as transmitter substances in nerve terminals which are termed serotoninergic, adrenergic, noradrenergic, and dopaminergic as appropriate. Demonstration of these is described in Sect.30.2.1. Serotonin is also found in large amounts in the enterochromaffin (argentaffin) cells of the intestinal epithelium and in carcinoid tumours derived from these cells. Adrenaline and noradrenaline are most abundant in the adrenal medullary cells and in phaeochromocytomas, the corresponding tumour.

18.1.5 Carotenes

These are exogenic pigments composed of hydrocarbons with long chains and conjugated double bonds.

18.1.6 Carbon

This is one of the most frequently encountered exogenic pigments. It is frequently found in the lungs and in the mediastinal lymph nodes, but is also found in the skin due to tattooing.

18.2 The Histochemical Properties of the Individual Pigments

A summary of the histochemical properties of individual pigments is given in Table 18.1.

18.2.1 Haem Pigments

Haemoglobin is difficult to demonstrate histochemically. See further Sects.18.4.4 and 31.6.1.

Porphyrin. This is characterized by its red autofluorescence (Sects.30.1 and 31.6.2).

Haemosiderin. Haemosiderin (Sects.18.1.1 and 31.6.3) is a non-reducing pigment characterized by a positive reaction for iron (Sect.17.7.4). It is also PAS positive.

Acid Haematins. These pigments (Sects.18.1.1 and 31.6.4) are birefringent and are soluble in alcoholic picric acid.

Bile Pigments (Sects.18.1.1 and 31.6.5) give several rather unreliable reactions (see Sects.18.4.6, 18.4.7, 18.4.8, and 18.4.11).

18.2.2 Lipofuscins

These pigments (Sects.18.1.2 and 31.6.6) are autofluorescent (Sect.30.1) and reducing (positive ferric ferricyanide reaction, Sect.8.3.1). In addition, they show acid fast basophilia (Sect.18.4.1), resistance to bleaching with H_2O_2, and varying degrees of sudanophilia (Sect.19.6.1). Although they are derived from fatty acids, lipofuscins are resistant to lipid extraction (Sect.19.3) and are retained in paraffin sections.

18.2.3 Melanins

These reducing pigments (Sects.18.1.3 and 31.6.7) are somewhat more easily bleached by H_2O_2 than lipofuscins (Sect.18.4.2), but this is an unreliable reaction. If present in larger amounts the most selective method appears to be the ferrous ion uptake method (Sect.18.4.9). The DOPA reaction (Sect.18.4.13) may sometimes be of value.

18.2.4 Biogenic Monoamines

These compounds (Sects.18.1.4 and 31.6.8, and 31.6.9) are most selectively demonstrated by induced fluorescence techniques (Sect.30.2) but may also, if present in sufficient amounts, be demonstrated by their reducing properties (Sect.8.3.1) and their ability to azo-couple with diazonium salts (Sect.18.4.10).

18.3 Determination of an Unknown Pigment

From Table 18.1 it is apparent that many of the reactions are common to several pigments. It may therefore be difficult to select the first reaction. Table 18.2 provides an algorithm to make this task somewhat easier. The identity of a pigment determined using this key may be confirmed by selecting an alternative reaction according to Table 18.1.

18.4 Individual Reactions for Pigments

Several important pigment reactions are discussed below. Some other reactions have been considered in more appropriate sections (e.g. ferric-ferricyanide (Schmorl's, Sect.8.3.1) and argentaffin reaction, Sect.8.3.2, PAS-reaction, Sect.9.2.1, and sudanophilia, Sect.19.6.1).

Table 18.1. Reactions for pigments and similar substances.

	haemo-globin	por-phyrin	haemo-siderin	acid haematin	bile pigment	Dubin-Johnson pigment	lipo-fuscin	melanin	sero-tonin	cate-chol-amines	caro-tenes	ascorbic acid	carbon
Fluorescence[1]	0	+	0	0	0	+	+	0	+	+	+	0	0
Specific light absorption	+	0	0	0	0	0	0	0	0	0	0	0	0
Birefringence	0	0	0	+	0	0	0	0	0	0	0	0	0
Basophilia	0	0	0	0	0	0	+	+	+	+	0	0	0
Acid fast basophilia	0	0	0	0	0	(+)	+	0	0	0	0	0	0
Acidophilia	+	0	0	0	+	0	0	0	0	0	0	0	0
Sudanophilia	0	0	(+)	0	0	(+)	+	0	0	0	0	0	0
9-phenyl-2,3,7-trihy-droxy-6-fluorone	+	0	0	0	0	0	0	0	0	0	0	0	0
Bathophenanthroline	0	0	+	0	0	0	0	0	0	0	0	0	0
Bleached with H_2O_2	0	0	0	0	0	0	+	+	+	+	0	0	0
H_2O_2-Perls' reaction	+	0	+	0	0	0	0	0	0	0	0	0	0
Perls' reaction	0	0	+	0	0	0	0	0	0	0	0	0	0
Ferric ferricyanide	0	0	0	0	+	0	+	+	+	+	0	+	0
Argentaffin	0	0	0	(+)	+	+	+	+	+	+	0	+	0
Chromaffin	0	0	0	0	0	0	0	0	+	+	0	0	0
Gmelin	0	+	0	0	+	0	0	0	0	0	0	0	0
Aldehyde-oxidation	0	0	0	0	+	0	0	0	0	0	0	0	0
PAS	0	0	+	0	+	(+)	+	0	0	0	0	0	0
Ferrous ion uptake	0	0	0	0	0	0	0	+	0	0	0	0	0
Azocoupling	+	0	0	0	+	0	0	(+)	+	(+)	0	0	0
Soluble in alcoholic picric acid	0	0	0	+	0	0	0	0	0	0	0	0	0
DOPA reaction[2]	0	0	0	0	0	0	0	+	0	0	0	0	0

+ : positive reaction, (+): sometimes positive, 0: negative reaction.

[1] See 30.1; 30.2.

[2] Must be performed on frozen sections.

Table 18.2. Strategy for the identification of pigments.

1. Demonstration of iron: Perls' reaction
   + = haemosiderin
   0 ⟶ 2
2. Demonstration of reducing properties: ferric-ferricyanide reaction
   + ⟶ 3
   0 ⟶ 6
3. Demonstration of serotonin: coupling with Fast Garnet GBC
   + = serotonin (adrenaline, noradrenaline (melanin))
   0 ⟶ 4
4. Demonstration of acid fast properties: prolonged Ziehl–Neelsen
   + = lipofuscin
   0 ⟶ 5
5. Demonstration of melanin: ferrous ion uptake
   + = melanin
   0 ⟶ 6
6. Demonstration of birefringent properties
   + = acid haematin
   0 ⟶ 7
7. Demonstration of bile pigment: diazonium salt coupling
   + = bile pigment
   0 ⟶ 8
8. Demonstration of masked iron: alkaline H_2O_2-benzidine[1]
   alkaline H_2O_2-Perls' reaction
   + = haemoglobin
   0 = exogenous pigment

+ : positive reaction; 0: negative reaction.
[1]: See 18.4.4.

18.4.1 Acid Fast Basophilia ("Long Ziehl-Neelsen") for Lipofuscin

This is a modification of the Ziehl-Neelsen method for acid fast bacteria. Even the faintest red staining is considered positive for lipofuscin or possibly Dubin-Johnson pigment. The colour of the pigment itself can sometimes make the reaction difficult to assess. This is particularly true of highly polymerized lipofuscin. A doubtful reaction can be verified using a lysochrome reaction (best with Sudan Black B), as lipofuscin is not extracted from the tissue during the preparation of paraffin sections. According to Lillie and Fullmer (1976), p.516, it is an advantage to use Night Blue (C.I. 44085) instead of Basic Fuchsin.

The mechanism is not known in detail but it is believed to depend on binding of a cationic dye to carboxylic groups arising from the oxidation of alkene groups in the lipids (Sect.9.1). Phenol is added to the staining solution (carbolic Fuchsin) and this gives a final pH of between 4 and 5 providing greater selectivity. Staining should be carried out at 60°C at which temperature lipofuscin is more permeable to the dye. Differentiation is carried out with 1% w/v HCl in 99% v/v ethanol at room temperature. Under these conditions the dye is not extracted from lipofuscin.

18.4.2 Bleaching with Hydrogen Peroxide

Both lipofuscins, melanins, and monoamine derived pigments can be bleached by treatment with strong oxidizing agents.

The period of time necessary for complete bleaching with H_2O_2 makes a crude differentiation between pigments possible. With 10% w/v H_2O_2 monoamine derived pigments are bleached within 1 h, melanins require between 24 and 48 h treatment, while bleaching of lipofuscins takes more than 48 h.

18.4.3 Perls' Reaction for Ferric Iron

See Sect.17.7.4.

18.4.4 Alkaline Hydrogen Peroxide-Perls' Reaction

The iron in haemoglobin can be demonstrated after demasking with alkaline H_2O_2, this dissociates it from the haem molecule and causes oxidation to the ferric condition. This method, however, frequently fails to produce the required result. The H_2O_2-benzidine methods of Lepehne and Pickworth are alternatives; N.B. benzidine is carcinogenic (Pearse, 1985, p.888). Lillie and Fullmer (1976), p.447 and p.487 suggest the use of either the leuco-Patent Blue-H_2O_2 method or reaction with 9-phenyl-2,3,7-trihydroxy-6-fluorone.

18.4.5 Chromaffin Reaction for Monoamines

In this reaction potassium dichromate is used as the fixative. This oxidizes the monoamines to brown pigments of unknown chemical composition. In Hillarp and Hökfelt's iodate method (1953) potassium iodate is used as an oxidizing agent. This makes it possible to differentiate between adrenaline and noradrenaline as adrenaline takes longer to react. The method can be used in the diagnosis of phaeochromocytomas, but it is not nearly as sensitive as the fluorescence methods (Sect.30.2).

18.4.6 Gmelin Reaction for Bile Pigments and Dubin-Johnson Pigment

A few drops of the reagent (equal parts of concentrated nitric acid and ethanol) are placed on the section which is observed under the microscope. Bile pigments change colour according to the sequence yellow, green, blue, violet-red. The method is capricious and the colour unstable.

18.4.7 Aldehyde-Oxidation Method for Bile Pigment

The majority of aldehydes can oxidize bilirubin to biliverdin. This also applies to formaldehyde and the characteristic green colour is seen with formaldehyde fixed material on unstained sections or sections stained with reagents such as van Gieson's Picrofuchsin.

The yellowish-brown colour which bile pigments acquire in Al-haematein Eosin stained sections is due to mixing of the green colour of the pigment with the red colour of Eosin.

18.4.8 Stein's Iodine Method for Bile Pigments

This method is designed to oxidize bilirubin in sections to the green biliverdin. As already shown above this is a rather doubtful procedure on formaldehyde fixed material. The method is of limited value.

18.4.9 Binding of Ferrous Ions to Melanin

Ferrous ions are bound complexly to melanin and can be demonstrated with potassium ferricyanide, giving rise to Prussian blue ("Turnbull Blue", Sect.17.7.4). Due to its intrinsic colour, melanin appears dark green to green-black, while the background can be stained bluish, especially if washing in water after ferrous sulphate treatment has been inadequate. For instance goblet cells are frequently stained rather strongly blue. The method is, however, one of the most selective for melanin (green to green-black), but the sensitivity is only moderate and is surpassed by far by the *ferric ferricyanide method* (Sect.8.3.1).

18.4.10 Azo-Coupling

This is discussed in detail in Sect.9.4.1. It provides selective and sensitive demonstration of serotonin. If, for example, the stabilized diazonium salt Fast Garnet GBC is used, serotonin is stained red to orange, while the background appears yellow due to the content of protein bound histidine and tyrosine. Accumulations of haemosiderin can sometimes give problems of interpretation as the high content of tyrosyl and histidyl residues produces a colour approaching orange after coupling (control: PAS-reaction). In addition, due to its content of tyrosine derivatives, melanin acquires a brownish-yellow colour.

The method is of value in the diagnosis of carcinoids, and also phaeochromocytomas which, due to their high content of adrenaline and noradrenaline, sometimes give a positive reaction if these monoamines have not been completely extracted during tissue processing. Adrenaline and noradrenaline in the adrenal medulla are not usually retained in sufficient quantities after tissue preparation to give a positive reaction, just as the amounts of monoamines in nerve terminals are not high enough to give a visible result with this method. A considerably more sensitive method is formaldehyde induced fluorescence (FIF, see Sect.30.2.1).

18.4.11 Azo-Coupling Reaction for Bilirubin

This reaction is best performed on frozen sections with diazonium salts freshly prepared from either ethyl anthranilate, (Raja, 1965; Desmet *et al.*, 1968; Lillie and Pizzolato, 1970), Safranin O or Methylene Violet (Lillie and Pizzolato, 1969). When reacting with the diazonium salt bilirubin is hydrolytically cleaved giving rise to unsubstituted α-carbon atoms in two of the pyrrole rings to which the diazonium salt is coupled forming two isomer azobilirubins:

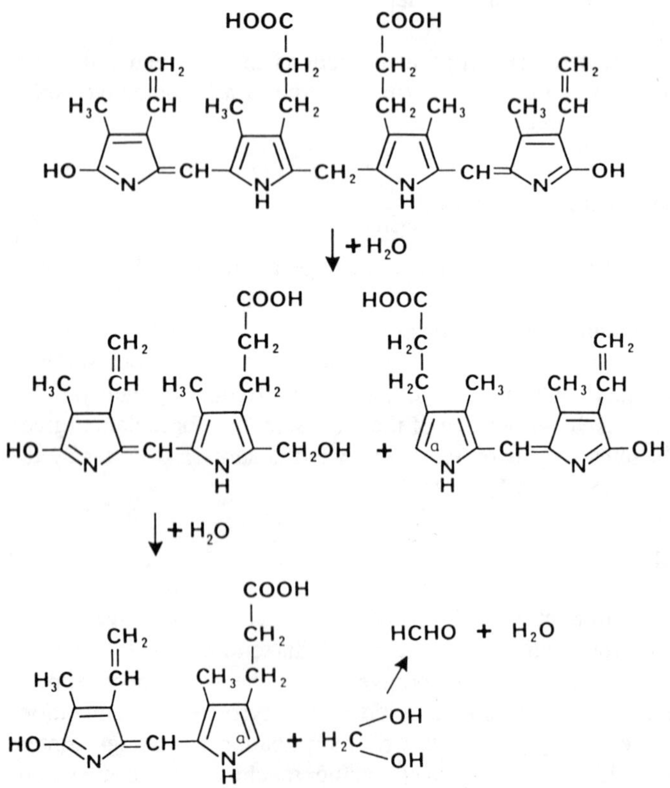

In our hands, however, these methods perform most erratically.

18.4.12 Solubility in Alcoholic Picric Acid

Characteristically, acid haematins are easily dissolved by treating sections with a saturated solution of picric acid in 99% v/v ethanol or 3% v/v ammonia in 70% v/v ethanol.

18.4.13 DOPA-Reaction

The DOPA-reaction (Lillie, 1956) is actually a reaction to detect the enzyme tyrosinase which, as shown in Sect.18.1.3, catalyzes the first steps in the synthesis of melanin. Cells containing tyrosinase (melanocytes) are able to oxidize DOPA (dihydroxyphenylalanine) to DOPA quinone after which the synthesis continues spontaneously to melanin. A positive DOPA-reaction therefore, shows formation of melanin, and it is the development of the brown-black pigment in cells which is registered. The reaction can be performed on frozen sections of fresh tissue but this does allow some diffusion. Formaldehyde fixation of small pieces of tissue for 1–4 h gives the best results. Usable results may still be obtained even after fixation in cold formaldehyde solution for up to a week.

Interpretation of the DOPA-reaction is rather difficult as:

1. Preformed melanin cannot be distinguished from that synthesized during the reaction
2. DOPA can be autooxidized in both alkaline and acid solution. The reaction should therefore be performed at pH 6.8–7.4
3. Other oxidases can oxidize DOPA and thereby initiate melanin synthesis. Leukocytes and red cells also give false positive reactions due respectively to their peroxidase content and the peroxidative activity of haemoglobin. Cytochromes may also have this effect, so that there is a theoretical possibility for reaction in non-melanin producing tumours

19 Lipids

P. Prentø

The definition, classification, and occurrence of lipids are discussed in Sect.2.1.4.

19.1 A Survey of Histochemical Reactions for Lipids

There are three central problems in lipid histochemistry: identification of lipids, their intracellular localization, and their native state in the cell. Most lipid histochemical methods do not allow reliable identification of lipids. The ability to interpret lipid reactions requires knowledge of the structure, biochemistry, and physiology of the tissue under examination as well as a clear idea of the appearance of different lipid accumulations with the various reactions available. Even with this basic knowledge great care and considerable imagination are necessary; for example, although often poorly understood, the pattern of metabolism, accumulation, and distribution of lipids in many invertebrates is very different from that occurring in vertebrates. *In vivo* a given lipid rarely or never occurs in isolation, is normally associated with other lipids, and may also be associated with non-lipids and metal salts. This of course complicates the interpretation of the histochemical pattern. The assignment of a histochemically demonstrated lipid to a particular class is therefore a conclusion arrived at on the basis of several different reactions. Only methods which are reasonably reproducible have been included here. These selected methods have been assembled into a key (Table 19.1).

19.2 Pretreatment

Sections of fresh frozen tissue should be used for histochemical investigation of lipids. Reactions should be performed shortly after sectioning as auto-oxidation of fatty acids can substantially change the physico-chemical properties of the lipids. If the sections are to be kept for more than a few hours they should be placed in a sealed container in a deep-freeze. Where fixation of other tissue components is required, buffered formaldehyde with added calcium chloride should be used as the

H. Lyon (Ed.)
Theory and Strategy in Histochemistry

Table 19.1. Histochemical methods for demonstration of lipids based on chemical and physical characteristics with notes on pretreatment.

Method	Reactive group	Lipid class	Staining result	Pretreatment[1]
	Demonstration according to chemical characteristics			
Ca-lipase	carboxylic ester	triglycerides	brown	A
Chromation-acid Haematein	choline + alkene	lecithin and sphingomyelin	blue/black	A
Gold-hydroxamate	carboxylic ester	phosphoglycerides	purple	C
Copper(II) acetate-rubeanic acid	carboxyl	free fatty acids	green	B
Modified PAS	vicinal hydroxyl	glycolipids	magenta	B
NaOH-chromation-acid Haematein	choline + alkene	sphingomyelin	blue/black	C
Nile Blue	phosphoryl sulphate carboxyl	phospholipids sulphatides free fatty acids	blue	B
Osmium tetroxide	alkene	unsaturated lipid	brown/black	A
Perchloric acid-naphthoquinone	3-hydroxy-Δ^5 (or $\Delta^{5,7}$) steroid configuration	cholesterol and cholesterol esters	blue/red	C
Plasmal reaction	alkene + ether	plasmalogen phospholipids	magenta	A
UV-Schiff	alkene	unsaturated lipid	magenta	A
	Demonstration according to physical characteristics			
Marchi reaction	hydrophobic character (+ alkene)	triglycerides and cholesterol esters	brown/black	A
Oil Red O	hydrophobic character	triglycerides and possibly cholesterol esters	red	B
OTAN	hydrophobic character (+ alkene)	predominantly triglycerides and cholesterol esters	brown/black	A
	hydrophilic character (+ alkene)	predominantly phospholipids	orange/brown-red	
Sudan Black B	hydrophobic and hydrophilic character	all lipids	black	B

[1] Pretreatment:
A: cryostat section of unfixed tissue, no postfixation.
B: cryostat section of unfixed tissue, postfixation 1 h in formol-calcium in refrigerator.
C: cryostat section of unfixed tissue, postfixation for at least 1 week in formol-calcium in refrigerator.

calcium immobilizes the phospholipids. Salt concentrations in excess of 0.1 mol/l should be avoided as phospholipids can appear to be hydrophobic under these conditions and thereby alter the results of procedures such as the OTAN reaction. The concentration of alkali metal ions should be kept as low as possible to avoid their extractive effect on phospholipids. Fixation should be avoided in material that has a high content of neutral lipid droplets (e.g. fatty liver) as this causes extensive confluence of the droplets. If fixation is essential then periodic acid (0.5%, 5°C, 5 min) can sometimes be effective.

Table 19.1 outlines recommended pretreatment for the different histochemical reactions for lipids (see Sect. 11.3.1 regarding cryostat sections).

19.3 Differential Extraction of Lipids

To enable control of the specificity of the histochemical method and to correlate histochemical and biochemical data on lipids, histochemical methods for lipids should in principle include staining results following differential extraction of lipids (Pearse, 1985, p.792).

The results of extraction experiments on tissue or frozen sections unfortunately do *not* fulfil the theoretical expectations and in addition are heavily influenced by preceding fixation (Edgar and Donker, 1957; Roozemond, 1969a; 1969b). Table 19.2 shows the solubility of *isolated, pure* lipids in organic solvents.

A lipid extraction may deviate from theoretical expectations in two ways:

1. Removal of insoluble lipids
2. Failure to extract soluble lipids

Table 19.2. Solubility of isolated, pure lipids.

Lipid class	Acetone	Methanol, ethanol	Diethyl ether	Chloroform	Chloroform-methanol 2:1
Triglyceride	C	(C)	C	C	C
Cholesterol	C	(H)	C	C	C
Cholesterol ester	C	(H)	C	C	C
Waxes	0 or H	0	0 or H	C or H	C
Phosphoglycerides	0	0 or C	0 or H	C	C
Lecithin	0	C	H	C	C
Cephalins, phosphoinositides	0	0	H	C	C
Plasmalogens	0	0 or C	0 or C	0 or C	C
Phosphosphingosides	0	0 or (C)	H	C	C
Sphingomyelin	0	C	H	C	C
Sphingocephalins	0	0	H	C	C
Glycosphingosides	0 or H	0	0	0 or H	C
Cerebrosides	H	0	0	H	C
Gangliosides[1]	0	0	0	0	C
Fatty acids					
Free fatty acids	C	C or H	C	C	C
Ca- and Mg-soaps	0	0	0	0 or (C)	C
Isoprenoids	C	C	C	C	C

C = soluble in cold solvent (0–20°C).
H = soluble in hot solvent (60°C).
0 = insoluble in the solvent.
( ) = usually soluble in the solvent.

[1] In addition non-protein bound gangliosides are soluble in water and mixtures of water and organic solvents.

Removal of Insoluble Lipids. Insoluble lipids may be removed by soluble lipids co-extracting the insoluble lipids completely or partially into the extracting solvent. This can occur both with pieces of tissue and with frozen sections. With pieces of tissue the circumstances are further complicated by the formation of a gradient between the extracting solvent and the aqueous phase of the tissue (water or a previously used extraction medium). As the solubility of a lipid in the gradient deviates from its solubility in the pure solvent a reproducible differential extraction is in practice impossible to carry out on pieces of tissue.

Failure to Extract Soluble Lipids. Soluble lipids may fail to be extracted due to binding to insoluble cell components. This may involve other lipids (particularly membrane lipids), proteins (notably phospholipids), or lipid-metal complexes (e.g. calcium and magnesium salts of cephalins, and fatty acids). The protein-bound lipids can be made soluble by treatment with protease (Tuqon and Adams, 1961; Adams and Bayliss, 1962) or by denaturing protein (absolute ethanol, picric acid, trichloroacetic acid, HCl etc.); disruption of metal-lipid binding can often be achieved with 0.1 mol/l HCl or 0.5% Na_2EDTA. EDTA treatment leads to Na-saponification and loss of free fatty acids.

Differential lipid extraction is not a routine procedure. Some general guidelines may, however, be given. As the solubility of lipids is increased with increasing temperature, solvents should *not* be used above 0–5°C, unless total lipid extraction is desired. Only solvents that are acceptable for use in biochemical lipid investigations should be used. Where there is doubt, the extracted lipids should be identified by chemical analysis or chromatography.

Extraction of homophasic lipids can usually be carried out without difficulty on sections fixed for short periods in aldehydes providing the material is fresh. For all other lipids fixation must *not* be performed before extraction. For differential lipid extraction, *only* fresh unfixed frozen sections should be used for starting material.

Lipid solvents, unless otherwise stated, should be completely anhydrous. Acetone, methanol, and ethanol may be dried over anhydrous $CaCl_2$. Frozen sections must be completely dry before they are placed in the solvents.

The procedure set out in Table 19.3 forms the basis for a rational differential lipid extraction. To counteract extraction and dislocation of phospholipids the frozen sections are first treated with a 1% $CdCl_2$ or $CaCl_2$ solution pH 7–7.5; see Table 19.3.

If a single step, total lipid extraction is desired then the water in the chloroform/methanol/water solvent should be replaced with 1 mol/l HCl. This makes calcium salts more readily extractable. Total lipid extraction may require preceding treatment with protease. The presence of water in chloroform/methanol aids the extraction of the hydrophilic lipids.

Table 19.3. Differential extraction of lipids. Procedure and results.

Procedure	Extraction of
1. Unfixed frozen section	
2. 1% $CdCl_2$, pH 7, 30 min; wash in water, 10 sec; dry; anhydrous acetone 0–5°C 1 h	triglycerides cholesterol cholesterol esters
3. 1 mol/l HCl, 30 min (possibly 0.1% trypsin, pH 7, 30 min followed by 1 mol/l HCl) wash in water, 10 sec; dry; anhydrous methanol 0–5°C 1–2 h	as above and: lecithin sphingomyelin free fatty acids
4. Chloroform : methanol : water 2 : 1 : 0.1, room temp. 1–2 h	as above and: cephalins, plasmalogens, glycosphingosides, and sphingocephalins

19.4 Masked Lipids

Lipid accumulations (droplets, granules) visible by light microscopy in most cells only constitute a small fraction of the total amount of lipid present. Lipids in membranes and disperse lipids are present in sections in concentrations that are too low to give visible staining with cytochemical methods. In addition, a portion of the lipid is present in a non-reactive form due to electrostatic and hydrophobic bonding to proteins (Cain, 1950; Hartmann and Fleck, 1952; Dixon, 1970; 1973). Lipids masked in this way can be made reactive by treating the section with protein denaturing agents such as ethanol, hydrochloric acid, concentrated sodium chloride solution, or with proteases.

These treatments cause disperse lipids to coalesce in varying degrees and they can often subsequently be demonstrated. In general membrane lipids also become more stainable after this treatment either because the lipids become more accessible following the demasking procedure or because the bonds between the reactive groups in lipid and protein are broken by denaturation.

Demasking leads to an increased solubility of intracellular lipids. This is probably due to more lipid being in an unbound state. Following demasking, gangliosides are usually soluble in water while lecithin is often soluble in lysochrome solvents (e.g. in 70% v/v ethanol). Similarly, cleavage of calcium soaps with acid frequently leads to a loss of fatty acids when lysochrome methods are applied.

The relationship between disperse and globular storage lipid (triglycerides, cholesterol esters) depends on the amount of lipoprotein and micelle-forming phospholipid in the cell. In fat cells and in other cells in certain pathological conditions (e.g. fatty liver) the amount of lipoproteins and phospholipids is low in relation to the amount of storage lipid, hence storage lipid forms globules. This process can take place very quickly and in a way corresponds to a natural demasking of storage lipid (Dixon, 1970; 1973).

19.5 Strategy for the Histochemical Investigation of Lipids

Before beginning histochemical analysis, it should be established which types of lipids are to be expected and, where appropriate, which types are of special interest to localize.

The biologist and the pathologist face different problems. The pathologist is working with relatively well-defined material in which deviations from the normal are to be determined; for example, the occurrence of triglyceride droplets in the liver or an abnormal accumulation of lipids in the central nervous system (sphingomyelin and cholesterol in Niemann-Pick's disease; cerebrosides in Gaucher's disease; etc., cf. Sect.31.7.3). The biologist, on the other hand, is usually working with, as yet, poorly understood material which, for example, is to be examined in relation to the physiology of nutrition (uptake, storage, and mobilization of lipid), or the physiology of reproduction (accumulation of lipids during the maturing of the ovum, changes in lipid hormone producing cells). The biologist is therefore principally interested in storage lipids and possibly also glycerophosphatides. Definitive identification of lipids in such investigations often requires chemical or chromatographical analysis of isolated material. This is especially true for investigations of invertebrates where the nature of stored lipid may be quite different from that found in vertebrates.

19.5.1 Control Material

A negative lipid reaction is only of value if the correct performance of the procedure has been confirmed using a positive control section. The following control material can be recommended: rat or mouse brain (all types of phospholipids, cholesterol); fatty liver (triglycerides, free fatty acids); rat adrenal gland (cholesterol, cholesterol esters).

All three tissues may be frozen into one common block and cut simultaneously. The cryostat sections may be stored in a deep-freeze, preferably at −70°C or below, for several months.

19.5.2 Identification of an Unknown Lipid

Table 19.4 is *not* exhaustive, but is sufficient for most histochemical investigations. As several types of lipids often occur together, it must be appreciated that a positive reaction for one type does not exclude the occurrence of others.

1. Lipid

	Lipid (including hydrocarbons and higher alcohols)
a Bromation-Oil Red O, 40 °C or Oil Red O, 40 °C and 60 °C	+
b Osmium tetrexide	+
c a and b after total lipid extraction with chloroform:methanol:HCl	0
	→ 2.

2. Hydrophobic or hydrophilic lipid

	Hydrophobic lipid (including hydrocarbons and higher alcohols)	Hydrophilic lipid
a Bromation-Sudan Black B, 40 °C or Sudan Black B, 40 °C and 60 °C	+	0
b Marchi method ($OsO_4/KClO_3$)	+	0
c a and b following extraction with acetone, 4 °C, 1 h	0	
	→ 3.	→ 4.

3. Hydrophobic lipid (storage lipid)

	Triglycerides	Sterols and sterolesters (cholesterol and cholesterol esters in animals)	Free sterols (cholesterol in animals)	Free fatty acids and calcium soaps	Free fatty acids and calcium soaps	Lipofuscin or phospholipids	Waxes, higher alcohols, hydrocarbons
a Calcium lipase	+						0
b PAN		+					0
c Digitonin-PAN			+				
d Nile Blue pH 1				+			0
e Copper(II)acetate-rubeanic acid					+	(+)	0
f b-e following HCl, 1 mol/l, 1 h - acetone, 4 °C, 1 h		0	0	0	0	+	

4. Hydrophilic lipid (heterophasic lipid)

	Phospholipids	Glycosphingosides (presumably cerebrosides)
a Nile Blue, pH 2	+	
b PAS		+
c a and b following HCl - acetone extraction	+	(+)
d a and b following chloroform:methanol:HCl extraction	0	0
	↓ 5.	

5. Phospholipids

	Phosphoglycerides	Acetal phosphatides (plasmalogens)	Lecithin	Cephalins	Sphingomyelin
a Gold - hydroxamic acid	+	+	+	+	+
b Plasmal reaction		+		0	
c Chromation - acid Haematein			+	0	
d NaOH - chromation - acid Haematein					+
c a-d following chloroform:methanol:HCl	0	0	0	0	0

+ positive reaction	(+) occasionally positive reaction
0 no reaction	↓ if the conditions are fulfilled: further to next step

19.6 Outline of Lipid Methods

19.6.1 Lysochrome Methods

Lysochromes differ from dyes proper in that they lack ionized groups. A lysochrome is practically insoluble in water, but readily soluble in alcohols, acetone, and liquid lipids. Lysochromes are soluble in alcohol/water mixtures but the saturation point is relatively low. The distribution of a lysochrome proceeds towards an equilibrium between the alcohol/water phase in the stain bath and the lipid phase in a section containing triglyceride droplets according to its relative solubility in the two phases. The solubility of the lysochrome in the lipid phase is always high compared to its solubility in the solvent (high partition coefficient), so the lysochrome is concentrated in the lipid provided the lipid phase and the solvent phase are immiscible. Following staining (10–20 min), the section is rinsed, "differentiated", by washing quickly with the lysochrome solvent. Critical factors in lysochrome staining are:

1. The concentration of the dye in the lysochrome solution
2. The lipid extractive or dislocative effect of the solvent
3. The staining temperature
4. The post-rinsing or differentiation step

These points are discussed individually below.

1). For ethanol-water mixtures, the ethanol content should never be below 70% v/v to ensure a sufficient concentration of the lysochrome dye (see Oil Red O). Below 60% v/v ethanol the solubility of the lysochrome is extremely low.

2). The ethanol content should not exceed 80% v/v as this leads to pronounced dislocation, or even extraction, of lipid, even with short staining times.

3). The staining temperature should always be kept as low as possible in order to minimize lipid extraction and dislocation. Except for tissue from mammals and birds, staining may generally be performed at 5–15°C, and even for mammals it is seldom necessary to use above 20–25°C, unless cholesterol or cholesterol esters are to be stained.

4). In order to permit differentiation without excessive loss of dye from the lipid droplets, neither lysochrome solution nor rinsing solvent should exceed an ethanol concentration of 80% v/v.

The organic solvents most often used for lysochrome solutions are ethanol (optimally 75% v/v), 2-propanol (60% v/v), and triethylphosphate (60% v/v). Of these triethylphosphate seems to be the most gentle with respect to lipid extraction and dislocation.

Visible accumulation of lysochrome implies that a substantial volume of lipid is present. If the lipid is present in a disperse form as opposed to droplets, the lysochrome cannot achieve a high enough concentration to give a visible result. Furthermore, if droplets are small, much of the lysochrome is washed out by the rinsing solvent. The possibility for a *false negative* reaction is therefore always present when using lysochrome reactions.

If the tissue section is pretreated with an acid protein coagulating agent such as picric acid or HCl, with urea, or with a protease the disperse lipid will be directly exposed to the aqueous phase and will sometimes coalesce into larger or smaller droplets. The thus "demasked" lipid can now be demonstrated with lysochrome methods. This approach does, however, give rise to artifacts of localization. Prior to lysochrome staining, the section may be pretreated for 5 min with cold 0.5% periodic acid. This stabilizes the section and improves the final morphology. At the same time it reduces the coalescence of the lipid droplets.

The most frequently used lysochromes are Oil Red O and Sudan Black B.

Oil Red O. This is a typical lysochrome. It normally only stains neutral lipids (triglycerides) but, at 60°C, cholesterol and cholesterol esters are also stained. Sometimes phospholipids are weakly stained as are fatty acids providing they have not been extracted by the lysochrome solvent. Commercial Oil Red O generally contains two components, one red and one orange yellow. The yellow component is more soluble at low ethanol concentrations than the red, and less soluble in triglyceride droplets. Thus, if the lysochrome solution is below 70% v/v ethanol, or only partly saturated, the yellow component tends to dominate and lipid droplets are stained orange instead of the bright red expected from saturated 70–75% v/v solutions. To ensure proper staining the ethanol content of the solvent should never be below 70% v/v and, during preparation, the temperature of the Oil Red solution should be kept at 70–80°C for several hours with constant or frequent agitation to ensure saturation or near saturation.

Very small triglyceride droplets may be stained more distinctly by use of Oil Red EGN (C.I. 26120), which is even more soluble in neutral fats than Oil Red O, and has a somewhat higher molar absorbtivity. Oil Red EGN offers no advantage for staining larger droplets.

Sudan Black B. This lysochrome stains all lipids and hydrocarbons if they are liquid. The staining of hydrophilic lipids is probably related to Sudan black being considerably more polar than the other lysochromes and the resultant possibility for binding in parallel to the phospholipid molecules in cell membranes. This organized disposition can be utilized by examining the section in polarized light where the Sudan Black B stained phospholipids show varying shades of red.

Sudan Black B contains two secondary amino groups which are responsible for the heterophasic properties of the compound. Unfortunately, the amino groups also make it possible for the molecule to be ionized thereby offering the properties of a cationic dye and possibilities for binding to non-lipids. Usually, this is of little importance but, if required, the lysochrome can, in theory, be made specific for lipids by esterifying the amino groups in Sudan Black B with acetic acid anhydride (Casselman, 1954).

Sudan Black B is sold as a mixture containing two main components (the ortho- and para-isomers of the same compound) and 8–11 coloured impurities (Marshall,

1977; Pfüller *et al.*, 1977). Sudan Black B is unstable, and older solutions (including stock solutions) contain degradation products which stain nucleic acids and proteins, but not lipids (Frederiks, 1977). It is therefore essential for reproducible and interpretable results, that the solution used, including any stock solution, is freshly prepared (not more than one week old). Even small differences in the concentration of organic solvent in the dye solution are of considerable importance for the concentration of the lysochrome. Ethanol should be used as 70% v/v and 2-propanol and triethylphosphate as 60% v/v solutions, and an araeometer should be used during the preparation. More lysochrome can be dissolved in both 2-propanol and triethylphosphate than ethanol, and this allows the staining time to be kept down to 5–10 min and minimizes the extraction of lipid.

In addition, triethylphosphate gives rise to considerably less confluence of lipid droplets than the other solvents and gives more stable solutions of lysochromes. Regardless of the solvent, staining should normally take place at room temperature as the extraction of lipid is considerably increased with increasing temperature.

Bromination. Bromination of the section prior to staining can considerably enhance lysochrome staining because firstly, it reduces the solubility of unsaturated lipids in lysochrome solvents and secondly, it effectively doubles the solubility of the lysochrome in the lipid. These two factors can render even very small lipid droplets visible. In addition, bromination leads to cholesterol and cholesterol esters becoming liquid at room temperature and thereby stainable. If large or even moderate sized accumulations of triglycerides are present, however, bromination should be discouraged as the increased fluidity promotes droplet confluence. With this reservation, the *bromination-Sudan Black B*, or possibly bromination-acetylated Sudan Black B, are probably the best routine methods for the demonstration of total lipid content (Bayliss and Adams, 1972).

Control Sections. For all staining with lysochromes it should be possible to extract the stain from a control section by brief immersion in absolute acetone or ethanol. In addition, preceding total lipid extraction should inhibit staining with Sudan Black B and extraction of neutral lipids should inhibit staining with Oil Red O. Extraction may not be possible if prolonged storage has allowed auto-oxidation and the formation of extensive cross-linking between the lipid molecules to occur.

Lipofuscin granules are natural auto-oxidation products of lipids. These lipid polymers take up lysochromes, frequently bind cationic dyes, but are not or only weakly affected by lipid solvents (Sect.18.2.2).

19.6.2 Nile Blue

According to Dunnigan (1968a) Nile Blue contains two dyes, a blue oxazine and a red oxazone, where the latter is an oxidation product of the oxazine. Nile Blue has a positive charge, but is otherwise a lysochrome that binds to phospholipids and free fatty acids including calcium soaps. The red oxazone is uncharged and reacts as a lysochrome. In concentrated solutions (1% and above) phospholipids stain red, fatty acids and their calcium soaps blue to red, and storage lipids more or less strongly red. In dilute solutions (less than 0.1%) staining with the oxazone is delayed, and both phospholipids and fatty acids are stained blue.

The selectivity of Nile Blue is pH dependent as for other cationic dyes. Below pH 2 staining of nucleic acids can be avoided, while phospholipids and fatty acids are still stained.

The oxazine-oxazone mixture cannot be used to distinguish between phospholipids and fatty acids with certainty. Moreover, Oil Red O is greatly preferable for staining storage lipids, thus it is really only the blue oxazine which is of interest from a cytochemical point of view.

Dunnigan (1968b) has described a reproducible and simple Nile Blue staining procedure with good selectivity for phospholipids. The method is based on treating the tissue section with 1 mol/l hydrochloric acid followed by acetone extraction applied to the dried section. This treatment removes RNA and fatty acids and leaves only phospholipids for Nile Blue staining at pH 2. To control co-extraction of phospholipids, an adjacent section should be stained following the hydrochloric acid treatment alone. The presence of calcium soaps can be controlled with Holczinger's copper acetate-rubeanic acid method for fatty acids and with the Alizarin Red S method for calcium salts (see Sect.17.7.1), providing appropriate controls are performed.

19.6.3 Copper (II) Acetate-Rubeanic Acid

Holczinger's copper (II) acetate-rubeanic acid (Holczinger, 1959) is a method for the demonstration of free fatty acids. In tissues these are predominantly found as calcium soaps and show a pronounced ability to bind metal ions, especially those of heavy metals. The most selective binding is achieved using a very low concentration of Cu^{2+} ions (0.005% copper (II) acetate in distilled or demineralized water). The bound copper ion is then demonstrated with rubeanic acid (dithiooxamide).

A copper (II) acetate-rubeanic acid method with high selectivity for free fatty acids has been worked out in accordance with the solubility properties shown in Table 19.5 (Elleder and Lojda, 1972).

In the modified copper (II) acetate-rubeanic acid method two sections are first treated with 1 mol/l HCl for 1 hour and dried. One (control) section is extracted with anhydrous acetone at 4°C for 20 min and both are then taken through the Holczinger method. Green to dark green colour in the test section alone indicates the presence of free fatty acids and/or calcium soaps.

19.6.4 Plasmal Reaction

The plasmal reaction (Hayes, 1949; Cain, 1949a, 1949b) demonstrates plasmalogens ("acetal" phosphatides).

Table 19.5. Reactivity with copper(II) acetate-rubeanic acid following extraction with hydrochloric acid and/or acetone.

	No pretreatment	1 mol/l HCl, 1 h	Acetone 4°C, 20 min	HCl-acetone
Acid proteins and sulphate containing proteoglycans	+	+	+	+
Phospholipids	+/0	+/0	+/0	+/0
Lipofuscins	+	+	+	+
Free fatty acids	+	+	0	0
Calcium soaps	+	+	+	0
Inorganic calcium and magnesium salts	+	0	+	0

The reactive group is an α,β-unsaturated ether bond between the α fatty acid chain and glycerol. This bond is most unstable and is hydrolyzed with mercuric chloride forming a fat-aldehyde-mercury addition product.

The fatty aldehyde can be demonstrated with Schiff's reagent which binds to the aldehyde group (Sect.9.7.1) or alternatively with reagents such as diphenylcarbazone which react with mercury. This latter approach, however, also demonstrates Hg bound non-specifically to alkene groups. Schiff's reagent is therefore preferred for the demonstration of fatty aldehydes.

The alkene groups in unsaturated lipids are relatively easily oxidized by atmospheric oxygen giving rise to Schiff-reactive aldehyde groups, "pseudoplasmal" reaction (Cain, 1949a). In fixed tissue and in older frozen sections a strong pseudoplasmal reaction can "drown" the plasmal reaction. For this reason unfixed, completely fresh, frozen sections should always be used.

With vertebrate material a control section which only has been treated with Schiff's reagent should be colourless apart from any lipofuscin present (Sect.18.2.2). In the case of invertebrate material a positive pseudoplasmal reaction *may* sometimes be natural.

19.6.5 Osmium Tetroxide

Osmium tetroxide (Adams, 1965, p.34; Adams *et al.*, 1967) selectively demonstrates alkene groups in unsaturated fatty acids, both free and in esters. OsO_4 should always be used in aqueous solution together with $CaCl_2$ or $CdCl_2$. As OsO_4 vapour is extremely toxic, any handling of the reagent should take place in a fume-hood. The process of addition of OsO_4 to the alkene group is described in Sect.9.1.1. The unsaturated storage lipids are oxidized to relatively insoluble products. The OsO_2 formed, "osmium black", remains in the lipid phase. Black, spherical accumulations are therefore a specific indicator for the presence of liquid storage lipids (triglycerides, cholesterol esters). Triglyceride accumulations containing only saturated, fatty acids do *not* react with OsO_4. This sometimes makes it necessary to perform a control with a lysochrome staining at elevated temperature. In heterophasic lipids the OsO_2 formed makes cellular membranes electron dense, giving the classical trilaminar appearance.

In unfixed frozen sections only alkene groups can reduce OsO_4 to OsO_2. The formation of "osmium black" is therefore a specific indicator for the presence of unsaturated lipid.

Marchi Methods. In the Marchi methods (Adams, 1958; 1960; 1965, p.36) it is possible by using OsO_4 together with another oxidant to distinguish between hydrophobic and hydrophilic lipids. Usually, potassium chlorate ($KClO_3$) is used as the other oxidant. OsO_4 then acts as previously described, however, when the lipid present is hydrophilic, $KClO_3$ penetrates and hinders the complete reduction to "osmium black". Instead, uncoloured addition compounds containing hexavalent osmium are formed. The result is that only hydrophobic lipids, i.e. fat droplets and possibly disperse triglycerides, are stained. (See however Sect.19.7, myelin methods). OsO_4 and $OsO_4/KClO_3$ are excellent methods for the demonstration of lipid as long as one remembers that any saturated short chain fatty acids present do not react, and that disperse hydrophobic lipid sometimes remains unstained with the Marchi method.

Sections stained with Marchi methods should be evaluated and any photomicrographs taken immediately after mounting as hexavalent osmium is gradually reduced with the formation of more "osmium black". This readily occurs with alcohol containing mountants.

OTAN Method. The OTAN method (Adams, 1959; 1965, p.40) is a further development of the $OsO_4/KClO_3$ method where the double oxidation is followed by treatment with α-naphthylamine (**o**smium **t**etroxide-**a**lpha-**n**aphthylamine); where hydrophilic lipids are present, α-naphthylamine forms a complex with hexavalent osmium giving rise to red to brownish red α-naphthylamine-OsO_4-alkene products of addition. The staining results are, however, very variable and furthermore, α-naphthylamine is very carcinogenic. A comparison of sections stained with OsO_4 alone, $OsO_4/KClO_3$, and Sudan Black gives far more information than OTAN.

19.6.6 Gold-Hydroxamic Acid Reaction

This method (Adams and Davison, 1959; Adams *et al.*, 1963) is specific for phosphoglycerides. As noted above the carboxyl ester bonds in phosphoglycerides are labile to alkali. In the presence of hydroxylamine a hydroxamic acid, or more correctly, a hydroxamate is formed during hydrolysis, which can form coloured complexes with different metals.

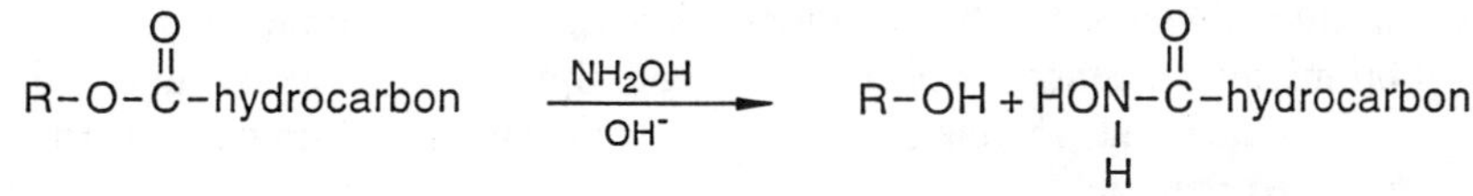

hydroxamic acid

If an ammmoniacal silver solution is used, the silver ions are reduced to free silver, and the phosphoglycerides are stained black. It is usual to follow this by toning with gold chloride.

As the reagents are insoluble in the lipid phase, triglycerides and cholesterol esters do not react. Sphingolipids do not contain carboxyl ester bonds, but the more alkali-resistant amide bond. The method is therefore specific for phosphoglycerides.

Treatment with NaOH makes it necessary to fix the section. This must not be too prolonged as lecithin forms lysolecithin during formol-calcium treatment.

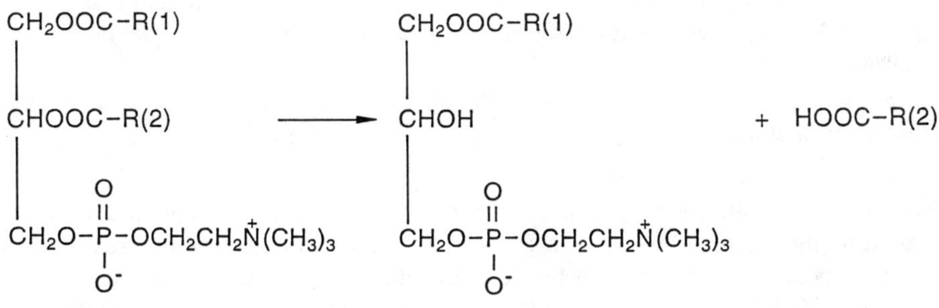

lecithin lysolecithin free fatty acid

19.6.7 Chromation-Acid Haematein

In this method (Baker, 1946; Elftman, 1954; Sinapius, 1961) the dichromate anion, $Cr_2O_7^{2-}$, oxidizes alkene groups in hydrophilic lipids relatively quickly, while hydrophobic lipids, due to the insolubility of the dichromate anion in the lipid phase, are oxidized slowly. During oxidation, cross-links are formed between the fatty acid chains and the hydrophilic lipids are rendered insoluble in organic solvents. Chromated tissue can therefore be embedded in paraffin and cut essentially without loss of phospholipids. During oxidation divalent chromium is formed, possibly as CrO, and this chelates with aminophosphate in the choline-containing phospholipids, lecithin and sphingomyelin.

On treatment of the chromated section with an acid Haematein solution, Haematein is bound to the divalent chromium which is concentrated in areas where dichromate has reacted with lecithin and/or sphingomyelin. Deposits of these lipids are thus stained black to dark blue. Contrary to other double-bond reactions chromation-acid Haematein can *not* be blocked by prior addition of bromine. Dichromate reacts with brominated alkene groups to form stable lipid-bromine-chromium addition compounds which bind Haematein in a way that cannot be distinguished from the non-brominated material.

Several variants of the chromation-acid Haematein method have been suggested. The best and most reproducible remains Baker's original method from 1946. It is, however, now mainly used on frozen sections instead of on pieces of tissue and the chromation step at 60°C is, under these circumstances, shortened to 2 h to avoid a partial chromation of hydrophobic lipids.

To preclude staining of non-lipid, (e.g. Haematein binding to iron or calcium deposits), a chloroform/methanol extracted adjacent section should be included or appropriate control reactions performed. For the specific demonstration of choline-containing phospholipids the differentiation step (sodium tetraborate/potassium ferricyanide) is an integral part of the method and must *not* be shortened.

Used on frozen sections, chromation-acid Haematein can be combined with Oil Red O staining, allowing demonstration of neutral lipids and phospholipids in the same section (Bourgeois and Hubbard, 1965). The treatment with acid Haematein can be substituted with Sudan Black and this results in all lipids being stained blue-black.

Some batches of Haematein and haematoxylin are not suited for use in the acid Haematein method.

NaOH-Acid Haematein. The precise mechanism of the chromation-acid Haematein method is unknown, but the presence of choline is clearly necessary. This makes it possible to distinguish between lecithin and sphingomyelin (Adams and Bayliss, 1963) as lecithin is hydrolyzed in strong alkali (2 mol/l NaOH for 1 h at 37°C) and only sphingomyelin remains for the reaction with chromation-acid Haematein. The method requires good preceding fixation.

19.6.8 UV-Schiff Reaction

In the UV-Schiff reaction (Belt and Hayes, 1956) ultraviolet irradiation for 3–4 h transforms alkene groups in unsaturated lipids to aldehyde groups and these can subsequently be demonstrated with Schiff's reagent. This oxidation method is considerably more specific, easier to perform, and more gentle than performic acid or peracetic acid oxidation. A high-pressure mercury lamp is a suitable source of light. When used on unfixed sections the method is specific for unsaturated lipids. Larger fat droplets will usually only be stained on their circumference. After similar irradiation, polyunsaturated fatty acids are fragmented and this sometimes results in diffusion of the aldehyde products of hydrophilic lipids. Microscopic examination should therefore take place immediately after the reaction.

A control section (no irradiation) is essential to rule out any possible "pseudoplasmal" reaction (see Sect.19.6.4).

19.6.9 Modified PAS Reaction

There are no specific direct methods for the demonstration of the glycosphingoside group as such. However, the PAS (periodic acid-Schiff's reagent) method (Adams and Bayliss, 1963) gives a positive reaction with the 1,2-glycol groups in the hexoses in *cerebrosides* and in protein-bound gangliosides. The majority of the gangliosides are not protein-bound and are extracted by water. As both polysaccharides and glycolipids can give a positive PAS reaction, the presence of glycolipid must be confirmed by lipid extraction and various blocking reactions. It is usually sufficient to perform a total lipid extraction and a "pseudoplasmal" reaction (see Sect.19.6.4) as a control for the demonstration of cerebroside. It is important not to oxidize for more than 5 min in periodic acid as this makes alkene groups reactive.

Recently Buk and Bayliss High (1986) have described a borohydride-periodate-Schiff sequence as a reliable method for the specific demonstration of sialic acid containing lipids *(protein-bound gangliosides).*

There is no satisfactory method for the demonstration of *non-protein bound gangliosides.*

19.6.10 Calcium Lipase Method

Triglycerides lack unique reactive chemical groups and can therefore only be identified unequivocally using enzymes. Adams *et al.* (1966) have developed a calcium lipase method. The triglycerides are cleaved into glycerol and fatty acids by pancreatic lipase. In the presence of calcium ions, insoluble calcium soaps are formed which are transformed into lead soaps on the addition of lead nitrate. Bound lead is visualized by reaction with ammonium sulphide which gives rise to the formation of brown to black sulphide.

Calcium salts and free fatty acids and calcium soaps present in the section prior to treatment also react with ammonium sulphide $(NH_4)_2S$. A control section, where treatment with lipase has been omitted, is therefore necessary.

As the enzymatic degradation of the lipid and soap formation takes place at the water-lipid boundary, larger lipid droplets will only be stained on their circumference.

19.6.11 Perchloric Acid-Naphthoquinone

The perchloric acid-naphthoquinone method (PAN, Adams (1961)) demonstrates 3-hydroxy-5-ene- and 3-hydroxy-5,7-diene-sterols, i.e. cholesterol, ergosterol, stigmasterol, and sitosterol. Of the demonstrable sterols, cholesterol is by far the most important in animals, while sitosterol is the most important in plants. Both free sterols and their esters react in the PAN method with the formation of blue or red pigments. The selectivity and rationale of the method is, as for other methods for sterols, very uncertain.

Crystalline cholesterol is normally stained blue with the reaction, while disperse or lipid dissolved cholesterol will be stained red. The colour is stable for several hours, and the method is relatively gentle to the tissue. For this reason the PAN method is much preferred to the previously used Schultz' methods, which use 5–8 mol/l H_2SO_4.

Carbohydrate deposits may sometimes give a weak red staining with PAN and an acetone extracted control section should therefore always be included.

Digitonin-PAN Method. If cholesterol and cholesterol esters are to be distinguished from each other, the digitonin-PAN method is often used (Adams and Bayliss, 1974). The sensitivity of this method is not overwhelming.

19.7 Myelin Methods

Most vertebrate axons are surrounded by a myelin sheath which predominantly consists of multiple layers of plasma membrane produced by Schwann cells. In consequence, myelin consists almost exclusively of protein and lipid, of which the lipid forms the greater part by far (70–80% dry weight). Methods for demonstrating myelin are therefore methods for demonstrating lipid. The most important lipids quantitatively are cholesterol, acetal phosphatides (plasmalogens), phosphatidyl cholines (lecithin), and sphingomyelins. Cephalins (phosphatidyl ethanolamines, phosphatidyl serines, and phosphatidyl inositols) and glycosphingosides (cerebrosides and gangliosides) also occur in large amounts.

The following lipid histochemical methods can be recommended for the routine staining of myelin: Chromation-acid Haematein, plasmal reaction, PAN method, and Sudan Black B.

In addition, Luxol Fast Blue methods (Salthouse, 1962) are useful. For this procedure, following fixation in cetyltrimethyl ammonium bromide, tissue is embedded in paraffin and the sections deparaffinized and stained with Luxol Fast Blue ARN.

Procedures using Nile Blue or OTAN, and various more traditional Haematoxylin methods are described in Lillie and Fullmer (1976), pp.576–579. While normal myelin is stained with osmium tetroxide alone, it is not stained with mixtures of osmium tetroxide and potassium chlorate as all the myelin lipids capable of being oxidized are hydrophilic and thus accessible to the chlorate anion. In *degenerating* myelin the lipid composition is gradually changed with an increase in hydrophobic or homophasic lipid (triglycerides, cholesterol esters), while the amount of hydrophilic or heterophasic lipids falls or even disappears.

Degenerating myelin can thus be identified by black staining with the osmium tetroxide-potassium chlorate mixture (Marchi method). With the OTAN method the normal red to red-brown colour of the myelin gives way to black staining due to reduction of osmium tetroxide to "osmium black".

20 Nucleic Acids

P.E. Høyer, A.K.N. Iversen, E. Schulte, H. Lyon

20.1 Methods for the Demonstration of Nucleic Acids

Some of the most important methods used for demonstrating nucleic acids are enumerated in Table 20.1, while Table 20.2 outlines a strategy for their identification using the classical histochemical reactions.

Table 20.1. Cytochemical methods for investigating nucleic acids.

Method	Demonstrates DNA	Demonstrates RNA	Selectivity	Quantitative method	Reference
Ultraviolet absorption	+	+	high	yes	28.8.4
Autoradiography	+	+	high	yes	Chapter 29
Affinity probe techniques					
BrdU-immunofluorescence	+	0	high	?	20.5
In situ hybridization					
RISH	+	+	high	?	20.6
NISH	+	+	moderate	?	20.6
Polymerase chain reaction	+	0	high	no	20.7
Basophilia					
Toluidine Blue	+	+	moderate	yes	6.1.1
Methyl Green-Pyronin	+	+	moderate	yes	6.1.5
Cuprolinic Blue	+	+	high	?	20.2
Chromium-Gallocyanin	+	+	high	yes	7.3
Feulgen	+	0	high	yes	9.9
Control methods					
RNase	0	+	high	yes	3.4.3
DNase	+	0	high	yes	
Extraction with acids	+	+	high		

Under well-defined conditions, cationic dyes such as Toluidine Blue can bind selectively to nucleic acids. One of the conditions is that pH should be sufficiently low for proteins to exhibit a positive net charge, i.e. pH not over 3 (cf. basophilia Sect.20.2).

Another approach is to use specific probes for the detection of the nucleic acids. These affinity probe techniques can be divided into:

H. Lyon (Ed.)
Theory and Strategy in Histochemistry

Table 20.2. Identification of nucleic acids in a tissue component which exhibits basophilia.

1. Demonstration of nucleic acid:	chromium Gallocyanin
	+ = nucleic acid or polyphosphate ⟶ 2
	0 = acid carbohydrates ⟶ Table 22.2
	0 = acid lipids ⟶ Table 19.4
2. Demonstration of DNA:	RNase followed by chromium Gallocyanin
	+ = DNA or polyphosphate ⟶ 3
	0 = RNA
3. Demonstration of DNA:	Feulgen's nucleal reaction
	+ = DNA
	0 = polyphosphate ⟶ 17.7.12

a. Radioactive, as used in autoradiography (Chap.29) or radioactive in situ hybridization (Sect.20.6)
b. Non-radioactive, as in the 5-bromo-2′-deoxyuridine immunohistochemical reaction (Sect.20.5) or non-radioactive in situ hybridization (Sect.20.6)

20.2 Basophilia

The use of Toluidine Blue for demonstrating nucleic acids is discussed in Sect.6.1.1, while the Methyl Green-Pyronin method is treated in Sects.6.1.5 and 28.8.4. The use of Cuprolinic Blue and Chromium-gallocyanin can also be included under this heading.

Cuprolinic Blue. The phthalocyanin-like dye Cuprolinic Blue, 'quinolinic phthalocyanin' (Scott, 1972c) is an analogue of Alcian Blue with the S-methylene tetramethyl-isothiouronium side groups removed and N-methylpyridine substituted for benzene in the central ring structure (see formula for Alcian Blue, Sect.6.1.2). The staining pattern of Cuprolinic Blue resembles that for Alcian Blue (Sect.6.1.2) but in addition the dye shows high affinity for nucleic acids, especially RNA (Scott, 1980). In the presence of magnesium chloride the dye has been used as a selective stain for RNA (see Critical Electrolyte Concentration, Sect.6.1.2), (Scott, 1972c; 1973b; Mendelson *et al.*, 1983; Tas *et al.*, 1983).

Chromium-gallocyanin. This metal complex dye (cf. Sect.7.3) binds to the phosphate groups of the nucleic acids.

20.3 Feulgen's Nucleal Reaction

The principles for this method have been given in Sect.9.9.

20.4 Application of Reactions for Nucleic Acids

The Feulgen nucleal reaction is the best method for in situ quantitation of DNA in nuclei. The Feulgen reaction in itself does not give any information as to the functional condition of the nucleus (cell).

In situ investigations on the localization and relative amounts of RNA were performed by Brachet (1940a; 1940b; 1942; 1953) and formed the basis for the first hypothesis on the exchange of information and transport:

nucleus (DNA $\rightarrow$ RNA) $\rightarrow$ cytoplasm (RNA $\rightarrow$ protein).

Brachet used the Methyl Green-Pyronin reaction coupled with RNase for the specific demonstration of RNA. This method is fast, reliable, and gives a reasonably good histological picture. For cytological details it is less well suited than Toluidine Blue or Cresyl Violet Acetate for example. The majority of cationic dyes are metachromatic and therefore not well-suited for quantitative studies. For this purpose Cr-gallocyanin is preferable.

Brachet's investigations showed that, in most cases, there was a functional connection between the amount of RNA in the cytoplasm, the size of the nucleolus and the content of non-histones in the nucleus for one cell type. Consideration of the level of protein synthetic activity should include these measurements as well as the ratio between the volumes of nucleus and cytoplasm (Sect.2.2.1). In normal circumstances, the following cell types contain high amounts of RNA and large nucleoli (usually):

- fibroblasts, chondroblasts and osteoblasts,
- plasma cells, both free and in active lymph nodes,
- nerve cells,
- exocrine protein secreting cells,
- epithelial cells continually being replaced (e.g. in the stratum germinativum of the skin, and in the intestines),
- embryonic cells undergoing rapid divisions,
- many tumours.

20.5 The 5-Bromo-2′-Deoxyuridine Method

Monoclonal antibodies (cf. Sect.26.1.4) specific for 5-bromo-2'-deoxyuridine (BrdU) provide a sensitive method for detecting DNA replication in single cells in a manner analogous to the use of tritiated thymidine (Gratzner *et al.*, 1975; Gratzner, 1982). It is thus possible to apply the usual techniques of immunocytochemistry (Chap.26) to the study of DNA synthesis in individual cells. The technique may, after incorporation of BrdU, be used for in vivo or in vitro studies of frozen or paraffin sections and for cultured cells, smears, cytospins, and chromosome spreads. The advantage of this method is that it avoids the technical difficulties of emulsion autoradiography and also that results may be obtained in much shorter times. Sensitivity can be increased by exposing cells to be labelled with BrdU simultaneously to 5-fluoro-2′-deoxyuridine, an inhibitor of thymidilate synthetase, thus increasing BrdU incorporation by decreasing competition from endogenous thymidine (Ellwart and Dörmer, 1985). It is necessary to denature cellular DNA to allow access of BrdU. This can be achieved by nuclease digestion simultaneously with the antibody incubation (Gonchoroff *et al.*, 1985). Nuclease treatment can be substituted using microwave irradiation (van de Kant *et al.* 1990).

Another approach was described by Apte and Puddle (1990). Their use of sodium ethoxide to remove plastic from tissue sections made a separate denaturation step of DNA unnecessary. The effect of different fixatives was compared. Best results were obtained with formaldehyde or Bouin's fluid.

Several authors (Hamada, 1985; Morstyn *et al.*, 1986; Apte and Puddle, 1990) have found a high correlation between the incorporation of ^{3}H-thymidine in DNA as detected by autoradiography and data based on the BrdU technique.

Frederiks *et al.* (1990) using BrdU-immunocytochemistry on isolated hepatocytes after partial hepatectomy in rats found the same labelling index of binuclear diploid, mononuclear tetraploid, and binuclear tetraploid cells. There did not appear a special role for mononuclear diploid cells in proliferation.

20.6 In Situ Hybridization

In situ hybridization histochemistry involves demonstration of specific nucleic acid sequences which may comprise, sequences with no identifiable gene products, or sequences of exogenous (e.g. microbial) origin. Estimates of the level of gene transcription may also be obtained by detecting mRNA. In some respects the technique may be considered analogous to immunohistochemistry (Chap.26) since both use highly specific probes for cell components and both require some means of generating a signal. In situ hybridization was first introduced to histochemistry by John *et al.*, (1969) and Gall and Pardue (1969).

Mechanism. The central procedure involves denaturation (melting) of double-stranded nucleic acid by heating to 80–100°C followed by hybridization (renaturation; reannealing) with selected complementary DNA or RNA probes (single nucleotide strands of 5–50 kB (bases × 1,000) at 65–75°C. By adding formamide to the hybridization buffer the temperature range for hybridization can be reduced to 37–50°C.

Visualization. This can be achieved by labelling the probe in one of the following ways:

1. RISH, radioactive in situ hybridization using suitable radioactive elements such as ^{125}I or ^{3}H followed by autoradiography (cf. Chap.29)

2. NISH, non-radioactive in situ hybridization using labels such as biotin. This can be detected as described in Sect.26.3.4

In situ hybridization can be achieved using cryostat sections or paraffin-, Lowicryl^{R}-, or methacrylate-embedded material. Cytology smears and metaphase chromosome spreads may also be used. Regardless of the sample preparation method, results are heavily dependent on very careful handling, fixation and pre-treatment. Gloves should be worn when handling specimens as DNase and RNase are present in epidermis. To avoid loss of sections during prehybridization (denaturation) and hybridization the use of slides coated with 3-aminopropyl-triethoxy-silane is recommended (Van Prooijen-Knegt *et al.* 1983; Warford, 1988).

Fixation. To ensure maximum retention of nucleic acids the material should be promptly frozen or fixed. Some delay does, however, seem to be admissible, particularly where demonstration of viral nucleic acid is intended as this has been achieved using autopsy material (Warford, 1988). For RNA, formaldehyde, preferably prepared from paraformaldehyde, is the fixative of choice both for cryostat and paraffin material. This fixative is also useful for DNA demonstration on paraffin-embedded material (McAllister and Rock, 1985). Clarke's or Carnoy's fixative can also be used for DNA (McAllister and Rock, 1985). These ethanol and acetic acid mixtures are preferred for cryostat, chromosomal, and cytology smear preparations.

Proteolysis. Proteolytic enzyme digestion with proteinase K, pepsin, and pronase is usually essential after aldehyde fixation but is often also desirable after protein precipitating fixatives. The proteolytic treatment must be carefully controlled as undertreatment yields suboptimal hybridization and over-digestion leads to dissolution of the specimen. Pronase shows considerable variations from batch to batch with regard to impurities and may contain DNase. The addition of glycine may to some extent prevent the adverse effects of this procedure and also tends to make the proteolytic treatment easier to control. It has, however, been found that the use of proteinase K is less prone to difficulties of this kind. After digestion, brief fixation with formaldehyde minimizes loss of nucleic acid during the subsequent steps.

Denaturation. The required temperature for denaturation depends on the relative amounts of guanine-cytosine and adenine-thymine base pairs in the nucleic acid sequence. Predominance of guanine-cytosine results in higher melting temperatures due to the presence of the third hydrogen bond in this base pair.

Hybridization. The detection of specific sequences can be accomplished using one of two main approaches:

1. By hybridizing a labelled sequence of nucleic acid to its complementary DNA/RNA sequence in the tissue section/cell smear
2. By hybridizing an oligonucleotide primer to its complementary sequence in a tissue section/cell smear, and in situ synthesizing a new DNA strand using the cellular DNA or RNA as a template while incorporating labelled nucleotides

In the first type of in situ hybridization the target nucleic acid sequence can be either nucleic DNA or nucleic/cellular RNA. Conceptually, the hybridization to the two types of nucleic acid is quite similar, but the technical details differ significantly. In both cases, in order to control hybridization conditions the labelled strand of nucleic acid, the probe, must have a known length. Whether the sequence of the probe has to be known in detail depends primarily on the problem. The stringency calculations only call for a knowledge of the approximate content of guanine-cytosine base pairs (cf. specificity).

In the second type of in situ hybridization called **p**rimed **in s**itu labelling (PRINS, Kock *et al.*, 1989), the target can be either DNA or RNA. The oligonucleotide primer is usually not longer than 25 bases as it is difficult to calculate the optimum temperature when the primer is longer. The DNA polymerase used varies with the nature of the template. When this is DNA, the Klenow fragment of E. coli, DNA polymerase I, is most frequently used, while with RNA as the template reverse transcriptase is usually employed.

Hybridization can only take place when the probe and target sequences are single stranded (following denaturation). The conditions used should be adjusted so that only complementary sequences can hybridize. Depending on the chosen probe and target, the resulting double strand can be DNA:DNA, DNA:RNA, or RNA:RNA. DNA:RNA hybrids have the advantage of having a higher optimum hybridization temperature.

Several reagents are usually added to the hybridization solution to ensure as specific a reaction as possible. Polyethylene glycol or polymers of dextran sulphate form networks in the solution from which the probe is excluded thereby increasing its relative concentration and the rate of reaction of the hybridization process. EDTA (Sect.15.4) blocks the activity of contaminating nucleases and the addition of formamide up to 50% effectively reduces the optimal temperature of hybridization (for DNA the temperature is reduced 0.65°C for each per cent formamide while the equivalent reduction is 0.38°C for RNA). In addition, Larsson and Hougaard (1990) found that hybridization temperatures between 40–45°C produce the best signal to noise ratio. Moreover, the inclusion of 50% formamide produces an enhanced signal to noise ratio in spite of higher background staining (Larsson and Hougaard, 1990).

Finally, denatured heterologous nucleic acid sequences, usually derived from salmon sperm DNA, are included. These form electrostatic bonds with positively charged components in the sample thereby reducing the opportunity for involvement of the probe in similar interactions. The effect of these blocking factors is maximized by introducing a prehybridization step using a solution composed as the hybridization solution without the addition of the probe. Although hybridization is usually complete in 3–5 hours, overnight incubation is commonly employed for convenience. The hybridization time should, however, be empirically tested as too long a hybridization time might reduce the sensitivity of the assay.

Specificity. The specificity of the hybridization reaction is influenced by the concentration of monovalent cations and hybridization temperature. Higher temperatures (within 25°C, or less, below the melting point) increase specificity while higher cation concentrations (low stringency) give greater opportunity for partial homologous annealing as mismatches are stabilized. In practice, with conditions of low stringency, one in situ hybridization procedure can demonstrate several related sequences such as those in HPV subtypes (human papilloma virus), while high stringency makes it possible to distinguish between different subtypes of HPV which differ only slightly in sequence.

Washing. Following the hybridization step washing is performed usually first as a low stringency wash (high concentration of monovalent cations and/or low temperature, and/or with or without low concentrations of formamide) to remove unbound or loosely-hybridized probe together with the other components of the hybridization solution. This is followed by a high stringency bath (low concentration of monovalent cations, and/or high concentrations of formamide, and/or high temperature) to "fine-tune" the specificity of probe hybridization.

Visualization. To visualize a biotinylated probe (NISH) treatment with avidin is required. This may, however, give rise to non-specific reaction as it can bind to negatively charged groups. This non-specific reaction may be reduced by pretreating the sections either with a dried milk product (Warford, 1988) or by making use of a buffer with high pH. Alternatively, avidin can be substituted with streptavidin. Antibody directed against biotin can be placed between the biotinylated probe and the avidin step. Antibody "trees" can be built by adding biotinylated antibody against Fc IgG. Other antibody "trees" are also possible as for instance biotinylated antibody against avidin. The choice of the subsequent biotinylated reagent depends on the required levels of sensitivity and precision for final localization. Fluorochromes, such as Fluorescein Isothiocyanate (FITC) or Rhodamine, give high sensitivity and precise localization but morphological correlation is difficult.

Enzymes such as alkaline phosphatase or horseradish peroxidase (HRP) give fairly good sensitivity but relatively poor precision after incubation with their substrates and suitable chromogenic reagents (cf. Sect.26.3.4). On the other hand, enzyme conjugates give permanent results and morphological information is easy to obtain.

Larsson and Hougaard (1990) found that detection of a biotinylated probe was best accomplished with monoclonal antibiotin antibodies and the alkaline phosphatase-anti-alkaline phosphatase (APAAP) system (cf. Sect.26.3.3).

Controls. Control reactions are necessary to ensure the specificity of the in situ hybridization reaction. These include:

1. Nucleases for removing DNA or RNA. This determines the class of nucleic acid to which the probe has hybridized. As RNase usually is contaminated with DNase and vice versa the unwanted enzyme must first be removed (cf. Sect.3.4)
2. When probes are carried on a vector such as a plasmid, it is essential to include a labelled vector (without probe) control to determine the extent to which direct vector hybridization is contributing to the result

3. An excess of unlabelled sequence added to the hybridization solution can be used to assess the specificity of hybridization
4. Hybridization carried out at a variety of stringencies (see above under hybridization solution) may give additional information on the specificity of the hybridization reaction
5. A positive control using a sample known to contain the target sequence will establish that the technique is working and also give some information on its sensitivity compared with previous analyses
6. Negative controls should include:
   a. a specimen known not to contain the target sequence
   b. a specimen containing the target sequence but hybridized with unlabelled probe

These reactions will demonstrate any non-specific interactions of detection reagents with the sample.

Application. In situ hybridization has proved to be an important tool in research work, but applications in diagnostic work are becoming increasingly common as appropriate nucleic acid probes are now becoming available commercially. At present the technique is chiefly used in the diagnosis of viral infection (Haase, 1986) and chromosome abnormalities (Cremers *et al.*, 1987).

A recent general review of in situ hybridization has been given by Warford (1988). For more detailed information on sensitive in situ hybridization techniques for detecting mRNA, see Bresser and Evinger-Hodges (1987). If fluorescence techniques are to be used, then the review by Bauman *et al.* (1984) should be consulted. Van den Brink *et al.* (1990) have demonstrated that microwave irradiation makes a very rapid label detection possible.

20.7 Polymerase Chain Reaction

The polymerase chain reaction (PCR) is an elegant method, developed by scientists at the Cetus Corporation, by which specific DNA sequences may be amplified. Using DNA extracted from the sample, oligonucleotide primers to initiate DNA synthesis at specific sites, and a heat stable DNA polymerase, a very large number of copies of a specific DNA fragment can be generated from within a complex mixture of DNA. The *primers* are oligonucleotide sequences (16–30 bases) complementary to DNA sequences flanking the DNA fragment that is to be amplified. They act as initiation sites for DNA synthesis. Two primers are prepared; these are complementary to sequences on opposite strands of DNA at either end of the fragment of interest and both are oriented in the 5′ to 3′ direction towards the fragment. The thus defined fragment can then be amplified by means of a heat resistant *DNA polymerase*, Taq DNA polymerase, originally isolated from the thermophilic bacterium *Thermus aquaticus* (Erlich *et al.*, 1988; Saiki *et al.*, 1988).

It should be noted that even though PCR can be used for demonstrating specific nucleotide in single cells, cell spreads, or even paraffin sections, the reaction is not cytochemical in the strict sense since the amplified sequence is not demonstrated in situ in the cell. The aim of this section is to provide an introduction to PCR in recognition of its potential as an adjunct to histochemical studies. For comprehensive reviews of PCR readers are referred to: Innis *et al.* (1989) and Erlich (1989).

PCR technology has an extensive range of actual and potential applications. As well as producing more target DNA for hybridization studies, the amplified nucleotide sequence can also be used to study DNA-protein interactions and for nucleotide sequence determination. Different primer sets can be used; these can enable differentiation of more subtle genetic differences such as virus subtypes present in tissue samples. Gene expression can be studied by detecting and even quantitating RNA transcripts after first making copy DNA (cDNA) using reverse transcriptase (Krawetz *et al.*, 1989; Gilliland *et al.*, 1989). Finally, PCR can itself be used for the production of probe DNA for in situ hybridization.

Mechanism. The polymerase chain reaction can be described as a three step cycling process: *Denaturation-Annealing-Extension* (Fig. 20.1).

The first two steps are, in principle, the same as those used in in situ hybridization with annealing taking place between the two primers and the flanking nucleotide sequences on either side of the target DNA fragment to be amplified. Both these steps take about 20 to 60 seconds. In the third step, extension, the temperature is raised to 70–75°C, the temperature optimum of TAq DNA polymerase, and the annealed primers are extended in the 5′ to 3′ direction.

The time necessary to secure full synthesis of the desired fragment, depends on the length of the fragment being amplified. Under standard conditions the recommended time is 1 min per kB. Depending on the nature of the DNA template, incorporation can approach 150 nucleotides/sec/enzyme molecule at 75–80°C (Innis *et al.*, 1989). The cycle can then be repeated, allowing re-annealing of the primers and further rounds of DNA synthesis. Between 25 and 35 cycles are employed for most purposes. Each cycle is achieved by stepwise changes in temperature without the need for addition of new reagents. The so-called "PCR machines" are therefore programmable heating/cooling systems that achieve the appropriate temperature-time profile required for the proposed reaction.

Although each cycle should theoretically double the amount of target DNA, in practice around ten-fold less than the expected quantity is produced. After an initial exponential amplification, a linear increase in DNA occurs with each cycle and the process becomes less efficient.

Identification of the Amplified Product. The amplified fragment is often produced in amounts that allow direct identification by gel electrophoresis. Alternatively a labelled probe-hybridization technique (either on a gel or in a dot-blot system) may be employed.

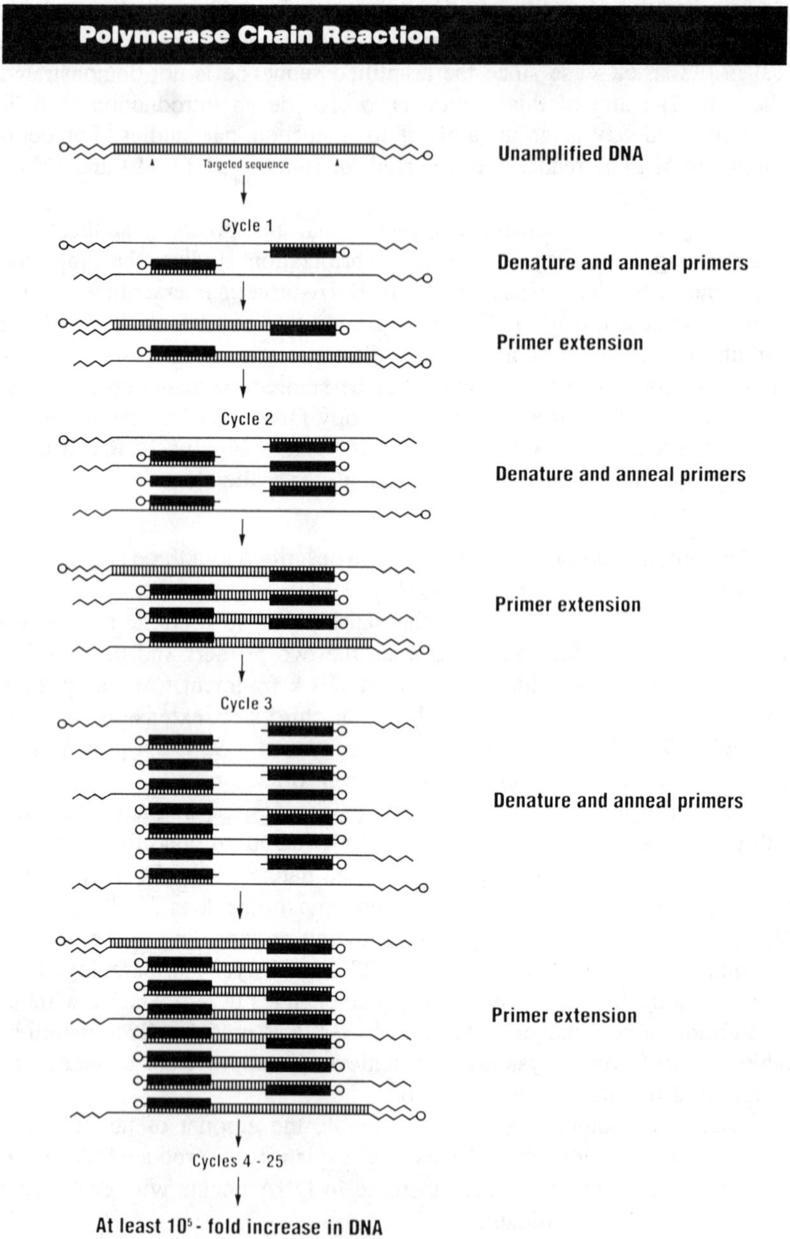

Fig. 20.1. Polymerase chain reaction technique (PCR) based on repeated cycles of high temperature template denaturation, oligonucleotide primer annealing, and polymerase mediated extension. Used with permission of Perkin-Elmer Cetus.

Source of Target Sequences. Target nucleic acid is extracted from the source material prior to amplification. The original source can be fresh frozen tissue/cells (DNA/RNA-cDNA) or formaldehyde fixed paraffin embedded tissue/cells (DNA). Unbuffered formaldehyde solutions or fixatives containing picric acid (Bouin) or mercuric chloride (Zenker, B5) cause severe degradation of DNA (Nuovo and Silverstein, 1988; Dubeau *et al.*, 1986).

If fresh tissue/cells are used, and if the fragment to be amplified is present as one copy per cell, only a few cells are necessary, and no particularly efficient DNA extraction is called for. If, however, more than 600 cells are necessary in order to get enough template sequences, DNA extraction must be performed (Innis *et al.*, 1989; Maniatis *et al.*, 1989).

If the source of DNA templates is formaldehyde fixed paraffin embedded tissue, three different ways to extract the DNA have been published:

1. Using xylene/ethanol (Dubeau *et al.* 1986; Shibata *et al.*, 1988)
2. By boiling the section(s) in a chelating agent (Chelex) (Singer-Sam *et al.*, 1989)
3. Using proteinase K/phenol/chloroform (Impraim *et al.*, 1987)

The yield of DNA is often only about 60–70% of what a similar amount of fresh tissue would give (Dubeau *et al.*, 1986).

Advantages and Disadvantages of the Method. The major advantage is clearly sensitivity. From as little as 10–100 copies of the target sequence, μg quantities of specific DNA can be synthesized and the amplification process itself can be automated.

The major disadvantage of the technique is its sensitivity to contamination and the tendency for PCR products to contaminate the laboratory. Almost every aspect of the procedure is open to contamination and many workers feel that a separate laboratory is desirable. Some of these problems may be reduced by UV-irradiating the PCR mixture prior to adding the sample DNA (Sarkar and Sommer, 1990).

For further details on PCR, the reader is referred to Innis *et al.* (1988) and Erlich (1989). An interesting account of the unusual origin of PCR is given by Mullis (1990).

20.8 Control Methods in Nucleic Acid Histochemistry

Nucleic acids may be extracted from the tissue either non-enzymatically or enzymatically. For non-enzymatic extraction trichloroacetic acid is usually preferred. This treatment primarily leads to the extraction of RNA but, depending on the temperature and the length of the treatment, DNA is also extracted to a greater or lesser degree. Hot trichloroacetic acid removes nucleic acids completely, after which the histones may be demonstrated with an anionic dye (Sects.6.2.5 and 21.4.2).

Perchloric acid and hydrochloric acid extract histones simultaneously with the nucleic acids. As conditions for the extraction of RNA vary from cell type to cell type and from fixative to fixative it is best to treat the tissue with the enzyme ribonuclease (RNase), which is specific for RNA. The purity of the enzyme is critical (Sect.3.4.3). Dichromate containing fixatives render RNA very resistant towards RNase. Aqueous formaldehyde solutions render RNA more resistant to RNase, while alcoholic formaldehyde solutions on the other hand seem to promote the degradation of RNA by RNase. It should also be noted that many salts (buffers) may have a direct extractive effect on RNA, especially at higher temperature.

In general, the same principles apply to deoxyribonuclease (DNase) as those outlined for RNase.

21 Proteins

H. Lyon, P.E. Høyer, P. Prentø

21.1 Introduction

Some proteins, such as the fibre proteins elastin and collagen, can be identified directly by their tissue location and their physico-chemical properties. Similarly, actin and myosin can be recognized in striated muscle by virtue of their organization into actomyosin, which appears cross-striated.

In general, selective demonstration of proteins according to amino acid composition is only possible for proteins which are present in a very high concentration at certain locations (Table 21.1). This approach depends on the protein concerned having an exceptionally high content of one or more of the amino acids that can be demonstrated histochemically (Cys, Trp, Tyr, Arg, etc.). Normally, individual proteins occur at low concentrations mixed with many other proteins. In most cases, selective demonstration of a given protein can therefore only be accomplished by using its biological properties such as *enzyme activity* (Chaps.23–25) or specific *antigenicity* (Chap.26).

21.2 Fixation

In general, an additive fixative such as formaldehyde is suitable as its effects are substantially reversible (cf. Sect.13.1, Fig. 13.1). The amino acid residue to be demonstrated is an important factor in making the choice (Sect.13.2). Finally, the use of albumin or gelatin as adhesives applied to slides should be avoided.

21.3 Demonstration

Two principles can be used for the demonstration of proteins:

1. Covalent binding to reactive amino acid residues giving rise to a coloured compound

H. Lyon (Ed.)
Theory and Strategy in Histochemistry

2. Binding of dye ions to proteins with a net positive or negative charge

Table 21.1 gives a survey of demonstrable amino acid residues in proteins and their occurrence.

Table 21.1. Demonstrable amino acids in proteins.

Amino acid	Frequency (mean number per 100)	Occurrence	Methods
Arg*	5	lysozyme, histones	NQS, PQ
Lys*	7	histones	oxidation-Schiff
His*	1	carboanhydrase	DAS-Azure A
Asp**	10	APUD-polypeptide hormones	
Glu**	11	APUD-polypeptide hormones	
Tyr	2	silk-fibroin, elastin	diazotization coupling
Trp	1	chymotrypsinogen, fibrinogen	DMAB – diazonium salt / nitrite
Cys-SH	2 (Cys-SH and Cys-S–Cys-S together)	neurosecretion, serum albumin	RSR, DDD, ferricferricyanide, maleimide
Cys-S | Cys-S		exocrine and endocrine granules in pancreas	reduction-RSR, etc. oxidation-cationic dye

Notes:
*: Basic amino acid residues. Can be demonstrated with reactions for acidophilia.
**: Acid amino acid residues. Can be demonstrated with reactions for basophilia.

Tyrosine occurs in practically all protein. The diazotization-coupling reaction (Sect.9.4.3) can therefore be used as a method for *general protein demonstration.* For this purpose, however, the azo coupling reaction (Sect.9.4.1) is preferred as this minimizes the importance of variations in the content of any individual amino acid. The azocoupling reaction also detects serotonin.

21.4 Reactions for Protein Bound Amino Acids

21.4.1 Covalent Reactions for Amino Acid Residues

Amino Groups. These occur in lysyl as primary ϵ-amino groups, in arginyl as primary and secondary amino groups in the guanidyl group, and as primary amino groups in the N-terminal part of proteins.

As stated in Sect.9.5, several methods are available for demonstrating amino groups. Lysine can be demonstrated selectively by oxidation-aldehyde reagent procedures (Sect.9.5.1) of which the chloramine T-Schiff is best for this purpose. Histones and globins are particularly heavily stained due to their lysine content, thus both tissue from the testis and red blood cells provide good test material.

Arginine. Arginyl residues can be demonstrated by the naphthoquinone sulphonate (NQS) method (Sect.9.6.1) and, for special purposes, the fluorescent reaction product of the phenanthrenequinone reaction (Sect.9.6.2). Arginine occurs in large amounts in histones and in the specific granules in eosinophils and Paneth cells.

Tyrosine. See Sect.9.4.3.

Histidine. See Sect.9.4.4.

Tryptophan. Tryptophanyl residues are best demonstrated using the dimethylaminobenzaldehyde (DMAB, Sect.9.4.5) followed by coupling to a diazonium salt. Formaldehyde may inhibit the reaction with DMAB as it forms a ring with the free α-position on the pyrrole group.

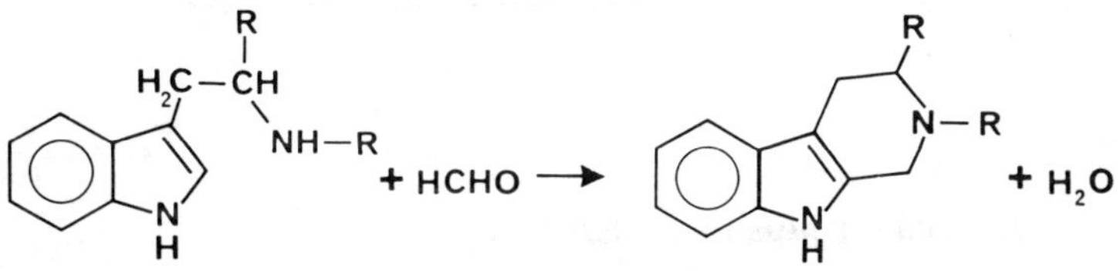

It is therefore recommended that fixation in neutral, buffered formaldehyde solution should not exceed three hours.

Tryptophan is not present in histones, but is abundant in certain proenzymes (trypsinogen, chymotrypsinogen, pepsinogen), thyroglobulin, gonadotropins, and glucagon. In addition, the granules of eosinophil leukocytes give strong positive reactions with methods for demonstrating tryptophan. These methods are also very useful for screening sections for fibrinogen and fibrin as both have a high tryptophan content.

Reactions for Protein Bound Cysteine and Cystine. Reliable histochemical distinction between -SH and -S-S- is not possible. A reasonable evaluation can, however, be achieved using trichloroacetic acid fixed material as follows:

–SH alone	1. RSR or DDD
–SH and –S–S–	1. reduction with thioglycollate 2. RSR or DDD
–S–S– alone	1. blocking of –SH with monoiodoacetic acid 2. reduction of –S–S– with thioglycollate 3. RSR or DDD

It is essential to maintain an acid pH following the reduction as even a slightly alkaline environment rapidly leads to reoxidation of a substantial proportion of the –SH groups.

The ferric-ferricyanide method (Sect.8.3.1) is the most sensitive method for SH groups, but less selective than Mercury Orange (Sect.9.3.2), the DDD-reactions (Sect.9.3.3), and the maleimide procedures (Sect.9.3.4).

Table 21.2. Identification of basic and acid proteins.

1. Demonstration of basic proteins: Fast Green FCF, pH 8
   + = unconjugated basic proteins, e.g. globins and lysozyme
   0 ⟶ 2
2. Demonstration of masked basic proteins (masked acidophilia) TCA + Fast Green FCF, pH 8
   + = conjugated basic proteins, e.g. in nucleohistones
   0 ⟶ 3
3. Demonstration of acid proteins: Toluidine Blue, pH 4
   + = acid proteins
   0 ⟶ 4
4. Demonstration of masked acid proteins (masked basophilia): HCl + Toluidine Blue, pH 4
   + = polypeptide hormones in APUD-cells
   0 ⟶ ?

TCA = trichloroacetic acid.

21.4.2 Reactions for Ionized Groups in Proteins

Identification of proteins with pronounced acid or basic properties can be performed as shown in Table 21.2.

Anionic Dyes at High pH (Basic Proteins); Relative Acidophilia. It is possible to demonstrate different basic proteins with reasonable selectivity and to discern their state by taking into account both the different conditions under which staining occurs and the localization of stained material (Table 21.3).

Table 21.3. Staining of proteins with anionic dye at high pH and varying pretreatments.

	Staining with anionic dye		
pH	8	8	10.5
pretreatment	none	TCA	none
Cytoplasm stained			
Certain globins (myoglobin in muscle cells and haemoglobin in red cells)	+	+	0
Ribosomal basic proteins	0	+	0
Dissociated ribonucleoprotein from ribosomes in certain pathological conditions	+	(+)	0
Proteins (lysozyme) in eosinophils and Paneth cells	+	+	+
Nuclei stained			
Dissociated nucleoprotein (histones) in certain pathological conditions. Mitotic chromatin and spermatozoa also frequently react	+	(+)	(+)
Histones and proteins after DNA extraction	0	+	+

Selectivity. If steric blockade or masking are disregarded the binding of anionic dye to protein bound amino groups will depend on pH, more or less as shown in the ionization curve in Fig. 6.7. Blockade of amino groups by nucleic acid and how this can be overcome by acid hydrolysis or enzymatic extraction were discussed in Sect.6.2.5.

Total Acidophilia. In principle using an anionic dye at low pH (in theory pH 1; in practice pH 2–3) this procedure detects all *available* amino groups. The main problems have arisen from blocking by negative charges and changes in the spectral characteristics of the dyes below pH 2. Although this reaction is rarely used, the method is very valuable for quantitating dissolved protein using protein in solution using Brilliant Indocyanine G (C.I. 42655; Serva Blue G; Coomassie Brilliant Blue G-250). See, however, also Sect.28.8.5.

Masked Basophilia; Cationic Dyes at Low pH. With dyes capable of staining metachromatically, the reaction is termed *masked metachromasia*. This approach has been developed into a procedure for demonstrating polypeptide hormones in cells belonging to the APUD system (Solcia *et al.*, 1968). These include calcitonin (C-cells in the thyroid), glucagon (A-cells in the pancreas), insulin (B-cells in the pancreas), and gastrin (G-cells in the stomach), all of which show a high content of glutamic and aspartic acids (acidic amino acids) or their amides. All are found in 0.2–0.4 μm diameter secretory granules. Treatment with HCl removes RNA and simultaneously converts the amides glutamine and asparagine to the corresponding acids. These can subsequently be demonstrated with a cationic dye at pH 4.

The procedure has been modified for use on Epon embedded tissue. This enables the identification of positive areas by light microscopy, and these can then be selected for ultrathin sectioning and electron microscopic examination (Pabst, 1985).

21.5 Demonstration of Elastin

Table 21.4 shows a comparison between collagen and elastin. The higher pI of collagen is due to a higher content of arginine, histidine, and lysine, while the principal non-polar amino acids in elastin are tyrosine, leucine, and valine.

Table 21.4. The chemical composition of collagen and elastin.

	pI	Polar amino acids	Non-polar amino acids
Collagen	6.0	40%	60%
Elastin	4.8	10%	90%

In accordance with the principles outlined in Chap.4, elastin shows a higher capacity for binding dyes by non-ionic bonds than collagen. Selective binding of anionic or cationic dyes to elastin should therefore be possible providing the ability of these dyes to form pure ionic bonds is suppressed. This approach is used in the methods mentioned under point 3 in Table 21.5.

Table 21.5. Methods for demonstrating elastin.

1. Autofluorescence (30.1)
2. Diazotization coupling reaction for tyrosine (9.4.3)
3. Use of dyes:
   Selected anionic dyes at high pH, e.g. Biebrich Scarlet
   Selected cationic dyes at low pH, e.g. Victoria Blue 4R
   Selected metal complex dyes, e.g. Verhoeff's Iron-haematoxylin
   Aldehyde Fuchsin without preceding oxidation
   Orcein
   Resorcin Fuchsin

Goldstein (1962) has shown that binding to elastin by both anionic and cationic dyes, including Aldehyde Fuchsin, Orcein, and Resorcin Fuchsin, is not hindered to any appreciable degree by the addition of sodium chloride to the staining solution. This finding makes it highly unlikely that ionic bonds are important in the staining of elastin by these dyes.

The importance of van der Waal attractions has been stressed by Horobin and Bennion (1973). One consequence of this is that most of the dyes involved can be dissolved in organic solvents without losing their ability to stain elastin. It should however be re-emphasized (cf. Sect.6.3) that, depending on the prevailing conditions, a dye binds to its substrate by all the bonding mechanisms available to it. Proctor and Horobin (1988) have presented further work on chemical structure and staining mechanisms of elastic fibre stains supporting this view.

Among the different methods for demonstrating elastin the Orcein methods can be singled out for their high selectivity and the technical convenience. In contrast, Verhoeff's Iron-haematoxylin is not recommended on account of its low selectivity, poor reproducibility, and its difficult differentiation step.

21.6 Demonstration of Collagen

Collagen is intermediate between globular proteins and elastin as regards staining properties. It predominantly binds large dye molecules which show a pronounced tendency to form van der Waal attractions in addition to possessing anionic groups. Horobin and Flemming (1988) in a structure-staining correlation analysis of anionic dye pairs emphasize that selectivity of one-bath anionic dye pairs is diffusion-rate controlled, involving the interaction of differentially permeable tissue sites (collagen being more permeable than muscle cytoplasm).

Van Gieson Methods. The van Gieson methods probably give the most selective staining of collagen fibres. The 25–50-fold excess of Picric Acid relative to the collagen dye in a van Gieson solution renders the cytoplasmic background "Picric Acid yellow". Puchtler *et al.* (1988) investigated a number of anionic dyes for their ability to stain collagen selectively when used together with Picric Acid, and concluded that dispersion forces (Sect.4.5.4) and hydrophobic bonding (Sect.4.5.5) were decisive, whereas ionic forces did not impart affinity. For the collagen dye it is advantageous to use long, planar molecules such as Sirius Red 4B (C.I. 28160) and Sirius Red F3B (C.I. 35780) as these become oriented parallel to the collagen fibre structure and increase birefringence. Red staining with "Picro-Sirius F3B" accompanied by a pronounced increase in birefringence is an unequivocal indication of the presence of collagen and reticulin fibres (Junqueira *et al.*, 1979; Montes *et al.*, 1980) providing the presence of amyloid can be ruled out (see Sect.21.7). In contrast to the above, the classical "Picro-Fuchsin" method does not demonstrate the thinnest collagen fibres and therefore excludes both reticulin fibres and basal membranes. Acid Fuchsin does not lead to an increase in birefringence associated with collagen fibres. Wolman and Kasten (1986) have produced a detailed discussion of the use of polarized light microscopy for studying the molecular structure of collagen and reticulin.

Trichrome Methods. The trichrome methods, in which phosphotungstic acid or phosphomolybdic acid are used in combination with several anionic dyes usually do not give staining of collagen and reticulin fibres that are as precise or as discriminatory as the van Gieson method. For an analysis of mechanism, see Horobin and Flemming (1982).

Collagen and reticulin fibres can be distinguished using silver methods (Sect.8.3.2). Here the omission of the toning process leaves reticulin fibres black and collagen fibres orange to flame red (Lyon and Prentø, 1973). Toning with gold chloride bleaches the collagen fibres without changing the colour of reticulin.

21.7 Amyloid

Amyloid is a glycoprotein which occurs as extracellular deposits in a number of pathological conditions (Francis, 1990). Protein comprises 95–98% of the amyloid. Clinically a distinction is made between:

a. Primary amyloid, which occurs spontaneously without any apparent connection with another disease. The condition affects tissue of mesodermal origin, especially muscles, heart, skin, and tongue
b. Secondary amyloid, which occurs in association with a number of chronic conditions, notably tuberculosis in the past but nowadays more commonly conditions such as rheumatoid arthritis. The characteristic sites involved are liver, spleen, kidney, and adrenal gland

c. Amyloid in multiple myelomas and other diseases of the immune system with the same distribution as (a)
d. Tumour associated amyloid is particularly associated with apudomas, i.e. tumours derived from cells in the APUD system

Ultrastructurally amyloid consists of irregularly deposited fibrils, each consisting of two filaments. X-ray diffraction analysis has shown that amyloid filaments possess a β-pleated sheet conformation.

Table 21.6 sets out the results of chemical analyses performed on amyloid. Three different kinds of proteins have been found.

Table 21.6. Types of protein found in different kinds of amyloid.

Type of protein	Clinical form of amyloidosis	Amino acids characteristically present	Amino acids characteristically absent
Immunoamyloid = kappa-lambda (or light chain) amyloid = AL protein	multiple myeloma, primary amyloidosis	tryptophan, tyrosine	
A protein amyloid = AA protein	secondary amyloidosis (chronic inflammatory conditions), familial amyloidosis (nephropathic form)	tryptophan, tyrosine	cysteine, threonine, proline
Prealbumin = AF protein	familial amyloidosis (neuropathic form), senile systemic amyloidosis, senile cardiac amyloidosis		
β_2-microglobulin = B2M protein	haemodialysis associated		
C protein amyloid = APUD amyloid	tumour associated		tryptophan, tyrosine

Although there are a number of chemical differences between the different kinds of amyloid, the characteristic β-pleated sheet conformation, the ability to bind Congo Red, and the fibrillar ultrastructure are, however, common to all.

Demonstration. The demonstration of amyloid in biopsies is important for the treatment of underlying disease. The diagnosis can be made on a biopsy of the rectum in 75% of cases. Deposits are characteristically found in the wall of small submucous vessels and sometimes along the epithelial basement membrane. Amyloid is seen as an amorphous material stained pink with Al-haematein Eosin and yellowish brown with Fe-haematein van Gieson. Variable degrees of positive reaction are seen with the PAS method. None of these appearances are diagnostic as tissue elements, such as hyaline and fibrinoid, give similar reactions. Table 21.7 outlines more selective methods for the demonstration of amyloid.

Table 21.7. Methods for the demonstration of amyloid.

Method	Appearance	Selectivity	Sensitivity
Congo Red	green by polarization microscopy	Specific	moderate
Thioflavine TCN	yellow to silver grey fluorescence	moderate	very high
Crystal Violet	"metachromasia"	low	rather low
Toluidine Blue O	red by polarization microscopy	low	high
Sirius Red F3B	green by polarization microscopy	moderate	moderate

Congo Red, Sirius Red F3B, Toluidine Blue O, and Thioflavine TCN are bound to the β-pleated sheet conformation of amyloid by hydrophobic bonds. In addition, it is known that the planar and linear dye molecules are orientated parallel to the amyloid fibrils (Glenner, 1981). Non-specific bonding of the dyes by ionic bonds is kept at a minimum by increasing pH of the staining solutions and by adding inorganic salts, e.g. NaCl, and organic solvents, e.g. ethanol. Some indication of whether an amyloid deposit is primary or secondary can be obtained by pretreatment with trypsin or (more effectively) an acid potassium permanganate solution (Wright *et al.*, 1977). Primary amyloid is resistant to these treatments and continues to give a positive reaction with Congo Red. Amyloid associated with multiple myeloma, despite the similarity of the protein to that found in primary amyloidosis, is frequently degraded. The mechanism involved in staining with Crystal Violet is probably similar but the reason for the reddish colour is uncertain.

The preferred method is staining with Congo Red. It should be noted that to achieve the optimum "apple green" colour seen by polarization microscopy, it is necessary to use tissue sections which are at least 10 μm thick.

22 Carbohydrates

H. Lyon, P.E. Høyer, P. Prentø

22.1 Introduction

Histochemically demonstrable carbohydrates comprise homoglycans, heteroglycans, and oligosaccharides bound to protein (glycoproteins) (cf. Sect.2.1.5). Homoglycans occur as glycogen in mammals and starch (amylose and amylopectin) and cellulose in plants. Heteroglycans are a component of protein complexes (proteoglycans). Together with certain glycoproteins they are collectively termed mucosubstances.

22.2 Demonstration

Fixation has been covered in Sect.13.4. The most frequently used methods for demonstrating carbohydrates are given in Tables 22.1 and 22.2. The periodic acid-Schiff reaction (PAS) is discussed in detail in Sect.9.2.1, while metachromasia and the Alcian Blue reaction are discussed in Sect.6.1.1 respectively Sect.6.1.2. In the following, modifications of these methods and supplementary methods will be discussed, while the enzymatic extraction methods are expounded in Sect.22.3.3.

Note that selective identification of acid carbohydrates using the AB/$MgCl_2$ and the PAS-reaction is only possible if the PAS- reaction is negative. For example, a mixture of a sulphate containing proteoglycan and a neutral glycoprotein will give the same staining result as sulphomucin. The histological localization and morphology of the stained material is therefore essential for the interpretation.

22.2.1 Culling's Modification of the PAS-Reaction

This method (Reid *et al.*, 1973; Culling *et al.*, 1974; 1976) depends on three processes:

1. Reduction of aldehyde groups induced by periodic acid oxidation (or in other ways) to primary alcohol groups with borohydride, BH_4, thereby blocking any subsequent reaction with Schiff's reagent

H. Lyon (Ed.)
Theory and Strategy in Histochemistry

Table 22.1. Reactions of histochemically demonstrable carbohydrates in formaldehyde fixed, paraffin embedded tissue.

	HOMOGLYCANS	PROTEOGLYCANS				GLYCOPROTEINS		
		neutral	acid			neutral	acid	
			polycarboxylates	polycarboxy-sulphates	polysulphates		sialomucins	sulphomucins
PAS	+	+	0	0	0	+	+	+
Am-PAS	0	+	0	0	0	+	+	+
MC pH 2	0	0	0	+	+	0	0	+
MC pH 5	0	0	+	+	+	0	+	+
AB 0	0	0	+	+	+	0	+	+
AB 0.2	0	0	0	+	+	0	0	+
AB 0.5	0	0	0	0/+	+	0	0	+
AB 0.7	0	0	0	0/+	+	0	0	0/+
AB 0.9	0	0	0	0	0/+	0	0	0
AB pH 1	0	0	0	+	+	0	0	+
AB pH 3	0	0	+	+	0/+	0	+	0/+
AF	0	0	0	+	+	0	0	+
LID	0	0	+	+	+	0	+	+
HID	0	0	0	+	+	0	0	+

+ : positive reaction; 0: no reaction; Am-PAS: amylase-PAS; MC pH 2: Toluidine Blue metachromasia at pH 2; MC pH 5: Toluidine Blue metachromasia at pH 5, non-dehydrated section; AB 0: Alcian Blue 0 mol/l $MgCl_2$; AB 0.1: Alcian Blue 0.1 mol/l $MgCl_2$; AB 0.2: Alcian Blue 0.2 mol/l $MgCl_2$; AB 0.5: Alcian Blue 0.5 mol/l $MgCl_2$; AB 0.7: Alcian Blue 0.7 mol/l $MgCl_2$; AB 0.9: Alcian Blue 0.9 mol/l $MgCl_2$; AB pH1: Alcian Blue pH 1; AB pH3: Alcian Blue pH 3; AF: Aldehyde Fuchsin pH 1–2; LID: Low iron diamine; HID: High iron diamine.

Table 22.2. Strategy for identification of carbohydrates.

1. PAS (1,2-glycol and 1-amino-2-hydroxyl groups):
   - a: PAS + → 2
   - b: PAS 0 → 3 or non-carbohydrate
2. Amylase – PAS (α-homoglycans):
   - a: PAS 0 = *α-homoglycans* (glycogen, amylose, amylopectin)
   - b: PAS + — [β-homoglycans (e.g. cellulose); other homoglycans (e.g. dextrans); glycolipids (in frozen sections); glycoproteins] → 3
3. Alcian Blue pH 3 without $MgCl_2$:
   - a: AB + — [PAS + = acid glycoproteins; PAS 0 = acid proteoglycans] → 4
   - b: AB 0 — [PAS + = *neutral glycoproteins* or see (2b); PAS 0 = *neutral proteoglycans** or non-carbohydrate] → x
   - x possibly sulphate containing carbohydrate → 4
4. Alcian Blue pH 5.7 with 0.2 mol/l $MgCl_2$:
   - a: AB + — [PAS + = sulphate containing mucins; PAS 0 = sulphate containing proteoglycans] → 6
   - b: AB 0 — PAS + = *sialomucins* (possible control with neuraminidase) → 5
     PAS 0 = *polycarboxylate proteoglycan* (e.g. hyaluronic acid (possible control with hyaluronidase)) or non-carbohydrate
5. Periodic acid-Thionin/KOH/PAS method (PAT/PAS):
   - a: PAT + (blue) = *non-acylated sialomucins*
   - b: PAS + (red) = *O-acylated sialomucins*
6. Alcian Blue pH 5.7 with 0.7 mol/l $MgCl_2$:
   - a: AB + — [PAS + = sulphomucins; PAS 0 = probably polysulphate proteoglycans] → 7
   - b: AB 0 — PAS + = sulphomucins
     PAS 0 = *carboxysulphate proteoglycans* (e.g. chondroitin sulphates, heparin, heparan sulphate)

Table 22.2. (*Continued*).

7. Alcian Blue pH 5.7 with 0.9 mol/l $MgCl_2$:

a: AB + — PAS + = sulphate rich sulphomucins
a: AB + — PAS 0 = *polysulphate proteoglycans* (e.g. keratan sulphate)

b: AB 0 — PAS + = sulphomucins
b: AB 0 — PAS 0 = carboxysulphate proteoglycans

PAS + , PAS 0: positive, respectively no reaction with periodic acid-Schiff.
PAT + : positive reaction with periodic acid Thionin-Schiff.
AB + , AB 0: positive, respectively no reaction with Alcian Blue.
* An example is pure chitin (poly-N-acetylglucosamine) which is PAS 0. However, chitin as a component of the exoskeleton in different invertebrates is never found in pure condition, but always together with glycoproteins which explains the positive PAS-reaction found in practice.

2. Protection of 1,2-glycols from attack by periodic acid by esterifying (O-acylating) one or both hydroxyl groups. Subsequent saponification with KOH may then restore the hydroxyl groups allowing reaction with periodic acid in the usual way
3. Production of a Schiff's reagent from Thionin. This gives a blue product with aldehyde groups

The steps and results for Culling's procedure for the demonstration of non-acylated and O-acylated sialomucins are shown in Table 22.3. The chemistry of sialic acids is covered in Sect.2.1.5.

Table 22.3. Culling's method for O-acylated sialomucins (SM).

	non-acylated SM	O-acylated SM
1. PAS	red	not stained
2. PAT	blue	not stained
3. PA/Bh/KOH/PAS	not stained	red
4. PAT/KOH/PAS	blue	red

PA = periodic acid; S = Basic Fuchsin Schiff; T = Thionin Schiff; Bh = borohydride, BH_4^-, reduction; KOH = potassium hydroxide saponification.

Procedures 1 and 2 are identical, except for the colour of the final product. Procedures 2 (1) and 3 can be performed on consecutive sections, or if the two kinds of sialomucins are present in approximately equal amounts in the section, it may be satisfactory to only perform procedure 4. In humans O-acylated sialomucins are not present in the gastric mucosa but appear in the intestines in increasing amounts towards the anus.

Volz *et al.* (1986) note that the use of the PAS reaction for the histochemical identification of sialic acids is complicated by oxidation of vicinal diols on other carbohydrate residues. One solution to this problem is to selectively oxidize with dilute periodic acid ("mild" periodic acid) (Weber *et al.*, 1975). Another solution is to use the periodic acid-phenylhydrazine Schiff method (PAPS) (Spicer, 1961). The mechanism of the latter method is that the aldehyde groups engendered by periodic acid are blocked with phenylhydrazine. Subsequent treatment with Schiff's reagent reverses the blocking of sialic acid monoaldehydes (Reid *et al.*, 1984). Volz *et al.* (1987b) showed that the use of 0.4 mmol/l in 1 mol/l HCl for one hour at 4°C leads to the selective visualization of sialic acids in the PAS procedure. This selectivity is a result of an increase in the rate of the oxidation of the sialic acid residues together with a decrease in the rate of oxidation of neutral sugars (vicinal diols located on hexose, 6-deoxyhexose, or N-acetyl-hexosamine residues of sialo- and sialosulphoglycoproteins).

At this juncture it is worth pointing out that an exhaustive range of additional sequential treatments have been proposed on the basis of the investigations of Volz, Reid, Park, and coworkers. To assist in identifying these new sequences, we have chosen to number them consecutively and to list the abbreviations used in alphabetic order.

AB1.0	0.1% w/v Alcian Blue in 0.1 mol/l HCl for 30 min at room temperature
Az	Azure A Schiff reagent for 6 hours at room temperature
Bh	0.1% w/v sodium borohydride in 1% w/v dibasic sodium phosphate (anhydrous) for 20 min at room temperature
DNPH	saturated solution of 2,4-dinitrophenylhydrazine in 1 mol/l HCl for 2 hours at 4°C
KOH	saponification with 0.5% w/v potassium hydroxide in 70% v/v ethanol for 15 min at room temperature
PA(2)	1% w/v (44 mmol/l) aqueous periodic acid for 2 hours at room temperature
PA(1)	1% w/v (44 mmol/l) aqueous periodic acid for 1 hour at room temperature
PA(1/2)	1% w/v (44 mmol/l) aqueous periodic acid for 30 minutes at room temperature
PA*	0.4 mmol/l periodic acid in 1 mol/l HCl for 1 hour at 4°C
S	Pararosanilin Schiff reagent for 1 hour at room temperature
T	Thionin Schiff reagent for 2 hours at room temperature

Volz *et al.* (1987a):

1. PA*-Bh-KOH-PA*-T-KOH-Bh-PA(1)-S
2. KOH-PA*-T-KOH-Bh-PA(1)-S
3. PA*-T-Bh-PA(1)-S-KOH
4. KOH-PA*-Bh-AB1.0-PA(1)-S
5. KOH-PA*-Bh-PA(1)-S

(Park *et al.*, 1987) - Based on the periodic acid-phenylhydrazine-Schiff reaction:

6. KOH-PA(2)-DNPH-Az-KOH
7. PA(2)-DNPH-Az-KOH
8. (PA*-Bh-KOH-PA(2)-DNPH-Az-KOH
9. (PA-DNPH-Az-KOH-Bh-PA(1/2)-S

Reid *et al.* (1988) have presented a new general method for the specific histochemical identification of O-acyl sugars in any epithelial glycoprotein. The term O-acyl sugar indicates the presence of 8- (or 9-) O-acyl sialic acids and an ester substituent(s) on all the potential vicinal diols of the hexose, 6-deoxyhexose, and N-acetylhexosamine residues (for nomenclature, see Sect.2.1.5).

10. PA-Bh-KOH-PA*-Bh-PAS

The initial Pa-Bh treatment renders vicinal diols located on either sialic acid or neutral sugars PAS unreactive. In the subsequent steps ester substituents are removed by the PA*-Bh sequence, and O-acyl sugars are stained with the PAS technique. With this method it has been demonstrated that O-acyl sugars occur in the epithelial goblet cell glycoproteins of adult human colon (Reid *et al.*, 1988).

The staining results expected from the ten new variants cited above are summarized in Table 22.4.

Table 22.4. Staining results of material containing epithelial glycoproteins using the more recent developments of Culling's methods.

Procedure (1)	O-acyl sugars	O-sulphate ester	Neutral sugar vicinal diol (2)	Sialic acid O-acyl side chain substitution		
				none (3)	C7	C8 or C9 (4)
1	M	0	M	0	T	T
2	M	0	M	T	T	T
3	0	0	M	T	T	0
4	M	B	M	0	0	0
5	M	0	M	0	0	0
6	Y	0	Y	A	A	A
7	0	0	Y	A	A	A
8	Y	0	Y	0	A	A
9	M	0	Y	A	A	M
10	M	0	0	0	0	0

(1) Numbers refer to the procedures cited in the preceding text.
(2) Neutral sugar (hexose, 6-deoxyhexose and N-acetylhexosamine residues).
(3) If 9-O-acyl sialic acids oxidize in the initial PA* steps of 1 and 2, then they will appear in the class of sialic acids without side chain substitution. If they do not oxidize under conditions used, they will be identified as 8-O-acyl sialic acids in 8 and 9 and will not stain in 7.
(4) For histochemical purposes 8-O-acyl sialic acids include sialic acids with two (C7C8, C7C9, C8C9) or three (C7C8C9) O-acyl substituents.
0 = no staining; M = magenta; T = blue colour obtained with Thionin Schiff; B = blue colour obtained with Alcian Blue; A = blue colour obtained with Azure A; Y = yellow.
Note that in methods 1–4 mixtures of the components will stain in various shades of purple. In methods 6–8 mixtures of the components will stain in various shades of green, while in method 9 mixed deposits will stain in a wide variety of colours depending upon the composition of the glycoproteins.

22.2.2 Iron Diamine Methods

Using N,N-dimethyl-meta-phenylene diamine together with a fairly low concentration of ferric ions (LID = low iron diamine) or respectively three times as high a concentration of ferric ions (HID = high iron diamine), it is possible to demonstrate acid carbohydrates and to some degree to differentiate between sulphomucins and sialomucins (Spicer, 1965). These methods use toxic amines and interpretation of the results achieved is difficult. Furthermore, it should be noted that in an analysis of the high iron diamine-Alcian Blue pH 2.5 procedure, Reid *et al.* (1989) conclude that the staining is non-specific as glycoproteins known to contain sialic acids fail to stain with Alcian Blue. Whether this effect is due to HID staining of anion groups other than sulphate or to the masking of the Alcian Blue staining by that of HID or some combination of both effects has not been established.

22.2.3 Colloidal Iron

A colloidal solution of ferric hydroxide is prepared by adding a small volume of an aqueous ferric chloride solution to a large volume of boiling water. In the resulting colloidal solution the individual ferric hydroxide particles are positively charged. After dialysis against water, during which small ions such as H^+, Cl^-, and any unreacted ferric are removed, the solution can be used as a reagent for polyanions.

pH is normally adjusted to 1.8. After washing in water or diluted acetic acid the sections are placed in a solution of potassium ferrocyanide. This gives rise to a precipitate of Prussian Blue (Sect.17.7.4) associated with sulphomucins, sialomucins, and proteoglycans. The selectivity of the method is fairly good, even though a weak background staining of nucleic acids and sometimes also proteins occurs.

22.2.4 Cuprolinic Blue

In addition to its use for staining RNA (Sect.20.2), Cuprolinic Blue has also been used for the staining of acid glycoproteins and proteoglycans (Scott, 1980; Van Kuppevelt *et al.*, 1984a; 1984b). With Cuprolinic Blue applied to semithin epoxyresin embedded material, Juarranz *et al.* (1987) have obtained metachromatic staining of goblet cell mucin, mast cell granules, and cartilage matrix (Sect.6.1.1). Electron microscopy of ultrathin sections of the same material showed an electron dense reaction in the same structures. Using Cuprolinic Blue combined with enzymatic digestion, Hussein *et al.* (1988) have demonstrated that heparan sulphate is the main glycosaminoglycan in the basement membrane of human gall bladder.

22.2.5 Alcian Blue-PAS and PAS-Alcian Blue

The sequence normally recommended is Alcian Blue-PAS. However, when using a double-staining technique it is essential that the two methods used do not affect each other qualitatively or quantitatively (cf. Sect.1.2.4), and if possible, the test should be repeated with the reactions in reverse order. Both for theoretical reasons and in practice results obtained with Alcian Blue-PAS and with PAS-Alcian Blue are not identical, and the only histochemically permissible sequence is Alcian Blue-PAS. In support of this viewpoint, using PAS-Alcian Blue Johannes and Klessen (1984) showed that some mucosubstances became Alcian Blue-positive even though they had been PAS-positive and Alcian Blue-negative with the Alcian Blue-PAS sequence. Yamabayashi (1987) confirmed these findings but still found the reverse sequence, PAS-Alcian Blue useful.

22.3 Blocking and Extraction

22.3.1 Acetylation

As stated in Sect.9.2.1 the first step in the PAS-reaction can be blocked by acylating the vicinal hydroxyl or hydroxyl and amino groups. The ease with which the groups are blocked, however, varies considerably as shown in Table 22.5.

Table 22.5. The influence of the length of acetylation for the blocking of the PAS-reaction of different tissue components.

	Acetylation time (h)			
Tissue component	2	6	16	20
Collagen, basal membranes, reticulin, lipofuscin, colonic "melanin", and brush border in tubules of the kidney	0	0	0	0
Intestinal mucins, cornea, corpus vitreum, and capsule of the lens	+	0	0	0
Glycogen, starch, cellulose, mucin in the stomach, and chromaffin in the adrenal glands	+	+	0	0
Cartilage matrix	+	+	+	0

0 = PAS-reaction prevented; + = PAS-reaction uninhibited.

22.3.2 Methylation and Methylation-Saponification-Sequence

These methods can be used for differentiating between different acid carbohydrates. The principle has been discussed in Sect.9.8.3 and a summary is given in Table 22.6.

Table 22.6. The effect of mild (MM) and drastic methylation (DM) combined with saponification (S) on the binding of cationic dyes (C) to acid carbohydrates.

	Acid proteoglycans			Acid glycoproteins	
Reactions	polycarboxylates	polycarboxy-sulphates	polysulphates	sialomucins	sulphomucins
C	(+)	+	+	(+)	+
MM/C	0	0	0	0	0
MM/S/C	(+)	+	+	(+)	+
DM/S/C	(+)	(+)	0	(+)	0

+ = stained; (+) = sometimes stained; 0 = not stained.

The value of these reactions is, however, limited in two respects. First, the reactive groups show very great individual variation in their willingness to react and second, the saponification reaction is very harsh on the tissue. Broadly, blocking is achieved quickest for sulphate in proteoglycans and sulphomucins followed by the carboxylate groups in proteoglycans, sialomucins and proteins. A longer methylation step is required to block the basophilia of lipofuscin and even longer for melanin.

22.3.3 Enzymatic Extraction

A number of enzymes can give important information in discriminating between the different histochemically demonstrable carbohydrates (cf. Table 22.2). For reliable interpretation it is essential that all enzyme preparations are of guaranteed high purity and activity (see Sect.3.4.3).

Table 22.7. Enzymes used for the extraction of carbohydrates.

Enzyme	Source	Substrate	Method of demonstration
α-amylase	ptyalin (saliva)	glycogen and starch	PAS
hyaluronidase	testis	chondroitin sulphate A and C and hyaluronic acid	metachromasia with Toluidine Blue pH 6
hyaluronidase	*Streptomyces hyalurolyticus*	hyaluronic acid	metachromasia with Toluidine Blue pH 6
chondroitin ABC lyase	*Proteus vulgaris*	chondroitin sulphate A, B, C	Alcian Blue, PAS, LID
chondroitin AC lyase	*Anthrobacter aurescens*	chondroitin sulphate A and C	Alcian Blue, PAS, LID
neuraminidase (sialidase)	*Clostridium perfringens*	sialomucins	Alcian Blue

For the histochemical investigation of carbohydrates in animal tissues, α-amylase, neuraminidase, testicular and bacterial hyaluronidase are the most important. The use of chondroitin ABC lyase and chondroitin AC lyase remains somewhat experimental. For plants, pectinases are of some interest. Most of the important enzymes together with the appropriate method for demonstrating their respective substrates are reviewed in Table 22.7.

Due to the problems associated with fixing sialomucins (Sect.13.4), it is important to have a control section which has been treated with enzyme-free buffer when interpreting treatment with neuraminidase.

22.4 Lectins

Lectins are proteins with a molecular weight between 200 and 300 kDa which can be extracted from various plants. They bind specifically to certain carbohydrates in a manner that often corresponds to the hydrogen and hydrophobic bonds formed between antigens and antibodies. It is possible to label lectins in exactly the same way as antibodies (see Sect.26.2). Examples of labelling include Fluorescein Isothiocyanate (FITC), ferritin, and horse radish peroxidase (HRP). It is also possible to label with ^{3}H-acetal groups (autoradiography, Table 29.1).

The specificity of the lectin must be determined by control stainings. The most important control is the inhibition of the staining by addition of high concentrations of carbohydrates which bind the lectin in question. For instance, concanavalin A is specific towards α-D-mannose and α-D-glucose. Dye binding can also be hindered by chemical blocking; for example, acetylation hinders the binding of concanavalin A.

The effects of different fixation protocols on the subsequent binding of different lectins have been evaluated by Allison (1987). There is no universally appropriate fixative, but the best results are usually obtained with 95% ethanol or fixatives containing mercuric chloride. Formol-saline is, however, adequate for the study of routine paraffin-processed tissue in many instances. There are nonetheless situations where frozen sections may be preferable.

Jeffrey *et al.* (1987) have stressed the importance of using purified enzyme preparations (notably trypsin) in the pretreatment of sections from formalin fixed paraffin embedded material for lectin histochemistry.

The use of lectins is thus still associated with many practical problems but is expected to assist in a more precise classification of tissue carbohydrates.

Part 5
Enzyme Histochemistry

23 Enzyme Histochemistry I: General Considerations

A.P. Andersen, P.E. Høyer, H. Lyon, P. Prentø

23.1 Biochemical Aspects

23.1.1 Definition

Enzymes are proteins or glycoproteins with selective, often specific, catalytic effects. They characteristically increase the rates of reactions by a factor of at least 10^7.

23.1.2 Specificity

The specificity of an enzyme results from the manner in which the substrate is bound to the enzyme.

The structural location at which the substrate is bound is called the *active* site of the enzyme. It is only a tiny portion of the entire enzyme but results from a complicated three-dimensional structure consisting of groups from several points in the linear amino acid sequence. This complex structure limits form and size of molecules that can be bound and thereby confers substrate specificity.

In many cells a series of enzymes function together in a multienzyme system. A compound is processed through a number of steps where the product from one process is the substrate for the next.

23.1.3 The Rate of Reaction of Processes Catalyzed by Enzymes

The rate of reaction depends on the following factors:

- Molecular activity of the enzyme
- Concentration of substrate
- pH
- Temperature
- Ionic strength
- Activators
- Inhibitors

H. Lyon (Ed.)
Theory and Strategy in Histochemistry

Prerequisites for the occurrence of a chemical reaction are:
1. Physical contact between the reacting molecules
2. Affinity between the reacting molecules
3. Sufficient energy to start the reaction

The energy required to initiate the reaction is called the *activation energy* (Fig. 23.1).

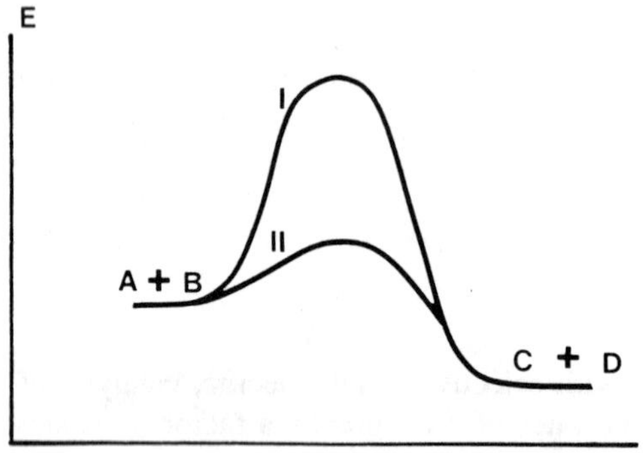

Fig. 23.1. The activation energy for the process $A + B \rightleftarrows C + D$
I non-catalyzed reaction
II catalyzed reaction
Energy (E).
Compounds A and B can be transformed to C and D, if the energy level of A + B is higher than that of C + D and E = activation energy is present.

For an enzyme-catalyzed process the activation energy is lower due to the formation of the enzyme-substrate complex. The affinity of the enzyme for the substrate and the rate of relevant molecular collisions is reflected in the dependency of the rate of reaction on the substrate concentration.

As the enzyme (E) is believed to form an intermediate product (ES) with the substrate (S), and (ES) is converted to enzyme plus product (P), the simplest equation for the equilibrium of the enzymatic process is:

$$E + S \underset{K_{-1}}{\overset{K_{+1}}{\rightleftarrows}} ES \overset{K_{+2}}{\longrightarrow} E + P$$

In 1913 Michaelis and Menten proposed an equation to describe the relationship between substrate concentration S and rate of reaction V with the constant K_M called Michaelis' constant.

$$V = V_{\max} \frac{[S]}{[S] + K_M} \quad \text{where } K_M = \frac{k_{-1} + k_{+1}}{k_{+1}}$$

This equation applies if the following conditions are fulfilled:
1. Steady state for ES, i.e. rates of formation and disappearance of ES are equal
2. $[E] \ll [S]$
3. $[ES]$ small compared to $\Delta[S]$ and $\Delta[P]$ in the same period of time

K_M is equal to the concentration (mol/l) of the substrate which gives half the numerical maximum velocity, $V_{\max}$.

If $k_2 \ll k_{-1}$, K_M is an expression for the affinity between enzyme and substrate.

In biochemical assays small K_M values (10^{-5} mol/l or less) mean high affinity while large K_M values (10^{-2} mol/l or greater) correspond to a low affinity.

Some examples of K_M values are given in Table 23.1.

Table 23.1. Comparison of K_M values for selected enzymes obtained by histochemical and biochemical assays (modified after van Noorden and Butcher, 1990).

Enzyme	Technique	Cytochemical assay K_M (mmol/l)	Biochemical assay K_M (mmol/l)	References
NADPH-ferri haemoprotein reductase	PVA	0.83	0.005	van Noorden and Butcher, 1986a
Alkaline phosphatase (tetrazolium salt method)	Acrylamide	0.05	0.05	van Noorden and Jonges, 1987
Alkaline phosphatase (tetrazolium salt method)	PVA	0.55	0.05	van Noorden and Jonges, 1987
α-Glucosidase	Conventional aqueous	0.68	0.64	Gutschmidt et al., 1979

PVA = polyvinyl alcohol.
Note that when using PVA in cytochemical assays, K_M is often considerably larger than that determined by biochemistry. This is probably due to a decrease in diffusion of the substrate in the viscous medium.

V_{max} is the rate of reaction when the enzyme is fully saturated with substrate. It is an expression of the molecular activity of the enzyme = "turnover number" = number of substrate molecules turned over in one minute per enzyme molecule.

Examples of maximum molecular activities are given in Table 23.2, and graphically depicted in Fig. 23.2.

Table 23.2. Number of substrate molecules turned over per minute per enzyme molecule (molecular activity) by different enzymes (determined biochemically).

Enzyme	Molecular activity (min^{-1})
Acetylcholinesterase	10^6
Chymotrypsin	10^4
Kinases	10^5
Peroxidase	10^3

$$V = V_{max} \frac{[S]}{[S] + K_M}$$

For low $[S]$ and with $[S] \ll K_M$ applies $K_M + [S] \cong K_M$

$$V = V_{max} \frac{[S]}{K_M} = k[S]$$

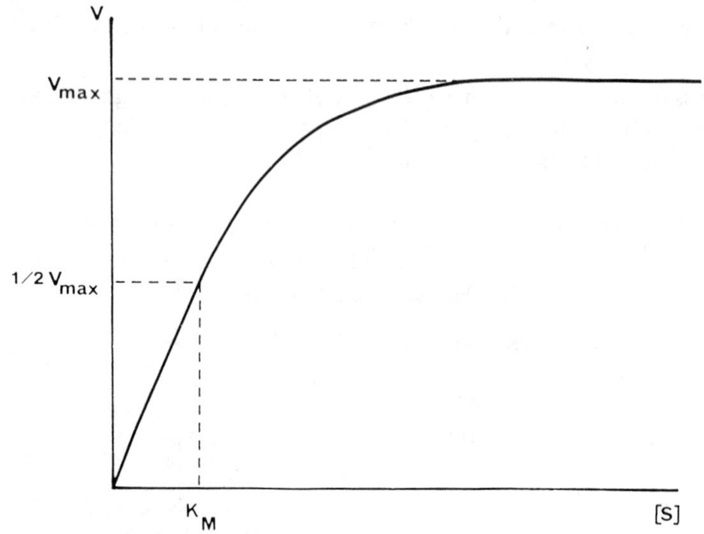

Fig. 23.2. Rate of the enzyme-catalyzed reaction as a function of substrate concentration. Reaction rate (V), substrate concentration ([S]), Michaelis' constant (KM).

At low values of [S] the reaction is of first-order (rate of reaction proportional to substrate concentration).

For high $[S]$ and $[S] \gg K_M$ applies $[S] + K_M = [S]$

$$V = V_{\max}\frac{[S]}{[S]} = K$$

For high values of [S] the reaction is of zero-order (rate of reaction is constant and independent of the substrate concentration).

K_M and $V_{\max}$ can be determined by measuring the rate of reaction (initial velocity) at different $[S]$ values.

Reaction Rate and pH. If the enzyme activity for a process is determined at different pH values a result similar to that shown in Fig. 23.3 will often be found.

The shape of the curve is determined by the following factors:

1. At extreme (high and low) pH values the enzyme is denatured
2. At the optimum pH there is optimum structure and charge of both enzyme and substrate

At the pH optimum for the process $E + S \rightleftarrows ES \rightarrow E + P$ the formation of ES complexes is at a maximum. If pH is varied from the optimum the charge distributions and consequentially the structures of enzyme and substrate are changed. This lowers the affinity of E and S for each other so that fewer ES complexes form thereby reducing the rate of product formation.

Reaction Rate and Temperature. Within certain limits, reaction rate increases by a factor of 2 for each 10°C increase in temperature (Q_{10}); however, enzyme

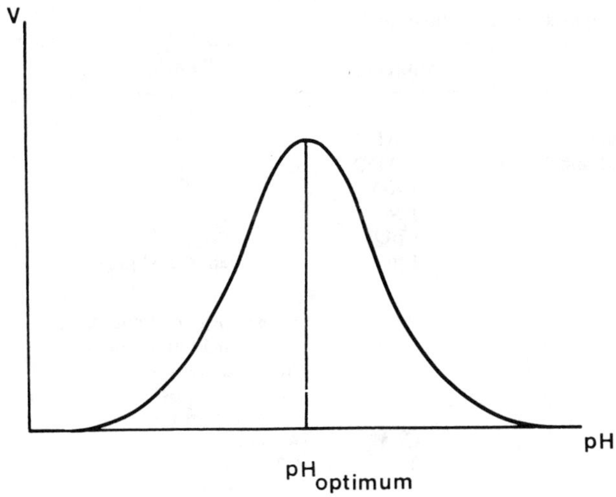

Fig. 23.3. Enzyme activity as a function of pH. Reaction rate (V).

denaturation also increases with temperature. The net substrate conversion therefore depends on the balance between reaction rate and denaturation. At temperatures > t_{max} the denaturing is greatest; t_{max} is not a constant value for the individual enzyme, as it also depends on time. With longer incubation times denaturation becomes more significant. The practical consequence of this is that for short incubation times higher temperatures can be used but for longer incubation times temperatures should be below the threshold for denaturation of the enzyme or be carefully monitored.

Reaction Rate and Ionic Strength. Low ionic strength does not inhibit enzyme activity and may even be activating. Certain enzymes need particular ions as co-factors. Greater ionic strength may inhibit activity either by blocking the active site or by denaturation.

Reaction Rate and Oxidants. Many enzymes contain thiol groups which are essential for their activity. Oxidation to disulphide bonds therefore results in reduced activity.

Cofactors. The catalytic activity of many enzymes is dependent on specific organic cofactors. These cofactors are called *coenzymes* if soluble and *prosthetic groups* if more strongly bound to the enzyme. A coenzyme can be a cofactor for several enzymes. Coenzymes can be classified according to the groups they receive from or deliver to the substrate (see Table 23.3).

Inhibition of Enzyme Activity. This can be irreversible or reversible.

Irreversible Inhibition. This results from denaturation of the enzyme or blockage of its active site. Denaturation occurs on exposure to high temperatures, large pH

Table 23.3. Some important coenzymes and their function.

Coenzyme	Abbreviation	Transfers
Hydrogen transferring		
Nicotinamide-adenine dinucleotide	NAD^+	Hydrogen
Nicotinamide-adenine dinucleotide-phosphate	$NADP^+$	—
Flavine adenine-dinucleotide	FAD	—
Flavine mononucleotide	FMN	—
Coenzyme-Q	CoQ	—
Lipoic acid	Lip.	Hydrogen + acyl groups
Group transferring		
Adenosine triphosphate	ATP	Phosphoric acid and AMP
Uridine diphosphate	UDP	Uronic acid and glucose
Pyridoxal phosphate	PLP	Amino groups
Tetrahydrofolic acid	FHY	Formyl groups
Biotin		Carboxyl groups
Coenzyme-A	CoA	Acyl groups
Thiamine pyrophosphate	TPP	C2-aldehyde groups
B_{12}-vitamin		Alkyl groups

displacements, high ionic strength, heavy metals, or organic solvents. Coagulant fixatives exert a strong denaturing effect.

Irreversible inhibition can also occur when small molecules bind permanently to the enzyme. The binding of aldehyde fixatives to amino groups often leads to inhibition, which in the case of glutaraldehyde is essentially irreversible. Where an enzyme needs free thiol groups to act, it can be irreversibly inhibited by iodoacetate which binds covalently to thiol groups

$$E{-}SH + CH_2ICOO^- \rightarrow E{-}S{-}CH_2{-}COO^- + H^+ + I^-$$

The irreversible aspect of the inhibition is due to the permanent alteration of the enzyme.

Reversible Inhibition. In contrast, reversible inhibition is characterized by the presence of an equilibrium involving enzyme and inhibitor (I).

It is customary to distinguish between two kinds of reversible inhibition, competitive and non-competitive.

Competitive Inhibition. A competitive inhibitor resembles the substrate and competes with the latter for the active site on the enzyme. This decreases the availability of the enzyme for the substrate thereby reducing the reaction rate. This form of inhibition can be abolished by increasing the concentration of substrate and Vt_{max} is therefore unchanged while K_M is increased as affinity is apparently decreased.

$$EI \rightleftarrows E \rightleftarrows ES \longrightarrow E + P$$

Non-Competitive Inhibition. Here, the turnover-number is decreased, i.e. Vt_{max} is decreased but K_M is unchanged. Binding of the inhibitor to the enzyme results in a lower substrate affinity.

$$\begin{array}{ccccc} E & \overset{S}{\rightleftarrows} & ES & \rightarrow & E+P \\ \uparrow\downarrow I & & \uparrow\downarrow I & & \\ EI & \overset{S}{\rightleftarrows} & EIS & & \end{array}$$

Non-competitive inhibition *cannot* therefore be abolished by increasing substrate concentration and this property allows distinction from competitive inhibition on an experimental basis.

23.1.4 Isoenzymes (Isozymes)

These are enzymes that differ in structure but catalyze the same process. They can show different patterns of activity and this is important for their biological function. One of the best examined isoenzyme systems is lactate dehydrogenase (Sect.25.4.1).

23.1.5 Proenzymes

A number of enzymes, including several of the proteolytic enzymes from the alimentary tract, are secreted in an inactive form called proenzymes. For example, pepsin is secreted as pepsinogen. At low pH pepsinogen dissociates into several peptides and is thereby transformed into active pepsin. In addition, pepsin itself catalyzes the transformation of pepsinogen to pepsin (autocatalysis).

23.1.6 Enzyme Classification

Enzymes are classified according to the reactions they catalyze. In 1955 the International Union of Biochemistry (IUB) appointed the Commission on Enzymes (Dixon and Webb, 1979).

According to IUB (1984) enzymes are classified in the following chief groups:

1. Oxidoreductases
2. Transferases
3. Hydrolases
4. Lyases
5. Isomerases
6. Ligases

Each main group is subdivided into sub- and sub-subgroups. The groups are numbered according to a system that allows for further additions.

The numerical classification system can be summarized as follows:

1. Each enzyme is given an EC number consisting of four digits
2. The first digit indicates which main group the enzyme belongs to

3. The second digit indicates the subgroup
4. The third figure indicates the sub-subgroup
5. The fourth digit is a serial number for the enzyme in its sub-subgroup

23.2 Histochemical Aspects

Many enzymes are completely or partly localized to certain cell organelles, e.g. acid phosphatase in lysosomes and succinate dehydrogenase in mitochondria (Sect. 31.11.1). Enzyme histochemical methods can therefore often be used as markers for certain organelles. Furthermore, to a certain extent these methods also enable cell identification as the relative occurrence of different organelles depends on the function of the cell concerned. This approach is applied in haematology where cytochemical methods allow distinction between cells of the myeloid and erythroid series (Sect.31.11.7).

Many pathological conditions give rise to histochemically demonstrable changes in enzyme distribution and activity in cells. Such changes may be due to organelle damage (e.g. increased permeability of lysosomes during hypoxia) and the effects of the biochemical lesion itself on enzyme activity. Examples of the latter include: the inhibitory effects of lead on δ-aminolaevulinate synthetase with resultant decrease in porphyrin synthesis and barbiturate induced production of the enzymes comprising the cytochrome P_{450} detoxification system.

At an experimental level a considerable number of publications report attempts to determine whether a cell has undergone a malignant change on the basis of altered enzyme activity (e.g. many malignantly transformed cells show increased succinate dehydrogenase activity).

23.2.1 Pretreatment of the Tissue

While all the standard principles of histochemical reactions apply (Sect.1.2) the following aspects may give rise to particular difficulty with enzyme reactions:

1. Immobilization of the enzyme in its original location
2. Deposition of the final reaction product corresponding to the enzyme under investigation
3. Adequate preservation of cell morphology

Where quantitation is desired a further requirement is that no activity should be lost during processing (Sect.28.1).

A very fine balance between fixation and retention of activity is necessary since many enzymes diffuse easily and nearly all oxidoreductases and several hydrolases are inhibited by even brief exposure to aldehydes.

The first steps in achieving reproducible results for the demonstration of any enzyme activity should be the rapid freezing of the fresh, unfixed tissue followed by preparation of cryostat sections. Depending on the enzyme to be demonstrated these sections are either used directly or after a brief fixation process.

It should be stressed that precautions aimed at achieving reproducibility are far more important than any attempts to simulate the physiological conditions for the enzyme reaction.

23.2.2 Incubation of Cryostat Sections

Protection of Tissue during Incubation. Some enzymes are tightly bound to membranes or other cell components (bound enzymes) while others are more or less freely diffusible (soluble enzymes). The latter may be found free in the cytoplasm, loosely bound to membranes, or enclosed in organelles. Membrane enclosed enzymes are not very diffusible under optimum conditions, but membranes may be damaged during processing and incubation.

Diffusion of the enzyme can be avoided in three different ways:

1. Preceding fixation of the section. Clearly this results in varying degrees of enzyme inhibition
2. Use of certain semipermeable films, usually dialysis membranes (McMillan, 1967; Meijer, 1980). The membrane is placed between the section and the incubation medium. This technique has been successfully used for the demonstration of certain hydrolases (e.g. acid phosphatase, β-glucuronidase, leucine aminopeptidase, E_{600}-resistant esterase) and a few oxidoreductases. Recently, Mihály *et al.* (1988) have used this approach for the demonstration of carbonate dehydratase (carbonic anhydrase). Agarose gel films containing the complete reaction medium have also been used. The reaction is simply initiated by placing the film over the tissue section (Pette and Brandau, 1962; Pette and Wimmer, 1980)
3. Addition of certain polymer compounds to the incubation medium. These include agarose, gelatin, polyethylene glycol, polyvinylpyrrolidone (PVP), polypeptides, carbohydrate polymers (e.g. Ficoll 400 and Dextran 10), and various kinds of polyvinylalcohol (PVA). A quantitative study of the effects of different grades of PVA on the activities of certain enzymes was carried out by Henderson *et al.* (1978). Reviews on the use of tissue protectants in enzyme histochemistry have been published by Altman (1980b) and van Noorden and Vogels (1989a)

Incubation Medium. The incubation medium is a solution containing the substrate and, as required, capture reagents, coenzymes, activators, inhibitors, and buffer salts.

Cofactors. As opposed to FMN and FAD, which are tightly bound prosthetic groups, NAD^+ and $NADP^+$ are diffusible and must therefore be added to the incubation medium. Too high a concentration of these coenzymes can sometimes lead to inhibition.

Activators. These are usually metal ions. In some cases they are bound to the substrate, in other cases to the enzyme molecule. Certain buffer anions can form a complex with the activating metal ion and block the activation.

Inhibitors. To determine whether a demonstrated enzymatic activity is due to a single enzyme or a whole group of enzymes it is usual to use one or several enzyme specific inhibitors. In the case of slowly acting inhibitors the section must be pretreated with the inhibitor before the substrate is added.

Buffer. The pH chosen for the histochemical reaction is often a compromise between the pH-optimum for the enzymatic catalysis and the pH-optimum of the capture reaction. As the enzymatic process nearly always changes the pH of the incubation medium it is necessary to maintain pH with the help of a buffer. Clancy (1987) has presented a computer program which enables the formulation of buffers and media of known pH and ionic strength, and the calculation of the thermodynamic acid dissociation constant of buffer substances.

Capture Reagent. Requirements for a good capture reagent or trapping agent are that it should react easily and quickly with the products of the enzymatic reaction and that this leads to precipitation of an amorphous coloured product at the location of the enzyme under investigation. At too low a concentration of the capture reagent coupling is difficult, while high concentrations frequently inhibit the enzyme.

Temperature. Material from humans and most warm-blooded animals is frequently incubated at 37°C to correspond with body temperature. This is a few degrees below optimum for most dehydrogenases. The temperature of the incubation medium is kept constant throughout incubation. To avoid a lag phase (with decreased initial velocity) both section and medium must be at 37°C at the beginning of the incubation.

If the reaction is only linear for a very short time it may, especially where determinations of K_M are intended, also be appropriate to perform the reaction at 20°C.

Incubation Time. For localization and semiquantitative assessment of enzyme activity, it is important that the incubation time should not exceed the minimum time required for the appearance of a visible reaction product. The semiquantitative aspect is directly assessed by the time it takes to obtain a visible reaction product at a particular location (Høyer and Andersen, 1970; Andersen and Høyer, 1974).

23.2.3 Posttreatment Procedure

The enzyme reaction is stopped by washing the section in distilled water, buffer, or fixative. If desired, the material may then be subjected to postfixation (e.g. formol calcium at 4°C for 15 min). The section is mounted in a hydrophilic mountant as the end products are usually soluble in organic solvents.

23.2.4 Definitive Identification of the Enzyme

This may be very difficult to achieve in tissue sections. Enzymes may be either substrate or group specific. Thus, glucose-6-phosphatase specifically catalyzes the hydrolysis of glucose-6-phosphate while acid phosphatase catalyzes the hydrolysis of most phosphate-monoesters including glucose-6-phosphate. Glucose-6-phosphatase activity cannot therefore be equated with the presence of the enzyme glucose-6-phosphatase. Identification must be attempted by:

1. Determination of activity with different substrates
2. The effect of specific inhibitors and activators
3. Biochemical analysis of the tissue, possibly after tissue fractionation
4. Immunohistochemical analysis with specific antibody directed at the enzyme in question

24 Enzyme Histochemistry II: Hydrolases

H. Lyon, P.E. Høyer, P. Prentø

Hydrolases or hydrolytic enzymes catalyze the breakdown of substrates (e.g. esters, glycosides, and peptides) using water:

$$AB + H_2O \rightleftarrows HA + B{-}OH$$

For the purposes of this account it is sufficient to note that the plasma membrane and lysosomes are prominent locations for hydrolytic enzymes.

Histochemical studies aim to determine both the location and the activity of an enzyme. In some cases, notably with lysosomal enzymes, assessments of the accessibility of the enzyme in its normal cytological location (membrane permeability) and metabolic context are of interest. Under physiological conditions lysosome membranes are relatively impermeable. When lysosomal enzymes are examined histochemically there may be a latent period after the incubation has been initiated before a drastic increase in lysosomal membrane permeability occurs. The length of this period may be a direct expression of the physiological condition of the membrane and the cell. In formal biochemical terms this lag is often referred to as *latency*. Investigations of latency may be carried out by comparing a series of sections incubated in media with or without added detergent (Fig. 24.1). Another approach has been described by Bitensky and Chayen (1977) in a study on leucyl-β-naphthylamidase in lysosomes in follicles of the guinea pig thyroid. The substrate used for the reaction can only sparingly penetrate the intact lysosomal membrane. The incubation medium is buffered to pH 5.0 and after a certain time an inflexion point is noted. This is taken to indicate that the acid medium renders the membrane more permeable for the substrate. The conditions pertaining to the permeability of the lysosomal membranes of thyroid follicle cells have formed the basis for developing a sensitive cytochemical bioassay for circulating levels of thyrotrophin (Bitensky and Chayen, 1977; cf. also 28.8.11).

24.1 Principles of Hydrolase Demonstration

The main principles in demonstration of hydrolases are:

1. Formation of a sparingly soluble coloured reaction product

H. Lyon (Ed.)
Theory and Strategy in Histochemistry

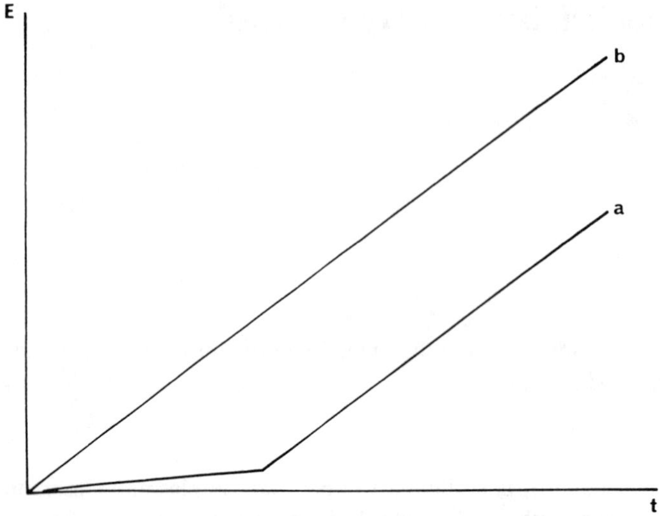

Fig. 24.1. The relationship between hydrolase activity (extinction) and incubation time.
a without detergent
b with detergent
Extinction (E), time (t).

2. Identification of the active enzyme (see Sect.23.2.4)

From a histochemical standpoint, hydrolases may be classified as substrate-specific – only demonstrable using their natural substrates, or group-specific – demonstrable using artificial substrates. Demonstration can be achieved using indoxyl ester, simultaneous coupling, post-coupling, and metal salt methods as described below. *Alkaline phosphatase is used as an example in all cases.* For each method at least one enzyme is mentioned for which it is either the most suitable or the only method available.

24.1.1 Indoxyl Ester Methods

In these methods an indoxyl phosphate is used as substrate. Alkaline phosphatase catalyzes the reaction shown below by removing phosphate from the substrate leaving an uncoloured leuco-indigo compound. Two molecules of the latter are converted into a coloured indigo compound by oxidation.

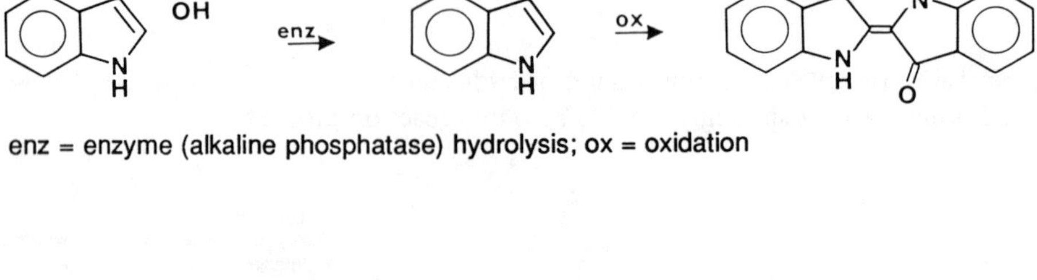

enz = enzyme (alkaline phosphatase) hydrolysis; ox = oxidation

Although the method has been used sparingly for the demonstration of alkaline phosphatase, reasonably good localization has been achieved with the calcium salt of 5-bromoindoxyl phosphate in the presence of a ferrocyanide-ferricyanide oxidation catalyst. The precipitated 5,5′-dibromoindigo is microcrystalline.

Problems with the method include the rather slow spontaneous oxidation of the leuco compound to indigo by action of atmospheric oxygen and the inhibitory effect of the ferrocyanide-ferricyanide reagent on a number of enzymes.

The indoxyl ester methods are often used to demonstrate carboxyl esterase activity and are very suitable for the quantitative demonstration of β-galactosidase (Lund-Hansen *et al.*, 1984).

With all indoxyl ester methods the liberated indoxyl couples very effectively with diazonium salts such as hexazotized Pararosanilin (simultaneous coupling 24.1.3). However, Fast Blue VB is the diazonium salt of choice (Lojda *et al.*, 1979, p.64). It yields the highest amount of azoindoxyl dye, allows precise localization, and does not decompose even after prolonged incubation. Moreover, the coupling velocity of the azoindoxyl procedure using Fast Blue VB is higher than that obtained in the simultaneous azocoupling methods.

The indoxyl-tetrazolium salt method for the localization of alkaline phosphatase activity was first described by McGadey (1970). In this method the substrate 5-bromo-4-chloro-3-indolyl-phosphate is hydrolyzed at alkaline pH with a concomitant release of hydrogen which can reduce tetrazolium salts. Tetranitro BT is the tetrazolium salt of choice, as it yields an amorphous or microcrystalline formazan of relatively high stability (Gossrau, 1978). The method has recently been optimized by van Noorden and Jonges (1987) who developed a specific and valid quantitative procedure. They showed that addition of polyvinyl alcohol to the incubation medium greatly improved the localization of the final reaction product in cryostat sections. The electron transfer from the substrate to tetranitro BT was significantly enhanced by 1-methoxyphenazine methosulphate which seems to indicate that electrons may be transferred by endogenous electron transport systems in the absence of 1-methoxyphenazine methosulphate. Addition of azide to the incubation medium enhanced the formazan production considerably by blocking the interference from oxygen. The specific reaction obeyed the Beer-Lambert law. A K_M of 0.05 mmol/l for aqueous media and a K_M of 0.55 mmol/l for polyvinyl alcohol-containing media were obtained. These values indicate that the indoxyl-tetranitro BT method is considerably more sensitive than any metal salt or diazonium salt method developed so far.

24.1.2 Post-Coupling Methods

These methods rely on the primary reaction product being sufficiently insoluble to remain at its primary location without diffusion during both the initial incubation and the subsequent performance of the visualizing reaction.

All post-coupling methods use a diazonium salt, and the final products are therefore azo dyes. These methods are particularly appropriate when long incubation times are necessary, and when the pH optimum of the enzyme is in the acid range. As distinct from the simultaneous coupling methods (Sect.24.1.3), a compromise between the pH optima for the enzyme and for the coupling reaction between diazonium salt and naphthol or naphthylamine is not necessary. Similarly, the enzyme reaction and the coupling reaction can both be performed at their respective optimum temperatures (e.g. 37°C and 4°C). Finally, any inhibitory effect of the diazonium salt on enzyme activity is avoided.

Despite these advantages and the fact that they are suitable for quantitative investigations (end-point measurements, cf. Sect.28.8.7), post-coupling methods are seldom used. The main reason for this is that it is necessary to use substituted naphthol compounds to achieve reasonable localization. These compounds have such a pronounced affinity for tissue proteins that coupling is inhibited and this may give rise to false negative reactions. In some cases, however, the approach may still offer significant advantages for particular enzymes (e.g. demonstration of N-acetyl-β-glucosaminidase, Robertson, 1980).

24.1.3 Simultaneous Coupling Methods

In these methods a diazonium salt is coupled to a product (e.g. naphthol) liberated by enzymatic activity. The reactions are:

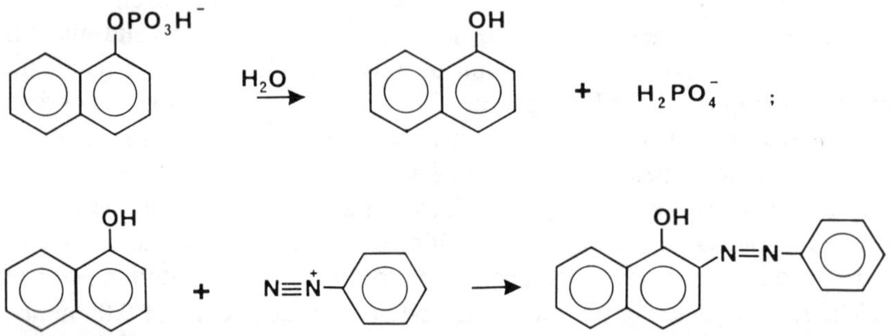

Diffusion of the primary reaction product before it reacts with the diazonium salt is an important problem that may prevent localization of the enzyme activity. Key factors influencing the final result are:

1. Specificity of substrate
2. Rate of substrate hydrolysis (naphthol compound)
3. Diffusion coefficient of the primary product (the naphthol)
4. Rate of the coupling reaction
5. Stability of the final reaction product

Simultaneous coupling methods are always a compromise between these factors. All diazonium salts lead to a greater or lesser degree of inhibition of enzyme activity. Unfortunately, a relatively high concentration of diazonium salt is necessary to achieve an effective coupling.

An advantage of the simultaneous coupling method is that the reaction can be followed directly and interrupted at different times. This makes it possible to resolve the time-course of the reaction as it occurs in different tissue components (continuous monitoring, cf. Sect.28.8.7).

The simultaneous coupling method for alkaline phosphatase has undergone a considerable development since it first appeared (Menten *et al.*, 1944) (see also azo coupling reagents Table 3.7 in Sect.3.3.8). Nowadays, effectively only substituted naphthols are used.

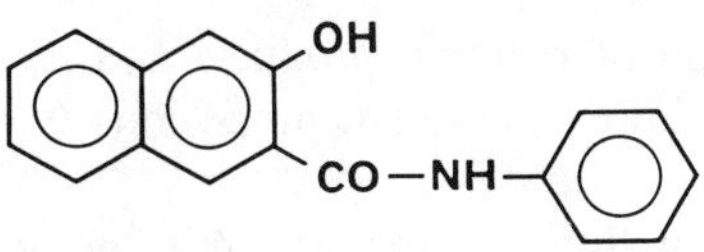

Naphthol-AS

As diazonium salts normally bind to histidyl and tyrosyl residues (Sect.9.4.1) as well as biogenic amines (Sect.18.4.10), it is essential that the incubation conditions should yield diazonium salt-naphthol and diazonium salt-tissue products that differ substantially in intensity and colour.

The naphthol-AS-phosphates are particularly suitable substrates for the demonstration of alkaline phosphatase activity (Table 3.7). These substrates show low solubility of the naphthol reaction products and high substantivity. Localization is very precise. Fast Red Violet LB or freshly prepared hexazotized New Fuchsin (C.I. 42520) are examples of appropriate diazonium salts. The latter is resistant to ethanol dehydration.

Simultaneous coupling methods also give good results with non-specific acid phosphatases, non-specific esterases, and some proteases.

24.1.4 Metal Salt Methods

These were the earliest methods developed for demonstrating enzyme activity. They were introduced for the demonstration of alkaline phosphatase by Gomori (1939) and Takamatsu (1939). The substrate is sodium β-glycerophosphate or another phosphomonoester and the incubation solution contains a high concentration of calcium ions which act as the capture reagent. Pb^{2+}, Cu^{2+}, and Co^{2+} are alternative capture reagents. Magnesium ions are added as an activator. The enzyme catalyzed reaction is:

$$\beta\text{-glycerophosphate} \rightarrow \beta\text{-glycerol} + \text{phosphate}$$

and the capture reaction with Ca^{2+} is:

$$3Ca^{2+} + 2PO_4{}^{3-} \rightarrow Ca_3(PO_4)_2$$

The section is treated with cobalt nitrate:

$$Ca_3(PO_4)_2 + 3Co^{2+} \rightarrow 3Ca^{2+} + 2PO_4{}^{3-} + 3Co^{2+} \rightarrow 3Ca^{2+} + Co_3(PO_4)_2$$

and then with ammonium sulphide:

$$Co_3(PO_4)_2 + 3S^{2-} \rightarrow 3Co^{2+} + 2PO_4^{3-} + 3S^{2-} \rightarrow 3CoS + 2PO_4^{3-}$$

This sequence relies on the fact that while calcium phosphate is only very slightly soluble at alkaline pH (pH 9.4), it is more soluble than cobalt phosphate at neutral pH which again is more soluble than cobalt sulphide.

The shortcomings of the method are diffusion of reaction products and non-specific binding of metal ions to the tissue. In addition, divalent metal ions (e.g. Pb^{2+}) inhibit the enzymes.

For demonstration of alkaline phosphatase by light microscopy this method is clearly inferior to the simultaneous coupling reaction (Sect.24.1.3) as artifactual blackening of chromatin is a constant finding. In general, however, for the substrate specific hydrolases, metal salt methods are the only possible approach.

24.1.5 Summary of Reactions for Hydrolases

The reactions taking place in the four methods described have been summarized below in order to clearly indicate the number of steps before the final, insoluble, coloured product is formed. (Precipitates underlined).

Indoxyl ester method:

$AB \rightarrow A + \mathit{B}$;	primary reaction product insoluble and stained

Simultaneous coupling method:

$AB \rightarrow A + B$;	the primary reaction product is soluble
$B + C \rightarrow \mathit{BC}$;	and unstained; the secondary reaction product is insoluble and coloured

Post-coupling method:

$AB \rightarrow A + \mathit{B}$;	the primary reaction product is
$B + D \rightarrow \mathit{BD}$;	insoluble and colourless; the secondary reaction product is insoluble and coloured

Metal salt method:

$AB \rightarrow A + B$;	the primary reaction product is soluble
$B + C \rightarrow \mathit{BC}$;	and colourless; the secondary reaction
$BC + D \rightarrow \mathit{BD} + C$;	product is insoluble, but colourless; the tertiary reaction product is insoluble and coloured

The choice of technique depends on the enzyme to be examined and on whether observation is to be by light or electron microscopy. If there are several possibilities, quality of localization is often the dominant influence in making the final choice.

Of the four approaches, it is only with the metal salt method that natural substrates can be used.

24.2 Pretreatment

Many hydrolases are diffusible and precise localization therefore requires the use of semipermeable membranes, colloid stabilizers, or fixation (see Sect.23.2.2). If fixation is chosen this should take place under carefully controlled conditions since enzymes are often either denatured or their substrate requirements altered or rendered less specific.

The ability of the different hydrolases to tolerate fixation is highly variable and an individual approach developed for each enzyme should be closely adhered to. In no circumstances should the recommended fixation time be exceeded.

24.3 Incubation

The incubation medium contains a number of other reagents in addition to the substrate (Sect.23.2.2).

24.3.1 Substrate

Artificial substrates should be readily soluble in their non-hydrolyzed condition and only slightly soluble and unable to form colloid particles or supersaturated solutions in their hydrolyzed condition. The substrate should also have a high diffusion coefficient and be easily hydrolyzed by the enzyme at a given pH, preferably over a wide pH range. It should not act as an inhibitor of the enzyme.

Considerations regarding the size of the substrate molecule may have bearing on the choice of substrate. Is the molecule able to penetrate the intact cell and organelle membranes? Should the membranes be opened, perhaps by fixing or drying for a short time?

24.3.2 Capture Reagent

Regardless of whether this is a diazonium salt or a metal salt the following requirements apply:

1. Ready solubility with a high diffusion coefficient
2. Easy and rapid coupling under the conditions of the reaction (pH, temperature, etc.)
3. Maximum substantivity of the reaction product formed
4. Preferably colourless
5. Unaltered by light
6. Stable under the various reaction conditions imposed (pH and temperature)
7. Minimal inhibitory effect on the enzyme catalyzed reaction

24.3.3 Cofactors and Activators

Metal ions are often necessary for, or increase, the activity of certain enzymes. For example, ATPase associated with myosin is activated by Ca^{2+} and the cell membrane "sodium pump" by Na^+ and K^+. In some cases a metal ions can be exchanged for another provided the charge is unchanged. For instance, in the histochemical demonstration of alkaline phosphatase both Mg^{2+} and Mn^{2+} are activators, while the natural cofactor is Zn^{2+}. It is worth noting that many of the stabilized diazonium salts contain Zn^{2+}, even though this is in suboptimal amounts.

24.3.4 Inhibitors

Metal ions can act as inhibitors (e.g. Be^{2+} for alkaline phosphatase). Chemical compounds which block $-SH$, $-NH_2$, or $-COOH$ groups frequently act as inhibitors, probably because these groups are often part of the active sites of the enzymes.

24.3.5 Post-Treatment Procedure

A problem with a number of methods using diazonium salts is the formation of nitrogen bubbles in the mounted slide. This seems to be due to a non-enzymatic degradation of the diazonium salt and can be counteracted partly by thorough washing, partly by the use of more stable diazonium salts, e.g. Fast Red Violet LB.

24.4 Controls in the Histochemical Investigation of Enzyme Activity

Where possible specific inhibitors are used but, failing this, heat inactivated sections. A further control is the omission of substrate.

24.5 Quantitation

In theory this can be achieved by determining λ_{max} for the final reaction product and measuring the mean integrated extinction (MIE) at this wavelength.

Clearly prolonged chemical fixation (e.g. 3–4 hours in formalin) would completely invalidate quantitative results, as would storage of the sections in gum acacia sucrose (cf. Sect.13.6 and Chap.28).

24.6 Demonstration of Selected Hydrolases

The following list of enzymes whose activity can be demonstrated histochemically is by no means complete but will include examples of enzymes acting on ester bonds (Sect.24.6.1), glycoside bonds (Sect.24.6.2), peptide bonds (Sect.24.6.3), and acid anhydrides (Sect.24.6.4).

24.6.1 Hydrolases Acting on Ester Bonds, EC 3.1

Included in this group of enzymes are carboxylic ester hydrolases, EC 3.1.1, and phosphoric monoester hydrolases, phosphatases, EC 3.1.3.

Carboxylic Esterases. Carboxylic esterases or carboxyl ester hydrolases, EC 3.1.1, catalyze the reaction:

$$R'{-}O{-}CO{-}R'' + H_2O \rightarrow R'{-}OH + HOOC{-}R''$$

where R′ is an alkyl or an aryl from an alcohol, phenol, or hydroxylated base such as choline. R" is an alkyl from a carboxylic acid with shorter or longer carbon chain. The distinction between the various esterases is confused, both in biochemistry and histochemistry, and generally the best strategy is to characterize any histochemical esterase activity by the substrate(s) (and co-factors) used, and the response to various inhibitors. Points worth noting are that no esterase is substrate-specific, that most often the substrate selectivity is low, that short-chain aromatic substrates such as α-naphthyl acetate are vigorously attacked by practically all esterases except pancreatic-type lipase, and that histochemical esterase substrates may also be attacked by some proteases (e.g. chymotrypsin).

A crude classification of histochemically localizable esterases can be achieved according to the nature and state of the substrate and the sensitivity of the enzymes to various inhibitors:

1. Non-specific esterases (these attack glyceryl and other esters of short-chain fatty acids and are especially abundant in the cytoplasm of liver, kidney, and pancreatic cells)
2. Cholinesterases (these hydrolyze fatty acid esters of choline and acetylcholine and are found in neuromuscular junctions, cholinergic synapses, and erythrocytes)
3. Lipases (these attack esters of long-chain fatty acids and are found principally in the pancreas)

A more detailed overview of the classification, substrates, products of hydrolysis, and inhibitors applicable to carboxylic esterases is shown in Table 24.1.

The judicious use of inhibitors can make it possible to identify individual enzymes belonging to the non-specific esterases or cholinesterases using naphthyl acetates (see Sect.24.1.2) or indoxyl acetates (see Sect.24.1.1) as substrates.

Table 24.1. Classification of carboxylic esterases with the products of hydrolysis and specific inhibitors.

Preferred name and E.C. no.	Substrate	Products of hydrolysis R′	Products of hydrolysis R″	Inhibitors
(I) Nonspecific esterases				
carboxyl esterase 3.1.1.1	carboxylic ester of aliphatic alcohol or aromatic alcohol or phenol	aliphatic alcohol aromatic alcohol	carboxylic acid	organophosphates (DFP, E600)
arylesterase 3.1.1.2	ester of acetic acid and a phenol	phenol	acetic acid	organic Hg compounds (PCMB)
acetylesterase 3.1.1.6	ester of acetic acid and an alcohol or a phenol	alcohol or phenol	acetic acid	
(II) Cholinesterases				
acetylcholinesterase 3.1.1.7	acetylcholine acetylthiocholine (H) phenylacetate (H)	choline thiocholine phenol	acetic acid	physostigmine organophosphates (DFP, E600)
cholinesterase 3.1.1.8	acylcholine acylthiocholine (H) phenylacetate (H)	choline thiocholine phenol	carboxylic acid	B.W. 284C51 physostigmine organophosphates (DFP, E600) ethopropazine hydrochloride
(III) Lipases				
lipase 3.1.1.3	triglycerides esters of polyhydric alcohols (other than glycerol) (H)	glycerol polyvalent alcohols	fatty acid	
phospholipase B 3.1.1.5	lysolecithine	glycerylphosphocholine	fatty acid	

(H) = histochemical substrate; DFP = diisopropylfluorophosphate, E600 = diethyl-p-nitrophenyl phosphate; PCMB = p-chloromercuribenzoate; B.W. 284C51 = 1,5-bis-(4-allyldimethylammoniumphenyl)pentan-3-one-dibromide.
Note: physostigmine = eserine.

It must, however, be stressed that the majority of the inhibitors are extremely toxic. The use of natural substrates together with metal ions as the capture reagent raises some further possibilities for distinguishing between the different esterases.

The term cholinesterases has unfortunately been used by histochemists to cover both *acetylcholinesterases*, EC 3.1.1.7, or true cholinesterases and *cholinesterases*, EC 3.1.1.8, or pseudocholinesterases. Both types are sensitive to organophosphorus compounds and are also inhibited by physostigmine (eserine) of which the latter does not inhibit the non-specific esterases. Both acetylcholinesterase and cholinesterase can be selectively demonstrated by use of acetylthiocholine which is not attacked by other carboxylesterases.

Acetylcholinesterase activity decreases with the chain length of the acyl group of the substrate and has the natural substrate acetylcholine. The enzyme can often be preferentially demonstrated by use of the histochemical substrate acetylthiocholine, but only at high substrate concentration which inhibits cholinesterase.

Cholinesterases show increased activity with increased chain length of the substrate acyl group, and, as stated above, are also inhibited to some extent when the concentration of acetylcholine or acetylthiocholine is high.

Lipases are able to attack long-chain (n more than 8) fatty acid esters. Most esterases show little activity towards long chain substrates. Bile salts are, as under in vivo conditions, required to disperse the lipase substrate (fats) so that it is accessible to the enzyme. In the absence of bile salts true lipases, or pancreatic-like lipases, show little or no activity. Several lipases are, however, known ("lipoprotein lipases") which hydrolyze long-chain esters in the absence of emulsifiers and many non-specific esterases attack long-chain fatty esters, although they do so more slowly than short-chain esters. Formerly, lipase activity was demonstrated using sorbitan esters ("Tween") but this method shows low selectivity and has been superceded by the naphthol-AS-nonanate-taurocholate method of Abe *et al.* (1964). The latter method permits selective demonstration of pancreatic-like lipases. As yet there are no published methods for the selective demonstration of lipoprotein lipases.

Proteases generally hydrolyze peptide bonds but several proteases also show significant activity against ester links. In fact, bromoindoxyl acetate and naphthol-AS-β-chloropropionate are useful substrates for the demonstration of cathepsin C and for chymotrypsin, as are naphthol-AS phenyl acetate, butyrate, and propionate. When demonstrating esterases, it may therefore often be wise to bear in mind the possibility that some, most, or all the enzyme activity may be due to a protease.

In this connection it should be noted that a particular aryl esterase, the *naphthol AS-D chloroacetate esterase*, shows several similarities with α-chymotrypsin (secreted by pancreatic exocrine tissue). Naphthol-AS-D chloroacetate esterase is found in neutrophils and myelocytes as well as in mast cells. It is also noteworthy that the enzyme activity is preserved after both formaldehyde and methanol fixation and following paraffin embedding (see also Sect.31.11.7).

Phosphoric Monoester Hydrolases, EC 3.1.3. These are enzymes that catalyze the hydrolysis of phosphate esters.

Alkaline Phosphatase, EE 3.1.3.1. Alkaline phosphatase comprises a group of enzymes which are group specific with a pH optimum of about pH 9.0–9.5. For maximum sensitivity, cryostat sections of freshly frozen tissue must be used. All methods for the demonstration of hydrolases can, as discussed in Sect.24.1, be used. For normal light microscopy a simultaneous coupling method may be used. However, for quantitative work an indoxyl-tetranitro BT method is preferable (van Noorden and Jonges, 1987). For electron microscopy a metal salt method must be used.

L-p-bromotetramisole is a specific inhibitor of some alkaline phosphatases particularly those in liver and bone, while leukocyte alkaline phosphatase is moderately inhibited and alkaline phosphatases in intestine and placenta are hardly not affected. L-phenylalanine shows exactly the reverse pattern of inhibition of alkaline phosphatases in different locations. These differences in inhibition are probably due to the presence of different isoenzymes in the different locations.

The enzymes are predominantly localized to cell membranes and are found in especially large amounts in transporting epithelia, e.g. intestinal epithelium.

Acid Phosphatase, EC 3.1.3.2. Acid phosphatase comprises a series of enzymes which are group specific with a pH optimum of about pH 5.0. Cryostat sections should be used as discussed under alkaline phosphatase.

Due to extensive diffusion of this enzyme a possible approach for achieving optimum localization is to use cryostat sections of tissue fixed in neutral phosphate buffered formaldehyde at 4°C. After fixation the material is washed thoroughly in water or preferably in gum acacia sucrose to remove phosphate ions. This technique can, however, not be applied in quantitative studies. Instead, tissue stabilizers (PVA, Perrild *et al.*, 1989) or semipermeable membranes (Stoward and Al-Sarraj, 1981a; 1981b) can be employed. With the latter technique in a simultaneous coupling method adsorption of hexazotized Pararosanilin onto tissue proteins gives rise to a weakly absorbing yellow product after the incubation has been terminated with formalin. After mounting this product is gradually transformed into a strongly absorbing purple material. This can be avoided by treating the section with 70% ethanol for 30 min at room temperature (Stoward *et al.*, 1981; 1982).

A metal salt method can be used for both light and electron microscopic investigations but one of the other methods is generally preferred for light microscopy. Molybdate and fluoride ions are effective inhibitors. If a metal salt method is used, precipitation of the lead salt must be avoided during the preparation of the incubation medium. Waters and Butcher (1980) have shown that this aim may be achieved by the use of acetate buffer at pH 4.7 that after maximal dilution is mixed with the other reactants. This has made possible the preparation of an incubation medium which contains the concentration of substrate, lead salt, and buffer originally proposed by Gomori (1950b).

Acid phosphatase is found in lysosomes and is particularly abundant in phagocytes.

5′-Nucleotidase, EC 3.1.3.5. This enzyme has been demonstrated by a metal salt method using natural substrates such as adenosine-5′-monophosphate with Mn^{2+}

ions as activator. Ca^{2+} are used as the trapping agent followed by substitution with Co^{2+} and visualization with sulphide as described in Sect.24.1.4. The reaction is preferably carried out on unfixed cryostat sections. The pH optimum of the enzyme is between 7.5 and 8.5. The reaction can with advantage be carried out at pH 8.3 minimizing activity due to acid phosphatase while alkaline phosphatase activity is inhibited by adding L-p-bromotetramisole to the incubation medium. 5′-Nucleotidase is itself inhibited by Zn^{2+} and Ni^{2+} ions.

Addition of PVA to the incubation medium results in better preservation of tissue morphology (Henderson *et al.*, 1980; Frederiks and Marx, 1988). A disadvantage with PVA is that total 5′-nucleotidase activity is significantly lower than the activity in unfixed sections incubated with an aqueous medium (Frederiks and Marx, 1988). The authors suggest that the activity of the enzyme is kept in the latent state by PVA.

The enzyme is bound to plasma membranes. An increased activity is observed in synoviocytes in rheumatoid arthritis (Henderson *et al.*, 1980).

Glucose-6-Phosphatase, EC 3.1.3.9. This enzyme can be demonstrated by the Wachstein and Meisel metal salt method using the natural substrate glucose-6-phosphate. The released phosphate couples with a lead salt (Wachstein and Meisel, 1956). Advantages and disadvantages of the method are described in Sect.24.1.4. The pH optimum of the enzyme is 6.0, and the reaction is usually carried out at pH 6.5. It is essential to perform control reactions as both alkaline and acid phosphatases will hydrolyze glucose-6-phosphate at this pH. These reactions include the addition of 1,5-sorbitan-6-phosphate, a specific inhibitor of glucose-6-phosphatase, to the incubation medium or the replacement of the specific substrate with β-glycerophosphate.

Glucose-6-phosphatase is used as a marker for microsomes in liver, kidney, and intestinal mucosa (de Duve *et al.*, 1955).

24.6.2 Hydrolases Acting on Glycoside Bonds, EC 3.2

From this group only examples of enzymes acting on O-glycosyl compounds, EC 3.2.1, will be given.

α-D-Glucosidase, EC 3.2.1.20. This can be demonstrated using a simultaneous coupling technique with 2-naphthyl-α-D-glucoside as the substrate and hexazotized Pararosanilin as the coupling agent. The pH used is 6.0 while pH optimum of the lysosomal enzyme has been reported to be in the range 4–5.

High activity is found in the brush border of enterocytes in the small intestine with peak activity towards the apex of the villi (Gutschmidt, 1981; Gutschmidt *et al.*, 1980a).

Turanose (3-O-α-glucosylfructose) inhibits lysosomal α-glucosidase. Other inhibitors are Hg^{2+} and Cu^{2+} ions (Barrett and Heath, 1977).

Natural substrates are glycogen and maltose. The enzyme is thus able to hydrolyze (α-1,4) and (α-1,6) glycoside linkages and therefore completely hydrolyzes glycogen to glucose. Hydrolysis of the α-1,6 linkage is rate-limiting.

Deficiency of the enzyme is present in the storage diseases type II glycogenosis or Pompe's disease.

β-Galactosidase, EC 3.2.1.23. This can be demonstrated with an indoxyl ester method. The technique has been improved by introducing a PVA technique so as to avoid diffusion of enzyme. Further, it was found necessary to perform air-drying of the material prior to incubation and to use 5-bromo-4-chloro-3-indolyl-β-D-galactopyranoside as the substrate at pH 4.0–4.1. Enzyme activity increases with the concentration of sodium chloride over the range 5 to 100 mmol/l but decreases above. The method relies on the oxidation of the primary product by ferro/ferricyanide (Lund-Hansen *et al.*, 1984).

The enzyme is inhibited by N-ethylmaleimide, N-acetyl-D-galactosamine, and heparin.

β-Galactosidase is a lysosomal enzyme which in the thyroid gland is involved in the processing of the carbohydrate residues of thyroglobulin (Perrild *et al.*, 1989). Deficient activity of the enzyme is present in the hereditary metabolic disease generalized or G_{M1}-gangliosidosis, one of the lipidoses (cf. Sect.31.7.3).

N-Acetyl-β-Glucosaminidase, EC 3.2.1.30. This enzyme can be demonstrated with a post-coupling reaction with naphthol-AS-BI-N-acetyl-β-glucosaminide as the substrate. Fast Garnet GBC is recommended as post-coupling agent. Optimum pH for the enzyme reaction is 4.5, while a pH of 6.2 is recommended for the post-coupling step. The reaction should be carried out as a PVA technique (Robertson, 1980). A specific inhibitor of the enzyme is 2-acetamide-2-deoxy-D-gluconolactone.

N-acetyl-β-glucosaminidase is a lysosomal enzyme. The activity of the enzyme is increased in synoviocytes in induced allergic arthritis, and it has a function in the processing of the carbohydrate residues of thyroglobulin (Perrild *et al.*, 1989).

β-Glucuronidase, EC 3.2.1.31. This can be demonstrated with a post-coupling technique with naphthol-AS-BI-β-glucuronide as substrate and Fast Blue B as post-coupling agent (Henderson, 1984). Activity is maximal in the region pH 4.3-5.0. A specific inhibitor is D-gluco-saccharo-1:4-lactone.

β-Glucuronidase is a lysosomal and microsomal enzyme found in granulocytes, particularly eosinophils. Further, the enzyme is found in fibroblasts, chondrocytes, and synoviocytes.

24.6.3 Hydrolases Acting on Peptide Bonds, Proteases, EC 3.4

The nomenclature of this group of enzymes is confusing and has probably not yet been finally settled. Proteases are divided into peptidases and proteinases.

Peptidases (exopeptidases, Table 24.2) only attack peptide bonds at or near the ends of peptide chains. Subclassification and designation is made according to their specificity rather than their mechanism of action (McDonald, 1985).

Table 24.2. Classification of some peptidases (exopeptidases).

Name	E.C. No.	Hydrolysis of
Aminopeptidases	3.4.11	single amino acids from the N-terminus of the peptide chain
Dipeptidases	3.4.13	dipeptides
Dipeptidyl peptidases	3.4.14	dipeptide units from the N-terminus
Peptidyl peptidases	3.4.15	dipeptide units from the C-terminus

Proteinases (endopeptidases, Table 24.3) catalyze the hydrolysis of internal bonds in peptide chains. Subclassification is here made on the basis of the composition of the active site of the enzyme (Barrett, 1980).

Table 24.3. Classification of some proteinases (endopeptidases).

Name	E.C. No.	Catalytic mechanism
Serine proteinases	3.4.21	active centre contains histidine and serine
Cysteine proteinases	3.4.22	active centre contains cysteine
Aspartic proteinases	3.4.23	pH optimum below 5 (acidic residue involved in catalysis)
Metallo Proteinases	3.4.24	a metal ion is involved in catalysis

Aminopeptidase A, EC 3.4.11.7. This enzyme may be demonstrated using a simultaneous coupling technique. The substrate of choice is at present α-L-glutamic acid-4-methoxy-2-naphthylamide with Fast Blue B as the coupling reagent (Lojda and Gossrau, 1980; Kugler, 1982a). The enzyme is activated by Ca^{2+} ions and inhibited by EDTA and 1,10-phenanthroline while organophosphates (E600, DFP), and organic mercuric compounds (PCMB) do not influence the activity. Important locations of the enzyme are endothelial cells of capillaries, brush border of epithelial cells in the small intestine, liver, and proximal tubules of the kidney (Lojda and Gossrau, 1980). In glomeruli the enzyme degrades any peptides leaking through the glomerular membrane, and it has been demonstrated that aminopeptidase A in this location is identical to angiotensinase A (Kugler, 1982a).

The enzyme is not bound very tightly to the membranes and up to 30% may leak out of fresh cryostat sections. To avoid this, pretreatment with acetone or acetone-chloroform has been used with reasonably success.

In a paper by Kurauchi *et al.* (1989) it has been shown that the activity of aminopeptidase A in placenta is stimulated by oestradiol and progesterone. Moreover, placental aminopeptidase A may degrade bioactive peptides such as angiotensin II (Mizutani *et al.*, 1985a) and oxytocin (Mizutani *et al.*, 1985b).

Aminopeptidase M or N, EC 3.4.11.2. This is a membrane bound enzyme. It is demonstrated by a simultaneous coupling technique using L-leucine-4-methoxy-2-naphthylamide or L-alanine-4-methoxy-2-naphthylamide and Fast Blue B. For electron microscopical demonstration Fast Blue B is substituted with hexazotized Pararosanilin.

The enzyme occurs in large amounts in brush borders of intestinal enterocytes, proximal kidney tubules, macrophages, and polymorphonuclear leukocytes. In a combined enzyme histochemical and in situ hybridization study, Norén *et al.* (1989) have shown that the onset of production of aminopeptidase M-mRNA probably takes place in a narrow zone of developing enterocytes at the transition zone between crypts and villi in the rabbit jejunum.

Dipeptidyl Peptidase I, EC 3.4.14.1. This enzyme contains cysteine in the active centre, and requires Cl^- ions for its activity. The addition of mercaptoethylamine-HCl simultaneously ensures that the thiol group is on the reduced form and that chloride is present.

The enzyme is demonstrated using a simultaneous coupling reaction with glycine-arginine- or proline-arginine-4-methoxy-2-naphthylamide as the substrates of choice and Fast Blue B as the diazonium salt (Sannes, 1988).

Dipeptidyl peptidase I has been demonstrated in for instance the placenta (Gossrau *et al.*, 1987).

Dipeptidyl Peptidase II, EC 3.4.14.2. This is demonstrated with a simultaneous coupling technique using L-lysyl-alanine-4-methoxy-2-naphthylamide as the most specific substrate and Fast Blue B as the coupling reagent. The enzyme shows a narrow pH optimum around 5.5 and is inhibited by DFP, Tris, and puromycin.

Dipeptidyl peptidase II is a readily diffusible enzyme occurring chiefly in lysosomes in for instance kidney, epididymis, testis, and spleen (Gossrau and Lojda, 1980). According to these authors freeze-dried celloidin-coated cryostat sections is a convenient way of treating the tissue. Celloidin coating may, however, decrease the activity of dipeptidyl peptidase II. An alternative approach is to postchelate with copper sulphate thus to some degree stabilizing the final reaction product. Microscopy, microphotography, or microspectrophotometry must, however, be carried out immediately as the copper chelate is only stable for a short time even if the sections are kept in the refrigerator.

In the proximal tubule of the kidney this enzyme could be involved in the lysosomal metabolism of reabsorbed proteins from primary urine (Jedrzejewski and Kugler, 1982).

Dipeptidyl Peptidase IV, EC 3.4.14.4. This can be demonstrated on unfixed or slightly fixed (methanol-acetone) cryostat sections with a simultaneous coupling technique with the reagents in a conventional aqueous solution. It has, however, been shown that the activity in blood and bone marrow cells is greatly improved by the addition of PVA to the incubation medium (van Noorden *et al.*, 1989). Morphology was found to be excellent and the localization of the final reaction product precise, enabling the recognition of organelles in cells in which enzyme activity is present. Furthermore, no loss of enzyme activity was seen due to fixation or long periods of drying of cell preparations.

Synthetic substrates with unsubstituted amino termini possessing L-glycine-proline-residues are preferred. Fast Blue B is used as the diazonium salt. The activity of the enzyme is inhibited by diisopropylfluorophosphate and mercuric chloride or specifically by the bacterial toxins, diprotins A and B, which are competitive inhibitors (Umezawa *et al.*, 1984). The enzyme has been demonstrated in plasma membranes in capillary endothelium in the cardiovascular, urogenital, and digestive systems and also for instance in T-lymphocytes (Gossrau, 1985; Lojda, 1985).

Cathepsin B, EC 3.4.22.1. This enzyme belongs to the class of cysteine proteinases. The necessity for reducing conditions during incubation as described above for dipeptidyl peptidase I is therefore self-evident. The recommended substrates are derivatives of N-benzyloxycarbonyl, abbreviated CBZ or just Z, for instance CBZ-L-alanine-arginine-arginine-4-methoxy-2-naphthylamide. A simultaneous coupling method with Fast Blue B as the diazonium salt is usually not very satisfactory and alternatively a fluorescent visualization system is used where 5-nitrosalicylaldehyde is substituted as the coupling agent (see for instance van Noorden *et al.*, 1987). Specific inhibitors are leupeptin or E-64 (van Noorden *et al.*, 1987).

Cathepsin B has been demonstrated in for instance gastrointestinal epithelium, macrophages, fibroblasts, chondrocytes, synoviocytes, and at the invasion front of carcinoma cells (Howie *et al.*, 1985; van Noorden and Vogels, 1986; van Noorden *et al.*, 1987; Graf *et al.*, 1981).

Methods have been described for demonstrating other proteinases such as cathepsin D (EC 3.4.23.5), cathepsin G (EC 3.4.21.20), kallikrein (EC 3.4.21.8) (Sannes, 1988; Garrett *et al.*, 1985). However, as described in Sect.24.1.3 the simultaneous coupling reactions for these enzymes are troublesome with difficulties in retaining the enzyme, inhibition due to the diazonium salt, slow rate of coupling, and instability of the formed azo dye product.

24.6.4 Hydrolases Acting on Acid Anhydrides, EC 3.6

Under this heading only enzymes acting on diphosphate bonds in compounds such as di- and triphosphates (EC 3.6.1) will be mentioned.

Adenosinetriphosphatases. Adenosinetriphosphatases, ATPases, comprise several substrate specific enzymes some of which are classified in Table 24.4.

The general reaction catalyzed by these enzymes is the dissociation of phosphate from ATP:

$$\mathrm{ATP} + \mathrm{H_2O} \rightarrow \mathrm{ADP} + \mathrm{orthophosphate}$$

Table 24.4. Classification of some ATPases.

Name	E.C. No.	pH optimum	Occurrence	Activators	Inhibitors
Ca^{2+}-dependent ATPase	3.6.1.3	7.0	sarcoplasmatic reticulum, mitochondria	Ca^{2+}, Mg^{+}	p-chloromercuribenzoate
Mg^{2+}-dependent ATPase	3.6.1.3	9.0	mitochondria, cell membranes	Mg^{2+}	p-chloromercuribenzoate
Myosin ATPase	3.6.1.32	9.4	muscle	Ca^{2+}	p-chloromercuribenzoate
Na^{+}/K^{+}-transporting ATPase	3.6.1.37	7.4–8.0	Na^{+}–K^{+} pump in membranes	Na^{+}, K^{+}, Mg^{2+}	VO_4^{3-}, ouabain, p-chloromercuribenzoate
Ca^{2+}-transporting ATPase	3.6.1.38	8.2	plasma membrane	Mg^{2+}, Ca^{2+}	p-chloromercuribenzoate

In general, the ATPases are very sensitive to fixation and fresh frozen cryostat sections should therefore be used for histochemical demonstration. As ATP is the only substrate, metal salt methods must be used.

Ca^{2+}-Dependent ATPase, EC 3.6.1.3. In the presence of Ca^{2+} this enzyme gives rise to $Ca_3(PO_4)_2$ that is only sparingly soluble at alkaline pH. Calcium is exchanged with cobalt by treatment with a solution of $Co(NO_3)_2$. For light microscopy the section is then treated with $(NH_4)_2S$ which leads to the formation of black CoS.

Mg^{2+}-Dependent ATPase, EC 3.6.1.3. This can be demonstrated by adding Pb^{2+} ions to the incubation medium together with the activator Mg^{2+}. $Pb_3(PO_4)_2$ is formed and for light microscopy this is converted to PbS. The specificity of this method has been severely questioned as Pb^{2+} concentrations of more than 1 mmol/l lead to spontaneous non-enzymatic hydrolysis of ATP. In addition, the method is encumbered with all the disadvantages mentioned in Sect.24.1.4.

Myosin ATPase, EC 3.6.1.32. This is in fact a part of the myosin molecule. In agreement with this, it has been demonstrated ultrastructurally in myofibrils. By varying pH, etc. it has been possible to classify muscle fibres into different types as described in more detail in Sect.31.11.6. The enzyme is inhibited by p-chloromercuribenzoate.

Na^+/K^+-Transporting ATPase, EC 3.6.1.37. Most of the disadvantages regarding the previously described enzymes are avoided with a new quantitative method for detecting Na^+/K^+-transporting ATPase (Chayen et al., 1981). Diffusion of enzyme can be prevented using a colloid stabilizer (Sect.23.2.2). The presence of free Pb^{2+} in the incubation medium is completely avoided by using a complex compound between ammonium citrate and lead. As lead has greater affinity for phosphate than ammonium citrate, the phosphate ions liberated by the enzymatic reaction bind to lead and precipitate as lead phosphate. Total and ouabain-sensitive Na^+/K^+-transporting ATPase activities are determined and by subtracting the latter from the former activity due to the specific enzyme is obtained. Ouabain is a specific inhibitor of Na^+/K^+-transporting ATPase. This new method has also been optimized with regard to all kinetic parameters.

Ernst (1972a; b) described a strontium capture method using the second phosphorylation step of this enzyme system with p-nitrophenyl phosphate as substrate. Firth (1987) compared the two methods described by Chayen *et al.* (1981) and Ernst (1972a; b). It was found that the methods yielded similar and appropriate patterns of activity distribution and inhibitor response. The indirect method gave preferable results, in that non-enzymatic background was lower and rate of reaction product accumulation considerably higher.

Ca^{2+}-Transporting ATPase, EC 3.6.1.38. This has been studied by El-Sherif *et al.* (1990) who with a lead salt technique studied the distribution of this enzyme on polyacrylamide gel films including cell homogenates of rat adenohypophysis. The films were fixed with a modified Karnovsky fixative containing sucrose and $CaCl_2$.

By substituting ATP with β-glycerophosphate, ADP, or AMP significantly reduced absorbance values were obtained. Maximum absorbance was obtained when 1% $CaCl_2$ was added to the fixative. Omission of Ca^{2+}, or surprisingly levamisole, led to considerably decreased activity.

The enzyme was in these experiments found to be located to the plasma membrane.

25 Enzyme Histochemistry III: Oxidoreductases

P.E. Høyer, H. Lyon

It is expedient to divide these enzymes into three groups:

1. Dehydrogenases
2. Peroxidases
3. Oxidases

Enzymes from all groups oxidize their substrates and simultaneously reduce a suitable acceptor (redox reactions).

The principles applicable to the demonstration of anaerobic dehydrogenases (Sect.25.1) are considered first as this field has received extensive attention and many of the methods are very well documented. Less-detailed accounts of peroxidase and oxidase methods are given in Sects.25.2 and 25.3 as many of the considerations appropriate to the dehydrogenases remain relevant.

25.1 Principles of the Cytochemical Demonstration of Anaerobic Dehydrogenases

Dehydrogenases (anaerobic dehydrogenases) catalyze the transfer of reducing equivalents from a specific substrate to an acceptor which is never oxygen but may be:

1. A prosthetic group (e.g. FAD)
2. A coenzyme (e.g. NAD^+)
3. Other acceptors

$$S_{red} + A \rightarrow S_{ox} + A{-}H$$

Transfer to a *prosthetic group* FAD (flavine adenine dinucleotide) or FMN (flavine mononucleotide). The prosthetic group is generally tightly bound to the enzyme molecule itself either non-covalently (e.g. NADH dehydrogenase) or covalently (e.g. succinate dehydrogenase).

Transfer to *coenzyme* (NAD^+ or $NADP^+$). Most dehydrogenases belong to this subgroup. In a few cases both NAD^+ (nicotinamide adenine dinucleotide) and $NADP^+$ (nicotinamide adenine dinucleotide phosphate) can be used, but in most cases the dehydrogenase specifically requires one of the two coenzymes. The electrons are then transferred via the respiratory chain to finally react with molecular oxygen.

H. Lyon (Ed.)
Theory and Strategy in Histochemistry

The respiratory chain comprises individual components that are arranged in a sequence of ascending redox potentials. The addition of a reagent with a suitable redox potential should enable capture of the reducing equivalents at a particular stage in the respiratory chain. Tetrazolium salts are the most frequently used capture reagents.

Using succinate dehydrogenase (SDH) as an example, electron transfer through the respiratory chain is shown in Fig. 25.1. The points at which electrons are believed to be captured by a number of commonly used acceptors are shown on the right hand side. It should be noted that the exact point at which the acceptors engage depends not only on their redox potential but also on such factors as steric conditions and pH. The sites of action of several inhibitors are shown on the left hand side of Fig. 25.1.

On reduction tetrazolium salts precipitate as virtually insoluble coloured formazan compounds (Sect.25.1.2).

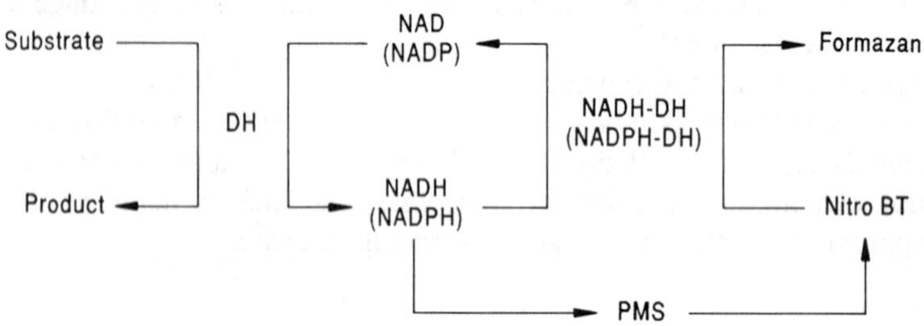

Fig. 25.2. Demonstration of NAD or NADP dependent dehydrogenases with tetrazolium salts as the capture reagent.

Fig. 25.2 shows the formation of either NADH or NADPH as intermediate products in the demonstration of NAD^+- or $NADP^+$-dependent dehydrogenases. The final transfer of hydrogen to the tetrazolium salt (see Sect.25.4) is catalyzed by NADH or NADPH dehydrogenases. Formazan is therefore precipitated, not at the site of the original dehydrogenase activity, but at locations corresponding to the NADH or NADPH dehydrogenase involved. These latter enzymes must therefore be close to the original dehydrogenase if staining is to give an accurate indication of its location. Thus, for enzymes that do not possess their own flavine prosthetic groups, the NADH or NADPH dehydrogenases are rate-limiting for formazan production.

It is, however, possible to bypass the NADH or NADPH dehydrogenases by adding a flavoprotein substitute such as phenazine methosulphate (PMS) (Fig. 25.2). PMS is able to act directly as an intermediate in the transfer of hydrogen to the tetrazolium salt. If the level of activity of the appropriate NADH/NADPH dehydrogenase is so low that it becomes rate-limiting in the demonstration of a particular dehydrogenase then a bypass with an intermediate carrier (Sect.25.1.2) is necessary. The need for this can be determined by performing a direct demonstration of NADH or NADPH dehydrogenase activity on a parallel section (Sect.25.4).

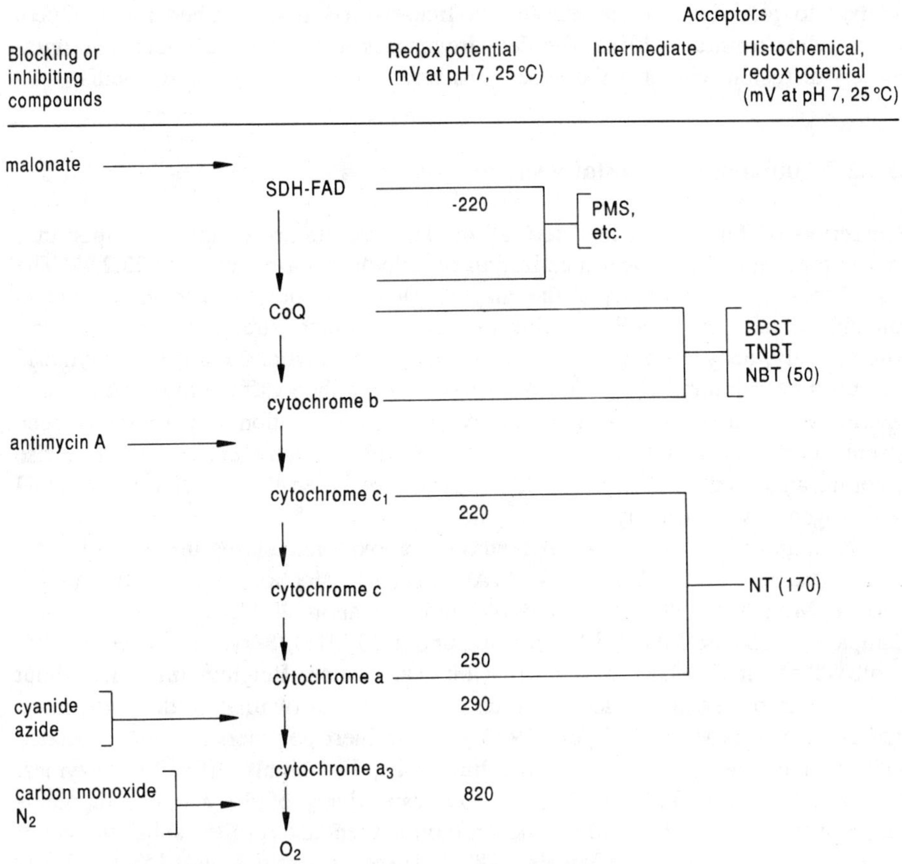

Fig. 25.1. The respiratory chain.

SDH-FAD	succinate dehydrogenase-flavine adenine dinucleotide
CoQ	coenzyme Q or ubiquinone
PMS	phenazine methosulphate
BPST	2-(2-benzothiazolyl)-3-(4-phthalhydrazidyl)-5-styryl-2H-tetrazolium
TNBT	tetranitro blue tetrazolium
NBT	nitro blue tetrazolium
NT	neotetrazolium

25.1.1 Pretreatment of Tissue

Cryostat sections of fresh frozen tissue are generally used for the demonstration of dehydrogenases (Sects.13.6 and 23.2.1). However, Murray *et al.* (1988) have described a method for the demonstration of dehydrogenases in freeze-dried or formaldehyde, glycol methacrylate resin-embedded tissue. Lactate dehydrogenase, NADH dehydrogenase, and NADPH dehydrogenase were all demonstrated in sections of formaldehyde-fixed resin-embedded tissue. In addition, glucose-6-phosphate dehydrogenase, glycerol-3-phosphate dehydrogenase (NAD^+-linked), and 3β-hydroxy-Δ^5-steroid dehydrogenase, but not succinate dehydrogenase, were claimed to retain maximum activity in freeze-dried tissue embedded in glycol methacrylate resin at −20°C. The dehydrogenases were accurately localized without any diffusion when the tissue sections were incubated in aqueous media.

25.1.2 Incubation of Cryostat Sections

Protection of Tissue during Incubation. Two factors are of critical importance for the precise cytochemical localization of dehydrogenases (cf. Sect.23.2.1). The first of these is the solubility of the enzyme. Dehydrogenase located in the cytosol can diffuse out of the cell resulting in a loss of demonstrable enzyme activity. Alternatively, dehydrogenase may diffuse out of one cell and reduce coenzyme in the substrate mixture. The reduced coenzyme may then diffuse into another cell where, even if there the enzyme activity under investigation had originally been absent, it can act as substrate for NADH/NADPH dehydrogenases and give rise to formazan formation (Høyer, 1980). It should be noted that the NADH/NADPH dehydrogenases are usually tightly bound.

The majority of cytochemical methods for oxidoreductases use a tissue protectant (tissue stabilizer) such as PVA (polyvinyl alcohol). PVA from Wacker Chemie, München, FRG (grade G04/140, mol. wt. about 20 kDa), Sigma Chemical Company, St.Louis, MO, USA (mol. wt. about 40 kDa), Serva, Heidelberg, FRG (mol. wt. about 22 kDa), or Janssen Chimica, Beerse, Belgium (mol. wt. about 22 kDa) can be recommended as tissue protectants for qualitative dehydrogenase studies (van Noorden and Vogels, 1989a). These inert polymers are not associated with spontaneous reduction of tetrazolium salts. For highly diffusible enzymes, 32% v/v, 18% v/v, 32% v/v, or 32% v/v respectively of the above compounds are prepared in a suitable buffer. The solution is then added to the substrate, coenzyme etc. (van Noorden and Vogels, 1989a). These authors concluded that 18% v/v PVA (Sigma, mol. wt. 40 kDa) is the tissue protectant of choice for quantitative cytochemical work.

Henderson *et al.* (1978c) recommend the following approach when a new grade of PVA is to be used in the demonstration of a soluble enzyme:

1. A solution containing the lowest concentration, which still retains the enzyme should be employed
2. The assay conditions should be made optimal at this concentration

Naturally, these principles also apply to the converse situation when the enzyme has not been investigated previously, but the laboratory is familiar with the grade of PVA to be used.

If the PVA is not of analytical grade, inhibiting impurities may sometimes be present (Butcher, 1982; van Noorden, 1988b).

It must be stressed that pipettes are far too inaccurate for dispensing the highly viscous medium. This should be carefully weighed (Høyer and Byskov, 1981; Robertson *et al.*, 1982b). The incubation itself can be performed in a "chamber" consisting of a ring of Perspex and a cover slip acting as "roof" (Fig. 25.3). The reaction is stopped by quickly transferring the section to ice cold, distilled water,

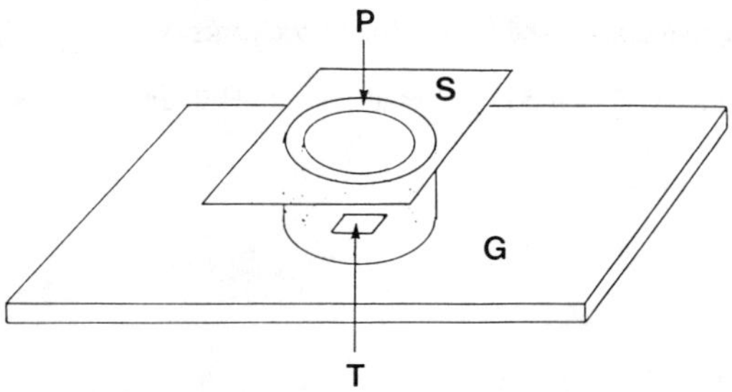

Fig. 25.3. "Perspex-chamber" used for the demonstration of enzyme activity.
S = cover slip T = tissue section
P = ring of Perspex G = slide

buffer, or fixative. Where microspectrophotometric end-point measurements are intended, it is essential that the mountant should not cause undue reflection.

Finally, it should be noted that excellent agreement between biochemical and cytochemical assays can be achieved by freezing in n-hexane (see Sect. 11.2.2) combined with the PVA-technique.

Tetrazolium Salts. Tetrazoles can have either of the following structures:

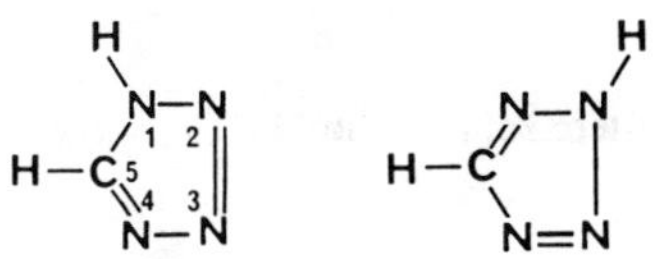

with hydrogen attached to nitrogen at position 1 or 2. Tetrazolium salts used in histochemistry are all quaternary derivatives of tetrazole with hydrogen in position 2. The tetrazolium ring itself is a resonance hybrid between two extreme forms:

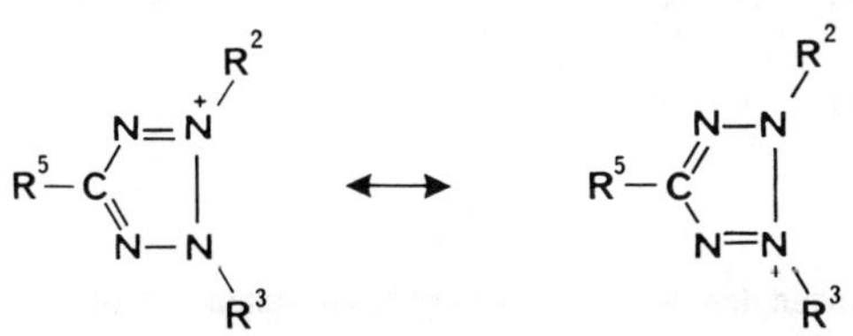

In practice, both monotetrazolium salts (e.g. BPST) and ditetrazolium salts (e.g. NBT, TNBT, NT) are used. In the latter compounds the two tetrazolium rings are linked at a position corresponding to the two quaternary nitrogen atoms:

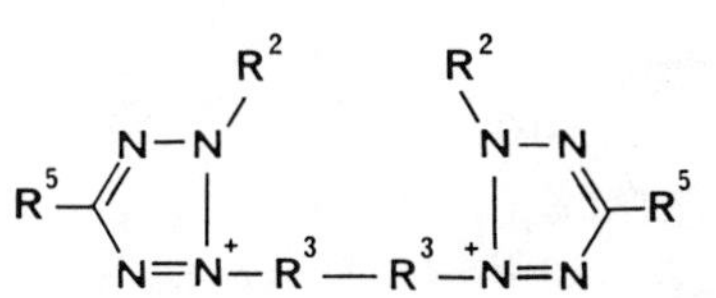

The tetrazolium salts used most frequently in current practice are shown in Table 25.1.

Table 25.1. Survey of the tetrazolium salts commonly used in enzyme histochemistry

MW = molecular weight; SN = systematic name; E′ = redox potential (mv at pH 7, 25°C; TN = trivial name

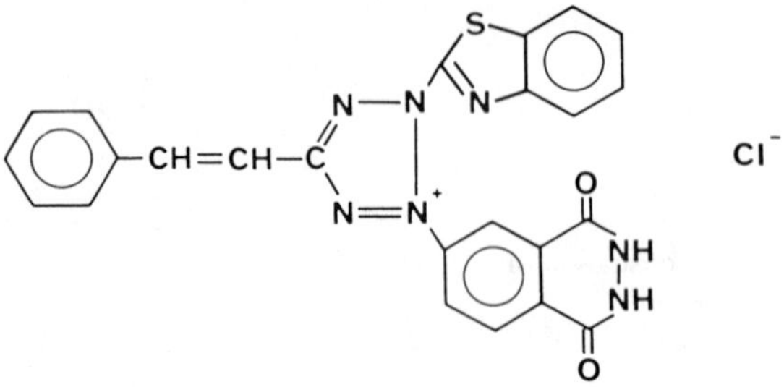

MW: 501.96; SN: 2-(2-benzothiazolyl)-3-(4-phthalhydrazidyl)-5-styryl-2H-tetrazolium chloride; TN: BPST

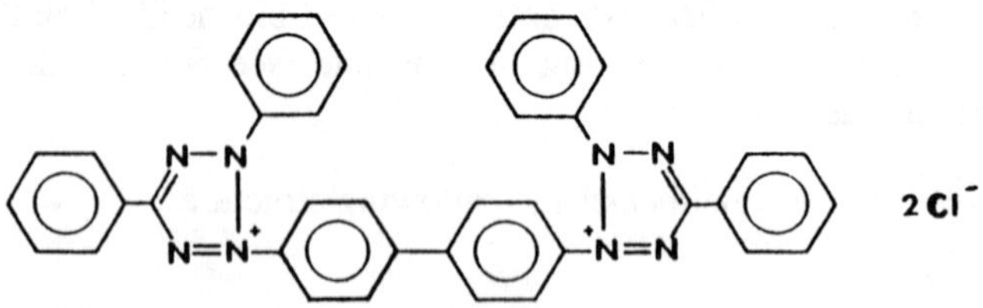

MW: 667.57; SN: 3,3′-(4,4′-biphenylene)-bis-(2,5-diphenyl-2H-tetrazolium chloride). E′ = +170 mV; TN: NT = neotetrazolium

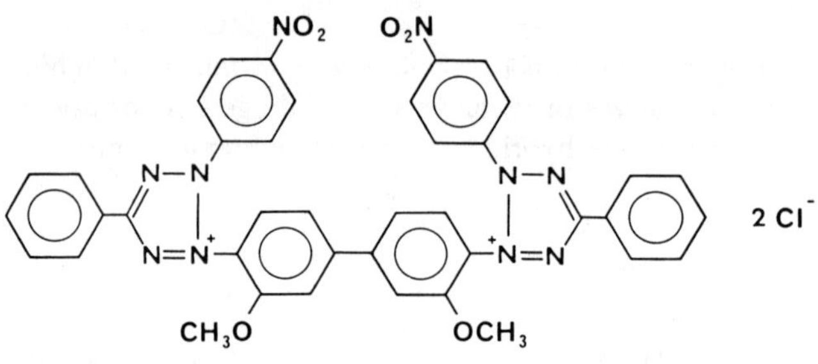

MW: 817.63; SN: 3,3′-(3,3′-dimethoxy-4,4′-biphenylene)-bis-(2-p-nitrophenyl-5-phenyl-2H-tetrazolium chloride). E′ = +50 mV; TN: NBT = nitro blue tetrazolium

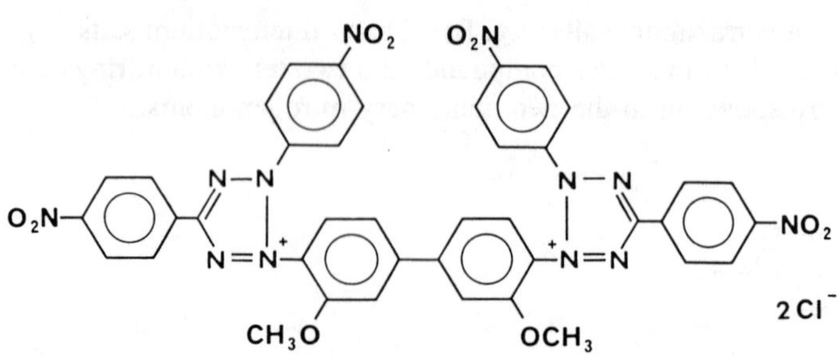

MW: 907.63; SN: 3,3′-dimethoxy-4,4′-biphenylene)-bis-(2,5-p-nitrophenyl-2H-tetrazolium chloride); TN: TNBT = tetranitro blue tetrazolium

The effect of reduction is shown below, using NBT as an example:

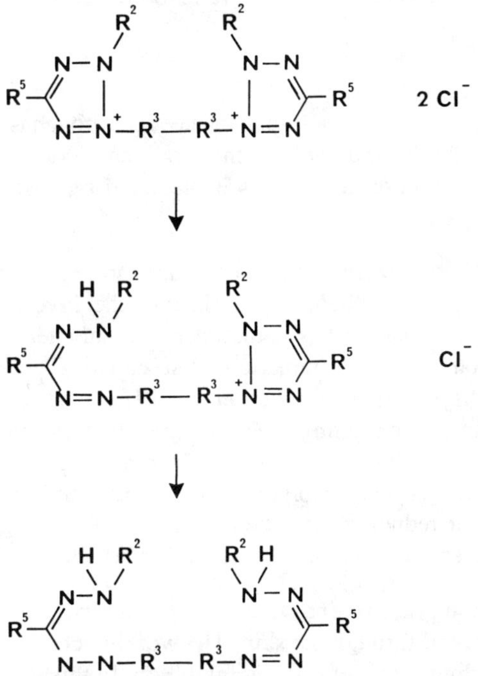

On reduction with a single hydrogen-equivalent, one of the tetrazolium rings is opened and thereby forms a monoformazan which, in the case of NBT, is red. Reduction of the second ring gives rise to a diformazan (blue for NBT).

The tetrazolium salt should:

1. be pure, stable in light, and easily soluble and colourless in water
2. be reduced reasonably quickly under histochemical conditions
3. have a suitable redox potential
4. not inhibit the enzyme
5. penetrate membranes readily

These points are discussed individually below.

1). Most tetrazolium salts are sufficiently soluble in water to allow maximum reaction rate and precise localization of formazans. They are colourless to pale yellow. Slow decomposition always occurs on exposure to light, so these compounds should always be stored in the dark. Moreover, according to van Noorden and Tas (1982a), incubation in darkness is a prerequisite as tetrazolium salts are photochemically reduced whether or not an exogenous electron carrier is present. It is particularly important to know the degree of purity as impurities may have a similar colour to that of the monoformazans. There is a considerable difference in the quality of reagents obtained from different suppliers. Buying from the right source, it is always possible to acquire pure BPST, NBT, TNBT, and NT. If, for some reason, impure tetrazolium salts must be used, it is necessary to purify to an acceptable level judged by chromatography (see Sect.3.3.8).

2). BPST, TNBT, and NBT are easily reduced in the test tube, while the NT reduction rate is much lower even under anaerobic conditions (van Noorden, 1988b). The relative order is as follows:

BPST $>$ TNBT $\geq$ NBT $>>>$ NT

In tissue sections, similar reaction rates for glucose-6-phosphate dehydrogenase activity were found with BPST, TNBT, NBT, and NT (Butcher and van Noorden, 1985). This seems to indicate that NT reduction in sections is mediated by tissue components (see also Sect.25.1.3).

3). The redox potential should allow the tetrazolium salt to capture the reducing equivalents from a position in the respiratory chain appropriate to the investigation. In consequence, NT (high redox potential) is sometimes the only tetrazolium suitable for certain investigations, e.g. determination of so-called Type I and Type II Hydrogen (Altman and Chayen, 1970; Chayen *et al.*, 1973a). Type I Hydrogen is believed to be involved in hydroxylating reactions (e.g. in steroidogenesis), whereas Type II Hydrogen is believed to participate in other types of biosynthetic reactions (e.g. fatty acid production). Moreover, for optimal activity NT must be used as the final acceptor of reducing equivalents when investigating NADPH-ferrihaemoprotein reductase (van Noorden and Butcher, 1986a).

4). Tetrazolium salts are potential enzyme poisons, and some may be carcinogenic. It should be noted that they can be absorbed through the skin. The working concentration should be as high as solubility allows without causing inhibition. In this way as precise a localization as possible is ensured, and the reducing equivalents are captured as effectively as possible. The inhibition caused by high concentrations of tetrazolium salt may often be diminished by inclusion of a polymer stabilizer such as PVA in the incubation medium (Fig. 25.4).

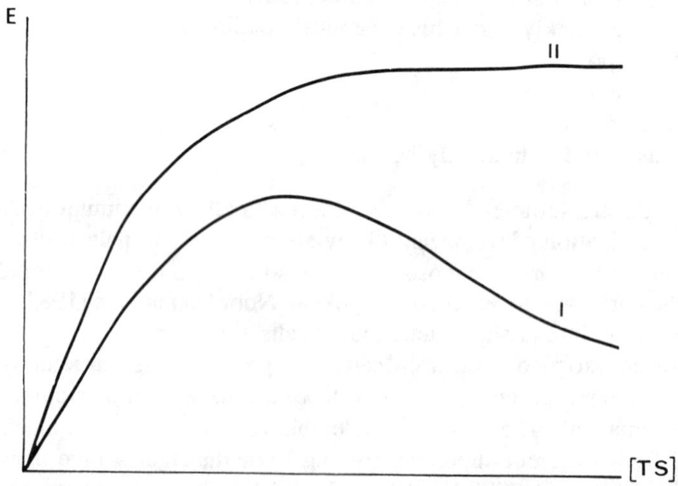

Fig. 25.4. Theoretical curve showing the effect of tetrazolium salt concentration ([TS]) on enzyme activity in a traditional aqueous medium (I) and in a medium containing an optimal amount of PVA (polyvinyl alcohol)(II). Extinction (E).

Although ditetrazolium salts possess two positive charges which are probably responsible for their relatively pronounced protein substantivity (Sect.1.2.3), this property does not seem to be the cause of the enzyme inhibition which sometimes occurs.

5). The relatively large planar ditetrazolium salts with two positive charges and polar groups often do not readily diffuse through intact membranes. To overcome this problem, passage through mitochondrial membranes is facilitated by brief fixation and the addition of DMSO. The use of a phosphate buffer, in place of others such as glycylglycine, has the same effect. Normally, however, the membranes are penetrated so quickly that the local intracellular concentration is not rate-determining.

Formazans should:

1. be poorly soluble and show a high substantivity for proteins.
2. not be soluble in lipids
3. be bound as very small and homogeneous particles
4. have distinct colours for which the absorption maxima in tissue are known
5. be stable to light

1). Substantivity for tissue proteins should be as high as possible. First, the formazan must almost instantly bind to proteins close to the enzyme molecule if exact localization is to be achieved. Second, low substantivity will increase the tendency to recrystallization and lead to formazan being removed when the incubation is stopped in distilled water. Diformazans show a relatively high substantivity. The large, planar molecules and hydrogen bonds are believed to be of importance in this respect. In general, monoformazans (from mono- or ditetrazolium salts) show less substantivity; thus BPST-formazan binding may be minimal in some cells. Under some circumstances, notably in certain tumour cells and in necrotic areas, even the diformazans can show decreased binding.

2). Solubility in lipids can lead to formazan diffusion from the site of formation to lipid droplets where unsaturated fatty acids can cause a non-specific reduction of tetrazolium salts. Of the tetrazolium salts in Table 25.1 only NT is significantly soluble in lipids. The formazans of all four salts have a certain affinity for lipid-water interfaces, and this may lead to accumulation in locations such as nuclear membranes and the surface of lipid droplets. This tendency is most pronounced for NT:

$$\text{NT} >> \text{NBT} > \text{TNBT} > \text{BPST}$$

The more lipid-soluble a formazan, the more pronounced is its tendency for both recrystallization and changes in its absorption maxima. In the case of BPST-formazan these problems can be completely avoided by chelating with Ni^{2+} or Co^{2+} (Altman *et al.*, 1979).

3). The smaller the formazan particles the more precise the localization is. In liver sections incubated to produce heavy formazan deposits, Altman (1976) found that the formazans of TNBT, NBT, BPST, and NT were granular in shape and had diameters of approximately 0.2, 0.5, 0.8, and 1.7 μm, respectively.

4). The individual formazans each have characteristic absorption maxima in tissue. BPST-formazan can be measured directly at its maximum (530 nm). However, when NBT, TNBT, and NT (ditetrazolium salts) are used, it is necessary to take into account the formation of mono- and diformazans (see Sect.25.1.4 and Chap.28).

5). Slow decomposition and recrystallization of formazans occur when the sections are stored in light. They should therefore be kept in the dark and preferably be examined shortly after mounting in a hydrophilic, non-extracting medium.

In conclusion, TNBT and NBT are generally preferred as they fill most of the requirements cited regarding both the tetrazolium salts and the resultant formazans. If the tissue is lipid-rich, TNBT should be used in place of NBT in order to achieve as precise a localization as possible. As shown by Horobin (1982b) it is possible by theoretical considerations of structural parameters to arrive at the same conclusions. The parameters of interest, which may be derived from the structural formulae of the tetrazolium salts, are the conjugated bond numbers and Hansch π values of molecular fragments (Horobin, 1982a, pp.237–246).

Seidler (1980), in a review on new nitro-monotetrazolium salts, concluded that these salts are particularly suited for quantitative histochemistry due to the uniformity of their amorphous reaction products. The substitution of nitro groups in tetrazolium salts (and the corresponding formazans) gives advantages of increased substantivity and reducibility along with improved localization and decreased solubility of the formazans in lipids. Of the monotetrazolium salts tested, 2,3-di(p-nitrophenyl)-5-phenyl-2H-tetrazolium chloride (2,3-p-DNTTC) was recommended for quantitative work as its formazan always appears in a blue form. Most of the other tetrazolium salts tested gave red formazans in normal tissue whereas varying amounts of blue formazans occurred in ischaemic lesions. Full details for the synthesis of 2,3-p-DNTTC are given by Seidler (1980). Unfortunately, this tetrazolium salt does not appear to be commercially available.

Potassium Ferricyanide as a Capture Reagent. Ferricyanide may be reduced in place of a tetrazolium salt. The principle of the method is that the ferrocyanide ion formed by reduction of the ferricyanide binds Cu^{2+} in the incubation medium. The product formed is often called "Hatchett's Brown". With succinate dehydrogenase (SDH) as an example the reaction takes place as follows:

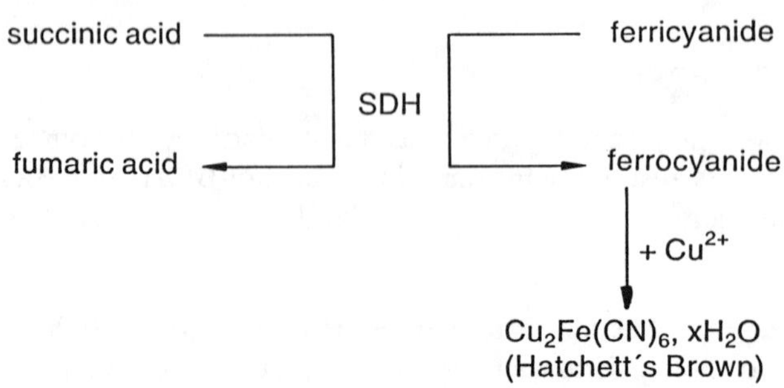

As most anaerobic dehydrogenases have a pH-optimum between 7 and 9, it is a disadvantage that copper ferrocyanide becomes increasingly soluble at pH values above approximately 7. A pH of about 7 has therefore often been used. At this pH the concentration of anions (especially hydroxyl ions) is sufficiently high for the solubility product of a number of copper compounds (i.e. copper hydroxide) to be exceeded. Non-specific precipitation can be avoided by adding an agent that forms complexes with Cu^{2+}. The dissociation constant of this compound must, however, permit rapid liberation of Cu^{2+} to ensure that the ferrocyanide ions formed are captured quickly and effectively.

At pH 7.2–7.3 either a potassium sodium tartrate buffer or a Tris buffer can be used to suppress non-specific precipitation of copper salts (Lukaszyk, 1971). In qualitative work observed by light microscopy, the brownish colour of the copper ferrocyanide is only slightly different from the colour of most tissues. The contrast can be increased with various post-treatments such as dithiooxamide which gives rise to a yellowish brown ethanol-insoluble complex. It is important to note that the ferricyanide ion and particularly Cu^{2+} can inhibit a number of enzymes.

Overall, potassium ferricyanide has been used most in EM cytochemistry as "Hatchett's Brown" has appreciably greater electron density than for example TNBT formazan (Hanker, 1975). In addition, metals from the transition groups (e.g. copper) can act as catalysts for the oxidative polymerization of certain aromatic amines (e.g. p-phenylenediamine or 3,3′-diaminobenzidine). In this way "Hatchett's Brown" is transformed to a reddish brown polymer, which after treatment with osmium tetroxide, forms an amorphous lipophobic product, Osmium Black, which is insoluble in organic solvents.

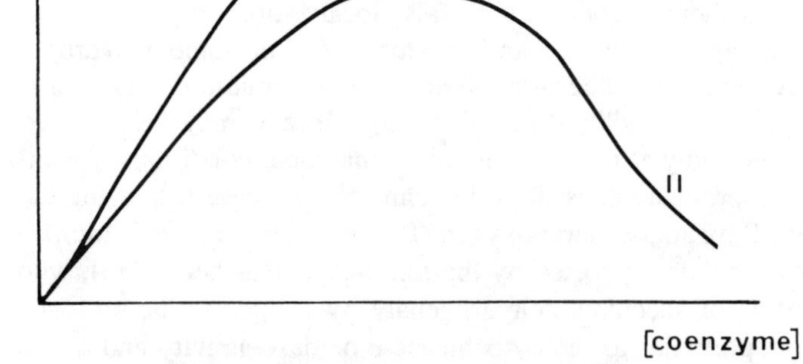

Fig. 25.5. Theoretical curve showing the dependence of enzyme activity on coenzyme concentration ([coenzyme]).
Usually high coenzyme concentrations result in zero-order kinetics (I) but inhibition may be seen with some enzyme-coenzyme pairs (II). Extinction (E).

Coenzyme. In quantitative work zero-order kinetics must be ensured (Fig. 25.5 and Chap.28). This usually means that the concentration of coenzyme should be of the order 1–2.5 mg/ml.

Occasionally, high concentrations of NAD^+ or $NADP^+$ can be inhibitory (Høyer, 1988). NADH and NADPH can still diffuse in gel media and, as noted in Sect.25.1.2 (Protection of tissue), this may result in the final product not having the same distribution as the enzyme.

If the dehydrogenase to be examined is dependent on coenzyme Q in the cytochemical system, it is important to determine whether the mitochondria have a rate-determining low concentration of this coenzyme (Andersen and Høyer, 1974).

Intermediate Electron Acceptors (Carrier Substances). It has been suggested that a number of compounds can be used to bypass the NADH/NADPH dehydrogenases. These may be termed flavoprotein substitutes and include:

- menadione
- phenazine methosulphate (PMS)
- 1-methoxyphenazine methosulphate, (mPMS)
- 2,6-dichloroindophenol
- Methylene Blue
- Thionin
- Meldola Blue

The presence of these compounds may increase the reaction rate so much that increasing amounts of H-equivalents are transferred directly to oxygen and not to the tetrazolium salt. This can be avoided by the addition of reagents, such as cyanide or azide, or the use of a nitrogen atmosphere (Fig. 25.1). The increased rate of reaction may also lead to the maximum number of tissue binding sites for the formazan being rapidly exceeded resulting in diffusion of formazan into the incubation medium or into the medium used to stop the reaction. Frequently these carriers cause an increased nothing dehydrogenase reaction (Sect.25.1.3). When PMS is used, incubation must either take place in the dark or in red light as it rapidly decomposes into compounds which may act as enzyme poisons. The reduced forms of both flavoprotein substitutes can, like NAD(P)H itself, diffuse even in the gel media and thereby contribute to false localization.

In addition, PMS may apparently have an inhibitory effect on some dehydrogenases when sections are incubated under normal atmospheric conditions (Hardonk, 1965; Robertson, 1979; Høyer, 1980; Butcher, 1983). Since it may be reversed by adding cyanide, but not amytal or by using strict anaerobic conditions, the inhibitory effect of PMS is probably caused by cytochrome-*c* oxidase catalyzing the oxidation of reduced PMS by atmospheric oxygen (Brody and Engel, 1964; McMillan, 1967). This view is further supported by the findings of Butcher (1983), who showed that the inhibition of succinate dehydrogenase by oxygen in the presence of PMS was directly proportional to the cytochrome-*c* oxidase activity and that it could be reversed by the addition of azide (see also Sect.25.1.2, The atmosphere).

Kugler and Wrobel (1978) introduced Meldola Blue and, using a sandwich gel technique, found it to be as efficient as an intermediate carrier as PMS. In contrast,

Henderson and Loveridge (1981), using a PVA technique, found Meldola Blue to be of little value. Stellmach and Severin (1987) found that the amount of formazan obtained after incubating living cells with Meldola Blue as the electron carrier was greater than that obtained with a number of the other intermediate acceptors including PMS, Methylene Blue, and menadione.

Butcher and Evans (1984) compared the effects of several intermediate electron acceptors using incubation media containing polyvinyl alcohol. Menadione was the least effective both in transferring reducing equivalents from the primary dehydrogenase to the tetrazolium salt and in preventing diffusion. PMS, Methylene Blue, and Thionin were all more efficient.

Considerable diffusion occurred when PMS was used, whereas diffusion was minimal with either of the thiazin dyes. Meldola Blue had little or no effect on formazan deposition. Kugler (1982b) suggested that the inefficiency of Meldola Blue in incubation media containing polyvinyl alcohol might be due to this dye being almost completely bound to the colloid stabilizer.Butcher and Evans (1984) agree with this view.

Van Noorden and Tas (1982a) compared the applicability of mPMS with PMS, Meldola Blue, and menadione using TNBT and polyacrylamide films into which either purified glucose-6-phosphate dehydrogenase or intact liver cells where incorporated. Neither intermediate compounds nor formazans diffuse in gel films of this kind. Consequently, it was not necessary to include PVA during incubation. In films containing purified enzyme, equimolar concentrations of PMS, mPMS, and Meldola Blue enhanced the transfer of electrons from NADPH to TNBT to a comparable degree. However, no enhancement of electron transfer was noted when menadione was used. They concluded that when menadione is used for the cytochemical assay of $NADP^+$- or NAD^+-dependent dehydrogenases in situ, it is not the primary enzymatic activity that is demonstrated but one of the compounds present in the cellular electron transport systems. This is explained by the fact that no electron exchange takes place between NADPH and menadione, whereas reduced menadione reduces TNBT (van Noorden and Tas, 1982a). The authors therefore suggest that menadione may be useful for the detection of succinate dehydrogenase as FAD/FADH has a higher redox potential than $NAD(P)^+/NAD(P)H$.

In films containing intact liver cells both PMS and mPMS enhanced formazan production. The "nothing" dehydrogenase activity (Sect.25.1.3) was less than 10% of the actual enzyme activity when mPMS was used. The background staining was somewhat higher with PMS, probably due to non-specific formazan production and a light pink staining of cells by PMS itself. With both menadione and Meldola Blue the non-enzymatic formazan production and background staining of the cells was quite high, and these substances were deemed "useless" in this context.

The background staining with Meldola Blue is partly due to binding of the intensely blue coloured compound to lipids and DNA (van Noorden and Tas, 1982b). Furthermore, mPMS is light-stable and photochemical reduction of tetrazolium salts is minimal. The authors therefore conclude that mPMS is the carrier of choice when extramitochondrial dehydrogenases are assayed (van Noorden and Tas, 1981a; 1982b). In the investigation of mitochondrial dehydrogenases PMS should, however, be used as mPMS cannot penetrate the intact mitochondrial membrane (Kugler, 1982b).

Buffer. In principle the pH optimum must be determined and used. The buffer should have reasonable capacity at this pH. For example, it is unacceptable to use phosphate buffer at pH 8. For further information on this and the ionic strength of buffers, see 23.2.2 and Clancy (1987). Buffers may have a direct effect on some enzymes or substrates. For example, Tris buffer can oxidize some steroid substrates (Ferguson and MacPhee, 1975) and maleate in Tris-maleate buffer inhibits urate oxidase (Angermüller, 1989). If a metal ion functions as an activator for the dehydrogenase a glycylglycine buffer may not be suitable due to its ability to form complex bonds. In general the ionic strength should be kept as low as possible, although it must be possible to keep pH constant throughout incubation. At pH > 9 spontaneous reduction of tetrazolium salts may occur. Both NAD^+ and $NADP^+$ are most stable at pH below 7, while the reduced versions are most stable at pH above 7.

Incubation Time. For quantitative work this must be chosen so that the amount of final reaction product due to the dehydrogenase under investigation varies linearly with time (Fig. 25.6). As shown by several authors (e.g. van Noorden and Butcher, 1986a) both the test reaction (plus substrate) and the control reaction (minus substrate) sometimes show levelling off of the reaction rate from the very start of the incubation. When, however, the control reaction is subtracted from the test reaction the resulting *specific* reaction is usually linear for a considerable period of time. One exception from this general rule has been described by van Noorden and Vogels (1989b).

Even in the linear zone, longer incubation times allow more extensive diffusion of NADPH and reduced carrier substances.

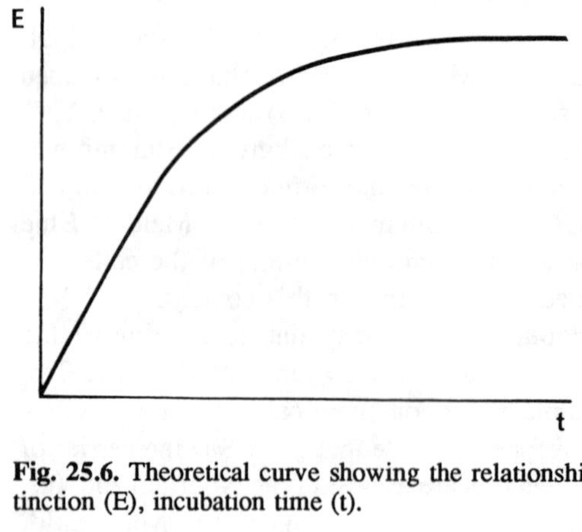

Fig. 25.6. Theoretical curve showing the relationship between activity and incubation time. Extinction (E), incubation time (t).

The Atmosphere. When using NT either the terminal part of the mitochondrial respiratory chain must be blocked (see Fig. 25.1) or a nitrogen atmosphere must be used. In the latter case the medium must be saturated by bubbling through with nitrogen.

Butcher (1978) has shown that reduction of NT *in vitro* only occurs under strict anaerobic conditions. Apparently, oxygen reduction is energetically far more favourable than NT reduction, both in the test tube and in non-carcinomatous tissue. NT reduction in nitrogen gave the localization pattern of the NADPH dehydrogenase, both in the presence and absence of intermediate electron carriers. This suggests that the transfer of reducing equivalents from the exogenous electron carrier to NT proceeds via cellular electron transport systems (van Noorden and Butcher, 1984).

In some cases it may be of diagnostic value to use a series of different oxygen tensions when using NT as the tetrazolium salt. Thus, the reaction for glucose-6-phosphate dehydrogenase in normal bronchial epithelial cells shows pronounced inhibition with high oxygen tension, while the same reaction in tumour cells from a bronchial carcinoma shows considerably less or no inhibition (Butcher, 1979). Similarly, in studies of human stomach and colon (Ibrahim *et al.*, 1983) and of human breast tissue (Petersen *et al.*, 1985) it was found that when incubation took place in an oxygen atmosphere, glucose-6-phosphate dehydrogenase activity could not be demonstrated in normal tissue whereas activity was retained in carcinomatous tissue. The components in carcinomatous tissue, which are responsible for this oxygen insensitivity, are not known at present.

Van Noorden (1988a) studied the direct effect of oxygen on the tetrazolium salt reduction mediated by NAD(P)H and PMS. The conclusions of this paper are given below. The first five relate to spectrophotometry using cuvettes containing cell and tissue free media with or without addition of pure enzyme, while the sixth conclusion was derived from work carried out on rat liver sections. The conclusions were:

1. Oxygen may interfere in the NBT and TNBT reduction, not only via flavine-containing enzyme systems (inhibited by azide), but also directly via tetrazolinyl radical intermediates. In the latter case oxygen inhibited competitively
2. In atmospheric air, 5 mmol/l NBT or TNBT is sufficient to exclude the oxygen effect via radical intermediates (cf. point 6)
3. Macromolecules (e.g. albumin, PVA, and tissue macromolecules) diminish the oxygen effect on NBT and TNBT reduction, probably by keeping the local concentration of tetrazolium salt high
4. BPST reduction appeared to be unaffected by oxygen, probably due to formazan production directly via divalent electron transfer and not via a radical intermediate
5. The inhibitory effect of oxygen on the reduction of NT was not significantly affected by increasing the concentration of tetrazolium salt
6. The maximum reaction rate and optimum localization of glucose-6-phosphate dehydrogenase in rat liver sections were obtained by using 5 mmol/l TNBT or NBT and 20% Sigma PVA. From the above it becomes apparent that when using

TNBT or NBT, any inhibition due to oxygen in atmospheric air in the absence of PMS can be overcome by increasing the concentration of the tetrazolium salt to 5 mmol/l and by adding macromolecules (e.g. PVA). If PMS is employed, it is necessary to block oxygen interference via flavoproteins by adding azide or cyanide (Butcher, 1983).

Temperature. Incubation is usually performed at 37°C as this is optimal for the majority of anaerobic dehydrogenases. If the reaction rate is extremely high, it may, however, be of advantage also to perform the incubation at lower temperature, e.g. 25°C. (See Sect.23.2.2).

25.1.3 Control Methods

Diffusion. By preincubating sections in different buffers it is possible to get an idea of the extent to which the enzyme diffuses in a traditional aqueous medium. So-called transfer experiments can be used to determine whether the concentration of PVA is sufficiently high to hinder diffusion of the enzyme (Henderson *et al.*, 1978b), while the "double section method" gives a control for diffusion of both the enzyme and of the reduced intermediary compounds (Andersen and Høyer, 1974; Høyer, 1980).

Diffusion of formazan may sometimes be detected by simply looking at the section immediately after its transfer to the medium used for stopping the reaction. Formazan diffusion may also be detected by chemical analysis of the medium used for stopping the reaction. Finally, formazan diffusion may also be detected by comparing continuous measurement of enzyme activity during incubation with end-point measurement. If for example the former method shows a linear increase of activity with time and the latter method shows a decrease of reaction rate with time, it may be concluded that diffusion of the final reaction product has taken place.

Inhibitors. Use of a specific inhibitor makes it possible to determine whether the activity found is due to the dehydrogenase investigated. For example, malonate is a specific competitive inhibitor for succinate dehydrogenase. Inhibitors may also be used to assess the occurrence of different isoenzymes. In this regard, urea can be used to inhibit the LDH 4 and LDH 5 isoenzymes of lactate dehydrogenase (see Sect.25.4.1). Moreover, inhibitors can be used to inhibit other enzymes that may disturb the investigated reaction; for example, the specific and irreversible inhibition of alkaline phosphatase with L-p-bromotetramisole and inhibition of alcohol dehydrogenase with 1,10-phenanthroline (Høyer and Andersen, 1977).

The "Nothing" Dehydrogenase Reaction. If incubation is performed without the substrate, one might expect no final product to be produced. However, even after a very brief incubation, some reduced product is detectable in nearly all tissues if optimal kinetic parameters (e.g. concentration of coenzyme, pH, temperature, etc.) are used. The reaction responsible for this product is, unfortunately, commonly referred to as the "nothing" dehydrogenase reaction.

In general, only a small component of this activity can be attributed to retained endogenous substrate for the dehydrogenase being examined. Other reasons include the presence of reactive thiol groups and alcohol dehydrogenase. Both of these latter influences can be completely blocked with N-ethylmaleimide (Sect.9.3.4). This reagent can, however, only be used to inhibit the non-specific reaction in exceptional cases as most dehydrogenases have thiol groups at or near their active sites. Alcohol dehydrogenase activity can also be suppressed by adding substances which bind to zinc at the active site (e.g. EDTA or 1,10-phenanthroline).

Van Noorden *et al.* (1985) investigating the nature of the "nothing" dehydrogenase reaction in rat liver sections were able to discern two components: an immediate formazan production at the beginning of the incubation period, and a much slower and almost linear increase with time after the first few minutes. They suggested that the first component of the reaction is due to lactate dehydrogenase oxidizing endogenous substrate and that this forms the main part of the "nothing" dehydrogenase reaction, at least in liver tissue. Furthermore, they suggested that the second component is due to thiol groups present.

In quantitative cytochemistry the "nothing" dehydrogenase reaction is not so important as it can be measured and subtracted, but it should preferably not exceed 10–15 per cent of the specific reaction (Høyer, 1980; Stoward,1980).

Other Methods. In the final analysis, however, comparison of the results of enzyme histochemical analysis with those obtained by other methods, such as biochemical and immunocytochemical procedures, is always valuable.

Another valuable approach is the use of model systems (Horobin, 1982a, pp. 229–234). Different materials have been proposed as the matrix for experiments of this kind (van Duijn, 1976; van Duijn and van der Ploeg, 1980), the most frequently used being polyacrylamide films (van der Ploeg and van Duijn, 1964).

Usually soluble fractions of isolated cells or pure preparations of macromolecules have been incorporated in the films, but van Noorden and Tas (1981b) have described a method for incorporating intact cells or pure enzymes in the polyacrylamide film. The films can be assayed biochemically as well as cytochemically. With model films linearity may be measured between increase of absorbance on the one hand and incubation time and film thickness on the other hand. Furthermore, absorbance per thickness of film (per unit of time) and the amount of enzyme incorporated can be studied (Beer-Lambert's law, cf. Sect.3.3.2). The polyacrylamide film technique is, however, not without problems. In the paper by van Noorden and Tas (1981b) on glucose-6-phosphate dehydrogenase these are shown to include:

1. A certain degree of enzyme inactivation must be expected during polymerization
2. Whereas NBT and TNBT can be used as final acceptors, the use of BPST and NT does not lead to formation of any formazan, possibly due to free radical residues, generated in the acrylamide during polymerization, reacting with the intermediary free radicals arising during tetrazolium salt reduction. Apparently the nitro groups in NBT and TNBT protect the intermediary radicals arising from these tetrazolium salts against interference from the free radicals in the polyacrylamide

3. TNBT measurements are most suitably carried out at 535 nm although TNBT is a ditetrazolium salt with an isobestic point (Sect.25.1.4) at 557 nm in tissue sections

The above conclusions emphasize the importance of optimizing matrix model systems. Other enzymes, apart from glucose-6-phosphate dehydrogenase, that have been studied in polyacrylamide films include peroxidases (van der Ploeg and van Duijn, 1964), alkaline phosphatase (van der Ploeg and van Duijn, 1968), acid phosphatase (Lojda *et al.*, 1967) aspartate aminotransferase isoenzymes (Papadimitriou and Van Duijn, 1970), esterases (Cabrini *et al.*, 1977), and acetylcholinesterase (Andrä and van Duijn, 1985).

25.1.4 Quantitation

Microspectrophotometric assays of dehydrogenase activities are based on the use of tetrazolium salts. With a monotetrazolium salt, BPST, measurements may be performed at the absorption maximum (Table 25.2). When ditetrazolium salts, (TNBT, NBT, NT) are employed, two reaction products may be precipitated simultaneously. It is therefore necessary to perform the measurements at a wavelength where both reaction products have the same molar extinction coefficient, the isobestic point (Table 25.2).

Alternatively, it is possible to analyze multiple components such as mono- and diformazans in tissue sections by using so-called component spectroanalysis (Araki *et al.*, 1987). For further details, see Sect.28.6.

Table 25.2. Absorbance maxima and isobestic points of some formazans in tissue sections.

Tetrazolium Salt	Absorbance maxima (nm) Monoformazan	Diformazan	Isobestic wavelength	Reference
BPST	530			Altman, 1976
TNBT	525	595	557	van Noorden et al., 1983
NBT	525	605	585	Butcher, 1978
NT	510	550	585	Butcher and Altman, 1973

Microspectrophotometric quantitation is further discussed in Sect.28.8.7. Detailed reviews of dehydrogenase cytochemistry are given by Wohlrab *et al.* (1979), Stoward *et al.* (1991), and Høyer *et al.* (1991).

25.2 Principles of the Cytochemical Demonstration of Peroxidases

Many of the peroxidases are iron containing haemoproteins, which catalyze the process:

$$DH_2 + H_2O_2 \rightarrow D + 2H_2O$$

where in histochemistry DH_2 can be amines, phenols, or the so-called leuco form of certain dyes.

Graham and Karnovsky (1966) introduced 3,3′-diaminobenzidine (DAB) as a substrate for ultrastructural studies. This is now the substrate of choice due to the insolubility in water and ethanol of the polymerized reaction product. A further advantage is its electron density rendering DAB suitable for electron microscopical studies. With added H_2O_2 as acceptor of reducing equivalents DAB is oxidized and polymerized forming indamine and/or phenazine coupling products (Seligman *et al.*, 1968). The precipitate may be amplified with OsO_4 forming Osmium Black.

25.3 Principles of the Cytochemical Demonstration of Oxidases

Oxidases (aerobic dehydrogenases) directly catalyze the transfer of hydrogen from substrate to an oxygen molecule which is either reduced to water or hydrogen peroxide:

$$S_{red} + O_2 \rightarrow S_{ox} + H_2O_2$$

$$2S_{red} + O_2 \rightarrow 2S_{ox} + 2H_2O$$

According to Angermüller (1989) two general principles may be applied for the cytochemical demonstration of oxidases. These are:

I. Reduction of a chromogen by the flavine adenine nucleotide (FAD) of the enzyme. The chromogen may be:
   a. A tetrazolium salt which on reduction forms a coloured, insoluble formazan. For details, cf. Sect.25.1.2 (Tetrazolium salts)
   b. Ferricyanide which after being reduced to ferrocyanide is captured by copper forming insoluble cupric ferrocyanide (Hatchett's Brown). For details, cf. Sect.25.1.2 (Potassium ferricyanide)

II. Detection of H_2O_2 generation
   A. Indirect methods:
      1. 3-amino-9-ethylcarbazole (AEC) method where AEC acts as hydrogen donor in the degradation of H_2O_2 by horseradish peroxidase (cf. Sect.26.2.2).
      2. 3,3′-diaminobenzidine (DAB) methods where DAB is the hydrogen donor in the degradation of H_2O_2 by either
         a. endogenous catalase (cf. Sect.25.4.2) or
         b. horseradish peroxidase (cf. Sect.26.2.2).
   B. Direct method:
      1. Cerium method. In this technique the oxidase transfers electrons from the substrate to molecular oxygen. As the generated hydrogen peroxide is broken down immediately by peroxisomal catalase (Sect.25.4.2), it is necessary to add a catalase inhibitor to the incubation medium. It is now possible for hydrogen peroxide to react directly with added cerium

ions ($CeCl_3$) forming the insoluble perhydroxide, $Ce(OH)_2OOH$. This compound is practically invisible by light microscopy but being electron dense can be visualized by electron microscopy. With this method it has been possible to improve localization of the reaction product by fixation with 0.25% glutaraldehyde in Pipes (piperazine-N,N′-bis(2-ethanesulphonic acid)) buffer at pH 7.4 (Angermüller, 1989).
A modification of this method for light microscopic application has been reported by Angermüller and Fahimi (1988a;b). The cerium perhydroxide is used either for oxidative polymerization and cyclization of DAB or for the formation of a lead precipitate in a two-step reaction with an alkaline lead citrate solution and ammonium sulphide. The DAB visualization procedure was further enhanced by adding nickel or cobalt ions. Gossrau *et al.* (1989) further improved the DAB-nickel and DAB-cobalt methods by adding very small amounts of hydrogen peroxide and incubating at 40 or 60°C. They presumed that H_2O_2 probably increased the potential for oxidation of DAB into its polymer in the presence of cerium perhydroxide and nickel or cobalt ions. No final reaction product was precipitated when the specific substrate, sodium azide, or aminotriazole were omitted or the inhibitor kojic acid was added thus indicating the specificity of the reaction. The authors conclude that the DAB-nickel-H_2O_2 one-step visualization procedure at 40°C is the method of choice.

25.4 Demonstration of Selected Oxidoreductases

Some of the oxidoreductases demonstrable using quantitative methods, are surveyed in Table 28.4.

25.4.1 Dehydrogenases

Selected examples are given below.

Lactate Dehydrogenase (LDH), EC 1.1.1.27. This enzyme reversibly catalyzes the oxidation of lactate to pyruvate with the participation of NAD^+. The enzyme is found in the cytosol of liver cells and in all cells where anaerobic glycolysis can take place (e.g. muscle). It is possible to distinguish between five different isoenzymes which each consist of four subunits on biochemical grounds. The subunits have a primary structure comprising one of two possible configurations which are designated type A (formerly M) or type B (formerly H) (Sjögren, 1984). The isoenzymes 1–5 have the quaternary structures B_4, AB_3, A_2B_2, A_3B, and A_4. In heart muscle LDH 1 and LDH 2 dominate, while LDH 5 dominates in liver cells. Jacobsen (1969) developed a cytochemical PVA technique for differentiating LDH isoenzymes by adding urea to the incubation medium.

LDH can withstand brief fixation with formaldehyde but normally cryostat sections of freshly frozen tissue are preferred for histochemical demonstration. Van Noorden and Vogels (1989b) determined the reaction rate of the enzyme using a quantitative assay. They were able to show that the addition of polymer compounds (Sect.23.2.2) to the incubation medium improves localization considerably . A method for simultaneous quantitative histochemical assay of lactate and succinate dehydrogenases in the same cell has been described by Stoward and Nakae (1988).

Glucose-6-Phosphate Dehydrogenase, EC 1.1.1.49. The enzyme catalyzes the reaction glucose-6-phosphate $\rightarrow$ 6-phosphogluconolactone (D-glucono-δ-lactone 6-phosphate) with the coenzyme $NADP^+$ as primary hydrogen acceptor. This initial and rate-limiting reaction in the pentose shunt (hexose-monophosphate shunt) leads on to the formation of ribose and subsequently deoxyribose. It is therefore a key enzyme in nucleic acid synthesis. The NADPH formed acts as an important hydrogen donor in pathways such as steroid and fatty acid synthesis.

The enzyme is found in the cytosol and high concentrations occur in liver, kidney, adrenal gland, and lactating mammary gland. Glucose-6-phosphate dehydrogenase does not tolerate fixation, and is therefore difficult to retain during incubation. For reasonable preservation, addition of polymer compounds such as PVA to the incubation medium is required (see Sect.23.2.2, Butcher, 1984, and Jonges and van Noorden, 1989).

An extensive review of the histochemistry of this enzyme has been given by van Noorden (1984).

Succinate Dehydrogenase (SDH) EC 1.3.99.1. This is a flavoprotein. The enzyme catalyzes the oxidation of succinate to fumarate during which reducing equivalents are transferred to the FAD moiety of the enzyme and thence to coenzyme Q and the respiratory chain (Fig. 25.1). Coenzyme Q itself may therefore be rate-limiting for the reaction. This problem may be overcome either by adding the intermediary electron carrier PMS (Sect.25.1.2) or by coating the slides with coenzyme Q before transferring the sections to the slides (Andersen and Høyer, 1973). SDH is tightly bound to the inner mitochondrial membrane and its presence may be used as a marker for this organelle (Sect.31.11.1). SDH is competitively inhibited by malonate and non-competitively by thenoyltrifluoroacetone (Old and Johnson, 1989). As the enzyme activity is very sensitive to fixation (Sect.25.1.1) only cryostat sections of freshly frozen tissue should be used. A kinetic study of this enzyme has been presented by van Noorden and Vogels (1989b).

NADH Dehydrogenase, EC 1.6.5.3 and 1.6.99.3. The first reaction in the mitochondrial respiratory chain is the transfer of electrons from NADH to ubiquinone-10 (coenzyme Q). This reaction is catalyzed by complex I (NADH dehydrogenase (ubiquinone) (EC 1.6.5.3) acting via flavoprotein (FMN) and iron-sulphur protein (FeS)).

Complex I can be degraded by various treatments to different types of NADH dehydrogenases which may use acceptors other than ubiquinone-10. The Enzyme

Commission (IUB, 1984) recommends that these types should be called NADH dehydrogenase (EC 1.6.5.3) without specification of an acceptor in brackets.

At present it is not clear which of the multiple NADH dehydrogenases are demonstrated in cells with cytochemical methods, and for this reason several synonyms have been widely used in the histochemical literature including *NADH-diaphorase*, NADH-ferricyanide oxidoreductase, NADH tetrazolium salt reductase, and NADH oxidoreductase. These terms are now considered inappropriate and *NADH-dehydrogenase* has been used throughout this text as a collective term for the oxidoreductase systems which catalyze the oxidation of NADH in cytochemical assays.

A substantial amount of NADH dehydrogenase activity is retained after brief formaldehyde fixation. Nonetheless, the use of cryostat sections of freshly frozen tissue is strongly recommended.

NADH may reduce tetrazolium salt non-enzymatically. This non-specific reaction becomes more pronounced above pH 8 and may therefore be reduced by lowering pH to 7.2–7.5.

It should be noted that when measuring the activity of an NAD^+-dependent dehydrogenase, a test, as to whether NADH dehydrogenase activity is rate-limiting for the overall reaction, should be made by performing the reaction at the pH optimum of the NAD^+-dependent enzyme (cf. Sect.25.1.2, Buffer). Furthermore, when tetrazolium salts with low redox potentials (e.g. NBT or TNBT) are employed, it may be necessary to use a relatively low concentration of NADH (about 0.5 mmol/l) in order to avoid too high non-specific formazan production in the medium. With NT an optimum concentration of NADH may, however, be used.

NADPH Dehydrogenase, EC 1.6.99.1. It is generally assumed that NADPH dehydrogenase activity demonstrated by cytochemical procedures represents the microsomal electron transport system. It is, however, not known which components of the system actually participate in the reaction.

According to the Enzyme Commission (IUB, 1984) at least six NADPH dehydrogenases of mammalian origin exist:

- $NAD(P)^+$ transhydrogenase (EC 1.6.1.1)
- NADPH-ferrihaemoprotein reductase (EC 1.6.2.4)
- NADPH-cytochrome-c_2 reductase (EC 1.6.2.5)
- NADPH dehydrogenase (EC 1.6.99.1)
- NAD(P)H dehydrogenase (quinone) (EC 1.6.99.2)
- NADPH dehydrogenase (quinone) (EC 1.6.99.6)

When NADPH dehydrogenase (*NADPH-diaphorase* is no longer applicable) is demonstrated cytochemically, any of the above enzymes may contribute to the result. Transhydrogenase can probably be excluded as insufficient NAD^+ is likely to be present or generated during incubation. Moreover, formaldehyde fixation completely inhibits this enzyme while only slightly reducing total cytochemical NADPH dehydrogenase activity (Stoward *et al.*, 1991).

NAD(P)H dehydrogenase (quinone), EC 1.6.99.2, was formerly called DT-diaphorase. It may accept reducing equivalents from both NADH and NADPH.

It is a two-electron transfer enzyme with menadione or vitamin K_3 as electron acceptor. It is inhibited by dicumarol (Raap and van Duijn, 1983).

NADPH-ferrihaemoprotein reductase, EC 1.6.2.4, was formerly called NADPH-cytochrome *c* (P-450) reductase. This enzyme is highly specific in only being able to accept reducing equivalents from NADPH. It is a one-electron transfer enzyme giving rise to semiquinones that are readily auto-oxidized forming superoxide radicals (van Noorden and Butcher, 1986a). These authors developed a quantitative cytochemical assay using a PVA medium that can be used for kinetic studies of reductase activity in tissue sections. In initial experiments 5 mmol/l TNBT was used as the final acceptor. With this tetrazolium salt (at pH 7.45) it was necessary to lower the concentration of NADPH to under 0.5 mmol/l to avoid too high formazan production in the medium. When 5 mmol/l NT replaced TNBT, it was possible to use concentrations of NADPH up to 5 mmol/l. The specific reaction (test minus control, cf. 25.1.2, Incubation time) was linear for at least 10 min. The specificity of the method was proved in two ways:

1. Activity is increased selectively in rat liver by phenobarbitone but not by 3-methylcholanthrene which induces NAD(P)H dehydrogenase.
2. It is completely inhibited by $NADP^+$ though unaffected by dicumarol, an inhibitor of NAD(P)H dehydrogenase (quinone), EC 1.6.99.2, activity.

Practical considerations concerning non-enzymatic reduction of tetrazolium salts, pH, and the addition of a tissue stabilizer are the same as for NADH dehydrogenase.

It should be noted that if the reaction for NADPH dehydrogenase activity is done on cells containing large amounts of non-specific alkaline phosphatase, artifacts may arise due to the removal of phosphate from NADPH and action of NADH dehydrogenase on the resultant NADH (Leeflang-de Pijper and Hülsmann, 1974). This problem can be solved by adding an alkaline phosphatase inhibitor. Either competitive (e.g. β-glycerophosphate) or non-competitive (e.g. L-p-bromotetramisole) inhibitors (Leeflang-de Pijper and Hülsmann, 1974; Høyer and Andersen, 1977) may be used.

25.4.2 Peroxidases

Peroxidases catalyze the oxidation of many different compounds (e.g. fatty acids and amino acids) using hydrogen peroxide as the hydrogen acceptor.

According to the Enzyme Commission (IUB, 1984) enzymes belonging to EC 1.11 act on hydrogen peroxide as acceptor. There is only one sub-subclass, EC 1.11.1, the peroxidases, that comprise several different enzymes of both animal and plant origin. From a histochemical viewpoint it may be convenient to subdivide peroxidases into endogenous and exogenous.

Endogenous Peroxidases. Examples listed by the Enzyme Commission (IUB, 1984) are cytochrome-*c* peroxidase (EC 1.11.1.5), catalase (EC 1.11.1.6), and iodide peroxidase (EC 1.11.1.8). Several other endogenous compounds, not classified by the Enzyme Commission, may, however, exhibit peroxidase or peroxidase-like activity. Examples are myoglobin, haemoglobin, and myeloperoxidase.

Endogenous peroxidases are often classified according to tissue of origin. Lactoperoxidases, reticuloendothelial phagocyte system peroxidases, thyroid gland peroxidase, reproductive tract peroxidase, mammary tumour peroxidase, and other tissue peroxidases are terms frequently used (Anderson *et al.*, 1979).

Exogenous Peroxidases. This term covers peroxidases used as probe tracer substances. Peroxidase extracted from horseradish (HRP) is of particular interest. HRP is widely used as a tracer in permeability investigations (Sect.27.2.3) and in immunohistochemistry (Sect.26.2.2). Histochemical demonstration using benzidine and its derivatives is described in Sects.26.2.2, 27.2.3, and 28.8.8. It should be noted that cytochrome-*c* oxidase, endogenous catalase, and haemoglobin may cause false positive reactions.

The peroxidase-like activity shown by haemoglobin may be suppressed by preliminary blocking with either H_2O_2 or by treatment with periodic acid (Pearse, 1980, p.235).

Cytochrome-*c* oxidase can be inactivated by fixation with formaldehyde, which HRP is relatively resistant to or inhibited by azide or cyanide. Catalase activity localized to peroxisomes can be suppressed if the hydrogen peroxide concentration of the incubation medium is kept under 3×10^{-3} mol/l. If quantitative determination of exogenous peroxidase (HRP) activity is to be made in probe assays, the plateau absorbance principle may be used (Sect.28.8.8).

In the majority of qualitative studies on "peroxidase" either glutaraldehyde or formaldehyde fixation has been used even though fixation must cause more or less pronounced inactivation.

In attempts to differentiate between the different peroxidases several different factors have been investigated, including: different fixatives, the influence of fixation temperature and time, concentrations of DAB and H_2O_2, and pH. See for instance Fahimi (1975), LeHir *et al.* (1979), and van Bogaert *et al.* (1980).

If quantitative demonstration of the activity of a specific endogenous peroxidase is intended, unfixed cryostat sections and optimal kinetic conditions for the enzyme in question must be used (Chap.28). Only a few endogenous peroxidases have been extensively studied cytochemically. An example, iodide peroxidase, is given below.

Iodide Peroxidase, EC 1.11.1.8, Thyroid Gland Peroxidase. This enzyme can be quantitatively demonstrated using the technique outlined in 25.2 (Ealey *et al.*, 1984; Perrild *et al.*, 1988). Optimal results are obtained after incubation for 20 min at 37°C and a pH of 8.0. Optimal concentrations of H_2O_2 are 0.15 mmol/l and of DAB 1.4 mmol/l. Maximum activity was found with a DAB concentration of 5.6 mmol/l but this gave precipitation also in the medium. To avoid this the concentration was reduced as indicated above even though the amount of reaction product formed was 2–3 times less. Inhibitors are 3-amino-1,2,4-triazole and methimazole with maximum inhibition at 10^{-2} mol/l.

When one I.U. thyroid stimulating hormone (TSH) was administered every 8 hours for two days maximal stimulation of the peroxidase in thyroid follicle cells was achieved after one day (Ealey *et al.*, 1984). Perrild *et al.* (1988) investigated the acute effects of low doses of TSH on the metabolism of guinea-pig thyroid segments maintained in non-proliferative organ culture. A sustained rise in peroxidase activity reaching 129% over control was observed after 30 min.

25.4.3 Oxidases

A subclassification of enzymes relevant to cytochemistry according to the recommendations of The Enzyme Commission (IUB, 1984) is followed below:

1. EC 1.1.3 act on the CH−OH group of donors with oxygen as acceptor

Glucose Oxidase, EC 1.1.3.4. This is a flavoprotein enzyme catalyzing the oxidation of β-D-glucose in the presence of oxygen to D-glucono-1,5-lactone and hydrogen peroxide. The histochemical importance of this enzyme is its use as a marker in immunohistochemistry. The glucose oxidase used for this purpose is of fungal origin thus preventing problems of background activity due to endogenous enzyme as glucose oxidase is absent from human and most animal tissues (Poulter *et al.*, 1987, Polak and van Noorden, 1983, p. 35).

An example of the application of glucose oxidase is given by Poulter *et al.* (1987) who used the enzyme conjugated to the monoclonal antibody, anti-HLA-DR. Activity was detected using β-D-glucose as substrate, NBT as acceptor of reducing equivalents, and PMS as intermediary carrier. The method was quantitated by microspectrophotometry, the reaction being found linear with time for at least 60 min.

If the one-step DAB-nickel-H_2O_2 visualization procedure (Sect.25.3) can be shown to be stoichiometric it may be a suitable alternative in quantitative immunocytochemistry using glucose oxidase as the label (Gossrau *et al.*, 1989).

(S)-2-Hydroxy-Acid Oxidase, EC 1.1.3.15. This is a flavoprotein enzyme with a wide substrate specificity that catalyzes the oxidation of α-hydroxy acids with oxygen generating the corresponding keto acid and H_2O_2. It is a peroxisomal enzyme present as two isoenzymes, one in liver, preferably metabolizing short-chain aliphatic α-hydroxy acids, and one in kidney, preferably metabolizing longer-chain aliphatic or aromatic α-hydroxy acids (Angermüller, 1989).

Xanthine Oxidase, EC 1.1.3.22. This is also a peroxisomal flavoprotein enzyme. In the catabolism of purines this enzyme catalyzes the oxidation of xanthine in the presence of oxygen generating urate, H_2O_2, and a superoxide radical. Allopurinol is a highly specific inhibitor (Angermüller, 1989).

2. EC 1.4.3 act on the CH-NH_2 group of donors with oxygen as acceptor

D-Amino-Acid Oxidase, EC 1.4.3.3. This is a peroxisomal flavoprotein enzyme catalyzing the oxidation of D-amino acids in the presence of oxygen and yielding the corresponding imino acids. The imino acid is then spontaneously hydrolyzed to the keto acid and NH_3. The enzyme is found in liver cells and cells of the proximal tubules in the kidney. Pipecolic acid or thiazolidine-2-carboxylic acid can be used as substrates for the cytochemical localization of the enzyme. Inhibitors are kojic acid and sodium benzoate (Angermüller, 1989).

Amine Oxidase (Flavine-Containing), EC 1.4.3.4. This enzyme, previously called monoamine oxidase, is a flavoprotein enzyme which in the presence of oxygen acts on primary amines and usually also on secondary and tertiary amines with small substituents. The products formed are an aldehyde, NH_3, and H_2O_2.

Two forms, A and B, have been described. It has been claimed that type A preferentially deaminates noradrenaline and serotonin while type B preferentially deaminates β-phenylethylamine and dopamine. Tyramine can equally well be deaminated by both forms of the enzyme. Type A is reversibly inhibited by brofaromine and moclobemide and irreversibly by LY 51641 and clorgyline while type B is irreversibly inhibited by 1-deprenyl.

A quantitative histochemical method has been developed by Frederiks and Marx (1985). Tryptamine was used as substrate and tetrazolium salts were employed as final electron acceptors. Highest activity was obtained using TNBT and BPST while lower activity was seen when using NBT. This is explained by the authors as possibly being due to NBT being soluble in the lipid component of the enzyme. No reaction product was formed in the presence of PVA probably due to binding of the substrate to PVA. At 37°C with TNBT, pH 7.7 and section thickness 4–12 μm, the reaction was linear with incubation time up to two hours. The activity was completely inhibited by p-chlorobenzoate. The enzyme is present in the outer mitochondrial membrane (Frederiks and Marx, 1985).

Kondradi *et al.* (1989) using a DAB technique with addition of nickel found activity of noradrenergic neurons of the locus coeruleus and serotoninergic neurons of the raphe nuclei, while, surprisingly, dopaminergic neurons of the substantia nigra gave no reaction whatsoever.

3. EC 1.7.3 act on diverse nitrogenous substrates with oxygen as acceptor

Urate Oxidase, EC 1.7.3.3. This is an enzyme of purine catabolism that catalyzes the oxidation of urate in the presence of oxygen yielding H_2O_2, CO_2, and allantoin. It is found in liver and kidney peroxisomes. Urate is the only substrate but high concentrations of this compound inhibit the enzyme. Inhibitors include chelating substances such as maleate that reacts with the copper atom at the active centre. Strong competitive inhibitors are oxonic acid and trisubstituted purines such as trichloropurine (Angermüller, 1989).

4. EC 1.9.3 act on a haem group of donors with oxygen as acceptor

Cytochrome-c Oxidase, EC 1.9.3.1. This enzyme is located in the mitochondrial cristae, is an enzyme of the terminal part of the respiratory chain (Fig. 25.1). Reduced cytochrome *c* generated by the transfer of electrons is reoxidized by this enzyme. The cytochemical method with 3,3′-diaminobenzidine (DAB) as an electron donor was introduced by Seligman *et al.*, 1968). Oxidative polymerization and cyclization of DAB gives rise to an insoluble indamine polymer localized in mitochondrial cristae.

An optimal assay of the quantitative microspectrophotometric determination of cytochrome-*c* oxidase activity in human skeletal muscle was shown by Old and Johnson (1989) to be achieved by using media containing 4 mmol/l DAB and 100 μmol/l cytochrome c. Reaction in individual muscle fibres was found to be linear for incubation times of more than 15 min. The reaction is linear with DAB concentrations from 1–6 mmol/l but to avoid precipitation of DAB from the solution a final concentration of 4 mmol/l was settled on. Optimum pH was found to be 7.0. Addition of catalase did not influence the reaction thus indicating that endogenous peroxidase did not contribute to the specific reaction. As shown in Fig. 25.1, azide and cyanide are terminal respiratory chain inhibitors. Without the addition of cytochrome *c* and in the presence of 5 mmol/l azide the non-specific reaction was found to be less than 5%. The rate of electron transfer is dependent on an adequate oxygen supply. To ensure aerobic conditions during incubation it was necessary to use a distance of at least 1 mm between coverslip and slide.

In any valid assay it is necessary to perform the reaction both in the presence and absence of exogenous cytochrome *c* (Henderson, 1979; Henderson *et al.*, 1978b). When the activity of cytochrome-*c* oxidase in normal synoviocytes was compared with that in synoviocytes in rheumatoid arthritis a 10–15 fold increase was found in the rheumatoid synoviocytes when exogenous cytochrome *c* was not added. In contrast no significant differences were observed when cytochrome *c* was added (Henderson *et al.*, 1978b).

5. EC 1.10.3 act on diphenols and related substances or ascorbate as donors with oxygen as acceptor

Catechol Oxidase, EC 1.10.3.1. This enzyme was previously called DOPA-oxidase or tyrosinase. This copper containing enzyme acts on catechols or substituted catechols. In the presence of oxygen, it converts L-tyrosine to L-dihydroxyphenylalanine (DOPA), which in turn is oxidized to a quinone. This substance is autooxidized to other products that polymerize into melanin (cf. Sects.18.1.3 and 18.4.13). For the cytochemical demonstration of this enzyme DOPA is the substrate of choice. Any wavelength in the range 400–700 nm may be selected for measurements of melanin as this compound does not possess an absorption maximum. Inhibitors of catechol oxidase are glutathione and the copper chelator diethyldithiocarbamate (Sect.17.7.3).

Whittaker (1981) has studied the reaction quantitatively using icecold 70% ethanol or absolute methanol fixed material. At pH 7.2 using DOPA as the substrate at a saturating concentration (4 mmol/l) the absorbance at 450 nm increased linearly from the onset of incubation up to nine hours.

Quantitating catechol oxidase activity at pH 7.2 in formaldehyde fixed single melanoma cells and measuring absorbance at 640 nm, Croce *et al.* (1988) identified two important parameters:

a. The rate of melanin production. This was found to be affected by many factors other than catechol oxidase activity (e.g. intracellular organization and distribution of the enzyme).
b. Lag-time. This was defined as the time required by intermediates to reach the critical concentration at which the polymerization process starts and melanin becomes measurable. This parameter was found to be a more direct expression of the enzyme activity.

As different levels of enzyme activity gave rise to different lag-times and rates of melanin formation, single measurements of the amount of melanin formed in single melanoma cells are insufficient for quantitation of catechol oxidase activity (Croce *et al.*, 1988).

6. EC 1.14 act on paired donors with incorporation of molecular oxygen and in subclasses 1.14.14–1.14.18 one atom of oxygen is incorporated into one donor, the other donor being some other compound, e.g. a reduced flavine or flavoprotein, a reduced iron-sulphur protein, a reduced pteridine, or ascorbate.

Monophenol Monooxygenase, EC 1.14.18.1. According to the Enzyme Commission (IUB, 1984) this enzyme belongs to a group of copper proteins that also catalyze the reaction of catechol oxidase (EC 1.10.3.1) provided only 1,2-benzenediols are available as substrates.

Schmidt (1988) has shown that monophenol monooxygenase is an excellent, highly specific marker enzyme of leukocytes. Histochemical detection of activity due to this enzyme, using DOPA as substrate, is thus a suitable means of demonstrating these cells in tissues and in various pathological conditions. Purulent processes are characterized by the presence of many leukocytes with an enzyme activity that is not resistant to methanol fixation, whereas proliferative, cell-mediated, immunological processes are marked by the presence of numerous cells with methanol-resistant enzyme activity (Schmidt, 1988).

Whether or not the monophenol monooxygenases of melanocytes, melanoma cells, and leukocytes are identical enzymes is not clear. Their almost identical behaviour with respect to different substrates and inhibitors does, however, support this view (Schmidt, 1988).

Part 6
Other Techniques

26 Immunohistochemistry

P.P. Clausen, M. Møller, B. van Deurs, O.W. Petersen

Immunohistochemistry is now a very substantial subject and it will only be possible to introduce the principles in this chapter. For more extensive coverage the reader is referred to the following: Bullock and Petrusz (1982; 1983; 1985), Polak and Van Noorden (1983; 1984), and Larsson (1988).

26.1 Definition

Immunohistochemistry and immunocytochemistry are defined as *in situ* demonstration of antigens in tissues and cells using antibodies. The antibodies are labelled directly or indirectly in order to visualize the location at which antigen binding has taken place using light or electron microscopy (regarding the latter, see Sect.27.3).

26.1.1 Antigens and Antibodies

A number of different molecules, mainly carbohydrates and proteins, induce production of antibodies when injected into vertebrate species. The immune system of the animal should not recognize these molecules as normal "self" components. Molecules that stimulate specific antibody production are referred to as *antigens*.

26.1.2 Antibodies

When an antigen is introduced into the organism it stimulates lymphocyte *activation*. Most antigens have several antigenic sites (epitopes) each of which may induce their own specific antibodies. As one lymphocyte only produces antibody of a single specificity, the response to an antigen with multiple epitopes entails activation and proliferation of several clones of lymphocytes. A proportion of the lymphocytes differentiate into secretory cells, plasma cells, which release antibodies (Fig. 26.1).

H. Lyon (Ed.)
Theory and Strategy in Histochemistry

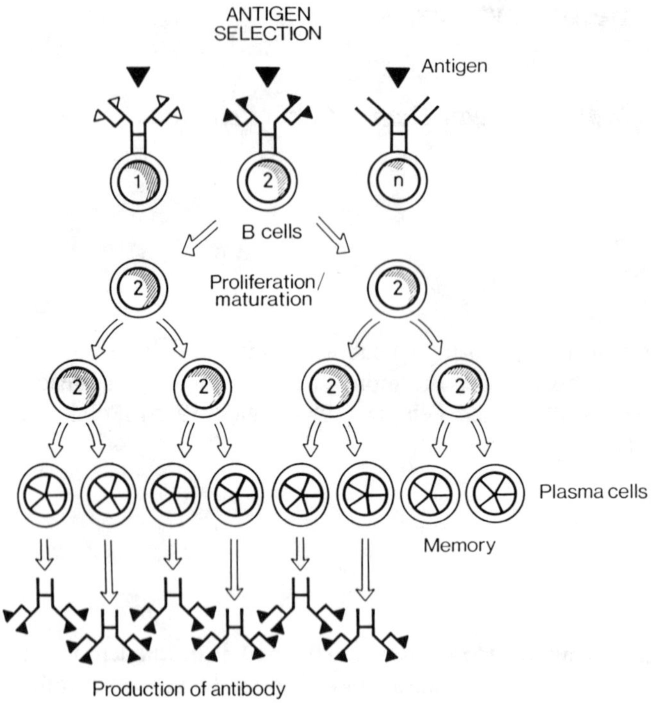

Fig. 26.1. Scheme showing three lymphocytes (1, 2, and n) of which (2) is "triggered" to produce antibodies. The triggered lymphocytes have receptors for the specific antigen on their surface.

The antibodies produced by plasma cells derived from an individual clone of lymphocytes are homogenous and specifically directed towards the antigen that evoked the initial plasma cell formation. The antibodies are secreted by the plasma cells and released to the blood stream where they form part of the serum protein class termed globulins (Fig. 26.2). The antibodies are therefore also called *immunoglobulins*.

Immunoglobulins are grouped into five classes:

- Immunoglobulin G (IgG), molecular weight 150 kDa
- Immunoglobulin A (IgA), molecular weight 160 kDa
- Immunoglobulin D (IgD), molecular weight 185 kDa
- Immunoglobulin E (IgE), molecular weight 200 kDa
- Immunoglobulin M (IgM), molecular weight 900 kDa

Most antibodies used for immunohistochemistry belong to either the IgG or, less frequently, the IgM class.

The essential structure of an IgG molecule is shown in Fig. 26.3.

The molecule is composed of two long or heavy polypeptide chains and two short or light chains. S-S-bridges form connections between the two heavy chains and interconnect the light chains with the heavy chains. Within the same molecule the two heavy and the two light chains, respectively, are identical. Only two types

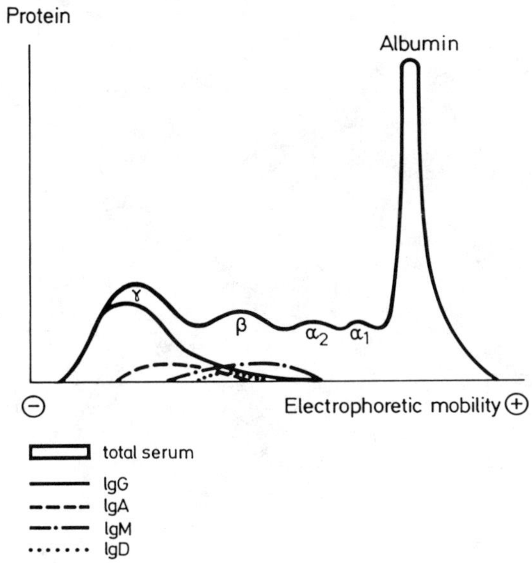

Fig. 26.2. Serum electrophoresis showing separation of serum albumin and different classes of globulin (α, β, γ).

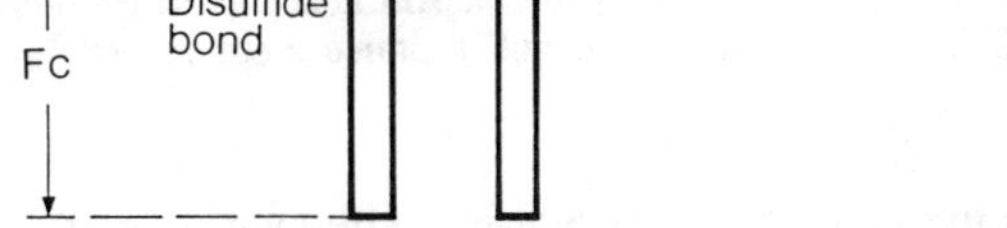

Fig. 26.3. Diagram illustrating the structure of the IgG molecule.

V_H	variable region of heavy chain	C_H	constant region of heavy chain
V_L	variable region of light chain	Fab	fragment antigen binding
C_L	constant region of light chain	Fc	fragment crystalline

of light chains are found: κ and λ. According to the five immunoglobulin classes, the heavy chains are named γ (IgG), α (IgA), δ (IgD), ϵ (IgE), and μ (IgM).

The enzymes papain and pepsin split immunoglobulin molecules into several components with different biological properties. From Fig. 26.4 it can be seen that digestion of IgG with papain results in the formation of two fragments termed Fab (fragment antigen binding) and one termed Fc (fragment crystalline). The binding between antigen and antibody takes place at one end of the Fab fragment, whereas the Fc fragment has other biological properties.

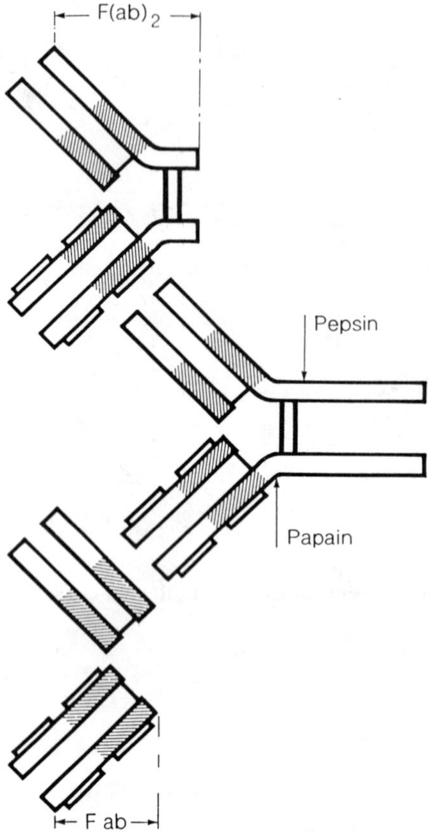

Fig. 26.4. Diagram illustrating the effects of the proteolytic enzymes pepsin and papain on the IgG molecule. Papain will produce two Fab fragments and pepsin will produce a single $F(ab)_2$ fragment.

Within both light and heavy chains there are different regions called variable and constant regions. The variable region contains the amino acid sequence and steric conformation responsible for the specific binding of the antibody to the antigen.

26.1.3 Antigen Antibody Reactions

Since antigens are usually large polypeptides or polysaccharides, only a small region of the molecule may actually participate in the antigen antibody reaction. Thus, each antigen molecule contains one or more regions comprising 4–8 amino acids or monosaccharide units called antigenic sites or determinants where specific antibody binding takes place. The larger the antigen, the more antigenic sites are present on the molecule. The sites themselves may be a sequence of 4–6 adjacent amino acids in the primary protein structure, or they may comprise more distant

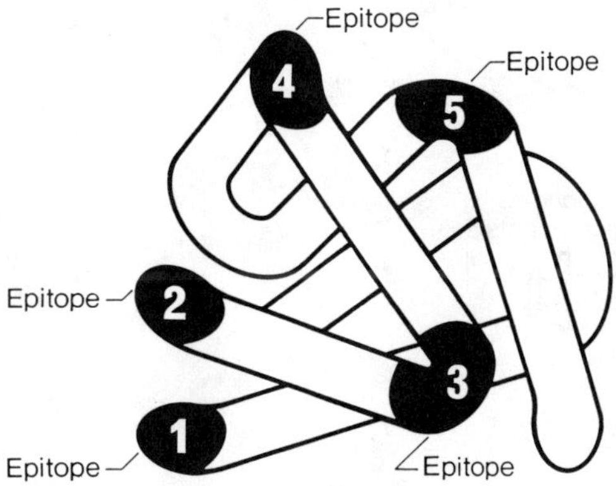

Fig. 26.5. Diagram illustrating the structure of a folded polypeptide. Five different epitopes are shown.

amino acids in the primary sequence brought together by the steric conformation of the molecule (Fig. 26.5).

The larger the antigenic site, the stronger is the binding between antigen and antibody. The binding is non-covalent, based on electrostatic forces, hydrogen bonds, van der Waal forces, and hydrophobic interaction. The reaction between antigen and antibody is reversible: AB + AG ⇄ ABAG, and the reaction can be described by the law of mass action:

$$AB + AG \overset{K}{\rightleftarrows} ABAG$$

If K is large, the equilibrium is displaced to the right and the antibodies bind strongly to the antigen and the antibody displays a high *affinity*. In practice we deal with multivalent antibodies, and in this case the term *avidity* is used instead of affinity of the antibody. In practical immunohistochemical work it is important to be aware of the fact that dilution of the antibody, for instance during prolonged washing between incubation steps, will shift the equilibrium to the left and thereby cause a weaker staining.

26.1.4 Production of Polyclonal and Monoclonal Antibodies

The classical way of producing antibodies is to inject the purified antigen subcutaneously into an animal. Most often rabbits, swine, goats, sheep, or guinea pigs are used. The animals are immunized several times with intervals of 2 or 3 weeks in order to obtain a high level of antibodies. When the serum from an immunized animal is isolated, it contains a mixture of different antibody molecules produced by different plasma cell clones, each of which reacts with only one antigenic determinant (Fig. 26.6).

An antiserum comprising multiple different antibodies towards the same antigen is referred to as a *polyclonal antiserum*. The crude antiserum is characterized

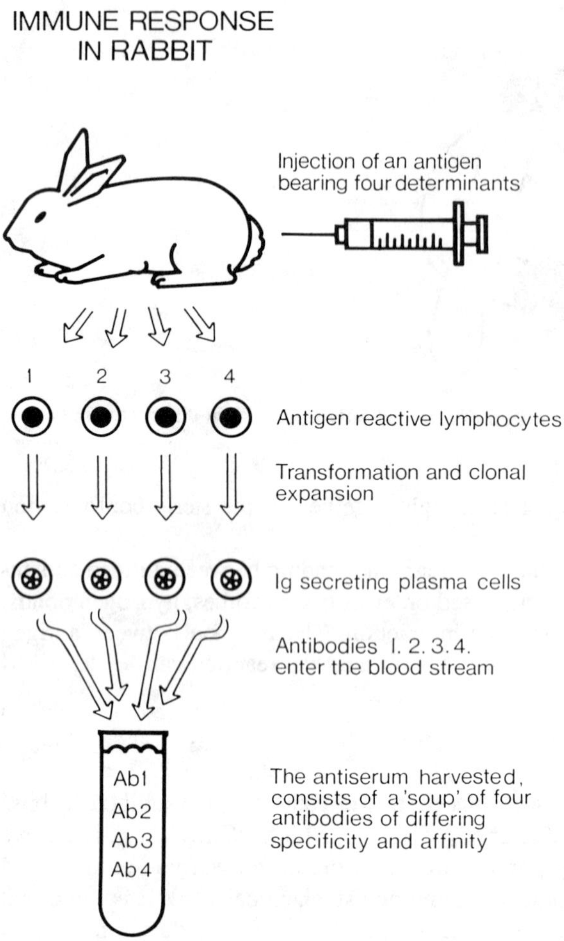

Fig. 26.6. Diagram illustrating the production of a polyclonal antiserum.

by the presence of both non-specific immunoglobulins of natural origin and the artificially induced specific immunoglobulins. For immunohistochemical staining the IgG-containing immunoglobulin fraction is often isolated. Even the purified immunoglobulin fraction will be a mixture of specific and non-specific immunoglobulins.

Further purification using antigen-affinity chromatography results in the isolation of specific antibodies. Theoretically, the use of affinity purified antibodies in immunohistochemistry should yield lower background staining, however, this is not always the case as some of the specific antibodies may be removed during purification.

An alternative way of producing antibodies is the "hybridoma" or "monoclonal antibody" technique. With this technique activated (immunized) B-lymphocytes are fused with malignant plasma cells (myeloma cells). This gives rise to a hybrid cell line which in theory is able to maintain an infinite production of specific immunoglobulins in cell culture (Fig. 26.7).

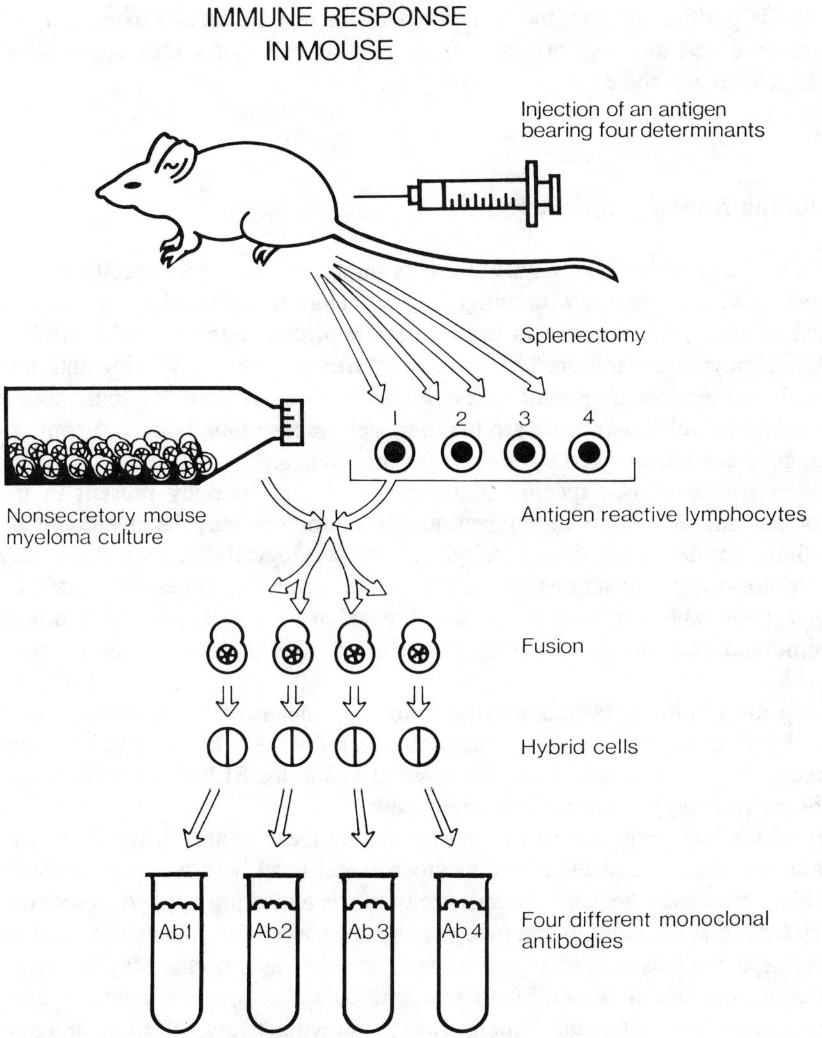

Fig. 26.7. Drawing illustrating the production of monoclonal antisera.

Briefly, mice are immunized with an antigen (not necessarily pure) and immune sensitized lymphocytes harvested from the spleen. These cells are fused with myeloma cells that are unable to secrete immunoglobulins themselves. Following this, only fused (hybrid) cells can grow in the specially formulated culture medium and some of these secrete immunoglobulins. The specificity of the antibody is determined by the original activated lymphocyte that took part in the fusion. The hybridoma cells are cloned and screened for antibody production. After test reactions with the purified antigen, the cell clones of interest are further expanded. As antibodies produced this way originate from a single cell clone, they are referred to as monoclonal antibodies.

26.1.5 Testing Antibody Specificity

Having raised and isolated an antiserum, it is important to test its specificity, i.e. the degree to which it reacts with antigens other than the one used for immunization. Lack of specificity may result from presence of contaminants in the original immunizing antigen preparation. These may give rise to their own specific antibody response. It is therefore important to use highly purified antigens as immunizing material when polyclonal antisera are being made. Traces of antibodies to contaminant antigens can be removed by absorption procedures.

Another source of non-specific immunoreactivity is antibody present in the serum of the animal before immunization. This problem may be overcome by using affinity purified antibodies. Finally, genuine serological cross-reactivity may also cause non-specific reactions. If amino acid or sugar sequences are shared by both the test and other antigens, antibodies directed at such sequences bind to both the specific and non-specific antigens. This sort of false positive reaction cannot be circumvented.

Testing for specificity is primarily performed by means of gel techniques such as crossed immunoelectrophoresis, immunoblotting (Western blots), and immunoprecipitation. In addition, radioimmunoassay (RIA) and the ELISA (enzyme-linked immunosorbent assay) technique are often used.

Even if the antiserum appears monospecific by these methods, its specificity still has to be confirmed at the immunocytochemical level before interpretation is valid. This is necessary because the specificity of the antiserum applied to sections may vary from that obtained in test tube reactions. Reasons for this include fixation and other aspects of tissue processing as well as other factors that may influence antigen conformation. It is helpful to test antibodies in a simple model system that reproduces many of these features while allowing many different antigens to be tested at the same time. Droplets of the antigens to be tested are applied to filter paper or nitrocellulose strips and immobilized by vapour fixation. The immunoreaction can then be performed in circumstances similar to the proposed tissue reaction and potential cross-reactions with a very large panel of different antigens evaluated.

The specificity of immunohistochemical staining is tested with control reactions as described in 26.5.

26.2 Labelling of Antibodies

A variety of labels have been developed to render antibodies visible by light or electron microscopy.

For immunofluorescence methods, a fluorescent molecule (fluorochrome) is coupled to the antibody and bound antibody located using fluorescence microscopy. In immunoenzyme methods, a suitable enzyme is used in place of the fluorochrome and an appropriate enzyme histochemical procedure (cf. Chaps.23–25) used to demonstrate the location of the coupled antibody. Some reaction products are osmiophilic and may therefore also be used at the electron microscopic level (Sect.27.3). Alternatively, for ultrastructural analysis, antibodies are often labelled with heavy metals using reagents such as ferritin or colloidal gold. These heavy metals are directly visible by electron microscopy. In contrast, gold particles are not generally visible by light microscopy (large gold particles appear pink) but the particles can be rendered visible using a silver intensification process.

Antibodies linked to a label are called conjugates and these will be discussed below.

26.2.1 Fluorochrome Conjugates

A fluorochrome is a molecule which absorbs light energy at a specific wavelength (excitation light) and subsequently emits light at a longer wavelength (emission light). The most commonly used fluorochrome coupled to antibodies is Fluorescein Isothiocyanate (FITC). FITC coupled to an immunoglubulin molecule has an excitation maximum at 495 nm and an emission maximum at 520 nm (apple-green colour). Another fluorochrome, Tetramethylrhodamine Isothiocyanate (TMRITC), has an excitation maximum at 568 nm and an emission maximum at 597 nm (orange-red region). TMRITC shows weaker fluorescence intensity than FITC but it fades less during exposure to ultraviolet light.

A disadvantage of immunofluorescence methods is that the background is dark, making it difficult or impossible to discern the structural pattern around the fluorescence image. Another problem is that fluorochromes fade so that the more prolonged the exposure to excitation the less the fluorescence emission of a given object. Fading can, however, be largely prevented by judicious choice of mounting media. The intensity of the FITC fluorescence is influenced by the mounting medium pH; an alkaline medium with a pH about 8 is optimal. A photometric analysis of antifading reagents was performed by Böck *et al.* (1985). Sodium azide and sodium iodide were found to increase the fluorescence intensity of FITC while the value of polyvinyl pyrrolidone, polyvinyl alcohol, 1,4-di-azobicyclo-(2,2,2)-octane, n-propylgallate, and sodium dithionite in this regard was questionable.

Some newer fluorescent tracers like Texas Red and Phycoerythrin show much less tendency to fading than FITC.

In contrast, localization of fluorochromes is easily detected in either sections or single cells in culture. Subcellular structures can be visualized, often with very high resolution. In addition, fluorochromes are very useful in double labelling immunocytochemistry, especially at the single cell level. This allows each fluorochrome to be visualized separately by switching between different filter systems within the fluorescence microscope.

26.2.2 Enzyme Conjugates

Several enzymes have been used for labelling antibodies (e.g. horseradish peroxidase (HRP), glucose oxidase, alkaline phosphatase, and β-galactosidase). Among these horseradish peroxidase (HRP) has been most widely used. It is a very stable enzyme, the enzymatic activity of which is hardly influenced by the coupling procedure. HRP has a molecular weight of 40 kDa and is coupled chemically to antibodies using glutaraldehyde or periodate.

The enzyme catalyzes the transformation of hydrogen peroxide to water if a hydrogen donor is available; 3,3′-diaminobenzidine tetrahydrochloride (DAB) is often used as hydrogen donor. During the catalytic process, DAB is transformed to a water-insoluble, brown product. As the reaction product is osmiophilic, DAB-labelled antibodies can be used for both light and electron microscopical investigations (Sect.27.3).

In the peroxidase-anti-peroxidase technique (PAP-technique, see Sect.26.3.3) three PAP molecules are bound to two IgG molecules in an immune complex. This approach increases the efficiency of the immunohistochemical detection system.

26.2.3 Ferritin Conjugates

Ferritin is an iron containing protein with a molecular weight of 500 kDa. Due to the high content of iron (each molecule possesses up to 4000 iron atoms) this molecule is electron dense and thereby visible in the electron microscope. (Sects.27.2.3, 27.3). The molecule is 10 nm in diameter. Using antibodies coupled to ferritin it is possible to quantify the number of antigens revealed in the section, as each "dot" seen in the electron microscope corresponds to one antigen molecule. A major disadvantage with ferritin is its high molecular weight. This leads to serious problems for antibody penetration into sections.

26.2.4 Colloidal Gold Conjugates

By chemical reduction of tetrachloroauric acid it is possible to make a gold sol containing particles of uniform diameter. Commonly used agents for this procedure include white phosphorus, citrate-tannic acid, or sodium borohydride. The gold particles can be prepared in different sizes (from 3–150 nm). The particles are negatively charged and bind easily to the positively charged groups of IgG or protein A (Sect.26.3.5). In the case of protein A, the probe is referred to as PAG. The colloidal gold labelled antibodies, like ferritin, show very distinct labelling of the immunostained structure. In contrast to the immunoenzyme methods, there is no diffusion of reaction products at the ultrastructural level. The colloidal gold conjugates have been especially popular for immunocytochemistry on ultrathin cryosections (Sect.27.3.2).

Colloidal gold particles can also be made visible for light microscopy by a silver intensification procedure in which silver ions are precipitated on to the colloidal gold particles using a photographic developer.

26.3 Immunostaining Methods

This section covers the theoretical aspects of staining methods using antibodies and conjugates on tissue sections.

26.3.1 Direct Staining Method

The simplest form of immunohistochemical staining involves a direct (one-step) procedure in which the labelled antibody binds directly to the antigen (Fig. 26.8). The procedure is fast, but the staining efficiency is low compared to the methods described below.

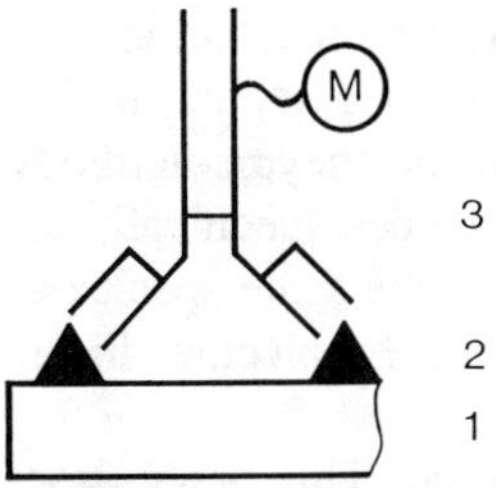

Fig. 26.8. The direct staining method.
1 = tissue section; 2 = antigen in tissue section; 3 = specific antibody labelled with a marker (M).

26.3.2 Indirect Staining Methods

Indirect staining (sandwich or two-step) methods use an unlabelled primary antibody. After a washing step, a labelled immunoglobulin directed against the primary antibody is added for a further incubation period (Fig. 26.9). This procedure may be repeated two, three, or more times, giving a higher signal-to-noise ratio and thereby increasing the staining efficiency of the method. A further advantage of the indirect method is that only a limited panel of antibody conjugates is needed for the demonstration of a wide variety of different antigens.

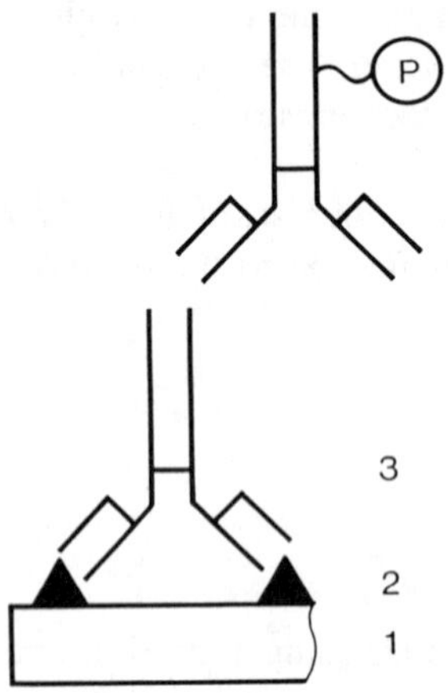

Fig. 26.9. The indirect staining method.
1 = tissue section; 2 = antigen in tissue section; 3 = specific antibody, raised in rabbits, against the antigen; 4 = antibody, raised in swine, against rabbit IgG. The swine antibody is labelled with horseradish peroxidase (P).

26.3.3 Methods Using Antibody-Label Immune Complexes

Another approach to increasing the sensitivity of immunohistochemical methods has been the production and application of complexes consisting of labels (often enzymes) bound to anti-label antibodies.

The complexes are used in the following way. First, the section is incubated with a specific antibody often made in rabbits. This is followed by incubation with an excess of a secondary antibody made against the primary antibody in another species (e.g. swine). Finally the material is incubated with an enzyme-antibody complex. This complex comprises an enzyme bound to antibodies specifically directed against it. If the antibodies in this complex are made in the same species as the primary antibody, the complexes bind to the free arm of IgG molecules in the second layer.

Immune complexes involving HRP have been widely used. This peroxidase-anti-peroxidase complex is referred to as the PAP-complex. It consists of three horseradish peroxidase molecules bound to rabbit antibodies directed against horse-

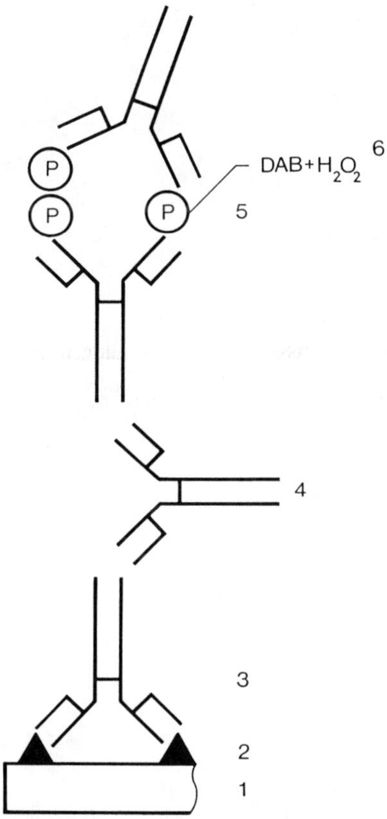

Fig. 26.10. The indirect PAP-method.
1 = tissue section; 2 = antigen in tissue section; 3 = specific antibody raised in rabbits against the antigen; 4 = antibody, raised in swine, against rabbit IgG; 5 = the PAP-complex; 6 = DAB substrate. P = horseradish peroxidase

radish peroxidase (Fig. 26.10). Alkaline phosphatase complexes have also been used (APAAP-complexes).

26.3.4 Biotin-Avidin Methods

In recent years, new marker systems have been developed which have increased the sensitivity of the immunohistochemical procedures. Biotin is a small water-soluble vitamin (molecular weight 0.244 kDa). This molecule binds strongly to avidin, a large glycoprotein (molecular weight 67 kDa) which is abundant in egg white. Each avidin molecule possesses four binding sites for biotin.

Antibodies conjugated to biotin in combination with avidin coupled to fluorochromes, enzymes, or colloidal gold represent a very sensitive detection system (Fig. 26.11).

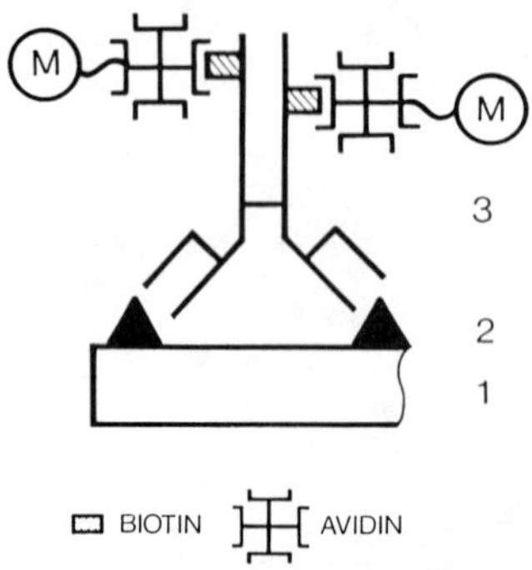

Fig. 26.11. Immunostaining using biotinylated antibodies. 1 = tissue section; 2 = antigen; 3 = biotin-conjugated antibody against 2.
Avidin conjugated with a marker (M, e.g. FITC or enzyme) binds to the biotin.

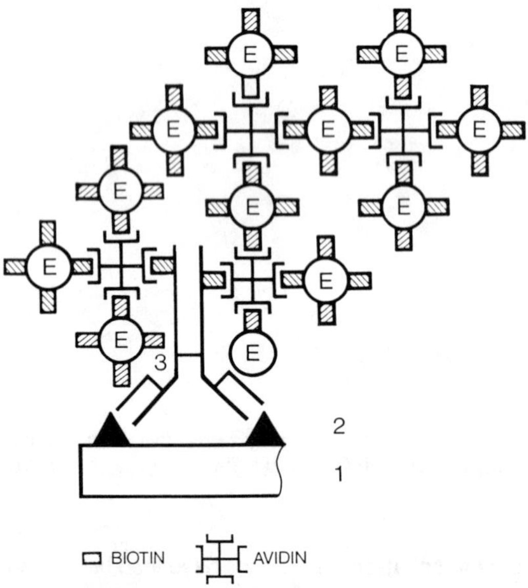

Fig. 26.12. Biotin-avidin staining with a biotinylated enzyme (the ABC-technique).
1 = tissue section; 2 = antigen; 3 = biotin-conjugated antibody against 2.
A biotin-conjugated enzyme (E = e.g. HRP or alkaline phosphatase) binds to 3 via an avidin bridge.

Even higher sensitivity can be obtained by using biotinylated IgG and unconjugated avidin followed by a biotinylated enzyme, e.g. HRP or alkaline phosphatase (the ABC-technique) (Fig. 26.12).

Streptavidin (molecular weight 60 kDa), a protein extracted from the bacterium Streptomyces avidinii, is often used instead of avidin due to its lower non-specific binding at neutral pH. Boon *et al.* (1990) have demonstrated a considerable reduction in incubation times with the ABC technique when applying microwaves (cf.Sect.12.3.12).

26.3.5 Protein A Methods

Protein A, a protein with a molecular weight of 42 kDa, can be extracted from the wall of the bacterium *Staphylococcus aureus*. This protein shows high affinity for the Fc portion of IgG molecules of many animal species including man, rabbit, guinea pig, swine, and dog (Fig. 26.13). Protein A can also easily be coupled to colloidal gold particles at pH 5 (the isoelectric point of the protein). Protein A-gold conjugates are therefore widely used in post-embedding immunocytochemistry and for immunocytochemistry on ultrathin cryosections.

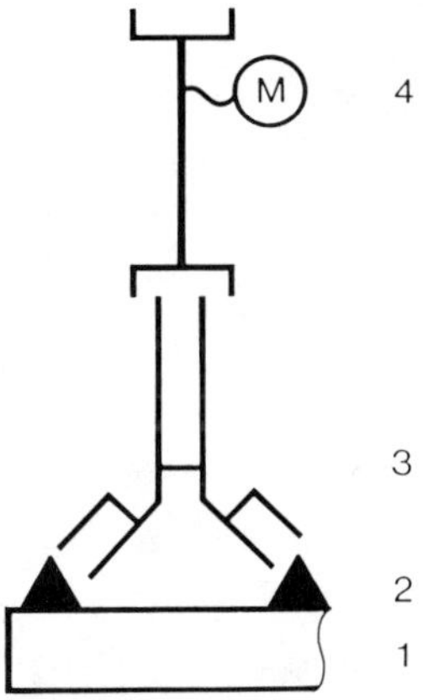

Fig. 26.13. The protein A method.
1 = tissue section; 2 = antigen in tissue section; 3 = specific antibody against the antigen; 4 = protein A bound to the Fc-part of the IgG molecule. Colloidal gold (M) is bound to the protein A molecule.

Protein A may also be used in the PAP-technique as a linking reagent between the primary antibody and the PAP-complexes (Fig. 26.14). In contrast to the ordinary PAP-technique, the antibodies in the primary antibody and in the PAP-complexes do not need to be from the same species.

26.3.6 Labelling of Two Antigens within the Same Tissue Section

In recent years an increasing number of antibodies of high avidity and specificity have been raised. Interest in demonstrating two or more different antigens within the same tissue section or cell has increased in parallel. Considerable effort has therefore been directed towards developing reliable double incubation techniques.

In theory, the simplest way to demonstrate two different antigens in the same section is to use direct antibodies conjugated with different markers (e.g. fluorochromes or enzymes). Since, however, labelled primary antibodies are rarely used, most interest has been concentrated on indirect methods.

In the "sequential incubation method", the section is incubated first with specific antibody followed by visualization with PAP-complexes. Next, all antibodies

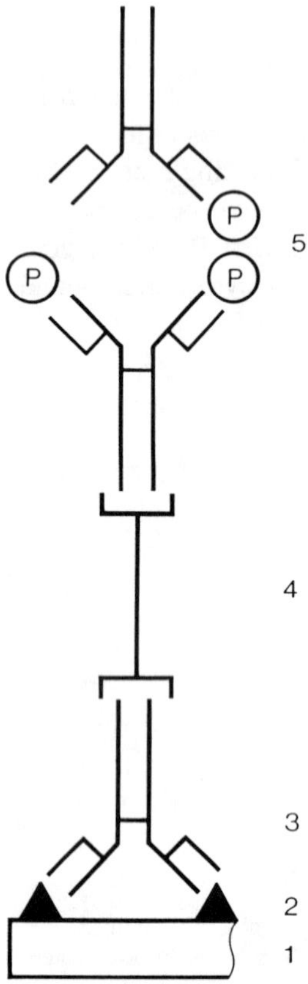

Fig. 26.14. The PAP-technique with protein A as an intermediate layer. 1 = tissue section; 2 = antigen in tissue section; 3 = specific antibody against the antigen; 4 = protein A; 5 = PAP-complex.
P = horse radish peroxidase.

are eluted in glycine/HCl buffer, or in acid potassium permanganate, and this is followed by incubation with a new primary antibody. It is essential that the antibody used for detection of this second primary antibody carries a different label from that used in the detection of the first antigen. FITC, TMRITC, alkaline phosphatase, and HRP have all been used in this sort of procedure.

In a modified version of the "sequential incubation method", the primary antibody is visualized by the PAP-method using DAB as substrate. By a further incubation in cobalt chloride the brown DAB reaction product is changed into a black substance and the enzymatic activity of HRP irreversibly inhibited. This approach enables the use of DAB again in the second round of incubation without confusing the results of the first reaction. Clearly, this method cannot be used to demonstrate different antigens within the same cell because the brown reaction product related to the second cell antigen would be obscured by the black reaction product from the first round of incubation.

In the *"multiple immunoenzymatic staining method"*, elution of the first round of antibodies is not necessary. If the primary antibodies have been raised in different species (e.g. rabbit, goat, sheep, or swine), or if monoclonal antibodies of different subclasses are used, the first incubation is performed in a solution containing both primary antibodies at the same time. The localization of the primary antibodies is then achieved using secondary antibodies directed against the different species or subclasses. This approach also requires separate labels for the two secondary antibodies (β-galactosidase has become increasingly popular as a marker in this area).

26.4 The Choice and Evaluation of an Immunohistochemical Staining Technique

The choice of an immunohistochemical method from the extensive range of options depends on the purpose of the investigation.

Immunofluorescence can only be used at the light microscopy level. The structural resolution and brightness of the staining reaction is often better than that obtained by immunoenzyme techniques. Fluorescence techniques are therefore still preferred for the demonstration of delicate patterns such as the linear immunoglobulin deposits in dermato- and nephro-immunopathology (Sect.32.4). When immunofluorescence is applied to cell cultures, the dark background enhances the contrast and makes detection of stained structure easier. Furthermore, the quality of resolution makes fluorescence techniques excellent for demonstrating more than one antigen within the same section or even within the same intracellular compartment.

Enzyme-based methods have the advantage of allowing counterstaining of the section. For electron microscopic work, enzymes and gold techniques are the techniques of choice at present.

Despite several relevant publications, there is no general agreement concerning the relative sensitivity of the different staining methods, see for instance De Jong *et al.* (1985) and Scopsi and Larsson (1986a). The direct staining method is commonly regarded as less sensitive than the various indirect and multistage techniques. However, when these latter techniques (e.g. PAP and biotin avidin) have been compared, the results seem to be more dependent on the reagents used, and to some extent on the investigator, than on the choice of immunocytochemical method. It is therefore recommended that each laboratory conducts its own tests for the relative sensitivity of the different methods, and chooses the method found to be most convenient for the intended purpose.

26.5 Immunohistochemical Controls

The reliability of an immunohistochemical procedure depends on the specificity of all the reactions and reagents including antibodies and conjugates. Several factors may be responsible for false positive and false negative reactions. Generally, two kinds of specificity are considered: antibody specificity and method specificity.

The specificity of the antibody is tested using various gel techniques (ELISA, Western blot, etc.), and immunocytochemical model systems (Sect.26.1.5).

The specificity of the staining method is tested by means of a panel of control incubations.

To exclude false negative reactions, a section known to contain the antigen of interest can be stained together with the experimental sections.

Testing of false positive reactions is performed by:

A. Incubation with a known non-immune serum as the first layer
B. Incubation with primary antiserum preabsorbed with the specific antigen
C. Incubation with pre-immune serum as the first layer
D. Omission of the primary and secondary antibodies
E. Development of the enzyme reaction in the absence of antibodies (only for enzyme techniques)
F. Observation in a fluorescence microscope of an unstained section testing for possible autofluorescence (only for immunofluorescence)

Apart from false positive reactions due to direct cross reactivity of the antibodies with other antigens, most non-specific reactions reflect non-specific adsorption of the antibodies or conjugates. See also Scopsi and Larsson (1986b).

Unintentional background staining is generally reduced by a combination of the following measures:

a. Application of the primary anti-serum in as high a dilution as possible (this should give the best signal-to-noise ratio)
b. Pre-incubation of the sections with non-immune serum to block the non-specific adhesion
c. Addition of detergents (Triton X-100, Tween-20) to the incubation and washing solutions

26.6 Tissue Processing

26.6.1 Fixation

Fixation of tissue prior to immunohistochemistry is a very critical step. In principle, all fixation procedures decrease the sensitivity of the immunoreaction by interfering with the structure of the antigen. On the other hand, some antigens are extracted during incubation unless the cell and tissue components are immobilized by fixation.

It is therefore normally necessary to use some kind of fixation for retaining the antigens. Fixation also improves the structural preservation.

Most antigen molecules in tissue are proteins or peptides consisting of long chains of amino acid residues. Antibodies generally "recognize" 3 to 6 of these residues (an antigenic "site"). Dehydrating fixatives (methanol, ethanol, and acetone) and the precipitating fixatives (picric acid, mercuric chloride, potassium dichromate, and acetic acid) are preferable from a theoretical standpoint because they denature the proteins by coagulating the secondary and tertiary structure without interfering with the primary structure. The overall structural preservation of the tissue achieved with these fixatives is, however, often unsatisfactory.

In contrast, structural preservation is excellent using the cross-linking fixatives (e.g. paraformaldehyde, acrolein, and glutaraldehyde). The structural alteration implicit in the action of these fixatives (Sect.12.2.1) suggests that they may interfere with the antigenicity of proteins. This problem can be partially circumvented by strict control of the fixation time (e.g. 4% paraformaldehyde in 0.1 mol/l phosphate buffer for 1 to 12 hours).

If glutaraldehyde is used for fixation, the concentration should be kept as low as possible (0.1%) although binding of antibodies raised against dopamine, noradrenaline, and serotonin has been demonstrated even after fixing with 5% glutaraldehyde.

One way of circumventing the decrease in antigenicity caused by formaldehyde fixation is to pretreat the section with proteolytic enzymes before immunostaining. This enzyme pretreatment probably breaks the cross-links between the fixative and the reactive groups of the antigenic sites. The three most commonly used enzymes are trypsin, pepsin, and a non-specific protease named "Pronase". Whereas most antigens are "demasked" by treatment with trypsin and "Pronase", some antigens require treatment with pepsin in order to obtain a satisfactory demasking.

In contrast, some antigens are destroyed by enzyme pretreatment. It is therefore important to determine the extent to which enzyme pretreatment enhances staining before it is used routinely. A further disadvantage of the enzyme pretreatment is that sections easily detach from the slides. This problem can be solved by coating the glass slides with gelatin or poly-L-lysine.

It should be stressed that protease pretreatment should be applied only when tissues have been fixed in formalin (paraformaldehyde). Enzyme treatment of sections fixed in coagulative fixatives usually causes total "digestion" of the section.

Finally, it is important to relate the time of enzyme treatment to the fixation time. Too long an enzyme treatment on sections fixed for a short period of time may give weaker staining reactions, due to digestion of the tissue material. Too short an enzyme treatment of tissues fixed for longer periods of time will give weak reaction due to insufficient demasking of antigenic determinants.

26.6.2 Embedding

Many immunohistochemical reactions are performed on paraffin embedded material. All embedding procedures decrease the sensitivity of the immunohistochemical reactions, and some reactions cannot be performed on this type of material. Alcohol dehydration and in particular heating during the paraffin embedding, can denature the tissue antigens. Careful control of the paraffin temperature and the use of low melting point paraffin is therefore advisable.

26.6.3 Cryostat Sections

Many immunohistochemical reactions can be performed on lightly fixed or unfixed cryostat sections. It is important that the tissue is cryoprotected before freezing (e.g. with 20% sucrose). The freezing should be performed in a CO_2-expansion cooler or in acetone cooled by solid CO_2. The sections (5–20 μm) must be placed on clean glass slides. Incubation of free-floating sections increases the sensitivity of the immunoreaction, but they need to be around 40 μm thick before they can be effectively handled.

Following cryostat sectioning, specimens are dried for 12–24 hours. This makes them stick to the gelatin layer on the slide.

26.6.4 Cytological Material

Cytological material such as cell smears, imprints, and cytocentrifuge specimens can be stained by immunohistochemical staining methods in the same way as tissue sections.

Following preparation of the cell smears; the slides are air- dried for 2–18 hours and fixed in acetone or mixtures of acetone, methanol, and formalin. Air-dried slides may be stored for several days at room temperature before staining. If longer periods of storage are required, the slides should be wrapped in aluminium foil and stored at -20°C. Following fixation, sections are hydrated in buffer for 5 min before staining.

26.7 Quantitation

Quantitation at the light microscopical level is further discussed in 28.8.8.

Quantitative work at the electron microscopical level has been made possible by colloidal gold labelled antibody techniques. These methods are described in 27.3.2.

27 Ultrastructural Cytochemistry and Immunocytochemistry

B. van Deurs. M. Møller, O.W. Petersen

A number of histochemical reactions can, with certain modifications, be usefully applied at the ultrastructural level. The value of this approach is that chemical or biochemical information can be related to cell ultrastructure. Although it is not in the general scope of this book to cover all reaction types in ultrastructural cytochemistry a few examples will be mentioned below (Sect.27.2) while ultrastructural immunocytochemistry will be covered in more detail (Sect.27.3).

27.1 Problems in Ultrastructural Cytochemistry

More detailed expositions on the issues raised below may be found in Weakley (1981) and Glauert (1977).

27.1.1 Embedding

Routine electron microscopy (EM) depends on the preparation of suitably thin sections (20–80 nm). To achieve this, it is necessary to embed the specimen in plastic (e.g. Epon; Sect.14.4.2) before cutting. Since many histochemical reactions cannot be performed on sections of material embedded in hydrophobic plastic resins, EM histochemistry is often carried out *before* embedding and sectioning the sample. In some cases, however, it may be valuable to embed in a hydrophilic plastic resin (Sect.14.4.2).

27.1.2 Background Staining

Another problem inherent in EM cytochemistry is that all "stains" or "reaction products" have to be *electron dense*. Hence, they either contain or have strong affinity for heavy metals such as lead, osmium, and uranium. This often leads to problems with non-specific background staining.

H. Lyon (Ed.)
Theory and Strategy in Histochemistry

27.1.3 Fixation

Fixation is very important in EM cytochemistry. On the one hand it is necessary to use a fixative which adequately preserves structural details while on the other hand it is obvious that excessive fixation with glutaraldehyde, for example, precludes many histochemical reactions, particularly those based on enzyme activity.

Glutaraldehyde, sometimes in combination with formaldehyde, is the principal fixative for EM. Formaldehyde alone is generally not considered adequate for the preservation of fine ultrastructural details. The addition of as little as 0.1% glutaraldehyde to a formaldehyde fixative (e.g. 2% formaldehyde) gives satisfactory structural preservation. For checking the influence of glutaraldehyde on a cytochemical reaction, sections of a well-defined thickness (e.g. 25–50 μm) are prepared by "tissue chopping" (Chap.10) or by cryostat sectioning (Sect.11.3.1). The sections are then exposed to glutaraldehyde for various periods of time and subsequently subjected to the enzymatic reaction and further processing for EM observation. In this connection it is appropriate to compare the degree of ultrastructural preservation with the degree of enzyme activity (the evaluation of both will probably be highly subjective), since an "inverse relationship" is often present.

An alternative approach is to "stabilize" a tissue section with formaldehyde (or cut a cryosection), carry out the cytochemical reaction (as briefly as possible), then postfix the specimen with glutaraldehyde and complete processing for EM. This approach is widely used in enzyme cytochemistry as it gives the most reliable results.

27.1.4 Diffusion of Reagents

To make diffusion of the cytochemical reagents into the samples as efficient as possible it is essential that the thickness be kept to a minimum. It should be stressed that much thinner sections can be produced with an ordinary cryostat than by using a tissue sectioner ("Tissue Chopper") or a Vibratome (Chap.10). Nonetheless, some EM cytochemical reactions (for instance, various peroxidase reactions) are readily performed on 50–100 μm chopper or vibratome sections.

If the substrate molecules are very big (e.g. $>$ 100 kDa) even 50 μm sections present considerable diffusion problems. In order to produce a uniform reaction throughout the section, it is important to reduce the thickness. Such problems are pronounced in immunocytochemistry (Sect.27.3) where high molecular weight antibodies are used. Penetration or diffusion into a sample can be facilitated by adding membrane perturbing agents (e.g. Triton-X, DMSO, or saponin) to the incubation medium. Due to the disruption of cell membranes during the freezing process, penetration of molecules takes place more readily into cryostat sections than into vibratome sections; the morphological preservation is, however, often poor in frozen sections. It is hoped that the introduction of *ultracryotomy* will provide solutions to several of these problems.

Quantitative studies in ultrastructural immunocytochemistry are extremely difficult. Enzymatic reactions may vary so much between experiments that it is impossible to achieve any degree of reproducibility. Generally speaking, therefore, ultrastructural cytochemistry is a qualitative discipline although, under some circumstances, quantitative work is possible. In any event, if results are interpreted with due care and appropriate controls included, these techniques can yield very important information which could not have been obtained even with the most advanced techniques in light microscopy.

27.2 Some Major Reaction Types

Additional details and reaction types are given in Weakley (1981) and Glauert (1977).

27.2.1 Carbohydrates

Carbohydrates on the cell surface with terminal sialic acid (sialoglycoproteins, etc.) can be demonstrated with a number of cationic "dyes" such as Ruthenium Red, Alcian Blue, and colloidal iron. Excellent ultrastructural localization of anionic sites on the cell surface can be obtained with cationized ferritin, a polycation of around 500 kDa, which due to its iron core, can be located directly by EM.

Various lectins (Sect.22.4) conjugated with gold or horseradish peroxidase are useful for demonstrating carbohydrate components on the cell surface. For instance, concanavalin A binds to mannose, and Ricinus communis lectin, a toxic 60 kDa protein from castor beans, binds to terminal galactose. The specificity of binding can be checked by preincubating with unlabelled lectin or the sugar residue to which it binds (e.g. 0.1 mol/l lactose in the case of Ricinus communis lectin).

The PASM method (Sect.9.2.2) for detecting 1,2-glycol groups by light microscopy can also be modified for EM work. Following oxidation with periodic acid the free aldehyde groups reduce silver methenamine to metallic silver. This produces a deposit of electron dense silver grains at the reaction sites and can therefore be used to detect various carbohydrate-containing molecules such as intracellular glycoproteins in the Golgi complex, in secretory vesicles, and in lysosomes (lysosomal hydrolases are glycoproteins). The reaction is performed directly on grid-supported ultrathin sections and usually produces a great deal of non-specific staining. This can be reduced by treating the sections with chromic acid between the periodic acid and the silver methenamine steps. Chromic acid acts by further oxidizing some of the aldehyde groups generated by the periodic acid step to carboxyl groups.

27.2.2 Receptors

It is also possible to detect more specific components of the plasma membrane, such as receptors for polypeptide hormones and growth factors. This can be done by making conjugates of the hormone with native ferritin, colloidal gold, or horseradish peroxidase. Controls for binding specificity are carried out as described in the preceding paragraph. When using ligand conjugates one should be aware that the conjugation may change the behaviour of the native ligand. In many cases the accuracy of a ligand conjugate can be evaluated by using ultrastructural immunocytochemistry (Sect.27.3).

27.2.3 Enzyme Reactions

The principles for these reactions have been discussed in detail in Chap.24 (Hydrolases) and Chap.25 (Oxidoreductases). For further references see Glauert (1977).

In studies of endocytosis and intracellular transport *exogenous peroxidases* are frequently used as *markers* or *tracers*. The principle is that the cells are exposed to peroxidase molecules for various periods of time and, following fixation (for which glutaraldehyde can be used), the specimens are incubated in media containing H_2O_2 and 3,3′-diaminobenzidine (DAB) and finally postfixed in OsO_4. The electron dense reaction product is readily visible in the EM, and it is therefore possible to obtain very detailed information about the pathways and compartments involved in intracellular transport of an endocytosed protein. Exogenous peroxidases are also used as tracers in studies of capillary (or epithelial) permeability. For this purpose a series of peroxidases with increasing molecular weights are available:

microperoxidase	1.9 kDa
cytochrome c	13 kDa
myoglobin	17.8 kDa
horseradish peroxidase	40 kDa
lactoperoxidase	82 kDa
myeloperoxidase	160 kDa
catalase	240 kDa

27.3 Immunocytochemistry

Ultrastructural immunocytochemistry, i.e., immunocytochemical labelling or staining techniques applied to material processed for examination in the electron microscope, is potentially a very powerful technique in cell and molecular biology. It should enable investigators to determine the precise subcellular location(s) of a particular antigen and to estimate the amount of that antigen in a specific compartment. In order to attain this idealized goal, many factors have to be considered and technical problems solved.

While the detection of antigens on the free surface of samples such as cells in monolayer cultures or single cells in suspension is fairly straightforward and can even be performed on living cells (at 4°C), detection of antigens in tissue samples such as biopsies, particularly intracellular antigens, is technically demanding.

Techniques for the ultrastructural detection of *intracellular* antigens can be subdivided into *preembedding* and *postembedding* techniques. In the first case, immunocytochemistry is performed on pieces of tissue, or on cells in culture, *before* the specimens are embedded and further processed for electron microscopy. In the second case, tissue samples or cells (cell pellets) are first embedded, ultrathin sections cut, and the immunocytochemical reactions are then done directly on the thin sections.

The "label" used in ultrastructural immunocytochemistry must be electron dense. Two types of label are of interest, enzymes and metal formulations. The enzyme peroxidase (horseradish peroxidase, HRP) is used extensively (Kuhlmann, 1977). When DAB is used as substrate, the polymerized oxidized diaminobenzidine, enhanced by treatment with OsO_4, provides an electron dense reaction product which is readily observed by EM, provided the amount of label (HRP) exceeds a critical level. The "metal" labels include ferritin and colloidal gold, the latter being most popular at present. Gold particles are normally used in sizes from around 3 nm up to 12–15 nm (larger sizes should not be used). In general, the larger the size, the more homogeneous the preparation with respect to size, the easier it is to see at low magnifications by EM. Either a (secondary) antibody directed towards the primary antibody or, most often, protein A (a glycoprotein extracted from the capsule of the bacterium *Staphylococcus aureus*) are adsorbed to the gold particles. In the case of protein A-gold, the probe is referred to as PAG. Colloidal gold particles of different sizes can be used for double labelling experiments (see below).

27.3.1 Preembedding Immunocytochemistry

Since the immunocytochemical reaction has to be performed on whole cells and organelles, the membranes need to be permeabilized. This can be achieved by fixation in a hypotonic fixative containing saponin (about 0.05% w/v) and/or a subsequent treatment with saponin in hypotonic buffer. Even though saponin alone must be assumed to generate "holes" in the membranes, the use of a hypotonic fixative/buffer seems necessary in most cases for optimal permeabilization. Following incubation of the permeabilized specimens (small tissue blocks, cell pellets, or cells in monolayer culture) with the primary antibody or the Fab fragment thereof, the specimens are carefully washed. Thereafter they are incubated with a secondary antibody (e.g., rabbit anti-mouse, if the primary antibody was a mouse anti-human, etc.) conjugated to HRP (or another marker). After further washing the specimens are processed for peroxidase activity ($DAB-H_2O_2$ incubation) and then prepared for electron microscopy (embedding in Epon, sectioning, and contrasting).

Excellent results can be obtained with this technique. The main advantage is that, once embedded in Epon, the block can be used again and again for making numerous sections, even to the extent that serial sections can be processed for three-dimensional reconstructions. The disadvantages of the technique are:

1. A gradient of immunolabelling is obtained from the exposed surface of the specimen (from the cell surface for cultured cells) to the centre due to the decreasing degree of penetration of antibodies
2. Various intracellular compartments may be permeabilized and rendered accessible to antibodies to differing degrees. These factors make it critical to consider for each section its original location within the tissue block, particularly when any form of comparative study is proposed
3. Since HRP is the most commonly used label in preembedding techniques, quantitation cannot be performed. It should be stressed that the amount or density of the HRP reaction product cannot in any way be used to interpret the observations in quantitative terms. Even when using a particulate marker (e.g., ferritin) quantitation is very difficult due to the penetration problems mentioned above (points 1 and 2)

27.3.2 Postembedding Immunocytochemistry and Cryoimmunocytochemistry

Before exposure to immunological reagents, the cells or tissue are embedded in either Epon^{R}, Lowicryl^{R}, LR White^{R}, or other hydrophilic resins, or ice (by freezing) for sectioning. Because the sections are very thin, most of the intracellular compartments are "opened" as far as exposure of antigens to antibodies is concerned (cf. permeabilization in the *preembedding procedures*). Typically the sections are collected directly on grids for electron microscopy and then treated with antibodies, etc.

In general, immunocytochemistry on sections of *Epon-embedded* specimens, performed after etching the section surface in order to remove the hydrophobic plastic around the antigens, is not considered a satisfactory technique although satisfying and convincing results have sometimes been obtained especially when large amounts of antigen are present in the sections.

Embedding of specimens in *hydrophilic resins* such as *Lowicryl* (or LR White) is often used for postembedding immunocytochemistry and, in general, good or even excellent results can be obtained, depending on the amount of antigen and the (sub-)cellular structure to be studied. These hydrophilic plastics destroy less antigens during the embedding procedure and probably allow a certain degree of penetration of the antibodies. However, the sensitivity of the immunoreaction is decreased as compared to the ultracryosections described below.

In principle, the best approach involving the least perturbation is to freeze the cells or tissue, to cut *frozen ultrathin sections* and use the thawed cryosections for immunocytochemistry while they are hydrated. A brief description of the principles of applying immunocytochemical techniques to ultracryosections in ultrastructural work is given below under the following headings:

a. Fixation and cryoprotection
b. Freezing
c. Sectioning
d. Immunocytochemistry
e. Quantitation

Fixation and Cryoprotection. Small samples of tissue, or cell pellets from cell cultures are used. The fixative can be formaldehyde (1–8%) or formaldehyde-glutaraldehyde mixtures (with 0.1–2.5% of the latter) in a "standard" buffer, depending on the antigen to be localized. In some cases the use of more specialized fixatives may be warranted. The effect of various fixatives (as well as fixation time and temperature) on antigenicity should be examined by light microscopical immunocytochemistry (immunoperoxidase or immunofluorescence). Following fixation and washing, the specimens are cryoprotected by incubation in 1.8–2.3 mol/l sucrose for at least 15 min depending on the size of the specimens. Several changes of sucrose are recommended.

Freezing. The cryoprotected specimens are mounted on special metal stubs (silver or copper), the size of which depend on the *cryoultramicrotome* to be used, and frozen in liquid nitrogen. When frozen on the stubs specimens can be stored indefinitely in liquid nitrogen-containers, and be used for sectioning as required. In this way, in principle, a "library" of specimen samples can be obtained. This can be used for controls, comparisons, etc., just as is the case with Epon-embedded material for example.

Sectioning. For cutting ultrathin cryo-sections a special *cryoultramicrotome* has to be used. Several models are commercially available and they all seem to be useful, selection depending on personal preference and experience.

In most cases a glass knife is used for sectioning. The glass knife is made on a special machine ("Knifemaker", LKB) which differs from the "Knifemaker" designed for "routine" (plastic) sectioning in several significant respects.

The preparation of a good knife is probably the most important single step of the ultracryosection immunolabelling procedure. It takes priority over the titer, purity, and specificity of the antibody. If the knives are coated with a thin layer of phosphotungstate, they cut better sections and will stay sharp for several months.

Diamond knives manufactured specifically for ultracryosectioning are available and may be an appropriate alternative to the demanding preparation of good glass knives.

The frozen sections can be removed from the knife edge and sorted (under a microscope) with an eyelash brush for example. The best sections (those having a golden-to-purple interference colour on the knife edge) are grouped closely together and collected in a drop of 2.3 mol/l sucrose using a small loop (e.g. platinum). The diameter of the loop should be smaller than that of the grid (less than 3 mm) to make it easy to centre the droplet on the grid. When the loop and the sections are removed from the cryochamber of the microtome, the sections thaw and must therefore be kept hydrated for the rest of the procedure

Immunocytochemistry. The grids bearing sections are placed on droplets of buffer (e.g. phosphate buffered saline—PBS) with the sections downward, and kept there until enough grids have been made. Thereafter the grids are transferred from droplet to droplet with fine forceps or with a loop as exemplified below for a typical incubation routine (principal steps):

1. PBS with glycine
2. PBS with foetal calf serum (FCS)
3. PBS/FCS with first (primary) antibody
4. PBS/FCS
5. PBS/FCS with PAG (protein A-gold)
6. PBS
7. H_2O
8. Methyl cellulose with uranyl acetate (on ice)

In step (8) the sections are finally "embedded" in methyl cellulose (Methocel) and contrasted at the same time. After about 10 min the grids are removed from the Methocel droplet using a loop, and excess Methocel is removed with filter paper so that only a thin "film" is left, containing the grid. The interference colour of the Methocel film should be gold to pink. More detail and variations in the procedure may be found in for instance Polak and Varndell (1984), p.71.

Double labelling is performed in the same way. After step 5, the sections are carefully washed and the grids then transferred to a droplet of protein A to prevent non-specific binding of the next antibody. The second primary antibody is then applied, the section washed, and PAG (gold particle size different from the first PAG) applied, followed by washing, etc. In double labellings the two gold sizes should be well separated, e.g., 5 and 10 nm, or 6 and 12 nm

Quantitation. While in principle the immunogold labelling allows quantitation of the relative or even absolute amount of antigen, in practice this is rarely possible. The major problems are associated with the low labelling efficiency of the technique. In order to obtain a 100% labelling efficiency, that is, 100 antigen sites give 100 gold particles in a particular section, there must be complete penetration of antibody and PAG throughout the section, and full accessibility to the antigen sites. This is normally not the case. The depth of penetration of the "immunolabelling" is unknown but published data indicate that it is very poor with labelling efficiencies of the order of a few per cent. In consequence, the antigen density (e.g., a membrane-associated molecule per μm^2 of membrane) has to be rather high to provide significant gold labelling (above the background level).

To make calculations of the total amount of a certain antigen in various intracellular compartments, at least the following should be available: a standard reference system giving a linear relation between the number of gold particles and the number of antigen molecules in the same type of section, or biochemical data on the absolute amount of antigen within the cell, as well as stereological data on the surface area and/or volume of the respective intracellular compartments. Even then, assuming that the labelling efficiency is largely the same for all the compartments of interest, counting of gold particles over the various compartments and calculations of the theoretical "amount" of gold particles per compartment per cell should be interpreted with care and can only lead to a rough estimate of the relative intracellular distribution of the antigen. On the other hand, the immunogold labelling approach may be the only technique by which an estimate of the amount of a particular protein in various intracellular compartments may be obtained. This estimate, despite its limitations, may still be of considerable significance to cell and molecular biology. In addition to these considerations it is clear that a number of routine controls should be included (and quantitated), for example, non-specific gold particles over structures which are known not to contain the antigen, number of gold particles over the structures of interest following incubation protocols where the primary antibody has been omitted, etc.

28 Quantitation in Histochemistry

P.E. Høyer, L. Kayser, M.R. Barer, H. Lyon

The purpose of this chapter is to provide an introduction to the techniques used in quantitative work and to cite key examples of staining procedures for which quantitation has been attempted. More detailed accounts of the instrumentation can be found in Pearse (1980), pp.384–395. Autoradiography is considered in Chap.29.

Quantitative histochemistry is primarily concerned with determining the amount and distribution of histochemical reaction products. Most of the physical methods applied in attempting to achieve this fall within the discipline of quantitative microscopy. Although a full discussion of this topic is beyond the scope of this text it should be appreciated that several techniques can yield quantitative information about the chemical composition of cells and tissues without recourse to staining; for example, cytochrome P450 (Altman *et al.*, 1975; Gooding *et al.*, 1978; Chayen *et al.*, 1979) and haemoglobin (Morselt, 1978) may be measured directly.

Quantitative methods in histochemistry may also serve to provide insights into the mechanism of staining and confirmation of the quantitative nature of the staining reaction. These aims may be realized by studying the reactions in control sections and model staining systems as well as by comparing histochemical assay results with those obtained by conventional biochemical procedures.

While this account concentrates on techniques applied to sections or single cells, it is important to appreciate the existence and value of other quantitative approaches to the study of biochemistry at the microscopic level such as flow cytometry and X-ray microanalysis.

The approaches to quantitation discussed in this chapter are surveyed in Table 28.1. They will be reviewed in turn (Sects.28.2–28.7) after some general considerations (Sect.28.1).

Table 28.1. Classification of quantitative methods in histochemistry.

Quality determined	Term
Ratio of intensities of incident and transmitted light	Absorption photometry (28.2)
Intensity of light emitted from an excited fluorophore	Fluorimetry (28.3)
Ratio of intensities of selectively reflected and incident light	Reflection contrast photometry (28.4)
Retardation of light in the section	Interferometry (28.5)
Transmittance, absorbance, or reflectance spectra	Spectral analysis (28.6)

H. Lyon (Ed.)
Theory and Strategy in Histochemistry

In order to correlate structure and chemical composition, it is clearly necessary to determine the dimensions of different tissue components and reaction products. This topic, *morphometry*, is concerned with determining the *dimensions* of individual cellular and histological structures or enumeration of granular reaction products such as gold particles or silver grains. It is relevant to quantitative histochemistry insofar as the approach allows for quantitation of the targets of histochemical demonstrations within a defined structure. Conversely, the choice of staining technique clearly influences the definition of structure. A short overview of morphometry and the related subject, *stereology*, is given in Appendix B.

28.1 General Considerations

28.1.1 Prerequisites

Ideally all histochemical techniques should be sensitive, precise, specific, reproducible, and valid. As these ideals are not achievable, sources of error should be defined. A clear view of these terms underpins the more specific considerations applicable to individual techniques.

A. A procedure is said to be *sensitive* if its sensitivity (Sect. 1.2.3) is close to or equal to that achievable using other *in vitro* methods for determining the concentration of the substance to be investigated

B. The results of a *precise* technique give an end product distribution (Sect. 1.2.4) that closely corresponds to the *in vivo* distribution of the investigated compound. There are three important prerequisites for precision:
   1. The tissue morphology must be the best achievable
   2. Separate experiments using different methods should confirm that the specific end product is only deposited in structures which contain the compound under investigation
   3. None of the essential components of the staining reaction (final or intermediate) should diffuse from the primary reaction site and/or bind nonspecifically to other tissue components

C. A technique is *specific* if all end product results from the presence of the investigated compound. If interfering compounds are present their contribution to the end product should be small and quantifiable

D. A technique is *reproducible* if it gives the same result when used by different workers and when it is applied on several different occasions. Clearly, the protocols used must be identical and the biological material used should be highly comparable

E. Finally, a procedure is *valid* if the amount of end product formed per unit volume in a given region of the tissue is related in a known manner to the concentration of the compound in question in that locality. The reactivity of the compound should not be inhibited (for instance by fixation) or masked (for instance by proteins), and the compound should not be lost from the sections or cells (for instance by diffusion)

28.1.2 Errors

It is convenient to distinguish three sources of error. These are associated with:

A. The specimen (sampling or processing).

B. The principles of the histochemical procedure adopted and its performance

C. The instrument itself or its inappropriate use

A and B have already been addressed to some extent in the appropriate chapters; Some additional notes on these and the third source of errors are provided later in this chapter.

A specimen error which may influence both validity and reproducibility may occur when cryosections of fresh frozen tissue are used (for instance in quantitative enzyme histochemistry and immunohistochemistry). In these cases a constant and reproducible speed of section cutting is critical (Sect.11.3.1; see also Butcher, 1971). Table 28.2 clearly shows that even if the micrometer screw is adjusted to a certain μm value a high cutting speed will give thinner and a low cutting speed thicker sections.

Table 28.2. The effect of cutting speed on section dimensions. Liver tissue cut with the micrometer screw adjusted to 12 µm (from Altman, 1980; reproduced by courtesy of the Royal Microscopical Society, Oxford).

Cutting speed (sections/min)	Area of section (cm^2)	Volume of section (mm^3)	Section thickness (µm)
10	0.114	0.174	15.3
25	0.146	0.171	11.7
50	0.178	0.174	9.8

This variation also depends on the type of tissue. For quantitative work microtomes should be set to give sections that fall within the range yielding a linear relationship between absorbance and section thickness (Fig. 28.1).

28.2 Absorption Photometry

Absorption photometry can be used for the quantitation of all histochemical reaction products that absorb light (ultraviolet to infrared) including native reaction products such as NAD^+. The basis of absorption cytophotometry is measurement of light intensity in incident and transmitted light at a particular wavelength using a beam of light directed at portions of the sample containing the reaction product. The usefulness of absorption photometry depends on whether the chromophore

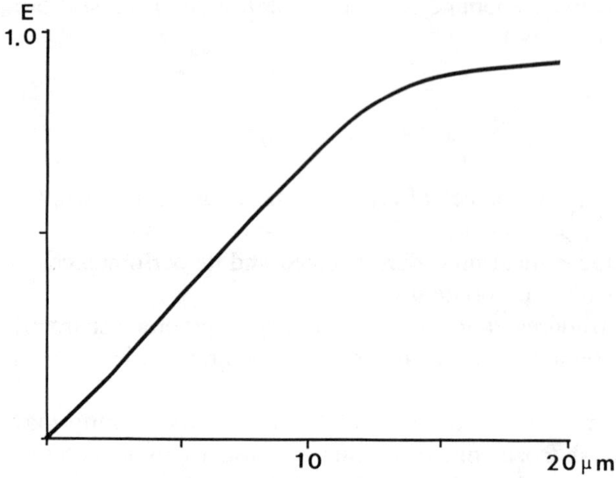

Fig. 28.1. The dependence of extinction (E) on section thickness (μm).

conforms to the Beer-Lambert law (cf. Sect.3.3.2):

$$E = \log \frac{I_o}{I} = \epsilon \times C \times d \quad (28.1)$$

E = absorbance; ϵ = molar extinction coefficient; C = concentration of solute in mol/l; d = light path in cm; I = intensity of light passing through the section; I_o = intensity of light reaching the section

Equation 28.1 can be transformed by introducing the mass, $M = CV$ (volume = V = Ad, where A = surface area) and by substitution this gives:

$$\begin{aligned} &E = \epsilon \times M/Ad \times d = \epsilon \times M/A \quad \text{or} \\ &M = EA/\epsilon \end{aligned} \quad (28.2)$$

As the area over which measurements are made can be defined and ϵ is known for several final reaction products in tissue, it is possible to express absorbance values in units of mass.

28.2.1 Absorption Cytophotometry

Absorption cytophotometry involves directing light of a specific wavelength via the condenser system of the microscope to areas in the specimen that contain the light absorbing compound. The wavelength of the incident light is selected according to the absorption spectrum of the chemical substance to be measured. The point at which absorption is at a maximum (λ_{max}) is usually chosen. The transmitted light passes through the microscope to a photomultiplier. This consists of an evacuated glass sphere, the inner surface of which is coated with a metal film (e.g. selenium). In response to light, this coat liberates electrons which are attracted to a positively charged metal plate inside the glass sphere. This photoelectric effect results in a current that is directly proportional to the intensity of the incident light.

As the precipitated chromophore is not uniformly distributed over the area, A (equation 28.2), it is necessary to divide the area into a large number of small regions of area, a. Otherwise, a very serious error may occur, distributional error (Sect.28.2.2). In some commercial instruments the smallest area that can be measured equals the resolution of the light microscope (0.1–0.2 μm), so ensuring that the field measured is always optically homogeneous.

Most instruments make individual measurements on a very small area of the section (e.g. a circular area with a diameter of 0.2–0.5 μm; see also Sect.28.2.3). Measurements over larger areas are achieved by taking multiple readings, either by moving the section in small jerks between measurements (scanning stage), by moving the illuminated area in the microscope in a similar pattern (flying spot), or by image analysis (Sect.28.2.3).

The central measurement provided by these instruments is the mean integrated extinction (MIE) or as it is also called mean integrated absorbance (MIA). This is given as relative values and is effectively the sum of all the individual absorbance values obtained (corrected for any overlap between adjacent measurements) divided by the number of small regions measured:

$$E = \text{MIE} = \frac{e_1 + e_2 \ldots e_n}{n}$$

where e_1, e_2 ... are the individual absorbances measured. Substitution in equation 28.2 gives:

$$M = \frac{e_1 + e_2 \ldots e_n}{n} \times \frac{na}{\epsilon} = \frac{(\text{MIE})A}{\epsilon} \tag{28.3}$$

At this step MIE is expressed in arbitrary machine units. To make it possible to compare results obtained with different machines or with the machine under different circumstances, it is necessary to convert these relative values into absolute values.

This can be achieved by measuring the relative absorbance of a series of neutral density filters with known absorbance values, thus producing a calibration graph. For the final determination of mass, M, we now have to determine the total area, A. In the case of image analysis the instrument can be programmed to automatically measure A. With other instruments, A is defined by an illuminated "mask" which may be varied within very wide limits. If the area of the object to be measured completely fills out the mask, it is not absolutely necessary to determine A, as this equals the size of the mask; however, if this is not the case, the area of the object must be determined.

28.2.2 Errors in Absorption Cytophotometry

Absorption cytophotometry may be subject to errors stemming from:

1. Distributional error and residual distributional error
2. Glare and diffraction errors
3. Non-linearity of the instrument and out-of-range error
4. Chromatic error
5. Out-of-focus error
6. Departure from Beer's law
7. Departure from Bouguer-Lambert's law

Many of these errors are dealt with in detail by Goldstein (1981), van Noorden and Butcher (1986b), and van Noorden (1989).

Distributional Error and Residual Distributional Error. In principle, if a parallel beam of monochromatic light passes through a homogeneous solution, equation 28.1 should be valid. Histochemical reaction products are, however, generally distributed heterogeneously which means that the mean absorbance cannot simply be calculated from one determination of the transmittance of the entire microscopic field, as this would give rise to the so-called distributional error. It is therefore, as mentioned in Sect.28.2.1, necessary to divide the area into a large number of small regions of area and to calculate the mean integrated absorbance (MIE) of all these regions. An example may clarify why this is necessary.

If the histochemical reaction product is evenly distributed throughout the microscope field and transmits 50% of the incident light, then according to equation 28.1:

$$E = \log \frac{100}{50} = 0.301$$

If, however, the same amount of reaction product is only distributed over half of the microscope field then the absorbance in this part of the field would be $0.301 \times 2 = 0.602$. The corresponding transmission (I) may be calculated from equation 28.1:

$$0.601 = \log \frac{100}{I} \quad \text{or} \quad I = 25\%$$

The amount of light transmitted through the entire microscope field would therefore be:

$$\frac{100 + 25}{2} = 62.5\% \text{ corresponding to } E = \log \frac{100}{62.5} = 0.204$$

instead of 0.301.

In Table 28.3 some further examples are given of the distributional errors arising when absorbance of a heterogeneously distributed final reaction product is determined by simple photometry without scanning.

From Table 28.3 it appears that the more unevenly the product is distributed and the higher the true absorbance of the microscope field is, the more serious will be the distributional error using simple photometry.

As shown in Table 28.3 the resultant error (distributional error) can be very substantial. This problem is approached by using a measuring spot that is sufficiently small to make the area illuminated for one measurement effectively homogeneous

Table 28.3. Distributional error arising due to heterogeneously distributed final reaction product when absorbance is determined by simple photometry (without scanning). Amount of incident light (I_0) is taken to be 100%.

True absorbance of microscope field	Amount of transmitted light (I%)	Absorbance determined by simple photometry	Distributional error in %
0.1	79.4	0.100	0.0
0.1	79.4	0.089	11.0
0.1	79.4	0.041	59.0
0.2	63.1	0.200	0.0
0.2	63.1	0.156	22.0
0.2	63.1	0.045	77.5
0.3	50.0	0.300	0.0
0.3	50.0	0.204	32.0
0.3	50.0	0.046	84.7

with respect to the distribution of measured product. With small heterogeneously distributed granules it may therefore be necessary to use a scanning spot as small as 0.1–0.2 μm in diameter.

Although scanning effectively reduces the magnitude of the error, the remaining or residual distributional error may still be important. It is dependent on the shape and absorbance of the object and on the size of the smallest step of the scanning raster in relation to the object size. The largest corrections for this error are to be expected when measuring small, intensely stained objects differing considerably in size and shape. Duijndam *et al.* (1980b) derived theoretical methods for correcting the integrated apparent absorbance values for the systematic errors due to residual distributional error, diffraction error, and glare error when using a scanning stage microspectrophotometer. Duijndam *et al.* (1980a) in a later paper tested their theoretical methods in practice and found a good correlation between theory and practice.

Glare and Diffraction Errors. Glare error (stray light) is caused by optical imperfections (dirt, etc.), particularly in the substage optics. It can be minimized by ensuring that all optical parts are scrupulously clean, by eliminating surfaces (oil instead of a dry objective), and by closing the field diaphragm appropriately.

Diffraction error on the other hand is not due to the optics of the system but is due to the wave characteristics of light (Duijndam *et al.*, 1980b). This error is dependent on the shape and absorbance of the object and on the step size to object size ratio and can be quite substantial for small and intensely stained objects.

Duijndam *et al.* (1980b) mention three ways of minimizing the effects of residual distributional error, glare error, and diffraction error in their conclusion. All these errors tend to give lower mean integrated apparent absorbance values as compared to the true absorbance values. The remedies suggested are:

1. To incorporate corrections for the errors in a computer scanning program

2. To use a confocal scanning microscope with a laser beam as the light source. Due to its coincident illuminating and measuring diaphragms, this instrument is essentially free from glare and diffraction errors
3. To obtain corrected results directly by using the two-wavelength scanning procedure program (Bicoscan) developed by van der Ploeg *et al.* (1979) employing the principles of two-wavelength scanning developed by Patau (1952) and Ornstein (1952)

Non-Linearity of the Instrument and Out-Of-Range Error. Microspectrophotometric measurements can exhibit systematic errors caused by non-linearity in the photodetectors, the associated electronics, or the mechanical design of the instrument. With modern instruments these errors should normally not be of any significance, but their possible presence should always be checked for by measuring the absorbance values of neutral density filters of known value.

Out-of-range error is potentially a much greater source of error and can occur due to failure to recognize the inbuilt instrumental cut-off in linear response at high absorbance values (van Noorden and Butcher, 1986b). It should be realized that even when the mean integrated absorbance is lower than the cut-off value there may be some small regions with absorbance values greater than the cut-off value resulting in an underestimate of the true absorbance value. For example, the apparent mean integrated absorbance should therefore not normally exceed about 0.7 when using the Vickers M85 microspectrophotometer (Chayen, 1980).

Chromatic Error. Chromatic error may occur if a wide spectral band is used for the measurements. This error will be most pronounced when the chromophore has a narrow absorption band. Chromatic error may consequently be reduced by using either an interference filter, a filter monochromator, a prism monochromator, or a grating monochromator with narrow spectral bandwidths.

Out-Of-Focus Error. Out-of-focus errror has been studied by Bitensky (1980) who found that, using a 100x oil immersion objective, a scanning spot of 0.2 μm, and a 10 μm thick section, the greatest error due to defocusing by ± 4 μm was $\pm 10\%$. This error apparently diminished with increasing intensity of the reaction product.

Departure from Beer's Law. Beer's law is fulfilled if absorbance is proportional to concentration of the chromophore. Although Beer's law applies to many of the dyes and coloured compounds that are used for, or arise during, histochemical reactions, there are many exceptions. For example, the law is not obeyed if intermolecular association occurs as in metachromasia; nor does it apply if the chromophore is present in high concentration. It is therefore essential to check that the relationship holds for each compound to be measured. This can be done by eluting the reaction product from the specimen and measuring the concentration of a sufficiently diluted solution with a spectrophotometer.

Departure from the Bouguer-Lambert Law. The Bouguer-Lambert law is fulfilled if absorbance is proportional to path length. As shown in Fig. 28.1, a plot of absorbance versus section thickness is initially a straight line which levels off at a certain critical thickness. Levelling-off may be due to out-of-range error (see above) or a particular problem encountered in enzyme histochemistry. For example, linearity is only retained up to a section thickness of 14 μm for acid phosphatase (Stoward, 1980) and up to about 16 μm for aminopeptidase (Felgenhauer and Glenner, 1966). Stoward (1980) has suggested that in order to maintain the validity of any quantitative enzyme histochemical technique, sections should not be thicker than about 12–14 μm. There are, however, also examples where linearity can only be maintained with even thinner sections: 10 μm for oestradiol 17β-dehydrogenase (Høyer, 1988), 8 μm for 20α-hydroxysteroid dehydrogenase (Robertson *et al.*, 1982b), and 5 μm for 3β-hydroxy-Δ^5-steroid dehydrogenase (Robertson *et al.*, 1982a).

The most likely explanation of this phenomenon is that linearity is limited by the rate at which substrate can diffuse into the section. At a certain critical section thickness the supply of substrate is no longer sufficient. Ideally, when the regression line (Fig. 28.1) is extrapolated towards zero, it should pass through the origin; however, it may deviate to either side. If the positive Y-axis is crossed, this suggests a non-specific reaction (Stoward, 1980), while, if the positive X-axis is crossed, this indicates a failure to overcome the critical supersaturation concentration of the final reaction product (van Duijn, 1974).

Extensive reviews of these and other errors in absorption cytophotometry may be found in Caspersson (1940), Glick *et al.* (1951), Swift and Rasch (1956), Chayen and Denby (1968), Atkin (1970), Mayall and Mendelsohn (1970), Piller (1977), Fukuda *et al.* (1978), Bitensky (1980), and Goldstein (1981).

28.2.3 Instruments for Absorption Cytophotometry

It is far beyond the scope of this chapter to describe these instruments in any detail. They can in principle be classified into three groups:

1. The *flying spot* principle. A stationary object is scanned by a moving light beam either in the image plane (Barr and Stroud GN5: smallest measuring spot 0.25 μm) or in the object plane (Vickers M85/85A/86 series: smallest measuring spot 0.2 μm)
2. The *scanning stage* principle. A moving object scans across a stationary measuring beam (Zeiss SMP 01; SMP 05; MPM series; Leitz MPV2: range of smallest measuring aperture in the specimen plane is 0.2–0.5 μm with a minimum step width of 0.25–0.5 μm)
3. The *image analysis* principle. The object is by means of a video camera converted into a television image where each point is measured by densitometry. (The minimum effective diameter of each image point is 0.1 μm). A video camera can act as a two-dimensional photometer. Two different principles of video cameras are used. These are:

a. Image tube based or solid-state video cameras, e.g. charge coupled devices (CCD-cameras)
b. slow scan solid-state cameras

Video cameras of the first kind produce images with a speed of 25 or 30 frames/second and can be used with conventional image analysis systems. The limitations of these video cameras are a limited range of linear relationship between input and output signals, geometric distortions, and a medium resolution (512×512 pixels). The advantage is low cost compared to slow scan cameras. Slow scan cameras do not use television standards but produce images with a speed of 4–5 frames/second. The advantages of slow scan cameras are a high resolution (greater than 1000×1300 pixels), a large range of linearity, and lack of geometric distorsion. The disadvantages are high cost and the need for special computer and image analysis systems. It is, however, to be anticipated that as resolution for the slow scan cameras further increases and their cost decreases, their use will become more and more widespread. This is all the more probable when the high speed of analysis together with the possibility for combining the analysis with morphometric measurements are taken into consideration.

The use of video cameras in microscopy is described in detail by Inoue (1986) and the use of solid state imagers by Aikens *et al.* (1989). Because of ongoing changes in the supply of commercial instrument systems, no specific instruments will be mentioned.

28.2.4 Other Absorbance Based Methods in Quantitative Cytochemistry

Two further quantitative techniques are relevant at this juncture, staining in artificial gel systems and quantitation of staining by elution combined with conventional absorption photometry. Both techniques are valuable for investigating quantitative aspects of staining processes.

Staining in Artificial Gels. This is achieved by incorporating the material to be stained into a gel (e.g. agarose), fixing, staining, and finally quantitating the coloured end product using a conventional spectrophotometer with a gel-scanning attachment. The approach is particularly suitable for investigating the mechanism, stoichiometry, and specificity of staining processes. Its main drawback is that the molecules stained are not in the same state or context as they would be in a histological or cytological specimen.

Stain-Elution. This was in fact one of the first methods applied to quantitation of histochemical reactions. The preparation is stained in the usual manner and may be inspected to determine the distribution of stain on a qualitative basis. Elution of the dye or coloured end product is then achieved using a known volume of a suitable solvent and absorbance determined in the extract. The amount of primary reactant present in the section as a whole can then be inferred taking into account many of the considerations discussed in 28.1. Clearly, great care must be taken to ensure the comparability of sections subjected to this procedure.

Recently the application of the stain-elution technique to cells cultured *in vitro* in *microtiter* plates has raised some interesting possibilities (Barer *et al.*, 1986a; 1986b). Firstly the use of clonally derived cells and distribution of identical aliquots into wells of the microtiter plates provides high replicate numbers of an essentially identical biological substrate for staining (up to 5000 identical monolayers can easily be prepared at one sitting). Fixation, staining, and elution are all performed in the microtiter plate and procedures are therefore limited to reagents compatible with the plate plastic. Secondly the microtiter plate (an 8 by 12 array of approximately 300μl flat-bottomed wells) provides a very convenient format for studying the quantitative effects of varying most aspects of a staining procedure (fixation, staining time, reagent concentration, etc.) on the formation of cell associated end product. This aspect is greatly facilitated by the existence of dedicated multi-channel microtiter pipetting devices.

Finally, relatively inexpensive equipment is available to obtain absorbance readings of eluted stain in situ (ELISA or microplate readers); a standard plate reader would read 96 wells in under 60 seconds and a more sophisticated instrument could provide values at up to 5 different wavelengths simultaneously and transfer the readings to a computer for further analysis.

The microtiter plate system is a somewhat heretical form of cytochemistry in that it provides no indication of the subcellular location of staining (some information can be obtained by direct inspection using an inverted microscope). Its use is not confined to investigating quantitative aspects of staining processes, in fact it was developed in order to study the effect of biologically active substances on cells cultured *in vitro*. These effects are characterized in terms of changes in the integrated staining properties of cell monolayers (Barer et al 1986a; Barer, 1987).

28.3 Fluorimetry

When molecules emit light solely due to high temperature, they are said to emit incandescence. All other forms of light emission are called luminescence. The emission of light as a result of absorbed radiation is called photoluminescence. Fluorescence occurs when light is emitted following absorption of light from an outside source of energy. Fluorescence is a type of luminescence in which light is emitted from molecules 10^{-8} seconds or less after they have absorbed light. If the delay is about 10^{-6} seconds, it is termed delayed fluorescence, while a delay of greater than 10^{-6} seconds results in phosphorescence. In fluorescence the re-emitted light is of a longer wavelength than that absorbed. This shift in wavelength is called Stoke's shift (Fig. 28.2).

The fluorescence microscope differs from the light microscope in being equipped with two filter systems. There are also differences as regards the light source,

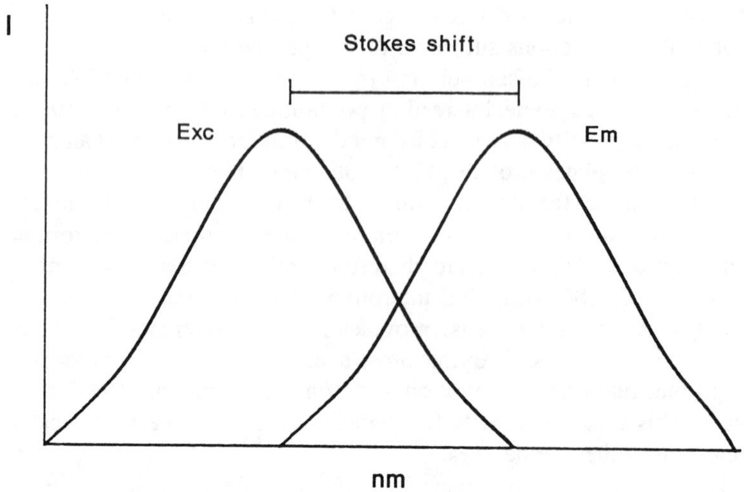

Fig. 28.2. Excitation (Exc) and emission (Em) spectra for a fluorochrome. I = relative intensity; abscissa = wavelength (nm)

illumination, and condenser systems as well as objective and ocular lenses. For details, see for instance Taylor and Salmon (1989).

The filter systems are excitation (or primary) and barrier (or secondary) filters. As excitation ideally should occur at a specific wavelength an interference filter is used as the excitation filter and is inserted between light source and specimen. To protect this filter, and also the specimen, it is necessary to position a heat protection filter between the light source and subsequent functional filters. Moreover, neutral density filters are often used for reduction of excitation light energy to avoid fading of fluorochromes. The barrier filter system is placed between specimen and detection system and should ideally remove all light with shorter wavelength than the light emitted from the specimen. Several different kinds of illumination may be used (e.g. full aperture and dark field). The most commonly used are, however, dia-illumination and epi-illumination.

The intensity of the emitted light depends on the numerical aperture of the objective and on the magnification as follows: $I_{\text{dia}} \approx NA^2M^{-2}$ and $I_{\text{epi}} \approx NA^4$ where I = intensity of emitted light in dia-illumination or epi-illumination; NA = numerical aperture of the objective; M = magnification.

There are a number of advantages of incident or epi-illumination. These include: a far greater light intensity than trans-illumination at objective magnifications of more than about 10× (cf. above), much less absorption of the emitted fluorescence (Ploem *et al.*, 1974), the possibility of using an inverted microscope, and finally the possibilities for combination with conventional trans-illumination methods such as phase contrast, polarization, and differential interference contrast.

In epi-illumination the objective functions as both condenser and objective lens. Light from the light source passes into the microscope at a right angle to the axis through objective, specimen, and detection system and having passed the excitation filter meets the chromatic beam splitter (dichroic mirror) placed at an angle of 45° to the light beam. This mirror has an interference coating that has a high reflectance at 45° for the wavelength transmitted by the excitation filter, but not for the longer wavelength passed by the barrier filter. Conversely, the dichroic mirror transmits the wavelength passed by the emission filter to the detection system, but not the wavelength passed by the excitation filter.

Fluorescence Measurements. Microfluorimetry is a technique for measuring the intensity of emitted fluorescence light. A linear relationship between the intensity of fluorescence and the amount of fluorescent molecules in the specimen only holds for low concentrations (Fig. 28.3). Van Oostveldt and Bauwens (1990) found that a linear relationship between integrated fluorescence intensity and integrated absorbance is only present if local absorbance values are kept below 0.1.

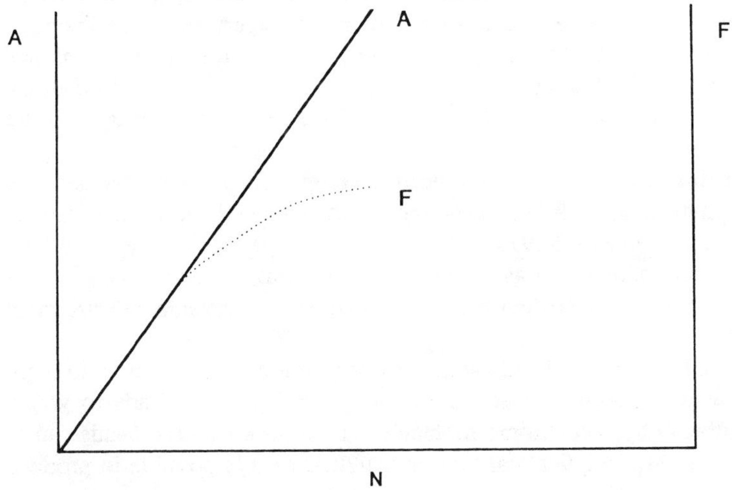

Fig. 28.3. Absorbance (———) and fluorescence (······) as functions of number of molecules (N). A = absorbance, F = fluorescence.

The advantages of microfluorimetry are:

1. The technique is very sensitive and using equivalent equipment microfluorimetry is several magnitudes more sensitive than absorption cytophotometry
2. The technique has a high resolution
3. In microfluorimetry there is no distributional error
4. Two or more fluorochromes can be used at the same time
5. The technique may be applied to living cells

Some errors in microfluorimetry are:

1. Fading or photobleaching arises as the result of photochemical decomposition of the fluorescent compound by the energy absorbed during excitation
2. Quenching results from the interaction of the fluorophore with its surroundings or with other fluorescent molecules. This effect can be kept at a minimum by keeping time of excitation low

3. Reabsorption of fluorescent light may occur if excitation and emission spectra overlap
4. Inner filter effect occurs if the upper layer of fluorophores in the specimen receive more excitation light than lower layers, that is, when the section is too thick

Both fading and quenching lead to reduction of the quantum efficiency which is the efficiency of conversion of absorbed photons to luminescent photons. Both these errors can be minimized by keeping excitation time short. All modern microspectrofluorimeters are equipped with electronic shuttle systems for standardizing excitation time, and with laser-induced fluorescence the necessary excitation time can be reduced to a few nanoseconds.

Reabsorption of fluorescent light and inner filter effect both cause deviations from the linear relationship between local absorbance and fluorescence light intensity. These errors occur at higher local absorbance values and are under conventional observation conditions dependent on the numerical aperture of the objective of the objective lens. Van Oostveldt and Bauwens (1990) have, however, shown that the use of a confocal imaging system eliminates the influence of the numerical aperture on the relationship between the integrated fluorescence intensity and the integrated absorbance.

A number of different standards have been developed for system checks or for calibration of analytical data. Advantages and disadvantages have been reviewed by Ploem (1970) and Sisken (1989).

The following alternatives are available for calibration:

1. Standard cells. Nucleated chicken erythrocytes fixed in glutaraldehyde are most frequently used
2. Polymer beads. For instance beads with various fluorescent dyes bound to them covering a range of excitation and emission wavelengths or beads to which different numbers of fluorochrome molecules are attached. These beads can in principle serve as an analytical calibration standard as it is possible to produce a calibrated curve by ultraviolet spectrophotometry
3. A solution with a known concentration of Fluorescein diacetate may serve as a system standard
4. Solid standards include fluorescent phosphorous crystals, uranyl glass, and Storna^{R} polymer blocks.

System checks can for example be used for determining the quality of the optical system, the stability of the light source and detector system, the exact focus, the linearity of the system, temporal variations in excitation intensities and/or output of the detection system, and for determining the influence of the age of the light source.

For calibration of analytical data either solutions or beads are used (Ploem, 1970; Sisken, 1989).

28.4 Reflection Contrast Photometry

Reflectance is defined as the ratio between the intensities of reflected and incident light. It may be calculated by using the formula:

$$^{x}R = \frac{(n-N)^2 + k^2}{(n-N)^2 + k^2}$$

where R = specular reflectance at normal incident light, n = refractive index of reflecting material, k = absorption coefficient of reflecting material, N = refractive index of the medium into which the light is reflected. The prescript x of R indicates this medium, such as immersion oil, water, etc. (e.g. $^{\text{oil}}R$) (Piller, 1977, p.135).

The refractive index (n) depends, among other factors, on the wavelength. It is referred to as *normal dispersion* when n increases with the frequency of light and *anomalous dispersion* when n decreases.

Ploem (1975) proposed two contrast-enhancing devices which fundamentally improved this approach for biological work:

1. An aperture central field-stop inserted in front of the collecting lens of the lamp housing in order to limit unwanted reflection of the central beam of light
2. An objective with high numerical aperture and a rotatable quarter wavelength plate mounted on the front lens. Such objectives were then produced by Leitz (Patzelt, 1977) and Zeiss (Pentz and Schulle, 1980) and the method was specified as reflection contrast microscopy

Apart from adding these two devices to a fluorescence microscope with epi-illumination it is necessarily to replace one of the fluorescence blocks containing a dichroic mirror and the fluorescence filters with a polarization block containing a 50% dividing mirror and two fixed and crossed polarizers (Cornelese-ten Velde *et al.*, 1989). To avoid reflection at glass-air interfaces, immersion oil with a refractive index equal to glass is used between the objective and the specimen, that is, no coverslip is used (Cornelese-ten Velde *et al.*, 1988).

In reflection contrast microscopy bright objects are seen against a relatively dark background, and in this respect the image may superficially resemble that seen with fluorescence microscopy. Contrast with the latter procedure is due to luminescence that occurs when a molecule, after absorbing radiation, emits a quantum of lower energy (van der Ploeg and van Duijn, 1979). The emission is almost always of longer wavelength than the exciting radiation.

In reflection contrast microscopy of stained biological material, the image is, however, either due to spectrally selective reflection or interference after reflection. *Spectrally selective reflection* may occur in many types of stained material, for instance in Feulgen stained cell nuclei or Giemsa stained chromosomes, and is due to anomalous dispersion (van der Ploeg and van Duijn, 1979). Selective reflection of light takes place at wavelengths near the absorbance peak of the chromophore bound to macromolecules embedded in a not completely matching medium. *Interference after reflection* is, however, the main reason for the whitish yellow appearance of the polymerized oxidation product of diaminobenzidine (DAB) in reflection contrast microscopy and is due to the high refractive index of this compound (Cornelese-ten Velde *et al.*, 1988).

Reflection of light by bound chromophores may be used for the visualization of microscopic material that may be difficult to distinguish using other methods. Reflection contrast microscopy is a sensitive method for detecting small amounts of the insoluble polymerized DAB oxidation product formed by the peroxidase reaction with DAB and hydrogen peroxide as cosubstrates, allowing the detection of details in the image having reflectance values of no more than 0.01% (Cornelese-ten Velde *et al.*, 1988). As reflectance of the polymerized DAB oxidation product is mainly due to interference of reflections, the stain must be present in a very thin layer, as in ultrathin sections, in the upper layer of semithin sections of Lowicryl^R embedded material, or on preparations of chromosomes (Cornelese-ten Velde *et al.*, 1988; 1989). Cornelese-ten Velde and Prins (1990) have, using post-embedding, stained different types of antigens in Lowicryl^R embedded material with immunogold for electron microscopy and with subsequent silver enhancement for reflection contrast microscopy. In principle, this makes it possible to use both reflection contrast microscopy and electron microscopy for visualizing the colloidal gold or silver enhanced gold particles on the same ultrathin section.

Accordingly, reflection contrast microscopy is a valuable tool in immunohistochemistry and in situ hybridization histochemistry when very small amounts of oxidized DAB are present, for example detection of single copy DNA sequences in chromosomes and quantitation of cell membrane receptors or other constituents in different compartments of the cell (Landegent *et al.*, 1985; Cornelese-ten Velde *et al.*, 1988; 1989).

Cornelese-ten Velde *et al.* (1989) compared the absorbance of oxidized DAB with its reflectance. Slides coated with peroxidase-conjugated immunoglobulins served as a model system. In addition, optimum conditions for the detection of single copy DNA sequences in metaphase chromosome preparations were established by using in situ hybridization and reflection contrast photometry. The intensities of reflected and incident light were measured using a 100% reflecting mirror in the object plane for calibration. The authors found a linear relationship between the amount of peroxidase and the reflectance determined for oxidized DAB. It was concluded that quantitative immunoperoxidase studies with the reflection contrast microscope are feasible.

28.5 Interferometry

Interference microscopy is another approach that can be used in quantitative histochemistry. Measurements are based on quantitating the retardation of light that has passed through the section relative to light that has not. The relative phase changes in the transmitted light give rise to addition and subtraction phenomena. The degree of retardation depends on the thickness and refractive index of the section, the latter depending predominantly on the local protein concentration in the section. This approach can therefore be used to measure local protein concentrations and section thickness, though it has largely been superceded by more specific techniques for the former application.

28.6 Spectral Analysis

The spectra of dyes and tissue bound chromophores can be used in several different ways. These include simple comparisons, colorimetry, spectral subtraction, and component spectral analysis.

The simplest approach to spectral analysis is to record transmittance, absorbance, or reflectance spectra of the chromophores in the tissue or cells under study. The general shape of the spectrum is observed and maxima, minima, and shoulders are determined. This makes possible the identification of a certain dye and can be used for determining the purity of dyes (cf. Sect.3.3.6). The absorption spectra of various blood cells stained with Azure B and Eosin were analyzed by Zipfel *et al.* (1984) who found that three absorption bands could be distinguished in the nuclei corresponding to DNA-bound monomers and dimers of Azure B and to the Romanowsky-Giemsa band corresponding to a complex of DNA, higher polymers of Azure B, and Eosin. In the spectrum of cytoplasm, absorption bands corresponding to monomers, dimers, and polymers of Azure B bound to RNA were identified (cf. Sects.6.1.1 and 6.3.2).

Further references to work of this kind of Romanowsky stained blood films are Marshall *et al.* (1981), Marshall and Galbraith (1984a), and Flanders *et al.* (1984). Galbraith *et al.* (1979) have analyzed the spectra of nuclei and cytoplasm of different types of cervical cells stained by the Papanicolaou technique. It is concluded that for automated image processing of this kind of material, the most suitable wavelength is 531 nm to maximize the contrast of cell against background and a wavelength in the range 560–605 nm to maximize the contrast of the nucleus against the cytoplasm. A further reference to work on the Papanicolaou stain is Galbraith and Marshall (1984).

Colorimetry. This attempts to quantitate the various components of the visual perception of colour. It therefore potentially provides the basis for understanding qualitative interpretation of stained material. It has not been used extensively for quantitating histochemical reaction products but the approach has been described in detail by Marshall *et al.* (1981), Marshall and Galbraith (1984a), and Flanders *et al.* (1984) studying Romanowsky stained blood cells and by Galbraith and Marshall (1984) studying Papanicolaou staining.

Visual perception is composed of geometric (site, size, shape, and texture), temporal (moving or at rest), and colour perception. The latter comprises three attributes: hue (red, blue, green, etc.), lightness (the same hue may appear lighter or darker), and saturation (proportion of colour). It is impossible for an observer to characterize differences in the colour perception quantitatively (Rietveld, 1986).

In colorimetry the colour measurements are based on the properties of the human eye and the conventions of the International Commission on Illumination (Commission International de l'Eclairage, CIE). Some of the standards defined by the CIE are the spectral tristimulus values X, Y, and Z of the CIE 1931 standard colorimetric observer and the chromaticity coordinates x and y. The relationship between these values is given by:

$$x = \frac{X}{X + X + Z} \quad \text{and} \quad y = \frac{Y}{X + Y + Z}$$

The chromaticity coordinates define the colour quality chromaticity (saturation and hue). The tristimulus value Y is equivalent to luminance (lightness), The tristimulus values X, Y, and Z may be calculated or computed by measuring the relative intensity of each spectral colour in the visible spectrum with a microspectrophotometer. Both reflectance and transmittance spectra may be used. In the dye and textile industries and also in criminological investigations, reflectance spectra are nearly always used as the samples usually are non-transparent. In cytology, histology, and sometimes in forensic medicine, on the other hand, colorimetry is normally based on transmittance spectra.

When performing colorimetric measurements in practice, a cell-or tissue-free area of the slide should be selected for determining the white standard values (Rüter *et al.*, 1982). The spectral transmittances thus become as independent as possible of the spectral characteristics of the measuring system (lamp, microscope, etc.). This ensures that the same results can be obtained irrespective of differing cytophotometric sensitivities in the same system (e.g. aging of the lamp). International agreement has, however, so far not been reached on an absolute standard making comparisons possible between measurements obtained with different instruments. The determination of the spectral white reference must be repeated for each measuring series.

All the colour values are indicated by reference to a standard illumination. For example, CIE Standard Illuminant D65 is an approximation of average daylight with a colour temperature of 6500°K, and Standard Illuminant A corresponds to the spectral power distribution of an incandescent lamp with a colour temperature of 2856°K. The power distribution is of decisive importance for the determination of the tristimulus values. It does, however, only participate by way of calculation, while the lamp, actually used for measuring the transmittance spectra of the specimen, is of no particular significance, as long as visible spectra are used. Only the instrument readings for the specimen and the white standard are determined with it; their ratio is then calculated separately for each wavelength, eliminating the intensity of the lamp. In fact, the tristimulus values of the specimen can be calculated with reference to several standard illuminants from a single transmission curve measurement.

A plot of the chromaticity coordinates, x against y, for all colours is called the CIE 1931 Chromaticity Diagram, and it is now possible to plot the calculated x and y values for a given specimen into this diagram. It should, however, be noted that equal distances on this diagram do not represent equal perceived differences in colour appearance. Several attempts have been made to transform the CIE 1931 Chromaticity Diagram by calculation to chromaticity diagrams with uniform chromaticity spacing. Standardized procedures have, however, so far not been agreed on. For instance, Marshall and Galbraith (1984b) employed four different sets of formulae to perform colour difference calculations for histological objects stained according to two different Romanowsky methods.

CIE coordinates contain less information than the spectrum from which they are derived. It is thus possible for two different spectra to give the same colour and coordinates, a phenomenon termed metamerism. More detailed spectral information may, however, be obtained by spectral subtraction or by component spectral analysis.

Spectral Subtraction Analysis. This is a technique that calculates sets of factors which are proportional to the quantities of the various component dyes which are bound to the cellular structures whose spectra are measured. The original method was described by Marshall *et al.* (1979) and later improved (Galbraith and Marshall, 1984) and has been applied to Papanicolaou staining (Marshall *et al.*, 1979; Galbraith and Marshall, 1984) and to Romanowsky staining (Galbraith *et al.*, 1980; Marshall *et al.*, 1981; Marshall and Galbraith, 1984a; and Flanders *et al.*, 1984).

Component Spectral Analysis. This is a method that makes it possible to separately obtain both the spectral patterns and the spatial distribution of different coloured components in tissue sections. Araki *et al.* (1987) have used a component spectrum analyzer for this purpose. This instrument measures the density images of a sample section at 25 nm intervals from 400 to 700 nm. The microscopic image is detected using a TV camera, digitized in a frame grabber, and stored in a microcomputer. The data are transferred to a large scale computer for further calculations. With this instrumentation, Araki *et al.* (1987) have determined the amounts of NBT mono- and diformazans formed by succinate dehydrogenase activity in central, intermediate, and periportal areas of rat liver lobules. The technique could probably be further refined by making it possible to perform the measurements with closer steps (1–5 nm) thus making possible direct determination of the isobestic point (cf. Sect.28.8.7). The sensitivity of the method might probably be increased by using a high power objective.

28.7 Analysis of Staining Kinetics

It is important to realize that a true thermodynamic equilibrium is rarely achieved in any of the commonly used staining methods (Goldstein, 1980). Rate (kinetic) factors affect the staining results decisively. These factors include diffusion of the dye in solution and/or substrate and the reaction between dye and substrate molecule. The latter has already been treated in Chaps.4–9. In Sect.12.3.11 factors governing the diffusion rate of fixatives are discussed. The arguments advanced here also hold true for the diffusion of dyes.

Bentley *et al.* (1979) have performed a microspectrophotometric study of Azure B Eosin staining of blood cells. The influence of stain formulation and staining technique was studied by varying stain concentrations, staining time, pH, metal salt contamination, dye contamination, buffer concentration, and fixation time.

Goldstein (1980) determined half-staining times for cytoplasmic RNA and nuclei with Azure A at pH 4 and 25°C using microdensitometry. Winzek *et al.* (1987) described the combination of a microspectrophotometer with a perfusion cuvette. The cuvette made it possible to ensure laminar flow of the staining solution in a completely closed system containing cytological or histological material and thus enabled continuous observation of a cell during the complete staining process.

A detailed review of factors pertaining to the kinetics of staining is given by Horobin (1982a), pp.56–110.

28.8 Some Applications of Quantitative Histochemistry

In principle, using absorption cytophotometry it should be possible to quantitate a substantial proportion of the histochemical methods discussed in the preceding chapters. Quantitation has, however, only been attempted with a few procedures and even then there have been serious problems with controls, standards, specificity, selectivity, and whether or not the reaction product measured can be relied on to give an accurate estimate of the target substrate.

Some examples of quantitative techniques are discussed below.

28.8.1 Metals and Metal Salts

Very little work has been done on quantitation of standard techniques applied to histological sections. An example is, however, the work by Wolters *et al.* (1983) who examined the effect of gradual, tolbutamine induced, degranulation and subsequent regranulation on the histochemically detectable zinc (dithizone, cf. Sect.17.7.5) and calcium (glyoxal-bis-(2-hydroxyanil) (GBHA), cf. Sect.17.7.1) content of B cells in rat pancreatic islets. The staining intensities were determined cytophotometrically and compared to the insulin content determined by staining with Aldehyde Fuchsin and by radioimmunoassay of insulin extracted from pancreatic tissue.

Although tissue-bound metal salts have received little attention, over the last ten years a number of fluorescent probes have been developed for the specific demonstration of individual diffusible ions in living cells. These techniques, which include methods for monitoring Ca^{2+}, H^+, K^+, Na^+, Mg^{2+}, and Cl^- in real time, are discussed in section 28.8.10.

28.8.2 Pigments

In principle, several methods could be made quantitative but this has not been attempted to any substantial degree. The ferric ferricyanide reaction (Sect.8.3.1) has been used for the quantitative assessment of glutathione (cf. Sect.28.8.10). It is, however, not specific for this compound as it also demonstrates ascorbic acid, cysteine residues, lipofuscin, melanin, and biogenic amines. This raises the possibility of quantitating some of the latter non-extractable substances in paraffin sections.

The biogenic amines can be demonstrated and quantitated by formaldehyde induced fluorescence (Sects.18.4.10, 30.2.1). A recent example has been given by Hougaard and Larsson (1990) who found a linear relationship between cellular polyamine concentration and formaldehyde-fluorescamine fluorescence yield.

28.8.3 Lipids

Staining with lysochromes can be combined with a subsequent extraction and chromatographic quantitation.

The acid Haematein method of Baker (cf. Sect.19.6.7) was applied to cryostat sections of human rheumatoid synovial membranes by Henderson *et al.* (1978a). Prior to the acid Haematein procedure sections were extracted by treatment with 1:3 (v/v) methanol:chloroform (cf. Sect.19.3). Measurements were performed at the absorption maximum of 575 nm and the non-specific staining measured over collagenous stroma was subtracted.

28.8.4 Nucleic Acids

Quantitation of cellular DNA has been used for studies on cell growth, tumour biology, and phylogenetic relationships. The DNA content of individual chromosomes can also be measured.

Measurements of DNA have been made with the Feulgen reaction, fluorescing Schiff reagents, autoradiography, with Cr-gallocyanin, with various cationic dyes such as Azure B and Methyl Green-Pyronin, and by ultraviolet spectrophotometry.

Photometric Quantitation of the Feulgen Reaction. By far the most frequently used quantitative technique for DNA is photometric quantitation of the Feulgen reaction (Sect.9.9). Several published procedures fulfil the prerequisites enumerated in Sect.28.1.1 (Coulton *et al.*, 1981; Henderson *et al.*, 1981).

Quantitative work with the Feulgen reaction should take into account, and in some cases actually utilizes, the fact that the rate at which a given segment of DNA is hydrolyzed depends both on its degree of association with DNA binding proteins and its relative state of condensation. Newly synthesized DNA (S-phase), which is only partially condensed (cf. Sect.31.8.1) is more susceptible to hydrolysis than "older" DNA (cf. Sect.31.8.3). Note, however, that differences in hydrolysis times of these "different" types of DNA only are apparent when using cold hydrolysis (5 mol/l HCl at 20°C; cf. Sect.9.9).

Errors have been considered in detail. Theoretical methods were developed for correcting absorption values determined with a scanning stage microspectrophotometer and these were subsequently tested in a practical examination of condensed, normal, and dispersed Feulgen stained chicken erythrocyte nuclei (Duijndam *et al.*, 1980a;b). They found a considerable degree of concordance between the predicted and experimentally determined magnitudes of systematic errors due to inhomogeneous distribution, diffraction, and glare; these sources of error are often pronounced when small, intensely stained objects are measured.

Standards that have been used include Feulgen stained nucleated chicken erythrocytes, neutrophil granulocytes, and lymphocytes (2n DNA) or spermatozoa (1n DNA).

In quantitative studies, the Feulgen reaction has been combined with several other methods for the simultaneous demonstration of dry mass, RNA, histones, non-histone proteins, enzyme activity, and immunoreactive substances (Welch *et al.*, 1979; Herzog, 1982; Olenev, 1983; Fukuda, 1983; Schiek *et al.*, 1987).

Fluorescent reagents. Two approaches have been used for the quantitation of DNA using fluorescent reagents. The first involves the use of *fluorescent Schiff reagents* prepared from dyes such as Acriflavine and Auramine O. These are applied after acid hydrolysis and are subject to similar considerations to those discussed under the Feulgen reaction. The second technique is the application of *fluorescent compounds that intercalate between adjacent base pairs* (e.g. ethidium bromide and propidium iodide). These reagents have primarily been used in flow cytometry but they can also be used for fluorimetric determinations on tissue sections and cytological preparations.

Few comparisons have been made between the results obtained using Feulgen techniques and DNA intercalating fluorochromes. Scherini *et al.* (1988) found that DNA content determined by fluorimetry of cells stained with Hoechst 33342 gave lower values than the Feulgen technique in cerebellar cells from rats treated with bleomycin. The authors put forth the hypothesis that the observed differences in part are due to differences in dye-DNA interaction, and that the reduction in Hoechst staining at least is partly due to a decrease in the content or accessibility of adenine-thymidine base pairs of DNA to the fluorochrome in cerebellar cells in animals treated with bleomycin.

Chromium-gallocyanin Method. The Cr-gallocyanin method introduced by Einarson (1932) is suitable for quantitating both DNA and RNA (cf. Sect.7.3). Selectivity can be obtained by pre-treatment with the appropriate nuclease.

Cationic Dyes. Methyl Green-Pyronin (Sect.6.1.5) has been used in several studies. The problems include, the purity of the dyes used, cross-over in absorbance spectrum between the two dyes, and uncertainties concerning the mechanism and stoichiometry of staining. Quantitation has been achieved by microspectrophotometry (Pellicciari and Fraschini, 1978), image-analysis (Lyon *et al.*, 1989) and by stain-elution (Barer *et al.*, 1986a). Lyon *et al.* (1989) in an image analysis study using a standardized Methyl Green-Pyronin procedure (Høyer *et al.*, 1986) have correlated the uptake of Methyl Green and the Feulgen staining intensity of cell nuclei. The absorption of cells was measured at the absorption maxima of the two dyes (630 nm Methyl Green and 550 nm Pyronin Y). The absorption of Pyronin was corrected for the residual absorption of Methyl Green at 550 nm according to Barer *et al.* (1986a) as

$$IOD_C(P) = IOD_C(550) - [IOD_C(630) \times IOD_M(550) / IOD_M(630)]$$

where IOD = integrated optical density; C = combined staining with Methyl Green and Pyronin Y; M = staining with Methyl Green alone. The coefficient of correlation of 0.978 shows that the IOD of Methyl Green parallels the IOD of the Feulgen stain.

Using a gelatin-RNA model system and spectrophotometry, Shea (1970) demonstrated that the binding of Azure B to RNA is stoichiometric and constructed a calibration curve. This was applied to determine the absolute amounts of RNA in biological material using the one-wavelength method and the type of cytophotometer described by Pollister and Ornstein (1959). Doebler *et al.* (1983) have also used Azure B combined with nuclease treatment.

Absorption of Ultraviolet Light. Unstained DNA and RNA have an absorption maximum in the ultraviolet region at 260 nm (Caspersson, 1940). Specificity can again be obtained by nuclease treatment. Appropriate instrumentation was developed by Caspersson and later by the Zeiss Company but cost has prevented their widespread use. In UV-microspectrophotometry the potential contribution from other natural chromophores such as reduced adenine nucleotides, pteridines, riboflavin, thiamine, carotenoids and porphyrins, and other haem pigments must be taken into account.

28.8.5 Proteins

Although methods depending on the formation of complexes or covalent bonds should be quantifiable, there is very little published work in this area.

Anionic Dyes. These have been used extensively. For some of these reagents, problems due to relatively low dye-substrate affinities have to be circumvented (Klunk *et al.* 1989a). One approach has been to omit the washing step after staining and to separate the staining solution from the substrate (e.g. cells) by filtration. Dye binding is then determined by measuring the *decrease* in dye concentration in the filtrate (Klunk *et al.*, 1989a). The problem can be avoided altogether when spectral changes are induced in the dye when it binds to certain proteins (e.g. Congo Red and amyloid-like proteins, Klunk *et al.*, 1989a; b).

At pH 2.8 (aqueous 1% acetic acid), Naphthol Yellow S, Light Green SF, and Orange II can all be used for quantitative staining of total protein (see Sect.21.4.2; Deitch, 1955; Oud *et al.*, 1984; Oud, 1986).

Nucleolar proteins have been quantitated using Phloxine B at pH 6.8. This procedure has been developed for automatic assessment of nucleolar size by image analysis (Boon *et al.*, 1988).

Amido Black 10B has been widely used for staining protein in chromatography (Lillie, 1977, p.142). Histologically the dye has been used as a stain for haemoglobin and as a highly selective stain for basement membranes, reticulum, and collagen where it is applied in Picric Acid mixtures of the van Gieson type (Sect.21.6; Lillie, 1977, p.142). Schauenstein *et al.* (1980) determined the molar extinction coefficient of complexes of Amido Black 10B and bovine serum albumin and went on to describe how a quantitative microspectrophotometric protein determination could be done on single cells. A modification of the method for quantitative microspectrophotometric determination of proteins in tissue sections was presented by Nöhammer (1984).

28.8.6 Carbohydrates

PAS Reaction. The applicability of the PAS (periodic acid Schiff) reaction to quantitative studies has been discussed extensively. Clearly the widespread acceptance of the Feulgen technique for quantitative work must have some bearing on these discussions. Moreover, both Hoogwinkel and Smits (1957) and Olsson and Dahlqvist (1967) have shown that glycogen solutions react stoichiometrically and that the reaction product conforms to the Beer-Lambert law.

Several studies demonstrate the importance of the relative availability of glycosyl units for oxidation. Halkjær-Kristensen and Ingemann-Hansen (1979) found that only 15% of the glycosyl groups were demonstrated in tissue sections, while Ovadia and Stoward (1971) showed that only 15–23% of the total glycogen content in liver sections as determined chemically could be oxidized by periodic acid. (For further discussion of this paper, see Sect.9.2.1). In molecular model experiments, Puchtler et al. (1974) showed that Schiff's reagent cannot diffuse freely through polysaccharides and that only every second glycosyl unit on the surface of glycogen granules can combine with Schiff's reagent for steric reasons.

Gahrton and Yataganas (1976) have tried to circumvent these problems by developing a system in which the total glycogen content of cells can be determined by reference to a microdroplet standard. Development of the microdroplet model allowed these workers to study the relationship between dry mass, as determined by X-ray absorption, and PAS staining using both standard and fluorescent Schiff reagents (Sect.28.8.4). Results using the two types of Schiff reagent gave a highly correlated linear relationship and the authors concluded that the microfluorimetry method could be used for glycogen determination in single cells providing certain procedural constraints are closely adhered to. These workers then went on to determine the mean value of glycogen per neutrophil leukocyte (13.3×10^{-12} g) in a healthy subject (Gahrton and Yataganas, 1976); this concurs with results obtained biochemically (Esmann, 1961).

Further quantitative studies using the PAS reaction have been published by Zoller *et al.* (1981) (microspectrophotometry) and Harkema *et al.* (1987) (automated image analysis). Frederiks *et al.* (1987a;b) used the PAS technique to assess glycogen phosphorylase activity by quantitating the amount of glycogen formed after incubation in a gelled medium containing glucose 1-phosphate as substrate in a semipermeable membrane technique.

Cationic Dyes. Quantitative work with cationic reagents that demonstrate *sulphate and carboxyl groups in acid glycosaminoglycans* has been sparse. An analysis of Alcian Blue and combined Alcian Blue-Safranin O staining procedures has been made by Tas (1977) using polyacrylamide films containing different glycosaminoglycans. When used alone, Alcian Blue staining followed the Beer-Lambert law making quantitation possible, while the combined Alcian Blue-Safranin O procedure could not be used as results were highly dependent on the amount of Alcian Blue bound to the glycosaminoglycan during the first step of the staining procedure. Safranin O can, however, be used alone for the microspectrophotometric quantitation of glycosaminoglycans in cartilage (Kiviranta *et al.*, 1985a).

Quantitative studies in this field include, Zoller *et al.* (1981) who used Alcian Blue pH 2.5, Berrisford *et al.* (1985) who used both PAS and Alcian Blue pH 2.5, and Dunham (1979) who used the Alcian Blue critical electrolyte concentration method (cf. Sect. 6.1.2). Finally, Harkema *et al.* (1987) have used automated image analysis of high iron diamine (Sect.22.2.2) stained material to quantitate sulphated mucins; however, it is uncertain whether this method fulfils all the requirements stated in Sect.28.1.1.

28.8.7 Enzymes

The advantage of quantitative enzyme histochemistry is that it enables localization and quantitation to be combined provided that the techniques for preparation and staining of cell and tissue samples have been shown to be reliable and valid (Glick, 1981).

Stoward (1980) has suggested some practical criteria for establishing the precision, reproducibility, validity, and specificity of quantitative histochemical techniques used for assaying the activities of enzymes in single cells and tissue sections. With a few minor abbreviations and modifications, these criteria are given below:

1. Precision
   a. Sections retain their morphology
   b. Specific final reaction product (FRP) is confined to the subcellular sites known to contain the enzyme from conventional biochemical studies
   c. The specific FRP must not diffuse

Some examples of how these criteria may be used are given by Fahimi (1980).

2. Reproducibility
   a. The mean values of measurable parameters do not vary significantly in repeated experiments
   b. The individual measurements of a given parameter within a preparation form a statistically-defined unimodal population

A method for testing these parameters is given by Stoward (1980).

3. Validity
   a. No enzyme is lost during incubation or if some is lost, the loss is small, constant, and known
   b. The specific FRP has its expected chemical composition
   c. There is a stoichiometric relationship between the amount of specific FRP in the specimens and the amount of primary reaction product formed by the enzyme
   d. The mean absorbance or fluorescence emission of the specific FRP is proportional to its mean concentration in the specimen
   e. Apart from fluorimetry (Sect.28.3), the mean absorbance of specific FRP is proportional to the thickness of the section up to a certain critical level when using zero order kinetics
   f. The reaction rate in a particular site is directly proportional to the specific activity of the enzyme at this site
   g. Mean absorbance or fluorescence emission of the specific FRP increases uniformly with incubation time (preferably linearly)
   h. When performing regression analysis of the absorbance- or fluorescence-incubation plots, the line should preferably extrapolate through the origin
   i. When the substrate concentration is above a certain level, the reaction rate should reach a constant maximum rate (zero order kinetics)
   j. The Michaelis constants should preferably be comparable to those determined biochemically. It should, however, be noted that the K_M values determined using PVA media often are an order of magnitude higher
   k. Enzyme modifiers (e.g. inhibitors) should have the same effect on rate of final FRP formation as known from biochemistry, and this should preferably be of the same order of magnitude
   l. Control experiments should indicate that the amount of non-specific final reaction product is small and preferably constant (not more than 5–10% of the specific FRP
4. Specificity. A technique is judged to be specific if:
   a. No specific FRP is formed in controls
   b. The reaction conditions that give rise to maximum formation of specific FRP in situ are the same or very similar to those used biochemically
   c. Modifiers (inhibitors) of the enzyme act in a similar manner as observed biochemically
   d. Potentially interfering enzyme systems have either been suppressed, eliminated or shown to be absent, or can be distinguished from the enzyme under study and measured separately

At present several techniques have been tested against these criteria. Examples are acid phosphatase (Stoward, 1980), the mixed-aggregation immunocytochemical technique (MAGIC) described by Wachsmuth (1980), and glucose-6-phosphate dehydrogenase (van Noorden, 1984).

These criteria have inspired much of the work on quantitative enzyme cytochemistry performed during the last ten years as exemplified by the fact that most of the papers referred to in Tables 28.4–28.6 to varying degrees have used these criteria.

At this point the reader is reminded that the *sensitivity* of the technique constitutes an additional criterion (Sect.28.1.1).

At the light microscopical level enzyme activity may be quantitated by absorption cytophotometry (Sect.28.2.1) or by microfluorimetry (Sect.28.3).

At present, even though microfluorimetry is a very sensitive technique, no fluorometric enzyme histochemical method adequately fulfils the above criteria. In particular, precise localization presents a problem as the available fluorogenic reagents lack sufficient substantivity and are not sufficiently insoluble in water (Raap, 1986).

Luminometry. The technique of luminometry is the quantitative measurement of luminescence (bioluminescence) (Sect.28.3). The advantage of this technique is great sensitivity, essential simplicity, and range of applicability (Glick, 1981). An example of its application is in reactions involving production or utilization of ATP that can be measured with the luciferin-luciferase firefly system.

$$\text{ATP} + \text{luciferin} \xrightarrow[\text{Mg}^{2+}]{\text{luciferin}} \text{adenyl luciferin} + \text{pyrophosphate}$$

$$\text{adenyl luciferin} \xrightarrow{O_2} \text{adenyloxyluciferin} + \text{light}$$

Measurement is therefore based on total light generated.

Much detail regarding these methods is given by Paschen (1990) who has presented new methods for some substrates of energy metabolism (ATP, glucose, and lactate) in thin tissue sections.

At the electron microscopic level either X-ray microanalysis of the final reaction product or microspectrophotometry of electron microscope negatives may be used. In the following only absorption cytophotometry is described.

Kinetic Measurements. There are two ways to make kinetic measurements; these are end-point measurements and continuous monitoring.

In end-point measurements incubation is performed for an appropriate certain time after which the reaction is stopped and the section mounted.

The advantages of this approach are that kinetic values can be obtained from as many areas as found necessary, and the approach is not limited to one-step procedures (i.e. post-coupling, metal salt, as well as plateau absorbance methods can all be used). If, however, the final reaction product diffuses into the solution used for stopping the reaction, continuous monitoring is the only possibility.

Table 28.4. Examples of quantitative cytochemical demonstration of hydrolases.

Trivial name	E.C. No.	Reaction type/substrate/capture reagent(s)	References
Esterase	3.1.1.1	simultaneous coupling/α-naphthyl acetate/p-nitrobenzene-diazonium tetrafluoroborate or Fast Violet B	Johnston and Ashford, 1980
Alkaline phosphatase	3.1.3.1	metal salt	Bader et al., 1984
		simultaneous coupling/naphthol AS-BI phosphate/Fast Blue B	Gutschmidt et al., 1980b
		indolyl-tetrazolium salt/5-bromo-4-chloro-3-indolyl phosphate/tetranitro BT	van Noorden and Jonges, 1987
Acid phosphatase	3.1.3.2	metal salt/β-glycerophosphate/$Pb(NO_3)_2$/H_2S	Yoffe, 1980
		simultaneous coupling/naphthol AS-BI phosphate/hexazotized Pararosanilin[1]	Stoward and Al Sarraj 1981a, b; Stoward et al., 1981
		post-coupling/naphthol AS-BI phosphate/hexazotized Pararosanilin[1]	Stoward et al., 1982
5′-nucleotidase	3.1.3.5	metal salt/adenosine-5′-phosphate/$CaCl_2$/$Co(NO_3)_2$/$(NH_4)_2S$	Henderson et al., 1980
Glucose-6-phosphatase	3.1.3.9	metal salt/glucose-6-phosphate/$Pb(NO_3)_2$/$(NH_4)_2S$	Hildebrand and Schleicher, 1986
α-D-glucosidase	3.2.1.20	simultaneous coupling/2-naphthyl-α-D-glucoside/hexazotized Pararosanilin	Gutschmidt et al., 1979
		simultaneous coupling/5-bromo-4-chloro-3-indoxyl-α-D-glucoside	Ruhnke and Gossrau, 1989b
β-Galactosidase	3.2.1.23	indolyl-ferricyanide/5-bromo-4-chloro-3-indolyl β-D-galactoside/ferricyanide	Lund-Hansen et al., 1984
N-acetyl-β-glucosaminidase	3.2.1.30	post-coupling/naphthol AS-BI N-acetyl-β-glucosaminide/Fast Garnet GBC[2]	Robertson, 1980
β-glucuronidase	3.2.1.31	simultaneous coupling/naphthol AS-BI glucuronide/hexazotized Pararosanilin	Henderson, 1984; Schofield et al., 1983
Aminopeptidase A	3.4.11.7	simultaneous coupling/α-L-glutamic acid-4-methoxy-2-naphthylamide/Fast Blue B	Lojda and Gossrau, 1980; Kugler, 1982a, b, c
Dipeptidylpeptidase II	3.4.14.2	simultaneous coupling/Lys-Ala-4-methoxy-2-naphthyl-amine/hexazotized Pararosanilin	Gossrau and Lojda, 1980
Dipeptidylpeptidase IV	3.4.14.5	simultaneous coupling/Gly-Pro-4-methoxy-2-naphthyl-amine/Fast Blue B	Ruhnke and Gossrau, 1989
Na^+/K^+-transporting ATPase	3.6.1.37	metal salt/adenosine-5′-triphosphate/lead ammonium citrate-acetate complex/H_2S	Chayen et al., 1981
		metal salt/p-nitrophenyl phosphate/$SrCl_2$/$CoCl_2$/$(NH_4)_2S$	Firth, 1987
Ca^{2+}-transporting ATPase	3.6.1.38	metal salt/adenosine-5′-triphosphate/lead citrate/ammonium sulphide[3]	El-Sherif et al., 1990

[1] Semipermeable membrane technique.
[2] PVA technique.
[3] Polyacrylamide gel film technique.
All other non-marked techniques used traditional aqueous incubation media.

Table 28.5. Examples of quantitative cytochemical demonstration of oxidoreductases.

Trivial name	E.C. No.	Primary acceptor	Function/location	References
Lactate dehydrogenase	1.1.1.27	NAD^+	glycolysis/cytosol	Evans et al., 1980; Stoward and Nakae, 1988; van Noorden and Vogels, 1989b
3-hydroxyacyl-CoA dehydrogenase	1.1.1.35	NAD^+	final step in β-oxidation of fatty acids to acetyl CoA/mitochondria	Chambers et al., 1982
6-phosphogluconate dehydrogenase	1.1.1.44	$NADP^+$	pentose shunt/cytosol	Altman, 1972; Jonges and van Noorden, 1989
Glucose-6-phosphate dehydrogenase	1.1.1.49	$NADP^+$	pentose shunt/cytosol	van Noorden, 1984; Jonges and van Noorden, 1989
20α-hydroxysteroid dehydrogenase	1.1.1.62 (formerly 1.1.1.149)	$NADP^+$ (NAD^+)	progesterone is converted to a less active metabolite/not established	Robertson et al., 1982b
3β-hydroxy-Δ^5-steroid dehydrogenase	1.1.1.145	NAD^+ ($NADP^+$)	synthesis of steroid hormones/smooth endoplasmatic reticulum, mitochondria (in some cells)	Robertson, 1979
Glycerol-3-phosphate dehydrogenase (mitochondrial)	1.1.99.5	FAD	transfers reducing equivalents into mitochondria/outer surface of the inner mitochondrial membrane	Lomax and Robertson, 1990
Aldehyde dehydrogenase	1.2.1.3 1.2.1.4 1.2.1.5	NAD^+ $NADP^+$ $NAD(P)^+$	oxidation of aldehydes, e.g. metabolites of biogenic amines or corticosteroids/mitochondria and cytosol	Chieco et al., 1986
Glyceraldehyde-3-phosphate dehydrogenase	1.2.1.12	NAD^+	glycolysis/cytosol	Henderson, 1976
Succinate dehydrogenase	1.3.5.1 or 1.3.99.1	FAD	citric acid cycle/mitochondria	Evans et al., 1980; van Noorden and Vogels, 1989b; Baker and Santer, 1990
Glutamate dehydrogenase	1.4.1.2 1.4.1.3[1] 1.4.1.4	NAD^+ $NAD(P)^+$ $NADP^+$	catalyzes the interconversion of L-glutamate and 2-oxoglutarate/mitochondrial matrix	Kugler, 1990b
NADH dehydrogenase	1.6.5.3 and 1.6.99.3	FMN	oxidation of NADH/mitochondria, (smooth endoplasmatic reticulum)*	Altman, 1972; Robertson, 1979

* In some cell types, for example steroid hormone producing cells.

[1] Enzyme present in warm-blooded animals.

Table 28.5. (*Continued*).

Trivial name	E.C. No.	Primary acceptor	Function/location	References
NADPH dehydrogenase	1.6.99.1	FMN	oxidation of NADPH/smooth endoplasmatic reticulum, (mitochondria)*	Butcher et al., 1979; Robertson et al., 1982b
Cytochrome-*c* oxidase	1.9.3.1	O_2	catalyzes the final step in the mitochondrial respiratory chain/mitochondrial inner membrane	Old and Johnson, 1989
Iodide peroxidase	1.11.1.8	H_2O_2	oxidation of iodide and formation of iodotyrosyl (thyroglobulin) and iodothyronine/rough endoplasmatic reticulum, Golgi complex, peroxisomes, luminal side of plasma membrane (thyroid)	Ealey et al., 1984

Table 28.6. Examples of quantitative cytochemical demonstration of enzymes other than hydrolases and oxidoreductases.

Trivial name	E.C. No.	Function	Technique	References
γ-Glutamyltransferase	2.3.2.2	cell membrane bound enzyme of the γ-glutamyl cycle of importance for the transport of specific amino acids (glutamine, cystine) into cells	conventional aqueous	Ruhnke and Gossrau, 1989
Glycogen phosphorylase	2.4.1.1	important in glycogen metabolism	semipermeable membrane	Frederiks et al., 1987b
4-Aminobutyrate aminotransferase (GABA transaminase)	2.6.1.19	catalyzes the transfer of the amino group of 4-aminobutyrate to 2-oxoglutarate	PVA	Kugler and Baier, 1990
Hexokinase	2.7.1.1	first and rate-limiting step in glycolysis	PVA	De Schepper et al., 1985; Lawrence et al., 1989; Kugler, 1990a
6-Phosphofructokinase	2.7.1.11	irreversibly phosphorylates fructose-6-phosphate to fructose 1,6-diphosphate in glycolysis	PVA	Butcher and Papadoyannis, 1979; Butcher, 1983
NAD^+ kinase	2.7.1.23	irreversibly converts NAD^+ and ATP to $NADP^+$ and ADP	PVA	Perrild et al., 1984
Creatine kinase	2.7.3.2	catalyzes phosphorylation of ADP by phosphocreatine	semipermeable membrane	Frederiks et al., 1987c
Ornithine decarboxylase	4.1.1.17	initial enzyme of the polyamine synthesis pathway	polypeptide	Dodds et al., 1990
Fructose-bisphosphate aldolase	4.1.2.13	catalyzes the splitting of fructose 1,6-bisphosphate to dihydroxyacetone and glyceraldehyde 3-phosphate in glycolysis	MAGIC*	Wachsmuth et al., 1975; Wachsmuth, 1980
Carbonate dehydratase	4.2.1.1	catalyzes the reversible conversion of CO_2 and H_2O to H_2CO_3	conventional aqueous	Loveridge, 1978
Glucose-6-phosphate isomerase	5.3.1.9	catalyzes the reversible conversion of glucose 6-phosphate to fructose 6-phosphate in glycolysis and gluconeogenesis	polyacrylamide gel film	De Schepper et al., 1985

* Mixed aggregation immunocytochemical technique.

With this latter approach an appropriate area in the section is selected. At zero time the incubation medium is either poured into a Perspex^R ring surrounding the section (cf. Fig. 25.3) or applied between slide and coverslip, preferably separated from each other by thin spacers. When using a Perspex^R ring it is necessary to use a water immersion objective with a long working distance (van Noorden, 1988b). Usually, the first measurement is made after 15 seconds and subsequent measurements may for example be made every 15 seconds. The "blank" value obtained after the first 15 seconds is subtracted from each subsequent measurement (van Noorden, 1988b).

This approach is limited to one-step reactions. If the microspectrophotometer is equipped with a computer-controlled fast scanning stage, it is possible to perform sequential monitoring of multiple fields in each section (Lomax *et al.*, 1989).

Incubation on the microscope stage of a cytophotometer can be performed at room temperature, in a hot room, or using a heated stage (Altman, 1978).

Measurements may be performed using an image analyzer or a microspectrophotometer. The latter may use either the flying spot principle or a fast scanning stage (cf. Sect.28.2.3). Usually, the measurement is made at the absorption maximum of the final reaction product. In case of high mean integrated absorbance values (MIE), it may, however, be necessary to make the measurements off-peak in order to avoid out of range error of the instrument (cf. Sect.28.2.2).

In the case of ditetrazolium salts measurements must be made at the wavelength (the isobestic point) at which the corresponding formazans have the same molar extinction coefficient (cf. Sect. 25.1.4).

To illustrate how absolute MIE (MIA) values can be converted to absolute units that may finally be compared to corresponding biochemical results, a simple example is calculated below:

Suppose that a specimen of diameter 10 μm has an absolute MIE of 0.15 at the isobestic point for NBT after 5 min incubation, then

$$A = 5 \times 5 \times \pi = 78.5 \mu\text{m}^2 = 78.5 \times 10^{-8}\text{cm}^2$$

and with ϵ at 585 nm = 16000 litre mol^{-1} cm^{-1} (Butcher, 1978) substitution in equation 28.3 gives:

$$M = \frac{0.15 \times 78.5 \times 10^{-8}}{16000} \frac{\text{cm}^2}{\text{litre mol}^{-1}\ \text{cm}^{-1}} =$$

$$0.74 \times 10^{-11} \frac{\text{cm}^2}{10^3\text{cm}^3\ \text{mol}^{-1}\ \text{cm}^{-1}} = 0.74 \times 10^{-14} \text{ moles} = 7.4 \text{ fmoles}$$

This is of course the amount of formazan formed and equivalent to the amount of hydrogen produced in the specimen per cm^2 per 5 min. The amount of hydrogen produced per cm^2 per min is thus 1.5 fmoles.

If the thickness of the section is known, it is now possible to calculate the hydrogen production as fmol/min/cm^3 provided that section thickness does not exceed the critical thickness (cf. point 7 in Sect.28.2.2). It should, however, be noted that even though a specific microtome setting (of say 10 μm) is used, the section thickness is not necessarily 10 μm. As already shown in 28.1.2 section thickness is a function of cutting speed (cf. Table 28.2). Even with a motorized cryostat section thickness does not necessarily correspond directly to the microtome setting. This is at least in part due to the considerable loss of water that takes place when the section, assisted by a kind of jet-effect, jumps from the cryostat knife and adheres to the slide (cf. Sect.11.3.1). A further loss of water may possibly occur when the slide is equilibrated to the temperature of the incubation medium. An illustrating example is given by Baker and Santer (1990). With a microtome setting of 20 μm the actual measured thickness, as determined by a surfometer, was found to be only slightly more than 5 μm.

Similarly, using interferometry, Anthony *et al.* (1984) found that the effective thickness of tissue sections was reduced by a similar order of magnitude (90%).

It thus appears that if section thickness has not been directly measured, it is better to give the quantities in fmoles per cm^2 per min with additional information on microtome type, setting, and cutting speed (strokes/min).

Several different methods may be used for determining section thickness. These include:

1. Weighing of serial sections (Troyer *et al.*, 1977)
2. Determination of protein content by micro-Kjeldahl analysis or by measuring Naphthol Yellow S stained sections (Lowry et al., 1951; Gutschmidt *et al.*, 1981)
3. Determination of DNA content (Butcher, 1968; 1971)
4. Use of a focusing technique with a Cejtronic probe attached to the microscope (Halkjær-Kristensen and Ingemann-Hansen, 1978)
5. Use of a surfometer (Pearse and Marks, 1974; Baker and Santer, 1990)
6. Interference contrast microscopy or interferometry (cf. Sect.28.5 and Anthony *et al.*, 1984)

The use of chemical quantitation of nucleic acid content to determine section thickness yielded a remarkable degree of agreement with the setting on the cryostat microtome when using a "normal" cutting speed (Butcher, 1971). The apparent discrepancy between the above results and those of Baker and Santer (1990) and Anthony (1984) might be explained by assuming that the latter authors have determined the thickness of the section after it has jumped from the knife and onto the slide. During this process the section is effectually flash dried on the slide. In contrast, Butcher (1971) really has essentially calculated the thickness of the section while it is still on the knife.

A direct comparison of the results of these papers is, however, difficult as the exact cutting speed has not been given. The remark "How thick is your section?" by Altman (1980a) is still relevant and we should perhaps add the further question: "Has section thickness been determined on the knife, after flash drying, after rehydration, during incubation, or after the histochemical reaction, posttreatment, and mounting have been done?"

To achieve comparability between the results obtained in cytochemistry and biochemistry it is necessary to define a common unit of comparison. Probably the most suitable approach is to relate the quantity measured to total protein content, or perhaps even better total DNA content, per unit volume or per cell.

Computer programs have been developed for cytophotometric measurements of double stained preparations allowing the simultaneous determination of protein content and enzyme activity. Proteins are stained with Naphthol Yellow S after the sections have been reacted for the enzyme activity. Thus a dual wavelength scanning cytophotometry program (Bicoscan) developed by van der Ploeg *et al.* (1979) may be employed for this purpose. This program will compute the corrected local absorbance values at specified wavelengths for each chromophore at each measuring spot and integrate these values over the total object to give separate totals for each chromophore. Van der Ploeg *et al.* (1979) originally used this program for simultaneous determination of DNA (Feulgen) and protein (Naphthol Yellow S).

Biochemistry normally correlates to total volume though from the cytochemical viewpoint correlation to total volume of cells would seem more reasonable when enzyme activities are concerned.

It would appear that image analysis is the easiest way of determining total volume of cells.

The importance of measuring the specific reaction rate cannot be sufficiently stressed. The specific initial reaction rate is the test reaction (complete incubation medium) minus the control reaction (medium without substrate). In most cases test reaction minus control reaction is a sufficient correction. Van Noorden and Vogels (1989b) found, however, that the test minus control reactions for lactate dehydrogenase were distinctly non-linear for all tissues tested. This appeared to be due to product inhibition by pyruvate generated during the reaction, and it was concluded that the appropriate control reaction in this case was the reaction obtained by adding sodium pyruvate to a final concentration of 20 μmol/l in the complete incubation medium.

Finally, it should be stressed that if an overall understanding of the key metabolic pathways is to be achieved the activities of flux-generating (rate-limiting) enzymes should be determined (Newsholme and Start, 1976; Newsholme *et al.*,1980). Unfortunately only a few quantitative methods exist for these enzymes at present.

Van Noorden and Butcher (1991) provide a recent general review of quantitative enzyme cytochemistry.

28.8.8 Immunohistochemistry

It is important to appreciate that immunohistochemical methods do *not determine biological activity*. For example, immunological quantitation of an enzyme may be confounded by the detection of few enzyme molecules with high molecular activity in one setting against many enzyme molecules with low molecular activity in another (cf. Sect.25.1.3). In contrast the *localization* of antigens is often of considerable diagnostic importance.

Immunohistochemical methods can, at least in principle, be quantitated by one of the following techniques:

1. Microfluorimetry
2. Counting of gold or ferritin particles using the colloidal gold or immunoferritin methods
3. Scanning and integrating microspectrophotometry

This section is, however, only concerned with the microspectrophotometric aspect. Only relatively few quantitative immunohistochemical studies of this kind have been published. Some examples are given in Table 28.7.

When performing immunocytochemistry it is imperative to have as much information as possible concerning the antigens and antibodies used in the study. While this has been universally accepted the establishment of relevant criteria has met with considerable difficulty. The Immunocytochemistry Editorial Sub-Board of the Histochemical Journal agreed by majority in 1985 (News and Views, 1985) that immunocytochemical manuscripts should fulfil the following criteria as a minimum:

1. The antibody used must be characterized as to the specificity of its affinity for the antigen in the cell or tissue under study. In addition, molecular characteristics (e.g. molecular weight) should be provided.
2. Any quantitative or semiquantitative statement must be justified by appropriate measurements.

In the same publication, (News and Views, 1985), Sternberger pointed out that the above criteria only consider serum-derived antibodies and ignore the existence of monoclonal antibodies where the establishment of specificity is redundant. Sternberger further points out that most specificity tests in fact are less sensitive than immunocytochemistry.

In response to the above, Montero, (News and Views, 1986), recommends that an additional criterion should be inclusion of the parameters of any fixation procedure used.

At this point it is worth remembering that, whether or not fixation is an integral part of the immunohistochemical method adopted, there are some further problems in the quantitation of antigens in tissue sections. These, according to Larsson (1988), pp.193–194, include:

1. Epitope preservation by fixation and post-treatment
2. Penetration of antibodies and detection reagents
3. Steric hindrance (including prozone phenomena)
4. Efficiency of the detection procedure

Moreover, it is necessary to optimize every step in the subsequent immunocytochemical procedure. In addition, it is extremely important to include relevant controls (cf. Sect.26.5). A very thorough review of control procedures at different steps of immunohistochemical methods is given by Larsson (1988), pp.19–36.

When enzyme labels are used in microspectrophotometric quantitation, it is a prerequisite that there is a stoichiometric relationship between the amounts of antigen and final reaction product.

Sternberger and Sternberger (1986) compared the peroxidase-antiperoxidase (PAP) technique (Sect.26.3.3) with the avidin-biotin complex (ABC) technique (Sect.26.3.4) and concluded that in principle the PAP method, unlike the ABC method, is suitable for making quantitative estimates of the concentration of an

Table 28.7. Examples of quantitative immunohistochemical studies.

Antigen demonstrated	Tissue	Primary antibody	Visualization	Quantitation technique	Reaction type	References
Class II HLA-DR	human tonsil	mouse monoclonal	MTT	microspectrophotometry	direct glucose oxidase	Poulter et al., 1987
Oestrogen receptor	human breast	rat monoclonal	DAB	microspectrophotometry	PAP	Petersen et al., 1987
Immunoglobulins	human synovial membrane	rabbit polyclonal	AEC	microspectrophotometry	PAP	Fritz et al., 1988
NADPH-ferrihaemoprotein reductase (EC 1.6.2.4)	rat liver	sheep polyclonal	DAB fluorescence	microspectrophotometry; microfluorimetry	a) PAP b) indirect FITC	Smith et al., 1983
Neurofilaments	rat brain	mouse monoclonal	DAB DAB	image analysis	a) ABC-peroxidase b) PAP	Sternberger and Sternberger, 1986
Metallothionein-1	rat placenta	rabbit polyclonal	DAB	microspectrophotometry	direct peroxidase	Roelfzema et al., 1989
Myelin basic protein	rat brain	rabbit polyclonal	DAB	image analysis	PAP	Sternberger et al., 1978
Tyrosine 3-monooxygenase (EC 1.14.16.2)	rat brain	rabbit polyclonal	DAB	image analysis	PAP	Benno et al., 1982a, b

antigen. The authors showed that when staining intensity was plotted against antibody dilution a flat curve was obtained with the ABC technique. In contrast staining intensity with the PAP technique showed an initial increase with progressive dilution of the antibody. Staining intensity then progressively decreased following an S-shaped curve. The initial increase in staining intensity has been explained by Bigbee *et al.* (1977) who noted that with the PAP technique decreased staining intensity or even false negative results might arise when a high concentration of antigen was combined with a high concentration of the primary antibody probably giving rise to steric hindrance. They believe that in this situation the secondary antibody may form cross-bridges to two antibody molecules from the first antiserum and therefore lose its ability to bind the PAP complex.

When measuring the activity of an enzyme it is normally a prerequisite that initial maximum rate kinetics are used. A plot of the time-course of product formation for an HRP catalyzed reaction is shown in Fig. 28.4.

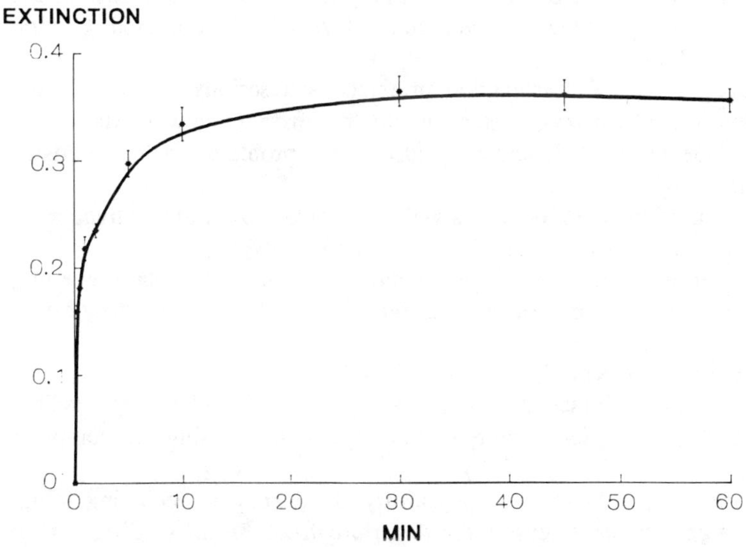

Fig. 28.4. Product formation (extinction) as a function of time for a horse radish peroxidase (HRP) catalyzed reaction. Pancreatic islets, rat.
Primary antibody: Specific islet cell cytoplasmic antibody (ICA)
Visualization: HRP conjugated protein A.
Cosubstrates: DAB 1mg/ml; H_2O_2 0.02%
(Marshall and Høyer, unpublished results, 1989)

In this example initial rate conditions only last a very short time (15–20 seconds) after which levelling off occurs. Several possible causes for this departure from linearity which are equally well-recognized in biochemistry, are listed below:

1. The reaction may be running out of substrate
2. The reaction may be approaching equilibrium
3. One of the products of the reaction may inhibit the enzyme

4. One of the components in the system is unstable under the assay conditions and is steadily being decomposed
5. An inhibitor of the enzyme may be present in the incubation medium
6. Loss of linear response by the microspectrophotometer (out-of-range error, cf. Sect.28.2.2)
7. The pH or ionic strength may change during the assay

These causes are elaborated on below:

1. In enzyme cytochemistry it is unlikely that the reaction will run out of substrate as the volume of incubation medium usually far exceeds that of the tissue. Theoretically, it could become a problem during continuous monitoring when the layer of incubation medium applied between slide and cover slip is thin
2. Equilibrium conditions do not occur with irreversible reactions, while for reversible reactions the probability of equilibrium is normally small as the reaction products diffuse from the enzyme site into the incubation medium. Equilibrium may, however, occur if one of the reaction products is unable to diffuse away, or does so very slowly, as may be the case if a PVA technique is used (Gordon and Robertson, 1986)
3. Inhibition due to one of the reaction products is a serious problem in the demonstration of 3β-hydroxy-Δ^5-steroid dehydrogenase, even when the use of a PVA technique is avoided, and may also cause problems in biochemistry (Caffrey *et al.*, 1979)
4. Instability of one of the components is well exemplified by heat inactivation of alkaline phosphatase (van Duijn and van Noorden, 1989)
5. The presence of an inhibitor in the incubation medium may explain levelling off in the case of HRP as one of the substrates (H_2O_2) has an inhibitory effect on the enzyme
6. Out-of-range error has been dealt with in Sect.28.2.2
7. Changes in pH taking place during the assay are easily checked. Possible changes in both pH and ionic strength can be calculated using the computer program suggested by Clancy (1987)

Apart from these seven factors, the possibility of the enzyme becoming "clogged" has been suggested as a reason for departure from linearity. Clogging of the enzyme site might be due to precipitation of the final reaction product here or to the cell and tissue matrix becoming obstructed. As noted by van Duijn and van Noorden (1989), no direct evidence for this phenomenon has, however, been presented.

On the basis of both theoretical calculations and practical experiments van Duijn and van Noorden (1989) point out that if plateau formation is due to gradual enzyme inactivation, it is possible to use plateau absorbance measurements as a parameter for enzyme activity, provided that the following requirements are fulfilled:

1. Enzyme inactivation should follow first order kinetics (Sect.23.1.3)
2. Other possible reasons for levelling off can be excluded
3. There exists a stoichiometric relationship between plateau absorbance values and amount of enzyme activity present

It has been demonstrated that these requirements are complied with in the case of catalase (Geerts and Roels, 1981), horseradish peroxidase (Nibbering *et al.*, 1986), and alkaline phosphatase (van Duijn and van Noorden, 1989)). As regards catalase and HRP, the inactivation process is probably due to reactive intermediates generated during the reaction of the enzymes with their substrates, whereas in the case of alkaline phosphatase heat inactivation was deliberately induced (van Duijn and van Noorden, 1989).

In conclusion, plateau absorbance values can in selected cases be used as a relative measure of enzyme activity instead of initial reaction rates (van Duijn and van Noorden, 1989).

Returning now to Fig. 28.4, this means that it is not necessary only to use the first part of the curve with initial rate conditions, but that any point on the curve may be selected for quantitating enzyme activity. Once the plateau has been reached, precise control of incubation time is not necessary. Before the plateau is reached, the same incubation time must, however, be used from experiment to experiment.

28.8.9 Glutathione

Smith *et al.* (1979) used the ferric ferricyanide method (Sect.8.3.1) to quantitate the distribution of reduced *glutathione* in cryostat sections of rat liver. The results obtained by microdensitometry were compensated for light scatter by subtracting the absorbance at the minimum absorption of Prussian Blue (475 nm) from the absorbance at the maximum absorption (675 nm). Although the Schmorl reaction is not specific for glutathione, the authors conclude that most of the reaction product formed is due to glutathione as there is a 70-fold greater concentration of this substance in the liver than of ascorbic acid.

28.8.10 Diffusible Ions

Information about the local concentrations of diffusible ions can be obtained using fluorescent probes or X-ray microanalysis. The fluorescent probes constitute an exciting new development that has received little application in conventional histochemistry. This is partly due to their dependence on intact, living cells as staining substrates.

By loading cells with the appropriate fluorescent probe *in vitro*, it is possible to make dynamic quantitative estimates of diffusible ions in living cells or tissues with a high degree of time resolution. The probes, which may also be termed indicators or dyes, are either introduced into cells by microinjection or by incubation with their respective membrane-permeable acetoxymethyl tetraesters. This latter approach effectively traps the probe within cells since cytoplasmic esterases produce tetra-anionic forms that are no longer membrane-permeable (Tsien, 1981; Grynkiewicz *et al.*, 1985).

The actual measurement of ion concentrations is based on either one of two phenomena arising from the binding of the probe to its specific ion; these are:

(1) a change in intensity of the emitted light and (2) a shift in the emission or excitation wavelength. By measuring light emission instead of absorbance, high sensitivity is achieved and ions can be measured in single cells. The equipment needed to measure the dynamic changes in ion concentrations at the single cell level comprises an epi-fluorescence microscope system connected to either a photometer or a video system. An inverted microscope is preferable or alternatively a water immersion objective can be used (Tanasugarn *et al.*, 1984). A high level of environmental control (temperature and CO_2) greatly facilitates the interpretation of results. Several investigations have been done at room temperature using HEPES buffered media without bicarbonate. These unphysiological conditions are to be avoided particularly when measuring intracellular pH (Thomas, 1989).

The photometer-based system is interfaced with a computer and this makes it possible to collect data at regular intervals and to calculate the intracellular concentrations based on the measured light intensity. While this system is cheaper than the video system and offers good time resolution, it only gives information about a single location (e.g. a cell or an organelle). The photometer approach and apparatus are described in detail by Tsien *et al.* (1985) and Cobbold and Rink (1987).

The video system is based on a very sensitive video camera (silicon intensified target (SIT) video camera) connected to an image analysis system. This system has a lower time resolution than the photometer system, but gives a visualization and a localization of the changes between cells or between different points in the cell. Video set-ups are described by Arndt-Jovin *et al.* (1985) and Poenie *et al.* (1986).

Measurements are made at either single or dual wavelengths depending on the probe used. The probes are classified as shown in Table 28.8.

Group I. Probes Changing Intensity of Light Emission on Binding to Ions. The concentration of the unbound ion is proportional to the intensity of emitted light from the probe after excitation of the probe at another wavelength. The intensity is measured at a single wavelength. Examples are quin-2 and fluo-3 (Fig. 28.5) for Ca^{2+}, and SPQ and SPA for Cl^-.

Group II. Probes Changing Excitation Wavelength. These probes change their excitation wavelength on binding to a specific ion. The emission wavelength remains the same but with an increasing concentration of the ion, the light emitted from the unbound probe will decrease at one excitation wavelength with a concomitant increase in light emitted from the bound probe at the other excitation wavelength. In principle, the concentration of the ion can be measured by registering the intensity of emitted light upon excitation of the bound probe. However, the change in excitation wavelength offers an opportunity of making a ratio between the intensity of emitted light during alternating excitation of the cells at the two excitation wavelengths, thus making the measurements insensitive to changes in the concentration of the probe, e.g. by leakage or photobleaching. For measurements with these probes more advanced equipment is necessary. The microscope must be equipped with a computerized filterchanger or two monochromators at the illumination site with equipment synchronizing the photometer or video equipment with

Table 28.8. Excitation and emission wavelengths of probes for the quantitative assessment of diffusible ions.

Probe	Group	Excitation wavelength(s) nm	Emission wavelength(s) nm	References
Calcium:				
BAPTA	I	254/274	363	Tsien, 1980
Quin-2	I	339	492	Tsien, 1980
Fura-2	II	360/340	505–12	Tsien et al., 1985
Indo-1	III	331 - 49	485/410	Grynkiewicz et al., 1985
Fluo-3	I	500–505	525	Tsien, 1988
Hydrogen:				
BCECF	II	508/?	530	Rink et al., 1982
SNARF-1	IV	579/518	640/587	Haugland, 1989
SNAFL-1	IV	537/479	623/543	Haugland, 1989
Potassium:				
PBFI	II	346/334	500	Haugland, 1989
Sodium:				
SBFI	II	346/334	500	Haugland, 1989
Magnesium:				
Furaptra	II	370/335	510	Raju et al., 1989
Chloride:				
SPQ[1]	I	320–50	445	Illsley et al., 1987
SPA[2]	I	400	490	Krapf et al., 1988

[1] High leakage, halftime: 10–30 min at 37°C.
[2] Lower leakage than SPQ, but a lower sensitivity for Cl^-.

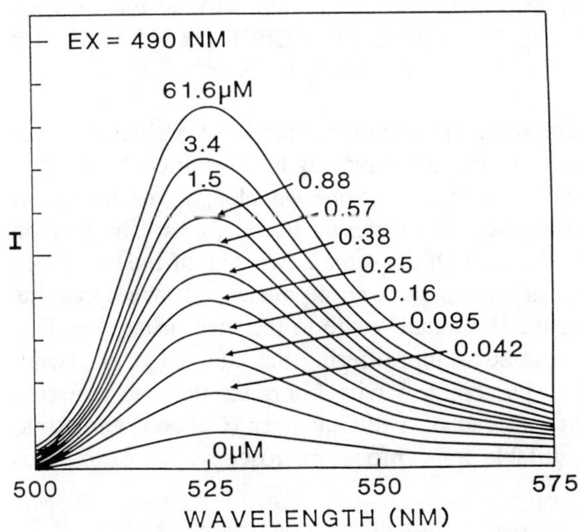

Fig. 28.5. Emission spectra of fluo-3 after excitation at 490 nm (EX = 490 NM) at various concentrations of calcium from 0 – 61.6 μM (μM = μmol/l of Ca^{2+}). Peak emission at 525 nm. I = fluorescence intensity (arbitrary units).
(Reproduced by courtesy of Molecular Probes Inc., Eugene, Oregon).

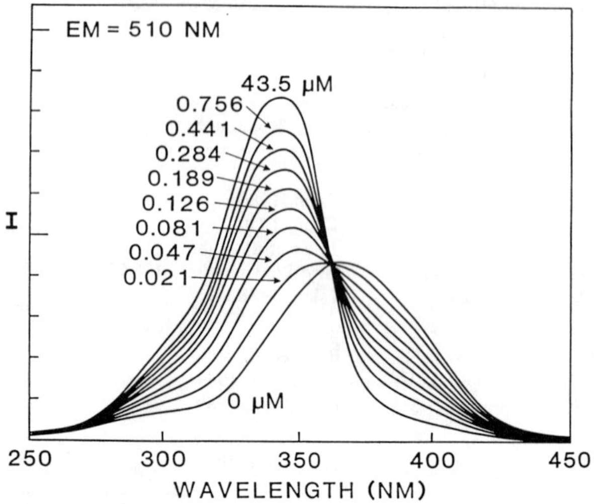

Fig. 28.6. Excitation spectra of fura-2 at various concentrations of calcium from 0 – 43.5 μM (μM = μmol/l of Ca^{2+}) obtained while monitoring the emission at 510 nm (EM = 510 NM). The excitation maximum shifts from 362 nm to 340 nm on binding of Ca^{2+}. I = excitation intensity (arbitrary units).
(Reproduced by courtesy of Molecular Probes Inc., Eugene, Oregon).

the changes of excitation wavelength. Continuous measurements are not possible since a ratio must be calculated from measurements at both excitation wavelengths. The speed of the filter change or change of monochromator is vital for the time resolution. Examples are fura-2 (Fig. 28.6) for Ca^{2+}, BCECF for H^+, SBFI for Na^+, PBFI for K^+, and furaptra, an analogue of fura-2, for Mg^{2+}.

Group III. Probes Changing Emission Wavelength. Instead of a change in excitation wavelength, these probes change the wavelength of emitted light upon binding to the specific ion. The cells are excited at one wavelength and the intensity of emitted light is measured at two other different wavelengths. The method is based on the same principle as above, with the same advantage of independence of probe concentration. In theory, measurements can be made with one apparatus by moving the filterchanger from the illuminating site to the measuring site. This can cause technical difficulties. Another setup, offering the advantage of continuous measurements, is to use two photometers with filters for the two emission wavelengths at the measuring site instead of a filterchanger (Cobbold and Rink, 1987). Probes in this group are suitable for confocal microscopy. An example is indo-1 for Ca^{2+}.

Group IV. Probes Changing both Excitation and Emission Wavelength. The probes in this group have the properties of both group II and III. The probes can be used in any of the described techniques (single or dual wavelengths). Examples of these probes are SNARF-1 and SNAFL-1 for determination of H^+.

Specific fluorescent probes for ions. The fluorescent probes have been extensively reviewed by Tsien, 1989. Commonly used probes are described below, whereas new and not so well documented probes are indicated in Table 28.3.

Calcium. So far, for the determination of diffusible ions, calcium probes have probably been the most frequently used. For additional information, see Cobbold and Rink, 1987.

BAPTA. This probe, BAPTA (group I), was the first rationally designed fluorescent probe showing high selectivity for calcium (Tsien, 1980). The intensity of light emission at 363 nm changes on binding to Ca^{2+}. The structure of BAPTA was based on the Ca^{2+}-chelator EGTA which in turn is a derivative of EDTA (cf. Sect.15.4). The stoichiometry of calcium binding is 1:1 with full saturation at 1 mmol/l Ca^{2+} and an effective dissociation constant K_d (cf. Sect.12.3.11) of 107 nmol/l. The wavelengths of excitation and emission are, however, too low to give BAPTA any practical value compared to the newer indicators, but BAPTA has served as a model for the other Ca^{2+} probes.

Quin-2. The probe Quin-2 (group I) is a derivative of BAPTA (Tsien, 1980) which emits at 492 nm on excitation at 339 nm (Tsien, 1982) and has a saturation level of 100 μmol/l. Quin-2 is pH insensitive at pH levels above 7.05. Leakage is less than 5%/h at room temperature. Quin-2 has unfortunately two major disadvantages:
1. K_d is low (115 nM) making Quin-2 unsuitable for intracellular Ca^{2+}-measurement of levels above 1 μmol/l as found in excited cells
2. Since it is the absolute intensity that is measured and not a ratio, the probe is sensitive to leakage, photobleaching, and to differences in cellular uptake

Fura-2. This probe (Fig 28.6) has several advantages compared to quin-2 (Grynkiewicz *et al.*, 1985; Tsien *et al.*, 1985):
1. It belongs to probe group II, changing excitation wavelength (from 362 to 340 nm) upon binding to Ca^{2+} with emission at 505–512 nm. To avoid loss of intensity in microscopes without quartz optics, measurements can be made using excitation wavelengths at 355 nm and 380 nm (Tsien *et al.*, 1985)
2. The intensity of emitted light is about thirty times stronger than that obtained with quin-2
3. The discrimination between Ca^{2+} and other divalent ions is better
4. The K_d is higher (224 nmol/l) than for quin-2

Indo-1. This probe has nearly all the advantages of fura-2 with a comparable increase in intensity (Grynkiewicz *et al.*, 1985). The main differences between the two probes are:
1. Indo-1 belongs to probe group III changing emission wavelength on Ca^{2+} binding (from 485 to 410 nm) on excitation at 331–349 nm
2. Photobleaching is much faster with this probe than with fura-2 (Tsien, 1988). K_d is 250 nmol/l

Fluo-3. This probe (Fig. 28.5) belongs to probe group I like quin-2, with the same disadvantages of not using the ratio technique. Fluo-3, however, offers other advantages such as a high K_d (400 nmol/l) and a wavelength of excitation in the visible spectrum (500–505 nm) making it suitable for confocal scanning microscopy (Tsien, 1988).

pH. Probes for H^+ make the determination of intracellular pH possible.

BCECF. The probe BCECF, belonging to group II, is an analogue of the pH sensitive 6-carboxyfluorescein (Rink *et al.*, 1982). The latter compound is not suitable for measurement of internal pH due to pronounced leakage. BCECF does not leak and has a pK_A (6.97) ideal for measurement of intracellular H^+ concentration. BCECF has an emission wavelength of 531 nm, and exhibits a shift in excitation wavelength upon binding to H^+ ions. Ratio measurements are possible when using excitation wavelengths of 440 and 500 nm.

28.8.11 Quantitative Detection of Hormones and the Effect of Various Compounds on Target Cells

The basis of *cytochemical microbioassays* is that a hormone acting on its specific target cell causes changes in the biochemical activity of that cell according to the physiological effect by which the hormone is recognized.

The basis technique for assaying hormones is outlined below:

1. The target organ is removed from a sensitive animal and cut into segments of a size that can survive *in vitro*. The segments are maintained in a non-proliferative culture medium (Trowell's T8 medium) for 5 hours. This procedure removes the target cells from the hormonal environment of the animal and allows them to recover from the trauma of excision
2. Initially, each segment was then exposed to one of a series of graded concentrations of a standard preparation of the hormone (to obtain a calibration graph) or to one of two dilutions (usually $1{:}10^2$ and $1{:}10^3$) of the test plasma. The segment is then frozen, sectioned in a cryostat at a suitable thickness (corresponding to the largest dimension of the cells), and finally reacted for the relevant biochemical activity (Chayen *et al.*, 1976)

It has subsequently been shown that similar responsiveness can be obtained by freezing and sectioning the segment after the 5 hour period of maintenance culture. The cells in the sections can then be exposed to the hormone, and the cytochemical reaction performed.

The maximum response of segments to picomolar concentrations of hormones occurs at 5–10 min while that of cells in sections is normally seen after 1–2 min. These assays have been shown to be highly reproducible, sensitive, precise, and specific. For a review, see Bitensky (1977).

The major advantages of this approach are according to Chayen (1980):

1. The cytochemical bioassays are "within-animal" assays and are therefore not disturbed by the considerable variability that exists between animals

2. The assays only measure the response in the target cells so that it is not confused by including non-responsive cells in the measurements
3. Cytochemical bioassays are extremely sensitive and are, for example, about one thousand times more sensitive than the equivalent radioimmunoassays for various polypeptide hormones (see Table 28.9)
4. Cytochemical bioassays measure the functional activity of molecules in contrast to radioimmunoassays. Cytochemical bioassays do thus not determine "big-ACTH", "big-gastrin", prohormones, or hormones that are otherwise changed metabolically or configurationally

On the other hand, cytochemical bioassays are not suitable for large scale screening assays as they are more time-consuming than radioimmunoassays. Moreover, these assays are very demanding on the skills of the investigator.

The advantages have nevertheless led to the World Health Organization recommending that these techniques be the international microassays of choice (World Health Organization, 1975).

Extensive reviews of the cytochemical approach to hormone assays are given by Chayen (1978; 1980).

Other examples of segment microassays include: The effect of pentagastrin and plasma gastrin on carbonic dehydratase in parietal cells of the guinea-pig fundus (Loveridge *et al.*, 1974); the influence of physiological or subphysiological concentrations of corticotropin on 3β-hydroxy-Δ^5-steroid dehydrogenase in cells of the fascicular and reticular zones of the adrenal cortex (Loveridge and Robertson, 1978); determination of the concentration of a sodium transport inhibitor in human plasma and urine by measuring the stimulation of renal glucose-6-phosphate dehydrogenase or the inhibition of renal Na^+, K^+-transporting ATPase (Fenton *et al.*, 1982); the acute effect of physiological concentrations of thyroid stimulating hormone on β-galactosidase, N-acetyl-β-glucosaminidase, and leucyl-β-naphthylamidase activities in thyroid follicular cells (Perrild *et al.*, 1989).

The concentration of certain immunoglobulins, the so-called thyroid growth stimulating immunoglobulins that occur in Grave's disease have also been investigated using the cytochemical bioassay technique (Bitensky *et al.*, 1974; McKenzie and Zakarija, 1977; Drexhage *et al.*, 1989).

In principle, not only segment assays and section assays, but also cultures of single cells, may potentially form the basis for assays for investigating the effect of any stimulating or inhibiting compound, such as hormones, drugs, toxic agents, cytokines, growth factors, or antibodies.

In conclusion, we agree with Bitensky and Chayen (1978), that quantitative cytochemistry is now a highly precise and remarkably versatile way of examining changes in cellular biochemistry induced by biologically active substances. It is an extension of conventional biochemistry to the cellular level which avoids many of the statistical problems inherent in conventional biochemistry; moreover, the technique is non-disruptive.

Table 28.9. Examples of microbioassays.

Peptide hormone assayed	Target cells	Biological function tested by quantitative cytochemistry	Sensitivity[1]	References
Corticotropin	zona reticularis cells of adrenal cortex	depletion of ascorbate determined by Prussian blue or silver nitrate reaction	5 fg/ml	Chayen et al., 1972; Loveridge et al., 1975
Luteinizing hormone	corpus luteum cells of ovary	depletion of ascorbate determined by Prussian blue reaction	10^{-4} mU/ml	Rees et al., 1973; Buckingham et al., 1979
Thyroid stimulating hormone	follicle cells of thyroid	lability of lysosomal membrane determined by testing the permeability for the substrate used for measuring leucyl-β-naphthylamidase activity	4×10^{-5} μU/ml	Chayen et al., 1980
Gastrin	parietal cells of stomach	production of H^+ indirectly measured by carbonate dehydratase	5 fg/ml	Loveridge et al., 1978; Loveridge et al., 1980
Parathyroid hormone	cells of distal convoluted tubules of kidney	transport of Ca^{2+} indirectly determined by glucose-6-phosphate dehydrogenase activity	5 fg/ml	Chambers et al., 1978; Goltzman et al., 1980
Corticotropin releasing hormone	cells of the anterior pituitary	corticotropin liberated		Buckingham and Hodges, 1977
Thyrotropin releasing hormone	cells of the anterior pituitary	thyroid stimulating hormone liberated	5 fg/ml	Gilbert et al., 1977

[1] According to Bitensky and Chayen (1978) and Chayen 1980.

29 Autoradiography

M. Møller, I.M. Krogh

In this chapter the general outlines of histochemical application of autoradiography are given. For more detailed information the reader is referred to Pearse (1980), pp.303–345 and Williams (1977). See also Sect.31.12.

29.1 Physical Principles of Autoradiography

The electromagnetic radiation emitted by radioactive isotopes affects photographic emulsions in the same way as light. During development, the resultant altered silver bromide crystals in the gelatin emulsion induce silver grain formation at sites corresponding to the incident radiation. Minute amounts of electromagnetic radiation can be detected because emulsions can be exposed over extended periods.

29.2 Application

When a molecule containing a radioactive isotope is injected into or administered to an animal it is absorbed and metabolized in the same way as a non-radioactive molecule. The absorption, metabolic translocation, and excretion of a specific molecule can therefore be followed by producing an isotope-labelled molecule and detecting its fate by autoradiography.

The isotope is often injected into the vascular system or peritoneal cavity. After a pre-determined period of time, the animal is sacrificed and the tissue fixed and processed for histology. Tissue sections are overlayed with a photographic emulsion and exposed in darkness to the radiation from the isotopes in the section for the desired time. After photographic development, fixation of the emulsion, and histological counter-staining, the autoradiograph can be studied by microscopy. The silver grains are localized in the emulsion above the histological structures containing the isotope.

29.3 Isotopes

Most elements (E) occur as several isotopes each of which possess the same chemical properties due to their identical complement of electrons but different physical properties, due to the differences in their nuclei.

An element is defined by its atomic number (Z) which corresponds to the number of positively charged protons in the nucleus. The element is also characterized by a mass number (A), which comprises the sum of the protons and the uncharged neutrons (N) within the nucleus. Neutrons have the same mass as protons.

Atoms with the same atomic number but different mass numbers are described as isotopes of the element concerned. If the number of neutrons is different from the number of protons, the isotope may be unstable and radioactive. Sooner or later such unstable isotopes transform into more stable isotopes by emission of ionizing radiation (α-, β-, or γ-radiation). The stable and most common isotope of carbon possesses a mass number of 12, formally written ${}^{12}_{6}C$ (${}^{A}_{Z}E$). The atomic number is often omitted because C alone specifies carbon. When denoting radioactive isotopes, however, the mass-number (A) is always included (e.g. 14C or sometimes C-14).

29.3.1 Ionizing Radiation

The type and the energy of the radiation is different for each isotope. Only some forms are suitable for autoradiography.

α-Radiation. This comprises helium atomic nuclei ${}^{4}_{2}He^{2+}$. All α-particles from a particular isotope are emitted with the same energy (monoenergetic radiation). Traces made by α-particles in a photographic emulsion appear as lines of identical length. This form of radiation is rarely used for autoradiography because it is only emitted by some heavy elements that are of little or no interest in biological systems.

β-Radiation. This consists of negatively or positively charged β-particles which are emitted by an isotope during transformation of the nucleus. β-Particles show a spectrum of energies ranging from a maximum value down to zero. The maximum energy (E_{max}) is characteristic for a particular isotope. Isotopes which emit β-radiation with low E_{max}, notably tritium (${}^{3}H$) but also ${}^{14}C$ and ${}^{35}S$, are very useful for autoradiography. H and C are clearly valuable because of their presence in all organic molecules.

Auger-Radiation. This is a monoenergetic soft β-radiation which is very useful for autoradiography. An internal conversion within the nucleus releases energy some of which is transferred to one of the electrons in an outer orbital. The radiation comprises the emitted electrons resulting from the energy transfer. The importance of Auger radiation in autoradiography has increased with the greater use of ${}^{125}I$. This isotope emits both γ-radiation as well as β-radiation by internal conversion. ^{125}I can therefore be used for combined biochemical and autoradiography experiments. The isotope is also easily incorporated into protein molecules.

γ-Radiation is a monoenergetic electromagnetic radiation which is emitted from a nucleus during nuclear transformation. It is often emitted in parallel with β-radiation. γ-radiation is not suitable for autoradiography because most photographic emulsions are not sensitive to it. It can, however, be detected with X-ray film.

29.3.2 Half-Life

Sooner or later unstable atoms transform into stable atoms by the emission of ionizing radiation. Although, *a priori*, it is not possible to calculate when emissions will occur, the time taken for half the atoms to decay into stable elements can be determined. This time is called the half-life ($T_{1/2}$) of the element and is characteristic for each isotope. The half-life is of ^{14}C is very long ($T_{1/2}$ = 5,730 years) while that of ^{3}H is shorter ($T_{1/2}$ = 12.3 years). Suitable isotopes for autoradiography do not have short half-lives. Table 29.1 reviews the characteristics of some biologically important isotopes.

Table 29.1. Isotopes and autoradiography.

Element	Radiation	E_{max}	Half-life
^{3}H	β^-	0.0186 MeV	12.3 years
^{14}C	β^-	0.156 MeV	5,730 years
^{22}Na	β^+, γ	0.54 MeV	2.6 years
^{24}Na	β^-, γ	1.389 MeV	15 hours $
^{32}P	β^-	1.710 MeV	14.3 days
^{35}S	β^-	0.167 MeV	87.2 days
^{36}Cl	β^-	0.714 MeV	3.1×10^5 years
^{42}K	β^-	3.52 MeV	12.4 hours $
^{45}Ca	β^-	0.254 MeV	165 days
^{55}Fe	Auger	0.006 MeV	2.7 years
^{125}I	Auger, γ	0.0056 MeV 0.034 MeV	60 days
^{131}I	Auger, γ	0.61 MeV	8.04 days
^{232}Th	α, γ	0.401 MeV 0.395 MeV	1.39×10^3 years

$ The half-lives of these isotopes are too short to be suitable for autoradiography.

29.4 Preconditions for Autoradiographic Experiments

Consideration must be given to both the physical and histological aspects of the technique:

1. The isotope must be retained and remain at the same tissue location during the histological procedure (fixation, dehydration, embedding, etc.). Several small hydrophilic molecules are extracted during these processes and their detection requires special procedures
2. It is essential to determine whether the radioactive isotope remains attached to the molecule being studied or whether it can be incorporated into another metabolite. Isotopes may also detach from the target molecule and bind to other molecules or lie free in the section. Controls can be performed to identify these pitfalls but their exact nature varies depending on the labelled molecule of interest
3. Sections should not contain substances that either sensitize the photographic emulsion directly (positive chemography) or that interfere with the sensitized bromide crystals in the emulsion so that they cannot be developed (negative chemography). (cf. Sect.29.7.2)

29.5 Light Microscopic Autoradiography

After tissue sections containing the radioactive isotope have been cut and placed on gelatinized glass slides, they must be de-waxed and placed in contact with an emulsion. The application of the photographic emulsion can be performed in two ways:

29.5.1 The Stripping-Film Technique

A stripping-film is a 5 μm thick photographic emulsion placed on a supporting layer of gelatin (approximately 10 μm deep) which is itself attached to a glass support. Working in a darkroom, the emulsion is stripped from the glass onto a water surface with the photoreceptive layer facing downward.

The slide bearing the isotope-containing section is then dipped into the water and the emulsion picked up in such a way that the section is covered by the stripping-film. After drying the slide is placed in light-tight boxes at 4°C for exposure. Exposure times can vary from few days to several months.

29.5.2 Liquid Emulsion Technique

Emulsions can also be made as gels. The gel is melted by heating to around 40°C and the emulsion diluted with an equal amount of distilled water. The glass slides bearing the labelled sections are dipped into the melted emulsion and placed on a horizontal table to dry. The thickness of the emulsion is dependent on the degree of dilution and the temperature of the emulsion. After drying the emulsions are exposed as in the stripping-film technique.

29.5.3 Emulsions

Photographic emulsions comprise silver bromide crystals in gelatin. The crystals consist of silver ions and bromide ions in a regularly spaced cubic crystal lattice. In order for the crystals to be sensitized by light (to produce a latent image) small defects have to be present in the lattice.

Our understanding of the photographic process in an emulsion is incomplete. It is proposed that light raises the energy of some of the electrons around the atomic nucleus so that they leave the orbit of the atom. The free electrons are thought to collect within the crystal in the defects (sensitivity specks) and reduce the silver ions in this part of the lattice to metallic silver. During development the entire silver bromide crystal is converted into metallic silver. The crystals which are not converted to metallic silver are removed during the fixation process.

The sensitivity and resolution of an emulsion is dependent on the size of the silver bromide crystals. Larger crystals have a greater chance of being hit by the emissions, and these sorts of emulsions are therefore the most sensitive. On the other hand, their resolution is low. It is possible for the manufacturers to increase sensitivity by adding certain "sensitizers". The principle producers of autoradiography emulsions are Ilford and Kodak. Ilford emulsions are designated by letters: G-emulsion (crystal size 0.27 μm), K-emulsion (crystal size 0.20 μm), L-emulsion (crystal size 0.15 μm). The letter is followed by a digit (e.g. L-4) indicating the sensitivity of the emulsion (from 0–5), where 5 indicates the most sensitive. The NTB emulsions from Kodak are also excellent, with crystal diameters of 0.20 μm and several grades of sensitivity.

29.5.4 Photographic Development and Fixation

During development the silver bromide crystals in the emulsion are reduced to metallic silver. As shown in Fig. 29.1 it is important to control the temperature and the development time in order to obtain the optimum balance between the development of grains initiated by specific radiation and the background.

29.5.5 Staining

A number of histological stains can be used after the autoradiographs have been developed. Stains have to pass through the gelatin layer of the emulsion and problems

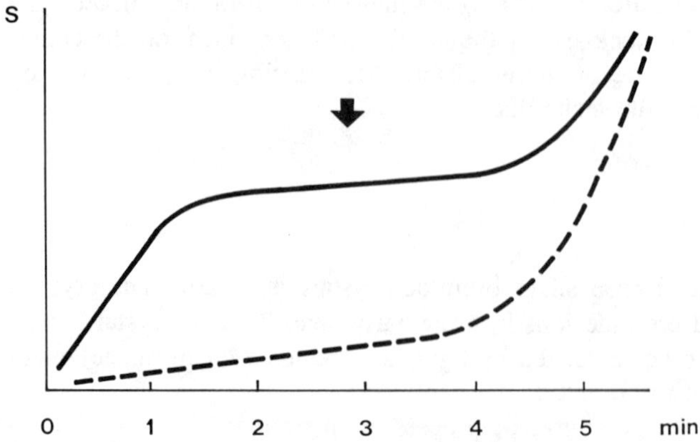

Fig. 29.1. The kinetics of development of specific grains compared with the background. Number of silver grains developed (S), development time (min).
(– – –) number of specific grains
(———) number of background grains
(⇓) optimum time for development

may arise with acidic reagents which may remove silver grains and with staining of the gelatin itself.

It is also possible to perform enzyme reactions through the gelatin layer but incubation times have to be prolonged. Eosin can be recommended for routine counterstaining because careful washing in tap water removes the stain from the gelatin before it is washed out of the section.

29.5.6 Autoradiography at the Ultrastructural Level

In principle, there are no fundamental methodological differences between autoradiography at the light microscopic level and at the ultrastructural level. It is important to appreciate that the resolution of autoradiography at the electron microscopic level is not much higher than at the light microscopic level. With tritium as the isotope, the resolution of light microscopic autoradiography is about 0.3 to 0.4 μm while at the ultrastructural level it is 0.1 to 0.2 μm. This EM-resolution is not very high compared to the resolving power of the microscope itself (0.2 nm). In spite of this, autoradiography at the EM-level remains very useful because labelling can be readily associated with specific organelles (e.g. mitochondria, endoplasmic reticulum, secretory granules and the Golgi-apparatus). The resolution of EM-autoradiography can also be increased by quantitative procedures (cf. Sect.29.8).

In EM-autoradiography a photographic emulsion is placed on a thin section which is in turn attached to a Formvar-coated grid. Ilford L4 or Kodak NTB emulsions are commonly used, the latter yielding slightly better resolution. The emulsion is either placed on the thin sections by use of a loop (Fig. 29.2)

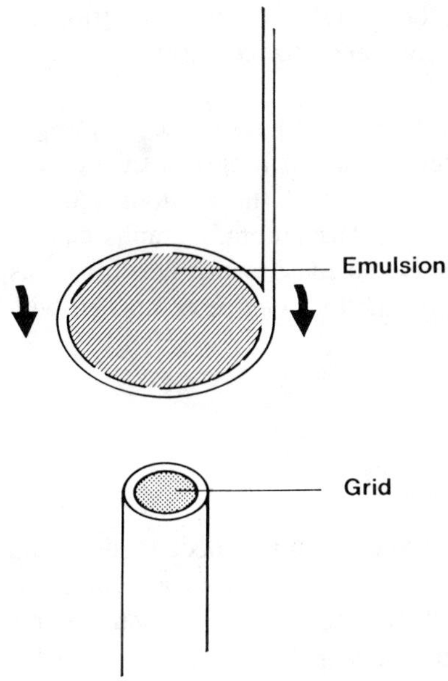

Fig. 29.2. The loop method.

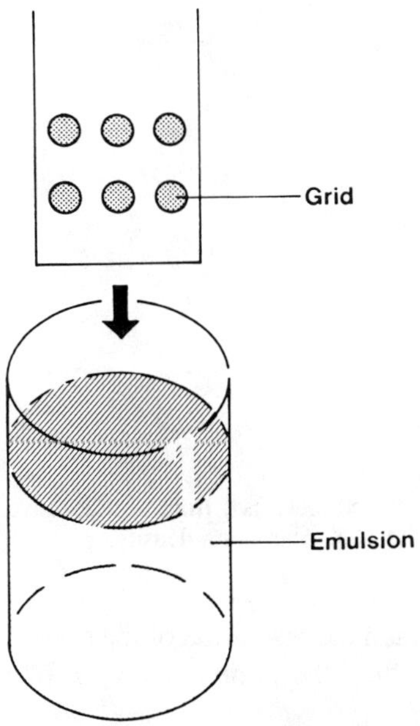

Fig. 29.3. The dipping method.

It is essential to obtain a monolayer of silver bromide crystals in the emulsion covering the thin sections. If the emulsion is too thick the resolution is drastically decreased. The thickness of the emulsion depends on the dilution and temperature of the emulsion and can be evaluated by the interference colour of the dried emulsion. The interference colour for a monolayer of an Ilford L4 emulsion should be purple. The thickness of the emulsion is also easily determined in the electron microscope.

or by dipping the sections in the emulsion. In the latter method the grids are placed on a collodion coated glass slide and the slide and attached grids are dipped into the emulsion (Fig. 29.3).

The autoradiographs are developed in light-tight boxes at 4°C containing a desiccant such as silica gel. The exposure times are much longer than for light microscopic work. This is due to the extreme thinness of the sections (80-100 nm) which contain very small amounts of isotope. The autoradiographs can be developed for 5 min in a D-19 developer at 20°C and fixed with thiosulphate as in the light microscopical procedure. After fixation sections can be contrasted with lead and uranyl acetate.

29.6 Resolution

The emissions from a radioactive isotope are random with respect to time and direction. Depending on the exposure time, each source produces a number of grains at different locations in the emulsion with the isotope at the centre. For a β-emitting source the distribution of grains would appear as shown in Fig. 29.4.

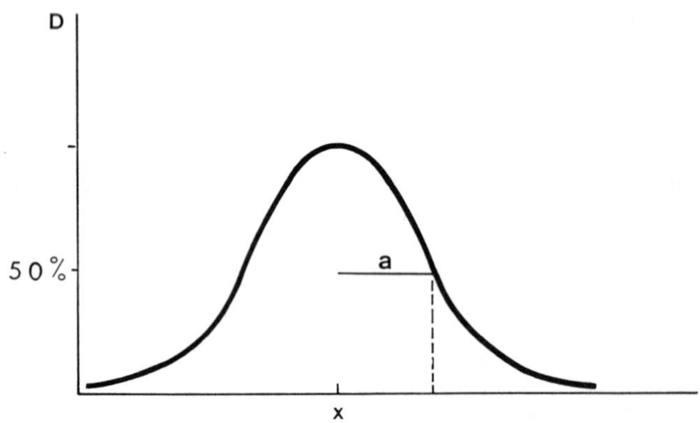

Fig. 29.4. Grain density (D) distribution as a function of distance (abscissa) from a radioactive source (x). Resolving power is the distance (a) from source to 50% peak grain density.

The curve shows a normal distribution within which the resolution of the system is defined as the distance (a) from the source at which the grain density falls to half of that detected over the source itself.

Because the resolution depends on several factors (isotope, grain size of the emulsion, thickness of the emulsion, developer) a related value based not on grain *density* but on total *number* of grains called the half radius (HR) is used instead (Williams, 1977, part II, pp.89-94). In a specific autoradiograph 50% of the grains generated by a point source should be located within a distance indicated by HR. The HR for ^{3}H detected with an Ilford L4 emulsion and developed in Microdol X is 280 nm. For ^{14}C the HR is 390 nm.

29.7 Artifacts

In every autoradiograph some of the grains are not caused by the presence of the isotope in the section. The reverse problem may also occur (i.e. sensitized silver bromide crystals disappearing or failing to give rise to silver grains by the end of development). Some of these artifacts will be discussed further below. They are, however, often unpredictable and careful autoradiographic controls are always required.

29.7.1 Autoradiographic Background

1. A high background is mainly due to extraneous radiation sources (e.g. cosmic and local radiation). This problem can be diminished by exposure in lead boxes, but this rather extreme precaution is not normally necessary
2. The emulsion may accidentally have ben exposed to light resulting in an increased background. It must be remembered that, even in a dark room, the amount of time the autoradiographs can be exposed to light is limited. Sparks from thermostats, nylon gowns and particularly adhesive tape can be a problem
3. The exposure and/or development times may have been too long
4. Contamination of the emulsion from glassware, chemicals, metal instruments (glass should be used throughout) as well as dust in the laboratory may result in an increased background
5. If the emulsions are too old a spontaneous increase of background will appear

29.7.2 Chemography

Chemography may be positive or negative.

Positive Chemography. This is caused by components of the section other than the specific isotope. These are often reactive molecules with reducing groups which can directly sensitize the crystals in the emulsion. The process is temperature dependent and can be decreased by exposing the emulsion at low temperatures. It is also possible to decrease the chemography by placing a layer of evaporated carbon between the sections and the emulsion.

Nuclear emulsions are sensitive to pressure and scratches and even fingerprints can produce grains. Lateral stress due to shrinkage during drying can also produce non-specific grains.

Negative Chemography. If the atmosphere during exposure is humid some of the sensitized grains lose this property and cannot be developed later on. This is called negative chemography.

29.8 Quantitation

A great advantage of autoradiography is that it can be quantitated with great accuracy. In fact, autoradiography can be compared with scintillation- or γ-counting with the emulsion serving as the detection system. In light microscopic work the grains over a particular histological structure (e.g. a cell) are counted and the number of grains per square μm calculated. After correction for the background, the results are compared statistically, often using a X^2-test.

Quantitation applied to EM-autoradiography is more complicated. As described in 29.7 grains related to one source are distributed at various distances from the isotope. It is therefore not possible directly to relate a grain to the structure located immediately beneath it. This problem can be overcome with a careful quantitative procedure. For example, if one wants to investigate whether an isotope is present in the mitochondria, the volume density of the mitochondria is measured (e.g. 3%). The number of grains overlying mitochondria is then determined and if more than the random predicted number (e.g. 3%) are located above this organelle then the isotope has been concentrated in mitochondria. The actual procedure is a little more complicated because the grains are not counted over the mitochondria but within the distance of a HD-radius (Sect.29.7) from the centre of the mitochondrion. A detailed description of this technique can be found in Williams (1977).

29.9 Radiation Safety

All countries have strict regulations with regard to storage and handling of radionuclides. These regulations generally follow the those laid down by the International Atomic Agency in Vienna (IAEA) in their booklet "Safe Handling of Radionuclides".

Radionuclides can be classified into four groups of radiotoxicity:

a. Very high
b. High
c. Moderate
d. Low

Isotopes used for autoradiography mostly belong to group d. Tritium and ^{14}C present absolutely no radiation hazard, however, ^{32}P and ^{125}I are in group c. Because ^{125}I can be concentrated in the thyroid gland this isotope should be handled in a fume hood. For further reading the booklet from IAEA is recommended.

30 Fluorescence Microscopic Methods in Histochemistry

M. Møller, H. Lyon

30.1 Autofluorescence

The majority of tissue components in unstained untreated sections show some degree of fluorescence. This is called primary fluorescence or autofluorescence. Such autofluorescence is particularly pronounced in plant tissues, while in animal tissue, collagen, elastin, and lipofuscin are noted for this property. Collagen and elastin (Sect. 21.5) show a blue-green fluorescence, while lipofuscin (Sect. 18.2.2) gives an orange fluorescence. The red autofluorescence of porphyrins (Sect. 18.2.1) may be useful in diagnostic work (Sect. 31.6.2). A number of drugs also fluoresce and it is therefore possible to follow their fate in tissues using fluorescence microscopy. Examples include tetracyclines (Sect. 15.9) and Acridines.

Background autofluorescence can be a nuisance in catecholamine work (Sect. 30.2) and immunofluoresence (Sect. 26.2.1). In these circumstances Cowen *et al.* (1985) suggest the use of Benzo Sky Blue (Pontamine Sky Blue 5BX, C.I. 24400) as a counterstain. This reduces autofluorescence.

30.2 Induced Fluorescence

Certain compounds, which are not fluorescent by themselves, can be converted to fluorescent products by suitable pretreatment. Formaldehyde is by far the most important reagent in this regard. It condenses with arylethylamines to form fluorescent isoquinolines, quinonoids and β-carbolines. Other useful reagents (see below) include glyoxylic acid, o-phthalaldehyde, hydrogen chloride, and acetyl chloride.

30.2.1 Formaldehyde Induced Fluorescence for Biogenic Amines

Biogenic amines can be divided into catecholamines and indolamines. The most important representatives of the catecholamines are adrenaline and noradrenaline which are localized to the nervous system and the adrenal medulla. Dopamine, which is found in the nervous system, may also be included in this group. The

H. Lyon (Ed.)
Theory and Strategy in Histochemistry

most important indolamine is serotonin which is found in the nervous system and in enterochromaffin cells (Sect.18.1.4). The formaldehyde induced fluorescence method (FIF) was first described by Eränkö (1955; 1967) and further developed by Falck and coworkers (1961; 1962). The method is very sensitive and highly specific. Due to the very high solubility of biogenic amines, it is usually necessary to perform the method on freeze-dried tissue (Sect.11.3.2). It is possible, particularly with serotonin, to obtain good results with standard paraffin-embedded material (Lyon *et al.*, 1982) or even better, material embedded in hydroxyethylmethacrylate (Lyon *et al.*, 1981) (Sect.14.4.2).

Mechanism. The material is treated with formaldehyde vapour which initiates a condensation reaction; with catecholamines the following occurs:

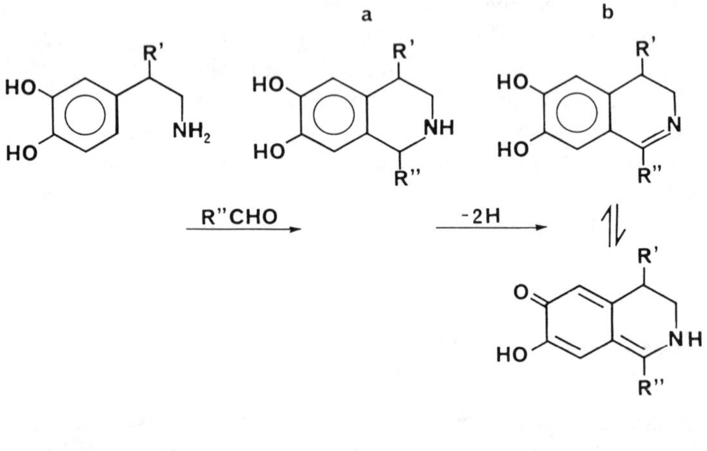

Dopamine: R′= H, Noradrenaline: R′= OH, Formaldehyde: R″ = H

The condensation is usually performed with formaldehyde (R″ = H), but can also be performed with vapours of acetaldehyde, glutaraldehyde, formic acid, and acetic acid all with excellent results.

With regard to adrenaline the dehydrogenation (autooxidation) (a→b) requires far more energy (Jonsson, 1967) than dopamine or noradrenaline.

For serotonin the fluorophore formation occurs in two steps which are similar to the steps for noradrenaline.

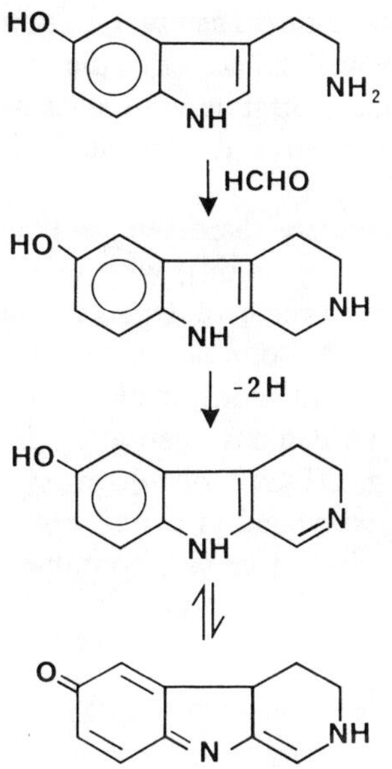

Selectivity. To obtain reasonable selectivity it is necessary to use a good quality fluorescence microscope and the best available filter combination. A combination of interference filters is strongly recommended. The excitation and emission maxima for biogenic amines reacted with formaldehyde are given in Table 30.1.

Table 30.1. Wavelengths for maximum excitation and the resultant emission maximum of formaldehyde induced fluorescence (FIF) of some biogenic amines.

Biogenic amine	Excitation (nm)	Emission (nm)
dopamine	410	470–480
noradrenaline	410	470–480
adrenaline	410	470–480
serotonin	385/410	520–540

An *interference filter* which permits light of wavelengths 405 ± 5 nm to pass through is recommended for the excitation or primary filter and secondary filters with a cut-off at 460 nm or 470 nm.

Catecholamines and serotonin can be distinguished because they emit light at different wavelengths (Table 30.1). To distinguish between the different catecholamines, however, it is necessary to plot both the excitation and emission spectra using a microfluorimeter.

Occurrence of Biogenic Amines. Noradrenaline is a synaptic transmitter in the sympathetic nervous system. The FIF method is therefore well suited for studies of this system and the tumours which it gives rise to. Also the central nervous system where dopamine and serotonin act as transmitters can with advantage be studied by this method.

FIF methods may also be used to demonstrate the widely dispersed peptide hormone-secreting cells (e.g. C-cells in the thyroid gland (calcitonin)) which collectively comprise the APUD-system (*A*mine and Amine *P*recursor *U*ptake and *D*ecarboxylation) (Pearse, 1968). In addition to their peptide hormone these cells produce a biogenic amine which can sometimes be directly demonstrated by fluorescence methods. Alternatively amine production can be demonstrated after (*in vivo* or *in vitro*) administration of an amine precursor (e.g. L-DOPA or 5-hydroxytryptophan) which is taken up by APUD cells and decarboxylated to the biogenic amine. These methods are also useful for the demonstration of certain hormone-secreting tumours in histopathology (e.g. carcinoids).

Sensitivity. This is very high indeed, e.g. noradrenaline in a neurone may be detected down to a concentration of 5×10^{-6} pmol/l.

30.2.2 Glyoxylic Acid Induced Fluorescence of Biogenic Amines

Glyoxylic acid can be used in place of formaldehyde to make fluorophores by condensation with biogenic amines. It may be used in vapour form, but tissue perfusion followed by freezing and freeze-drying (to sublimate the ice) is more usual. The reaction between the biogenic amines and glyoxylic acid is then initiated by heating the tissue. In a modification of this method (Lorén *et al.*, 1976), instead of heating, the tissue is exposed to formaldehyde vapour. The tissue block is then embedded in paraffin under vacuum and sections are cut and treated as for the formaldehyde method. Another modification (de la Torre and Surgeon, 1976) uses cryostat sections which are dried and subsequently treated with glyoxylic acid.

Mechanism. Glyoxylic acid reacts with biogenic amines as seen in the following scheme with noradrenaline as an example.

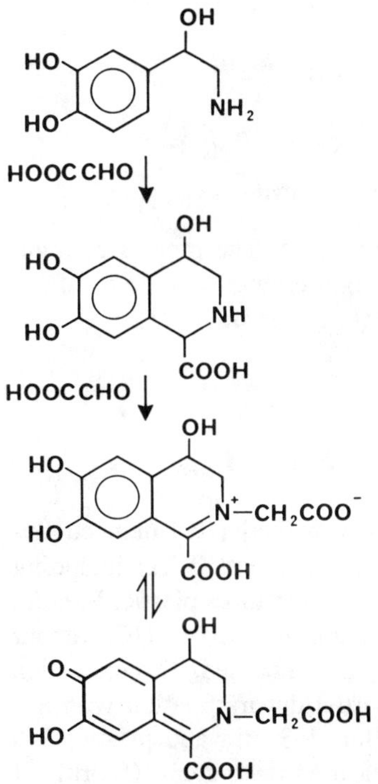

Fluorophore formation with serotonin occurs in two stages; a condensation with glyoxylic acid producing a weakly fluorescent product, followed by formation of a strongly fluorescent compound either by an autooxidative decarboxylation or by a further acid-catalyzed reaction.

Selectivity. The considerations are essentially the same as those mentioned in Sect.30.2.1. The substances reacting are with few exceptions (e.g. histamine) arylethylamines. Thus both this method and the FIF-method may demonstrate terminal tryptophan or DOPA (dihydroxyphenylalanine) residues in peptides and proteins as well as catecholamines and serotonin. These additional reactants have, however, different excitation and emission maxima and can therefore be easily distinguished using microfluorimetry. With fluorescence microscopy specificity can be determined by changing primary and secondary filters during observation.

Sensitivity. The glyoxylic acid method is even more sensitive than the FIF-method, noradrenaline can be detected down to a concentration of 10^{-7} pmol/l.

30.2.3 o-Phthalaldehyde Fluorescence Method for Histamine

The o-phthalaldehyde method (OPT) was first described by Håkanson *et al.* (1970). Deparaffinized sections are treated with the vapour of o-phthalaldehyde.

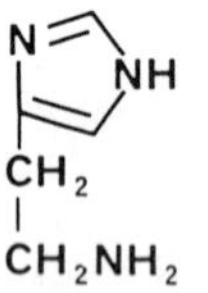

histamine

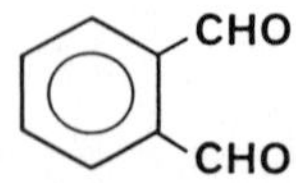

o-phthalaldehyde

Mechanism. A fluorescent condensation product is formed. The precise configuration is unknown. Using excitation at 365 nm, the light emitted is blue or, if the histamine concentration is particularly high, yellow (Cross *et al.*, 1971).

Selectivity. This is high for histamine.

30.2.4 Hydrochloric Acid Vapour-Induced Fluorescence for Dopamine

In the hydrochloric acid vapour-induced fluorescence method (HCl-induced fluorescence) the material is treated as for the FIF-method (Sect.30.2.1) including treatment with formaldehyde vapour and embedding in paraffin or plastic. Paraffin sections are partially deparaffinized after which they are exposed to HCl vapour for 15–30 sec at 25°C (plastic sections for 2–4 min) and mounted. The excitation spectra are then measured, followed by a second partial deparaffination with xylene, and further treatment with HCl-vapour (paraffin, 2–3 min and plastic, 3–6 min). After mounting, the excitation spectra are measured once again (Björklund *et al.*, 1968; 1972; Eränkö and Eränkö, 1971).

Mechanism.

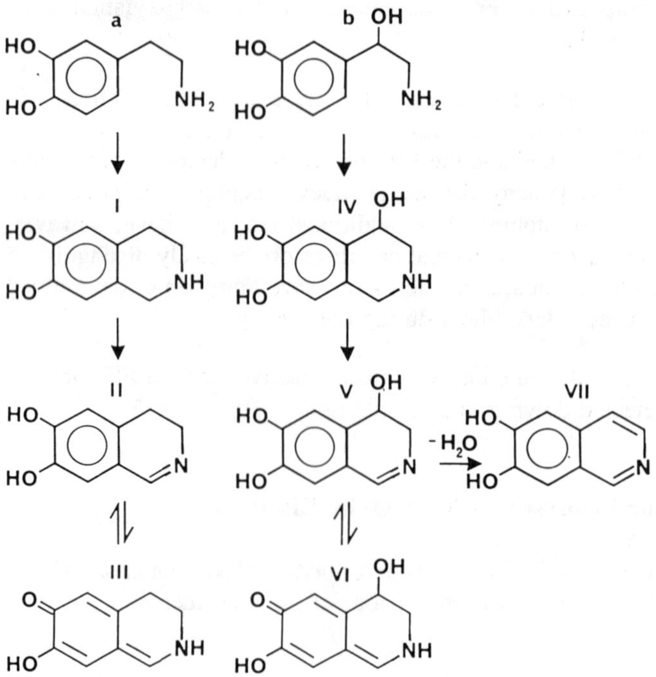

Atabrine Atabrine Mustard (R as in Atabrine)

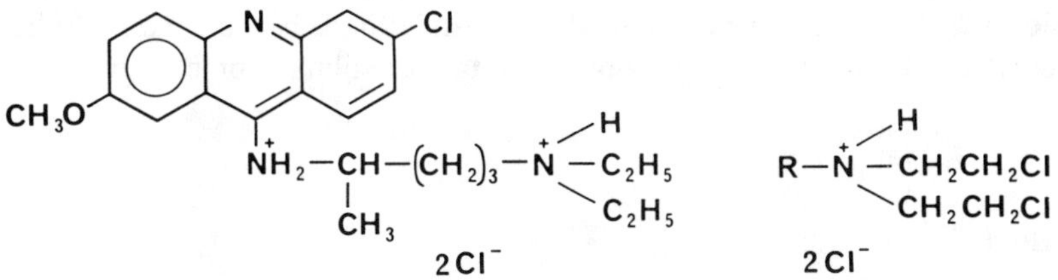

Dopamine (a) and noradrenaline (b) form the isoquinolines I and IV on condensation with HCHO. Dehydrogenation leads to the fluorescent dihydro-derivatives in their quinone form III and VI with excitation maxima at 410 nm. Short treatment with HCl-vapour leads to formation of the non-quinone forms II and V with excitation maxima at 370 nm. Further treatment with HCl leads to the transformation of V to VII with an absorption maximum of 320 nm; II cannot react in this way, and it thus possible to differentiate between dopamine and noradrenaline.

Selectivity and Sensitivity. These are high.

30.3 Direct Fluorochromy

This means staining with fluorescent dyes. Dyes may be subdivided into *diachromes* and *fluorochromes* (Sect.26.2.1) depending on whether they can seen by ordinary light or fluorescence microscopy. Many diachromes are also fluorochromes, e.g. Congo Red, Eosin, and Pararosanilin.

Fluorochromes are extremely sensitive stains. Compared to non-fluorescent dyes, very small amounts of dye can be demonstrated by fluorescence microscopy. Fluorescent dyes are very useful for screening procedures since the fluorescence is read against a dark background. Specific staining can often be rapidly detected at low magnifications even by inexperienced workers (Wachsmuth, 1988).

Staining can be quantitated using microfluorimetry (Chap.28).

Some examples of direct fluorochromy are given below.

30.3.1 Chromosome Banding

The antimalarial drug Atabrine (Quinacrine, Mepacrine) and the closely related mitosis inhibiting substance Atabrine Mustard (Quinacrine Mustard) are fluorochromes which stain chromosomes in mitotic cells and give rise to highly characteristic bands which make it easier to identify the individual chromosomes.

Staining is usually performed on material which has been pretreated with trypsin. Q-bands and QM-bands are referred to depending on which fluorochrome has been used. Similar results can be obtained with the Giemsa-method, G-bands.

It is possible to "reverse" the bands, so-called R-bands (dark bands become light and vice versa) by denaturation (e.g. with a strong salt solution or heat prior to staining).

30.3.2 Amyloid

Amyloid can be demonstrated with the fluorochrome Thioflavine TCN (Sect.21.7). The specific silver-white fluorescence is lost if the secondary filter is yellowish. Congo Red exhibits red fluorescence, but this is not specific for amyloid, and the section should instead be examined with a polarizing microscope.

30.3.3 Supravital Staining

Supravital staining must be performed with very low dye concentrations to avoid toxic effects. There are therefore considerable advantages in using fluorochromes such as Neutral Red, Acridine Orange, Congo Red, and Evans Blue for this purpose.

30.3.4 Schiff's Reagent

A Schiff's reagent prepared from Basic Fuchsin or Pararosanilin is fluorescent in itself. Fluorescent Schiff's reagents can also be made from Acriflavine or Auramine O.

30.3.5 Fluorescence Metachromasia

Fluorescence metachromasia is in principle identical to that seen by light microscopy (Sect.6.1.1) as the emitted fluorescence is shifted to a longer wavelength and the intensity of the fluorescence is diminished. Acridine Orange (C.I. 46005) and Coriphosphine O (C.I. 46020) are frequently used for this purpose. Both are green in their orthochromatic forms and red in their metachromatic forms. The advantages of using fluorescence metachromasia are that the number of binding sites in the tissue and the distance between these become apparent through changes in intensity of fluorescence and degree of metachromasia respectively. In particular, Acridine Orange has been used in the study of nucleic acids as DNA is stained orthochromatically and RNA metachromatically. Coriphosphine O has been used for masked metachromasia (Sect.21.4.2).

30.3.6 Procion Yellow MX

Procion Yellow MX is a fluorochrome belonging to the reactive dyes (Sect.3.3.8) that has been found to be an excellent intracellular tracer for neurons. The dye is introduced by electrophoresis and penetrates even the finest ramifications. The staining is resistant to the normal histological processing and can be combined with procedures such as FIF.

30.4 Indirect Fluorochromy = Immunofluorescence

(See Chaps.26 and 32).

30.5 Enzymatically Provoked Fluorescence

Enzyme activity in cells can be detected when the enzyme causes a change in fluorescence by its action on a substrate or a coenzyme. The redox kinetics of enzyme systems in which pyridine nucleotides participate can be followed using the fluorescence of NADH. In the case of hydrolases it is possible to use the fluorescence of the enzyme product (naphthol derivative) instead of coupling reactions. This makes increased sensitivity and optical specificity possible and facilitates microfluorimetric quantitation (Chap.28).

Part 7
An Introduction to Applied Histochemistry

31 Applied Histochemistry – An Overview

H. Lyon, E. Schulte, J. Visfeldt, E. Hasselager

In this chapter and Chap. 32, a number of applications for histochemical procedures are outlined. Cross references are made to the relevant sections earlier in the book. The overall organization follows that used for the preceding chapters. The sections can therefore be regarded as appendices applicable to specific earlier chapters. Each section is divided into subsections dealing with normal or pathological entities.

31.1 Tissue Processing

The requirements for the majority of histological and histopathological investigations are fulfilled by formalin fixed material stained with Al-haematein and Eosin. There are, however, several situations where supplementary histochemical staining provides information crucial to determining the correct diagnosis.

Where possible, the necessity for supplementary investigations should be anticipated and appropriate special methods of tissue preparation selected (Chap. 10).

Tissue preservation may be achieved by freezing (Chap. 11) or by chemical fixation (Chap. 12). If freezing is performed to a high technical standard (Sect. 11.2.2), preservation is excellent and the range of techniques that may be performed is essentially unlimited. In contrast, chemical fixation inevitably limits this potential (Chap. 12).

31.2 General Oversight Stains

The morphological evaluation of cells and tissues requires that key structures are demonstrated by staining. The chosen methods should give reproducible results irrespective of the source of material, manner of fixation, or thickness of the preparation.

The most frequently used general oversight stains are "Haematoxylin"-Eosin, Romanowsky-Giemsa, and Papanicolaou. For oversight staining of plastic embedded material Toluidine Blue O may be used.

H. Lyon (Ed.)
Theory and Strategy in Histochemistry

31.2.1 "Haematoxylin"-Eosin

Probably the most frequently used method in histology and histopathology is "Haematoxylin"-Eosin, H-E, or more correctly Al-haematein-Eosin. Using the metal complex dye Al-haematein (Sect.7.2.2) nuclear staining is highly reproducible almost regardless of fixation and tissue processing. Depending on the staining conditions many other tissue components can be stained (e.g. mucin, Sect.7.2.2). Cytoplasm and matrix proteins are counterstained with Eosin. The results of this are, in contrast to Al-haematein staining, highly dependent on fixation and other aspects of tissue processing (Sect.6.2.4) as well as staining pH (Sect.6.3.1), and post-treatment (Sect.16.2).

31.2.2 Romanowsky-Giemsa Staining

Sequential (May-Grünwald-Giemsa) or one step procedures (Giemsa) may be used. Many other variations have been described including Wright's stain and MacNeal's stain. The binding of Azure B and Eosin to chromatin giving the *Romanowsky-Giemsa effect* and the related Azure A-Eosin method are discussed in Sect.6.3.2.

These methods give excellent differentiation between different leukocyte series. They are the methods of choice for haematological preparations, smears of "cytological" material and sections with a high content of leukocytes (e.g. lymph nodes). The Romanowsky-Giemsa can also be used for the investigation of chromosomes (Sect.30.3.1).

A detailed description of a standardized *Azure B-Eosin staining method* is given in Appendix A.

31.2.3 Papanicolaou's Method

This is a sequential trichrome method developed for the staining of cytological smear preparations and "millipore"-preparations. In addition to a specified protocol of fixation and washes in water and ethanol, three staining solutions are applied:

1. Al-haematein
2. Phosphotungstic acid/Orange G
3. Phosphotungstic acid/Light Green SF/Bismarck Brown/Eosin (Sect.21.6, trichrome methods)

Papanicolaou's method is described in detail in Lillie and Fullmer (1976) p.721. Wittekind *et al.* (1979; 1982; 1983) recommend that Light Green SF be substituted with Fast Green FCF (for reasons of stability), that Bismarck Brown be omitted (it doesn't stain anything anyway!), and that Al-haematein should be substituted with Thionin.

31.3 Demonstration of Ionized or Ionizable Groups

The basic principles for the demonstration of ionized or ionizable groups are discussed in Chap.6 where also the definitions for acidophilia and basophilia are given.

31.3.1 Basophilia

For investigating basophilia or azurophilia Toluidine Blue O is most frequently employed, but also Azure A and Azure B can be used. All three dyes can exhibit a reddish-purple metachromasia on binding to polyanions (Sect.6.1.1). Toluidine Blue O may be used as a general stain either as a quick method in cryostat diagnostics or for the staining of "semithin" (1 μm) plastic sections for the survey and selection of areas of tissue for electron microscopy (Sect.14.4.2).

Basophilic structures comprise *chromatin* (DNA, Sects. 20.2, 20.4, and 30.3.1 (G-banding technique)), *nucleoli*, and *ribosomes* (RNA). A particularly high content of ribosomes usually explains cytoplasmic basophilia (Sect.20.4). Examples are lymphocytes, macrophages, plasma cells, monocytes, erythroblasts, platelets, exocrine epithelial cells in the pancreas, the majority of the chromophil cells in the pituitary gland, and the perikaryon of nerve cells (Nissl bodies). All of these are cells with a high protein synthesis. On degeneration cells lose their cytoplasmic basophilia, while basophilia may be increased in certain inflammatory conditions, for instance epithelial cells in the crypts of the large intestine in chronic ulcerative colitis.

Basophilic Granules. These are found in mast cells and in the basophils in blood. They chiefly contain a proteoglycan in which heparin forms a part (Sect.2.1.5). This is of importance in identification as the granules are stained metachromatically even at low pH (particularly if preceded by deamination). Mast cells and basophils are encountered in increased amounts in inflammatory processes, particularly acute, allergic reactions. It should be noted that heparin is soluble in alcoholic fixatives. Therefore, in cytological specimens (blood and bone marrow smears) dye-fixation as in Giemsa stains is preferable to fixation in pure methanol (cf. Appendix A).

Azurophilic Granules. These usually represent *lysosomes* (Sect.2.2.8) and may be demonstrated in macrophages, monocytes, lymphocytes, as well as in neutrophils. In the neutrophils only about 1/3 of the granules are truly azurophilic, while the remaining 2/3 are stained weakly violet or not at all (neutrophilic) in a Giemsa stain (Sects.6.3.2 and 31.2.2).

Acid Polypeptide Hormones. These hormones of the APUD-system may be demonstrated in the cytoplasm of the cells synthesizing them using the masked basophilia or masked metachromasia reactions (Sect.21.4.2) (Solcia *et al.*, 1968).

Virus Particles. These may be found as basophilic inclusion bodies in for instance warts.

Bacteria. These frequently exhibit a more or less pronounced basophilia which is utilized in the Gram stain and the Ziehl-Neelsen stain (Sect.6.1.6). Further, the Giemsa stain may be used (Sect.31.2.2).

The fluorochrome Auramine O is used for the demonstration of acid fast bacilli (see further Sect.31.4.1).

31.3.2 Acidophilia

Acidophilia is determined using anionic dyes (Sect.6.2). Amongst the oversight stains, Eosin is the most commonly used, hence the alternative term, eosinophilia (Sect.31.2.1). More information concerning basic proteins may be obtained by staining with anionic dyes at high pH (frequently pH 8–10.5) (Sect.21.4.2). Fast Green FCF and Biebrich Scarlet are often used for this purpose.

Acidophilic structures include: Cells with acidophilic (eosinophilic) cytoplasm such as red blood cells with a high content of the basic protein haemoglobin (Sect.21.4.2), muscle cells with their content of myoglobin (Sect.21.4.2), and about a quarter of the chromophils in the pituitary gland. The cytoplasm of the parietal cells of the stomach is distinctly acidophilic. This is probably due to their high content of mitochondria containing the oxidoreductases of the respiratory chain; these are basic proteins (Chap.2 and Sect.25.1).

Induction. The cytoplasm of liver cells can show increased acidophilia on induction. This results from exposure to various metabolites and exogenous compounds and involves an increase in the amount of smooth endoplasmatic reticulum bearing the enzymes which decompose and/or detoxify these molecules.

Hyaline Degeneration. In hyaline degeneration of a cell, the cytoplasm becomes increasingly acidophilic as a result of organelle breakdown. The *Mallory body*, which can be demonstrated in liver cells in alcoholic hepatitis and seems to be predominantly composed of microtubules, is a prime example (Sect.2.2.9).

Askanazy or Hürthle Cells. In Hashimoto's disease the thyroid follicular epithelial cells transform into the highly acidophilic Askanazy or Hürthle cells which are characteristic of this condition. Their cytoplasmic acidophilia probably results from their high content of mitochondria.

Russell Bodies. These are strongly acidophilic cytoplasmic bodies occurring in plasma cells in chronic inflammation and myelomas.

Acidophilic (Eosinophilic) Granules. These are found in *eosinophil* polymorphonuclear leukocytes. They contain an arginine rich protein (Olsson *et al.*, 1977), the function of which is still unclear (Olsson *et al.*, 1977; Gleich, 1977). Increased numbers of eosinophils occur in type I immune reactions. They localize to areas where mast cell degranulation and the liberation of histaminase have occurred. Many eosinophils are encountered in ulcerative colitis, in nasal polyps associated with allergic rhinitis, and also in Hodgkin's disease.

Paneth Cells. These cells in the intestinal mucosa contain coarse acidophilic granules. These consist of the strongly basic, arginine rich enzyme *lysozyme* (muraminidase) which is able to degrade the material of the cell wall surrounding Gram positive bacteria (Sect.2.3.1).

Zymogen Granules. These acidophilic structures are seen in areas such as the apical part of the exocrine gland cells of the pancreas.

Acidophilic (Eosinophilic) Bodies. These are seen in acute viral hepatitis and probably represent dead cells.

Amyloid. This appears as an acidophilic material (Sect.21.7).

Oedema Fluid. With a sufficiently high content of protein, oedema fluid can be seen as an extracellular, light, acidophilic material.

Fibrin. This appears as acidophilic threads in fibrinous inflammation and in thrombi (Sect.2.4.7).

Fibrinoid. This stains as fibrin and is seen in different immune reactions such as in rheumatoid nodules and in the vessel wall in polyarteritis nodosa.

Viral Inclusion Bodies. These can be acidophilic, for instance molluscum bodies in molluscum contagiosum or Negri bodies in the cytoplasm in rabies.

31.4 The Demonstration of Microorganisms

31.4.1 Bacteria

The structure and chemical composition of bacteria is dealt with in Sect.2.3. The Gram and Ziehl-Neelsen methods are considered in Sect.6.1.6.

A number of other methods for the demonstration of mycobacteria have been described. Fluorescence microscopy of mycobacteria is possible using Auramine O and Rhodamine B (red-golden) or Thioflavine TG (blue-green fluorescence) in the acid fast staining method. The advantage of these methods compared with the classical Ziehl-Neelsen method is that the acid fast organisms are seen strongly fluorescent on a dark background and are therefore easily demonstrated using dry objectives (25× or 40×) with resultant larger visual field and time saving.

Bartholomew (1981) insists that the result should be confirmed with a standard Ziehl-Neelsen staining.

31.4.2 Fungi

Selective demonstration of fungi is dependent on the carbohydrates in the fungal cell wall.

The most important methods are the periodic acid-Schiff reaction (Sect.9.2.1), Grocott's argentaffin method (Sect.8.3.2), and Gridley's technique. The latter method depends on a combination of chrome trioxide oxidation, Schiff's reagent, and Aldehyde Fuchsin (Sect.6.1.3).

Wachsmuth (1988) has reported on the use of two fluorescent dyes, Calcofluor White M2R and Uvitex 2B, for the demonstration of fungi. Both dyes stain a large variety of common pathogenic fungi and do not stain bacteria. Background staining is rather pronounced with Calcofluor White but virtually absent with Uvitex 2B. Both fluorochromes stain elastic fibres.

31.5 Demonstration of Metals

The pathologist may wish to confirm the presence of calcium in lesions suggestive of dystrophic or metastatic calcification (Sect.17.7.1). With the exception of iron (cf. Sect.31.6.3), demonstrations of other forms of metallic deposit such as copper are rarely performed. Details are given in Chap.17.

31.6 Demonstration of Pigments

A strategy for the identification of an unknown pigment is given in Table 18.2. The key reactions are directed at demonstrating reducing properties (Sect.8.3.1) and iron (Sect.17.7.4). The occurrence and identification of individual pigments is discussed below.

31.6.1 Haemoproteins

Demonstration of the haem containing proteins, haemoglobin and myoglobin, is of particular interest in forensic investigations. The H_2O_2-Perls, benzidine H_2O_2, and leuco-Patent Blue-H_2O_2 reactions are proposed in Table 17.3, Sects.17.7.4 and 18.4.4).

31.6.2 Porphyrins

Large amounts of porphyrins accumulate in acute porphyria. Demonstration can be achieved either by observing red autofluorescence (Sect.30.1), or by the Gmelin reaction. The former is highly reliable while the latter is unreliable and gives results that may be confused with bile pigments (Sect.18.4.6).

31.6.3 Haemosiderin and Ferritin

Haemosiderin is normally found in the reticulum cells of bone marrow. Accumulations that are not associated with tissue damage are termed *haemosiderosis* (e.g. haemorrhage and haemolysis). *Haemochromatosis* denotes accumulation of iron accompanied by damage and may, for instance, be observed in the liver (cirrhosis).

Iron (Fe^{3+}) is demonstrated with Perls' reaction (Sect.17.7.4) or by chelate formation (Fe^{2+}) with Bathophenanthroline (Sect.17.7.4). In addition, the iron containing glycoprotein, aposiderin, can be demonstrated using the PAS reaction. Electron microscopy allows the demonstration of the electron dense ferritin molecules. Ferritin has also been widely used for tracer studies (e.g. to elucidate the endocytic cycle, Sect.2.2.8).

31.6.4 Acid Haematins

Acid haematins (Sect.18.1.1) usually occur as an artifact on fixation in blood containing tissue when unbuffered or insufficiently buffered formaldehyde-based fixatives are used. In malaria and schistosomiasis (Sect.18.1.1) analogous pigments are deposited in histiocytes. Identification may be achieved by demonstrating that the pigment is birefringent and that it can be dissolved by treatment with a saturated alcoholic solution of picric acid (Sect.18.4.12).

31.6.5 Bile Pigments

Abnormal accumulations of bile pigments can either be localized (haemorrhages) or universal. In the latter case the condition is termed *jaundice*.

Demonstration of bile pigments can be useful in the examination of liver biopsies. Although extensive extraction occurring during tissue processing makes the use of cryostat sections desirable, true bile thrombi are preserved in routine paraffin sections. Thrombi can be seen in Al-haematein Eosin stained sections but they are often more easily observed in sections stained with Picrofuchsin since this does not obscure the green colour of the bile. Methods for the demonstration of bile pigments are discussed in Sect.18.4.

31.6.6 Lipofuscin

Lipofuscin (Sect.18.1.2) may be seen in the Leydig cells of the testis and in tumours derived from these cells. Nerve cells contain increasing amounts of lipofuscin with age as do several cell types undergoing atrophy (e.g. heart muscle cells).

Lipofuscin can be demonstrated by its orange autofluorescence (Sect.30.2) and its acid fast basophilia (Sect.18.4.1). In case of a doubtful reaction the Sudanophilia of lipofuscin may be of value (Sect.19.6.1), as this is preserved even in paraffin embedded material. Both Sudanophilia and PAS positivity decrease with the age of the lipofuscin.

31.6.7 Melanin

Melanin (Sect.18.1.3) is a reducing agent and very small amounts can be demonstrated using the ferric ferricyanide method (Sect.8.3.1). With slightly larger amounts, the ferrous ion uptake method of Lillie is a valuable, highly selective method (Sect.18.4.9). The ability to produce melanin (e.g. in malignant melanomas) can be demonstrated with the DOPA reaction (Sect.18.4.13) (cryostat sections of fresh or formaldehyde fixed material are necessary). Immunohistochemical techniques can also be useful in the diagnosis of malignant melanomas.

31.6.8 Serotonin

The demonstration of serotonin (Sect.18.1.4) is of interest in determining the distribution of transmitter substances in neurons and in the characterization of carcinoid neuroendocrine tumours (apudomas).

The relatively large amounts of serotonin present in enterochromaffin cells (and often present in carcinoid cells) can be demonstrated with the azo coupling reaction (Sect.18.4.10). If the serotonin concentration is lower, the azo coupling reaction can be supplemented with a ferric ferricyanide (Sect.8.3.1), and an argentaffin (Sect.8.3.2), or a chromaffin (Sect.18.4.5) reaction. Where serotonin content is very low, (e.g. in some neuroendocrine tumours) an induced fluorescence method is required (e.g. formaldehyde – FIF) (cf. Sect.30.2.1).

31.6.9 Catecholamines

The demonstration of catecholamines (Sect.18.2.4) is useful for investigating the distribution of neurotransmitters and for characterizing the cells and tumours of the adrenal medulla.

Unlike serotonin, catecholamines are readily extracted from the tissue during processing, and the azo coupling reaction (Sect.18.4.10) is therefore considerably less reliable.

The argentaffin (Sect.8.3.2), chromaffin (Sect.18.4.5), and the ferric ferricyanide (Sect.8.3.1) reactions are frequently positive in phaeochromocytomas but the most reliable results are achieved using formaldehyde induced fluorescence on freeze-dried material (Sect.30.2.1).

31.6.10 Asbestos

The inhalation of asbestos fibres may lead to asbestosis which is a pneumoconiosis or lung fibrosis. The accumulations of asbestos may be seen without staining or, where they occur as iron-coated complexes within macrophages, they may be demonstrated by Perls' reaction (Sect.17.7.4). Asbestos is carcinogenic and may lead to the development of bronchogenic carcinoma and mesothelioma.

31.6.11 Silica

Inhalation of silica (silicon dioxide) is an occupational hazard in industries such as metal mining, foundries, pottery making, and sandstone and granite cutting. It may, often after up to 30 years exposure, lead to silicosis, a fibrogenic pneumoconiosis.

The silica crystals may be demonstrated by their birefringence.

31.6.12 Urate

Crystals of monosodium urate occur in deposits, in and around joints and tendons and several other sites, in patients with gout. Although monosodium urate is only slightly soluble in water, losses during tissue processing involving aqueous solutions may be extensive unless the urate is bound to protein. Identification can be achieved by demonstration of birefringence and by the argentaffin reaction with Gomori's silver methenamine method (Sect.8.3.2).

31.7 Demonstration of Lipids

Lipid classification is described in Sect.2.1.4 and pretreatment of tissue to be investigated is discussed in Sects.13.5 and 19.2. A survey of methods for demonstrating lipids is given in Table 19.1.

Histochemical demonstration of lipids is of interest in conditions where the lipid content is altered (Sect.31.7.1), in lipid containing tumours (Sect.31.7.2), and in lipidoses (Sect.31.7.3).

31.7.1 Changes in Lipid Contents

The formation of fat droplets in cells (fatty degeneration) is usually due to the accumulation of triglycerides. The droplets are usually seen as empty vacuoles in the cells in paraffin sections. Frozen sections are required for a positive demonstration using lysochromes (Sect. 19.6.1).

The degeneration of myelin can be investigated using the *Marchi* methods (Sect. 19.6.5). The use of OsO_4 with another oxidant enables differentiation between hydrophobic (black-stained) and hydrophilic lipids. In degenerative conditions (demyelination) the lipid content is more hydrophobic.

31.7.2 Lipid Containing Tumours

In general, methods for the demonstration of lipids are only occasionally useful in the diagnosis of lipid containing tumours. This is due to the fact that many malignant soft tissue tumours give a positive reaction, while some liposarcomas give a negative reaction.

31.7.3 Lipidoses

Lipid storage diseases or lipidoses are familial disorders of lipid metabolism characterized by enzyme defects (Sect.31.11.2). They are classified according to the nature of the accumulated lipids.

For example the diagnosis of type B Niemann-Pick disease can be made on bone marrow aspirates or liver biopsies where the defect in sphingomyelinase activity leads to abnormal deposition of sphingomyelin and cholesterol. With this and other lipidoses determining the exact distribution and nature of the lipid deposits enables an accurate diagnosis.

Sphingomyelin is stained weakly with Sudan Black B (Sect. 19.6.1) and appears red in polarized light. In the central nervous system the sphingomyelin stains blue with the chromation-acid Haematein reaction or NaOH-acid Haematein (Sect. 19.6.7) reaction. Cholesterol is demonstrated with the PAN method (Sect. 19.6.11) and gangliosides can be demonstrated with the modified PAS reaction, but the borohydride-periodate-Schiff method (Sect. 19.6.9) of Buk and Bayliss High (1986) is probably better.

31.8 The Demonstration of Nucleic Acids

Fixation of nucleic acids and nucleoproteins is discussed in Sect.13.3. The chemistry of DNA and RNA is considered in Sect.2.1.6, and the occurrence of nucleic acids and their participation in the composition of chromatin and ribosomes is described in Sects.2.2.1 and 2.2.2. A strategy for locating and identifying nucleic acids is given in Table 20.2.

31.8.1 Amount of DNA

The amount of DNA present in the nucleus changes with the cell cycle. In mammals, for rapidly dividing cells, the cycle lasts about 24 hours; the G1-phase (G = gap) takes about 10 hours (lowest DNA content), in the S-phase (S = synthesis), which lasts about 8 hours, there is a gradual increase in DNA to twice the G1 level and this level then remains constant through G2 (4 hours) until the end of the M-phase (M = mitosis) which takes about 2 hours. In more slowly dividing cells the G1-phase is prolonged.

Pathological variations include malignant cells where abnormal numbers of chromosomes may occur (aneuploidy) and where the amount of DNA per chromosome is often elevated.

31.8.2 Chromosome Analysis

Chromosome analysis has been useful in classifying leukaemias. The analysis is usually performed on a bone marrow aspirate either directly or after short term culture. Hypotonic treatment and staining allows "bands" to be visualized (Sect.30.3.1).

Clones. Chromosome anomalies in leukaemia diagnostics will often appear as clonal changes. An abnormal clone is defined as two or more metaphase cells with an identical structural rearrangement or extra chromosome, or three or more cells lacking the same chromosome.

Leukaemia. The first constant chromosomal aberration associated with a human neoplasia was an abnormal chromosome, termed the Philadelphia chromosome or Ph^1 in *chronic myeloid leukaemia.* It can be demonstrated in around 90% of patients with chronic myeloid leukaemia.

Banding techniques have been of some value in the classification of *acute leukaemias*. Immunological techniques are now much more widely used for this purpose (Sect.32.5.5).

31.8.3 Different Forms of DNA

The staining properties of DNA vary according to its physicochemical state. For example DNA in condensed chromatin (non-S-phase) is hydrolyzed more slowly by Feulgen hydrolysis (Sect.9.9) than newly synthesized DNA (S-phase, partially condensed) so that in cell smears the optimal hydrolysis times are 60 and 20 min respectively. This sort of differential characteristic can be used to assess the content of different forms of DNA and may be useful in detecting neoplastic and preneoplastic cells where, in addition to the normal forms of DNA, small amounts of a more labile form with an optimum hydrolysis time of 3–5 min are found (Husain, 1983).

31.8.4 Amount of RNA

The amount of ribosomal RNA in a cell usually parallels protein synthesis (Sect. 2.2.2). Thus, large amounts of RNA are found in the cytoplasm of cells producing proteins for export, such as liver cells (albumin), glandular cells (enzymes and mucus), and plasma cells (antibodies).

In haemolytic anaemias, where the need for new red blood cells is high, an increased number of young erythrocytes (polychrome erythrocytes or reticulocytes) is found in peripheral blood. These cells have, in contrast to mature erythrocytes, rather large amounts of RNA in their cytoplasm as haemoglobin synthesis is still taking place.

Polychrome erythrocytes give a faint bluish colour with the Romanowsky-Giemsa stain. Supravital staining with the oxazin dye Brilliant Cresyl Blue, yields blue granules and threads in the cytoplasm.

31.8.5 Methods for Demonstrating Nucleic Acids

These include Al-haematein Eosin (Sects. 31.2.1 and 7.2.2), Chromium-gallocyanin (Sect.7.3), Methyl Green-Pyronin Y (Sect.6.1.5), Acridine Orange (Sect.30.3.5), and the Feulgen nucleal reaction (Sect.9.9). Examples of fluorochromes that can be used for demonstrating DNA include Ethidinium bromide and Acridine Orange. On RNase treated sections, the former reacts with collagen and keratin in addition to DNA. A fluorochrome Schiff reagent prepared from Acriflavine instead of Basic Fuchsin (Sect.3.3.9) can be used in the Feulgen reaction. The resultant increase in sensitivity can be useful where limited material is available (e.g. in prenatal diagnosis).

A standardized Methyl Green-Pyronin Y method is detailed in Appendix A.

31.9 Demonstration of Proteins

Methods based on the enzymatic activity (Sect.31.11) or the antigenic properties (Chap.32) of the protein are discussed elsewhere. The value of the general oversight stains and the trichrome stains (Sect.21.6) is substantially due to differential staining

Table 31.1. Methods for the demonstration of specific amino acid residues.

Method	Amino acid residues	Proteins or cells with strong reaction	Cross references
Oxidation/aldehyde reagent	amino groups (lysyl)	histones, globins	9.5
NQS	arginyl	histones, granules in eosinophils and Paneth cells	9.6.1
Diazotization/coupling	tyrosyl	elastin	9.4.3 21.5
DMAB/diazonium salt coupling	tryptophanyl	fibrin, fibrinoid, pepsinogen, glucagon, granules in eosinophils, and Paneth cells	9.4.5
Ferric ferricyanide	thiol groups (cysteine)	keratin	8.3.1 21.4 2.2.10

of proteins. It is often desirable or useful to demonstrate a certain protein or cell by its unusually high content of a specific amino acid residue (Table 31.1).

31.10 Demonstration of Carbohydrates

Histochemical demonstration of carbohydrates is useful in diagnostic work involving skin (Sect.31.10.1), stomach (Sect.31.10.2), colon (Sect.31.10.3), prostate (Sect.31.10.4), and breast (Sect.31.10.5). It may also be helpful in the differential diagnosis of myxoid tumours (Sect.31.10.6) and malignant mesotheliomas (Sect.31.10.7). Finally, it is sometimes of interest in the diagnosis of glycogen storage diseases (Sect.31.10.8).

31.10.1 Carbohydrates in Skin Biopsies

Skin biopsies are dealt with in Table 31.2.

Table 31.2. Histochemical methods of value in the examination of skin biopsies.

Diagnosis	PAS	Methods argentaffin reaction	Toluidine Blue metachromasia	Alcian Blue pH1
Fungi	+ (stratum corneum)	+	0	0
Diabetic micro-angiopathy	+ (capillary walls)	0	0	0
Paget's disease	+ (epidermis)	0	0	0
Urticaria	0	0	+	+

31.10.2 Mucins in Stomach

The mucous secretory cells of the normal gastric epithelium almost exclusively contain neutral mucins, while traces of sialomucins and perhaps sulphomucins can be found in the mucous neck cells of the fundus. Sialomucins and sulphomucins are sometimes also found in the antrum and in the glands of the oesophageal-gastric junction.

Mucins in Carcinoma of the Stomach. The demonstration of malignant cells in the stomach is usually no problem in sections stained with one of the oversight methods, e.g. Al-haematein Eosin. In cases of undifferentiated or signet-ring cell carcinoma Alcian Blue pH 2.5-PAS can be of value (Sect.22.2.5).

Intestinal Metaplasia. Intestinal metaplasia of the mucosa of the stomach can be divided into three forms depending on the histological cell type and the type of mucin produced (Filipe, 1983).

Type I or complete intestinal metaplasia resembles the small intestine and secretes sialomucins which can be both N- and O-acylated. In type II the goblet cells secrete sialomucins which can be N-acylated, but not O-acylated, while the intervening epithelial cells contain neutral mucins and sialomucins in different proportions. In type III the goblet cells secrete either only sulphomucins or a mixture of sialo- and sulphomucins, while the epithelial cells contain sulphomucins in addition to neutral mucins and sialomucins. The histochemical demonstration of these features is summarized in Table 31.3.

Table 31.3. The results of some carbohydrate reactions on the mucosa of normal stomach and stomach with intestinal metaplasia.

Method	Mucous membrane normal	Mucous membrane with intestinal metaplasia: goblet cells type I	II	III	intervening epithelial cells type I	II	III
PAS	+	+	+	+	+	+	+
Alcian Blue pH 3	0	+	+	+	+	+/0	+
Alcian Blue pH 5.7 +0,2 mol/l $MgCl_2$	0	0	0	+	0	0	+
Alcian Blue pH 5.7 + 0.7 mol/l $MgCl_2$	0	0	0	+/0	0	0	+/0
Periodic acid/borohydride/potassium hydroxide/PAS	0	+	0	0	0	0	0

\+ = positive reaction.
+/0 = positive or negative reaction.
0 = negative reaction.

Several investigations have suggested that type I intestinal metaplasia usually occurs in benign conditions, while type II and III more frequently accompany carcinomatous changes, especially if the secretion of sulphomucins is pronounced (Filipe, 1983).

31.10.3 Mucins in Colon

Goblet cells in the lower two-thirds of the *colonic* crypts predominantly secrete sulphomucins, while the goblet cells in the upper one-third and surface epithelium secrete N- and especially O-acylated sialomucins (Filipe, 1983).

Inflammatory Conditions. In inflammatory conditions an altered secretion occurs associated with an increase in sialomucins and a reduction or absence of sulphomucin. The amount of O-acylated sialomucins is reduced in some pathological conditions.

Carcinomas. In the mucous membrane adjacent to carcinomas, the so-called *transitional mucosa* (Filipe, 1969), increased secretion of sialomucins is a constant finding. The relative proportions between N-acylated and O-acylated sialomucins are changed in both carcinoma, transitional, and also more distantly located mucosa. In addition, an increase in neuraminidase-sensitive sialic acids is found in both transitional mucosa and in the carcinomatous tissue. This is accompanied by a reduction in the amount of acylated sialomucin in the carcinoma. There seems to be a connection between the degree of changes in the mucin and the aggressiveness and prognosis of the tumour. Possibly, the sialic acids play an important role in the immunological reactions towards tumour specific antigens and in the control of tumour growth (Filipe, 1983). Methods for a sophisticated study of mucins are described in Sect.22.2.1.

Adenomas. In adenomas the loss of O-acylated sialomucins parallels the degree of dysplasia (Filipe *et al.*, 1981).

31.10.4 Mucins in Prostate Gland

Acid mucins are rarely present in prostate gland without neoplastic changes. Conversely, these mucins are found in two thirds of prostatic carcinomas. They are usually not present in the most poorly differentiated tumours (Jöbsis, 1983).

31.10.5 Mucins in Carcinoma of the Breast

In the breast both neutral mucins and sialomucins are found in the epithelial cells in normal tissue, in fibrocystic disease, and in carcinoma (Ormerod and Sloane, 1983). In carcinoma of the breast a combined Alcian Blue pH 3-PAS reaction will frequently demonstrate large globules with a blue rim and a magenta spot in the middle, so-called bull's eye appearance. These vacuoles are particularly associated with in situ and infiltrating lobular carcinoma.

31.10.6 Proteoglycans in Myxoid Tumours

Myxoid tumours or tumours of soft tissue may be classified as shown in Table 31.4.

Table 31.4. Classification of myxoid tumours.

	Without chondromatous elements	With chondromatous elements
benign	myxoma myxoid lipoma lipoblastoma	myxoid chondroma
malignant	myxoid liposarcoma myxoid, malignant, fibrous histiocytoma myxoid fibrosarcoma	myxoid chondrosarcoma

These tumours produce proteoglycans containing the following glycosaminoglycans: hyaluronic acid, chondroitin-4-sulphate, chondroitin-6-sulphate, and keratan sulphate. Identification is achieved, as shown in Table 22.2, using Alcian Blue pH 2.5 and the Alcian Blue-CEC method, and controls are made by pretreating the sections with hyaluronidase and chondroitinases ABC (Sect.22.3.3). The staining results and the content of glycosaminoglycans in myxoid and chondromatous tumours are given in Table 31.5.

Table 31.5. The histochemical characteristics for myxoid tumours and chondrosarcomas.

	AB pH 2.5	Hyaluronidase + AB pH 2.5	AB-CEC (mol/l $MgCl_2$)	Glycosaminoglycan type
Myxoid tumour without chondromatous elements	+	0	0.3	hyaluronic acid
Myxoid chondrosarcoma	+	+ 0	0.6	keratan sulphate, chondroitin-4-, and -6-sulphate
Benign chondromatous tumours	+	+ 0	0.6	keratan sulphate, chondroitin-4-, and -6-sulphate
Chondrosarcoma well differentiated	+	+ 0	0.6	keratan sulphate, chondroitin-4-, and -6-sulphate
Poorly differentiated	+	0	0.6	chondroitin-4-, and -6-sulphate

AB pH 2.5 = Alcian Blue pH 2.5.
AB-CEC = Alcian Blue-critical electrolyte concentration.
\+ = positive staining result.
\+ 0 = variable staining result.
0 = negative staining result.

31.10.7 Mucins in Malignant Mesotheliomas

In the differential diagnosis between malignant mesothelioma and adenocarcinoma, either primary (lung) or metastatic, the diagnosis mesothelioma is histochemically supported by:

1. Alcian Blue pH 2.5: positive reaction
2. Hyaluronidase/Alcian Blue pH 2.5: negative reaction
3. PAS: negative reaction

Thus a mesothelioma contains hyaluronic acid in contrast to an adenocarcinoma which contains sialomucins and/or sulphomucins.

31.10.8 Glycogenoses

Glycogen storage diseases (GSD) or glycogenoses are metabolic diseases. They are caused by enzyme defects (Sect.31.11.2) and are characterized by accumulation of glycogen. An example is Pompe's disease (GSD II) in which the defect involves acid α-glucosidase. An accumulation of glycogen is seen in lysosomes (Sect.2.2.8). This is most pronounced in basal ganglia, brain stem, and spinal cord, but also occurs in sympathetic ganglion cells, liver and spleen reticuloendothelial cells, hepatocytes, and in smooth, striated, and cardiac muscle cells. The glycogen deposit gives a positive PAS reaction and is accompanied by increased activity of acid phosphatase. The diagnosis is made on blood smears, where glycogen is demonstrated in the lymphocytes which are filled with vacuoles.

31.11 Demonstration of Enzyme Activity

Histochemical demonstration of the activity of various enzymes in diagnostic work may be of value for instance as "markers" of different cell organelles (Sect.31.11.1) or as an aid in the specific diagnosis of inborn errors of metabolism (Sect.31.11.2) and Hirschsprung's disease (Sect.31.11.3). The demonstration of enzyme activity is also helpful in the differential diagnosis of malabsorption (Sect.31.11.4), prostatic carcinoma (Sect.31.11.5), neurogenic muscle disorders (Sect.31.11.6), and haematological conditions (Sect.31.11.7).

31.11.1 "Marker" Enzymes

Cell organelles may be selectively demonstrated at both the light and the EM level by the presence of certain "marker" enzymes (Novikoff, 1976). Typical examples are given in Table 31.6.

Table 31.6. Marker enzymes for different organelles.

Organelle (cell type)	Enzyme
Endoplasmatic reticulum	nucleoside-diphosphatase, glucose-6-phosphatase
GERL and lysosomes	acid phosphatase
Peroxisomes and microperoxisomes	catalase
Golgi	thiamine-pyrophosphatase
Mitochondria	enzymes and coenzymes which oxidize 3,3′-diaminobenzidine
	NADH dehydrogenase, succinate dehydrogenase
Plasma membrane	Nucleoside-phosphatase, Na/K ATPase

31.11.2 Inborn Errors of Metabolism

A number of hereditary diseases are attributable to deficiencies in the activity of enzymes, many of which are lysosomal, concerned with the metabolism of lipids and carbohydrates. The defect leads to accumulation of metabolic products associated with the pathways in which the enzyme concerned participates, hence the term storage diseases. Reticuloendothelial cells and nerve cells are particularly affected (Lake, 1983).

The sphingolipidoses, mucolipidoses, and glycogenoses are all important storage diseases. Examples include: Niemann-Pick's disease, a sphingolipidosis involving a defect in sphingomyelinase (Sect.31.7.3), Mucolipidosis 2 or I-cell disease, a mucolipidosis with deposition of lipofuscin-like granules in vacuolated cardiac muscle cells accompanied by increased acid phosphatase activity, and Pompe's disease, a glycogenosis associated with a defect in acid α-glucosidase (Sect.31.10.8). The diagnosis is usually made by demonstrating the accumulated compound or compounds. Direct histochemical investigation of the enzyme defect is currently only possible in a few instances (e.g. loss of β-galactosidase activity in the lipidosis G_{M1}-gangliosidosis). Fresh frozen tissue is preferred, even though formaldehyde fixed material can often be used.

31.11.3 Enzyme Histochemistry in Hirschsprung's Disease

Classically, the diagnosis of Hirschsprung's disease is made by demonstrating the absence of ganglion cells in a 1–2 cm segment of the rectum immediately above the anus. This requires at least 50 serial sections from a very substantial biopsy. Lake *et al.* (1978) have, however, shown that the diagnosis can be made by demonstrating acetylcholinesterase positive nerve fibres in 10 μm cryostat sections from fresh frozen mucosal biopsies. In the normal rectum only a few fine nerve fibres are seen in the propria and muscularis mucosae. The ganglion cells appear well-stained with a stippled appearance. In Hirschsprung's disease, both short-segment (confined to the rectum) and long-segment (with changes in the sigmoid), nerve fibres in both propria and muscularis mucosae are more numerous and often also increased in size (Lake, 1983).

31.11.4 Enzyme Histochemistry of the Malabsorption Syndrome

The malabsorption syndrome is a collection of symptoms which may occur regularly or less predictably in connection with the intake of food. It includes disturbances in digestion, absorption, secretion, and mobility of the small intestine. When the disorder is due to a defect in the epithelial cells of the intestine (primary malabsorption syndrome) (Lojda, 1983) a partial or total reduction in the activity of disaccharidases can often be demonstrated. The defect can affect one or more enzymes (lactase, sucrase, or trehalase). Examples of combined enzyme defects include coeliac disease, tropical sprue, and intolerance of cow's milk protein.

31.11.5 Enzyme Histochemistry in Prostatic Carcinoma

Metastatic tumours can be identified as prostatic in origin by demonstrating prostatic acid phosphatase activity (Jöbsis, 1983). This isoenzyme is almost unique to epithelial cells of the prostate. It is distinguished from other acid phosphatase isoenzymes by its relative resistance to formaldehyde and its near complete substrate specificity for phosphorylcholine. Demonstration can be achieved using cryostat sections, fine needle aspirates, or imprints.

The activity of acid phosphatase in the prostate is more pronounced in benign epithelial cells than in carcinoma cells. Decreased amounts of androgen or therapy with oestrogen reduce activity. Enzyme activity is irreversibly inhibited by ethanol and acid decalcifying agents and therefore cannot be demonstrated in paraffin sections or in acid decalcified tissue. The antigenic properties of the enzyme are, however, usually preserved, and the enzyme can be demonstrated immunohistochemically (Sect.32.5.2).

31.11.6 Enzyme Histochemistry of Muscle Biopsies

Histochemical techniques can be used to distinguish between the different types of muscle fibres (Johnson, 1983). Cryostat sections of tissue frozen within 30 min of removal should be used and the pretreatment recommended by Dubowitz and Brooke (1973) should be considered. Correct orientation of the biopsy is critical as well-orientated cross-sections are essential for assessing fibre sizes. Cryostat sectioning should be performed at −20°C to −25°C and a 10 μm thickness is suitable for most purposes.

A listing of recommended staining methods (Johnson, 1983) is given in Table 31.7.

The enzyme profiles of different muscle fibre types are given in Table 31.8.

Table 31.7. Methods for histochemical assessment of muscle biopsies (according to Johnson, 1983).

Method	Purpose
Myofibrillar ATPase with acid preincubation	Determination of fibre type
Succinate dehydrogenase	Distribution and activity of mitochondria
Periodic acid Schiff	Demonstration of stored glycogen
Myophosphorylase	Assessment of glycolytic activity
Acid phosphatase	Assessment of lysosomal changes and degree of histiocyte infiltration

Table 31.8. Enzyme activity in different muscle fibre types.

	Type I	Type 2A	Type 2B	Type 2C
ATPase pH 9.5 after fixation in methanol-free formaldehyde solution	+	+ +	+ + +	+ (+)
ATPase pH 9.5 after preincubation in buffer pH 4.6	+ + +	+	+ +	+ + (+)
ATPase pH 9.5 after preincubation in buffer pH 4.3	+ + +	0	0	+ (+)
Succinate dehydrogenase	+ + +	+ +	+	+ + (+)
Myophosphorylase	+	+ +	+ + +	+ (+)

0: no reaction; + : weak; + (+): weak-moderate; + + : moderate; + + (+): moderate-strong; + + + : strong.

Striated muscle comprises approximately equal numbers of type 1, type 2A, and type 2B fibres. Normally, there should not be more than 3% type 2C in adults. The random distribution of fibre types in adults corresponds to the random distribution of the muscle fibres from different motor units. The fibres from one motor unit all belong to the same type, but as the motor unit territories overlap each other, a mosaic is formed.

Neurogenic Muscle Diseases. In these conditions, where the most prominent pathological change is denervation, there are many common characteristics regardless of where in the nerve the lesion is localized (central or peripheral). If only a few motor units are denervated the distribution of the atrophied muscle fibres is scattered. The atrophied muscle fibres include both type 1 and type 2 fibres as there is generally no selective effect of the denervation process on either type 1 or type 2 motor units. If, however, the denervated fibres are reinnervated by branches from one single axon the result will be a group of fibres of the same histochemical type. This phenomenon with *grouping* of fibres of the same type is a clear indication of reinnervation in neurogenic muscle diseases.

Quite frequently biopsies from well-compensated (i.e. reinnervated) disorders do not show any signs of fibre atrophy, and the diagnosis can only be made by the histochemical demonstration of fibre type grouping. During the process of reinnervation fibres change from one type to another if the reinnervating axons are of a different motor type from the original axons. Fibres, which are undergoing change from type 1 to type 2 or vice versa, pass through a transition stage. This can be demonstrated by the reaction for ATPase pH 9.5 after preincubation in buffer at pH 4.3. The activity of normal type 2A and type 2B fibres is normally completely inhibited by this while "transition" fibres, usually called type 2C fibres, retain activity (Table 31.8). In neurogenic muscle disorders demonstration of succinate dehydrogenase activity reveals that the normal homogeneous distribution of mitochondrial enzyme activity in the fibres is lost. Instead three concentric zones are observed with low activity in the inner zone, high activity in the middle zone, and normal activity in the outer zone. Fibres showing this distribution are called *target fibres*.

Muscular Dystrophies. These conditions comprise a heterogeneous group without common histochemical characteristics. The signs of denervation (grouping and target fibres) are therefore absent.

31.11.7 Enzyme Histochemistry of Haemopoietic Cells

Investigations of *bone marrow* and *peripheral blood* smears are based on staining with the Romanowsky-Giemsa method (Sect.6.3.2). Supplementary histochemical staining methods can facilitate differentiation of the haemopoietic cells and are of particular value in the subgrouping of *leukaemias*. These methods include Sudan Black B, Neutral Red, PAS, Toluidine Blue metachromasia, and reactions for myeloperoxidase, α-naphthylacetate esterase, chloroacetate esterase, alkaline phosphatase, and acid phosphatase (Hayhoe and Quaglino, 1980). It is important to perform the staining reactions as soon as the specimens have been prepared. Table 31.9 shows the results expected for the above methods applied to normal haemopoietic cells (after Smith, 1983).

The histochemical classification of leukaemias is beyond the scope of this chapter. While the techniques comprising Table 31.9 are all useful, immunohistochemical classification schemes are now pre-eminent in this field (cf. Sect. 32.5.5).

31.12 The Use of Autoradiography

Autoradiography is covered in Chap.29. For an excellent review on the applications of autoradiography, particularly in neurobiological research, see Pilgrim and Stumpf (1987).

Table 31.9. Staining pattern of the cytoplasm of normal haemopoietic cells.

Cell	Sudan Black	Neutral Red	PAS	Myelo-peroxidase	α-naphthylacetate esterase	Chloroacetate esterase	Acid phosphatase
Myeloblast	0–(+)	0	(+)	0– +	0–(+)	0– + + +	+
Promyelocyte	+ +	0	+	+ + + +	0–(+)	+ + + +	+ +
Polymorph	+ + +	+ + + +	+ + + +	+ + + +	+	+ + + +	0– +
Megakaryocyte	(+)	0	+ + +	0	+ + + +	0–(+)	+ + + +
Erythroblast	0	0	0	0	0	0	+ +
B-lymphocyte	0	0–(+)	0– + +	0	0–(+)	0	+
T-lymphocyte	0	0–(+)	0– + +	0	+ + + +	0	+ + +
Plasma cell	0	0	0– +	0	+ +	0	+ + + +
Monocyte	0– +	+ +	0– + + +	0– + +	+ + +	0– +	(+)– + +
Platelet	0	0	+ + + +	0	+ + + +	0	+ + +
Mast cell	0	0	+ – + +	0	0	+ + + +	+ + + +
Marrow histiocyte	0	0	0– + +	0	+ + +	0	+ + +

0: no reaction; + : weak; + + : moderate; + + + : strong; + + + + : very strong reaction.

31.13 The Use of Fluorescence Microscopy

Fluorescence microscopy (Chap.30) is used in the demonstration of autofluorescent compounds (e.g. elastin, vitamin A, porphyrins, lipofuscin), as a means of detecting antibodies (e.g. FITC-conjugated), and in the FIF reaction. In *apudomas* it is possible in many cases to demonstrate formaldehyde induced fluorescence (FIF, Sect.30.2.1) or glyoxylic acid induced fluorescence (Sect.30.2.2). This applies to all tumours which store biogenic amines (e.g. carcinoids) or which can take up L-DOPA or 5-hydroxytryptophan in vitro and convert these to dopamine and serotonin respectively (e.g. in calcitonin producing medullary carcinomas of the thyroid). Cryostat or paraffin sections can be used (Lyon *et al.*, 1981; 1982). In addition, fluorochromes are useful for screening procedures (Sect.30.3), for amyloid (Sect.30.3.2), mycobacteria (Sect.31.4.1), and fungi (Sect.31.4.2).

Other examples of the uses of fluorochromes are given by Mai *et al.* (1984) and Shennawy *et al.* (1984). The former group used Acridine Orange for histological examination of the central nervous system, while the latter group described the fluorescence of Aniline Blue WS (C.I. 42755) used as a stain for glycogen. Aniline Blue WS was also shown to be helpful in tracing pathological and autolytic changes in lysosomes, mitochondria, erythrocytes, and nuclei, and for demonstrating bacteria in tissue sections and smears.

32 Applied Immunohistochemistry

M. Vyberg, P.P. Clausen

This chapter represents an appendix to the book and in particular to Chaps.26 and 27. The discussion below on applications of immunohistochemistry in pathology does not cover all aspects of this large subject. For further information the reader is referred to for instance DeLellis (1981; 1984; 1988), Taylor (1986), Nadji and Morales (1987), and Seifert (1987).

32.1 Introduction

Initial enthusiasm for the application of immunohistochemical methods in diagnostic pathology has in recent years been replaced by a more critical attitude. Along with the widespread use of these methods in the diagnostic process it has become evident that the reliability of the results obtained and the diagnostic inferences drawn from the staining reaction are highly dependent on knowing the effects of tissue preparation and the cross reactivity of the antibodies used.

32.2 Use and Interpretation of Immunohistochemistry in Diagnostic Pathology

32.2.1 Choice of Fixation and Processing Methods

The basic prerequisites for obtaining reliable staining results with immunohistochemical staining methods are the use of fixation and processing methods of the specimen compatible with optimal preservation of the antigen under investigation (Sect.26.6).

As fixation in routine diagnostic pathology is usually carried out with neutral buffered formaldehyde solution, it is extremely important to know whether the antigen under investigation can withstand this process. Although many antigens have been shown to tolerate routine fixation and processing, some do not (e.g.

H. Lyon (Ed.)
Theory and Strategy in Histochemistry

membrane related immunoglobulin deposits and lymphocyte surface antigens), and cryostat sections are required for their demonstration.

False positive staining reactions due to inadequate tissue fixation and false negative reactions due to over-fixation, use of unsuitable fixatives, or too high temperature in the embedding procedure, are problems often encountered in diagnostic immunopathology.

In diagnostically difficult cases, as in the classification of undifferentiated tumours, it is therefore recommended that the unfixed material should be treated according to the flow chart in Table 32.1. If this scheme is followed, there will always be tissue available for the demonstration of any kind of antigen.

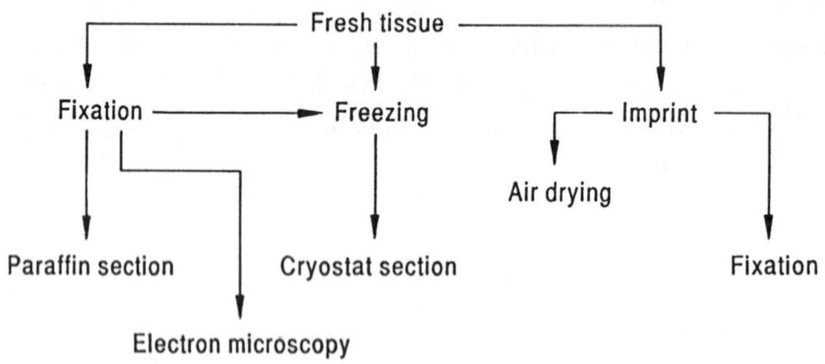

Table 32.1. Flow chart illustrating handling of fresh tissue in diagnostically difficult cases.

32.2.2 Choice of Primary Antibodies

The choice of primary antibodies is another important factor. In recent years huge numbers of commercial antibodies have been released on the market, and it is often very difficult to select from different antibodies to the same antigen.

As there is no general agreement on quality specification, individual laboratories have to make their own pilot studies in order to select the most suitable antibodies with respect to economy, specificity, and lack of non-specific staining and cross-reactivity.

32.2.3 Choice of Immunohistochemical Staining Technique

When selecting an immunohistochemical staining technique the most efficient and/or sensitive staining technique should be chosen (Sect.26.4). It should, however, be remembered that the use of a highly sensitive staining method may increase the risk of revealing cross-reactivity of the primary antibodies thus causing difficulties in the interpretation of the staining results. It is therefore necessary to perform a meticulous titration of the antibody adapted according to the staining technique and the degree of antigen preservation.

32.2.4 Diagnostic Conclusions

Only very few antigens can be considered as cell or tissue specific. The diagnostic conclusion drawn from an immunohistochemical staining should therefore not be based on only one staining reaction but on the results obtained using a panel of positive and negative reactions which will increase the diagnostic specificity of the staining results. In this context it is of great importance to correlate the staining results with the histomorphological picture, and with results obtained using other histochemical staining methods and/or electron microscopy.

In the following sections of this chapter the most common areas, in which immunohistochemical staining methods are used, will be summarized. Following a short description of the pathological problems, the typical staining profiles will be presented in tables.

32.3 Diagnostic Applications

Immunohistochemical staining methods are mainly used in diagnostic pathology within the following three major topics: immunological disorders (Sect.32.4), classification of tumours (Sect.32.5), and demonstration of microorganisms (Sect.32.6).

32.4 Immunohistochemistry of Immunologic Disorders

Essential elements in the pathogenesis of immunologic disorders are reactions between tissue antigens and antibodies to these or the formation of deposits of antigen-antibody complexes and complement fractions. The immunohistochemical demonstration of such tissue bound antibodies, antigen-antibody complexes, and complement fractions therefore forms an important part in the diagnosis and classification of these diseases. The disorders manifest themselves primarily in skin and kidney. The immunohistochemical reaction pattern in these organs can be roughly subdivided into linear and granular. The linear reaction pattern is due to antibody reaction with tissue antigen (Type II reaction), while the granular reaction is found in connection with deposits of immune complexes (Type III reaction). Furthermore, immunohistochemical staining methods are used for demonstrating circulating antibodies in the serum of patients in connection with autoimmune disorders.

32.4.1 Immunohistochemistry in Skin Diseases

Deposits of immunoglobulins and complement are predominantly present in the bullous skin diseases, lichen planus, lupus erythematosus, and in conditions where vasculitis is an essential element. A survey is given in Table 32.2 of the most important reaction patterns in skin disorders in which immunohistochemical demonstration of immunoglobulin and complement is especially important. Deposits may be encountered in conditions other than those shown in the table, thus deposits of C3 in the basement membrane in urticaria, IgM intercellularly in the epidermis in pemphigoid, lupus erythematosus, and lichen planus as well as in elderly people without skin disease. In connection with diseases accompanied by vasculitis deposits of IgG, IgM, and complement are seen. Granular deposits of IgA in vessel walls is a characteristic reaction in allergic purpura (Henoch-Schönlein).

Table 32.2. Typical reaction patterns in different skin disorders with deposits in the epidermis, basement membrane, and/or dermal papillae. One or more reactions may be absent. In the basement membrane granular deposits are seen in lupus erythematosus and dermatitis herpetiformis, while the deposits are linear in the other disorders.

	Epidermis intercellularly		Basement membrane				Dermal papillae (vessel walls)			
	IgG	C3	IgG	IgA	IgM	C3	IgG	IgA	IgM	C3
Pemphigus[2]	+	+	(+)[1]	−	(+)[1]	−	−	−	−	−
Bullous pemphigoid[2]	−	−	+	−	(+)	+	−	−	−	
Dermatitis herpetiformis	−	−	−	+		+	−	+	−	+
Lupus erythematosus[3]	−	−	+	−	+	+	−	−	−	−

[1] Pemphigus erythematosus.
[2] Reaction may also be seen in skin outside the lesion.
[3] Usually massive deposits in affected skin in discoid lupus erythematosus; in systemic lupus erythematosus deposits may also be seen in unaffected skin exposed to light in two thirds of the cases.

32.4.2 Autoimmune Disorders

An autoimmune disorder is characterized by the production of autoantibodies to an endogenous antigen with consequent injury to tissues or cells. The circulating autoantibodies can be demonstrated and are in some cases so specific that they solely react with cellular components in one organ while they in other cases may react with cellular components in several different organs. It is thus usual to distinguish between organ specific and non-organ specific autoimmune disorders, but in a number of these diseases there is a gradual transition from one to the other group. Frequently, several autoimmune disorders occur simultaneously in the same patient, for example autoimmune myocarditis together with pernicious anaemia. The key organ specific and non-organ specific autoimmune disorders are surveyed respectively in Tables 32.3 and 32.4, along with the antigenic specificity of the autoantibodies.

Autoimmune antibodies are demonstrated in the serum of patients by indirect immunohistochemical techniques. Serum from the patient is applied in increasing dilutions to cryostat sections of relevant tissue types, depending on the character of the disorder. After washing, incubation is performed with anti-human immunoglobulin which is fluorochrome or enzyme conjugated.

Table 32.3. Organ specific autoimmune disorders.

Disorder	Antibodies to
Hashimoto's thyroiditis	thyroglobulin
Primary hypothyroidism	thyroid microsomes
Grave's disease	TSH-receptors[1]
Pernicious anaemia	intrinsic factor, parietal cell microsomes
Idiopathic Addison's disease	cells of the adrenal cortex
Myasthenia gravis	acetylcholine receptors
Juvenile diabetes mellitus	pancreatic islet cells

[1] Thyrotrophin-receptors.

Table 32.4. Non-organ specific autoimmune disorders.

Disorder	Antibodies to
Systemic lupus erythematosus	DNA, nucleoprotein
Discoid lupus erythematosus Dermatomyositis Scleroderma	nuclear antigen, IgG
Rheumatoid arthritis	IgG

32.4.3 Glomerular Diseases of the Kidney

The identification of immunoglobulins and complement fractions is in some cases an important supplement to the light and electron microscopic classification of glomerulonephritis. The findings are, however, often difficult to interpret because the reaction patterns in different kinds of glomerulonephritis overlap to a large extent, and because deposits can occur both in other diseases and in kidneys without demonstrable disease.

While immunofluorescence investigations do not give substantial information regarding clinical symptoms, effect of treatment, or prognosis, the immunohistochemical findings substantially contribute to clarify the immunopathogenesis of glomerulonephritis. In general, two basic reaction patterns are found:

1. A linear reaction with a delineation corresponding to the basement membrane as an expression of the deposition of anti-glomerular basement membrane antibodies

2. Granular deposits peripherally or in the mesangium or in both as an expression of the deposition of circulating immune complexes

The most important immunohistochemical reactions in glomerular diseases are listed in Table 32.5.

32.5 Immunohistochemistry in the Diagnosis of Tumours

In the diagnosis of tumours, it is the task of the pathologist to decide whether a true neoplastic process is involved, whether the neoplasia is benign or malignant, and to classify the tumour process histogenetically. While the two first questions generally can be answered on the basis of the light microscopic evaluation of a specimen stained with Haematoxylin and Eosin, perhaps supplemented with a few other staining methods, the immunohistochemical methods are able to give important information regarding the histogenetic classification of tumours.

Antigens demonstrated in the diagnosis of tumours are most variable. Some of them are structural proteins taking part in the formation of the cytoskeleton of cells, the so-called intermediary filaments (Sect.2.2.9), others are secretory products such as hormones, and oncofoetal proteins, enzymes, and cell surface antigens.

Only very few of these antigens are really cell or tissue specific, but, as shown below, it is often possible to achieve a high degree of certainty regarding the histogenetic classification of tumours by the systematic use of antibodies directed against several of these antigens.

The choice of antibodies depends on the diagnostic problem as defined by the initial conclusions from light microscopic evaluation of a Haematoxylin and Eosin stained specimen. The choice will therefore be quite different depending on whether the tumour process is completely undifferentiated, appears to be a metastasis, or is more likely to be a primary tumour in a given organ. The first of these problems represents the basic diagnostic level where it is possible by a systematic selection of antibodies to arrive at a more specific diagnosis.

The following section will illustrate the use of antibody panels in the classification of undifferentiated tumours and tumours of unknown origin on the one hand, and in the classification of tumours developing from a number of well-defined organ systems on the other.

32.5.1 Immunohistochemistry in Undifferentiated Tumours

The primary task in connection with a histogenetic classification of undifferentiated tumours is to try to place the tumours in one of the diagnostic groups: Carcinoma, sarcoma, melanoma, or lymphoma.

Antibodies against keratin, vimentin, S-100, and leukocyte common antigen may be used to perform a primary immunohistochemical screening.

Table 32.5. Immunohistochemical reactions in glomerular diseases.

Diagnosis	Deposits of IgA	IgG	IgM	C_3	Fibrin
Minimal change	((+))MM	((+))MM	((+))MM	((+))MM	
Focal segmental glomerulosclerosis			+ SL	+ SL	
Focal segmental proliferative glomerulonephritis					
Henoch-Schönlein	+ MM	+ MM		+ MM	+ MM
Recurrent haematuria	+ + MM	+ MM		+ MM	
SLE	+ MM BM GR	+ MM BM GR	+ MM BM GR	+ MM BM GR	+ MM BM GR
Berger's disease	+ MM				
Diffuse membranous glomerulonephritis	(+)BM GR	+ BM GR	(+)BM GR	+ BM GR	
Diffuse mesangial proliferative glomerulonephritis					
Recurrent haematuria	+ MM	+ MM		+ MM	
Nephrotic syndrome	+ MM		+ MM		
Diffuse endopapillary glomerulonephritis		+ SE GR		+ SE GR	
glomerulonephritis		+ MM		+ MM	
Membranoproliferative glomerulonephritis					
I: mesangiocapillary		+ SEnGR	+ SEnGR	+ SEnGR	
II: dense deposit				+ + MM GR	
Crescentic, extracapillary					
Goodpasture syndrome		+ L BM		+ L BM	+ CR
Poststreptococcal and others		+ GR BM	(+)GR BM	+ GR BM	+ CR

Grading of reactions: ((+)) occasional weak; (+) occasional moderately positive; + moderately positive; + + strongly positive; MM: mesangial matrix; SL: sclerotic lesion; BM: basement membrane; GR: granular; L: linear; SE: subepithelial; SEn: subendothelial; CR: crescentic.

Keratin. This is a common designation for a complex group of cytoskeletal intermediary filaments comprising a total of at present 19 types with molecular weights between 40 and 70 kDa (Sect.2.2.9). Keratins are present in all kinds of epithelium, and the lack of keratin in benign or malignant tumours strongly suggests that the tumour is not an epithelial tumour. Typically, keratins with low molecular weights are present predominantly in more simple epithelia, while stratified epithelium also contains keratins with higher molecular weights. This distribution is to a certain degree retained in carcinomas, but, particularly in more undifferentiated tumours, the expression of keratins is very heterogeneous and it is therefore important to use keratin antibodies with a reactivity towards both low and high molecular keratins in the primary screening of tumours.

Vimentin. This is a cytoskeletal intermediary filament with a molecular weight of 58 kDa. It is typically present in mesenchymal cells and tumours developing from these. In malignant tumours a positive reaction is seen in tumours derived from mesenchymal cells but also in a number of carcinomas.

S-100. This is a calcium binding protein with a molecular weight of 21 kDa. S-100 is primarily present in the nervous system but may also be demonstrated in a number of different cell types outside the nervous system. There are three kinds: α-α, α-β, and β-β. In the differential diagnosis of undifferentiated tumours S-100 is especially useful as a marker for malignant melanoma. A positive reaction may further be seen in gliomas, neuroblastomas, malignant schwannoma, liposarcoma, chondrosarcoma, medullary carcinoma of the breast, and carcinoma of the sweat glands.

Leukocyte Common Antigen. Leukocyte common antigen (LCA) is a cellular membrane antigen which is found on the surface of most leukocytes, but in greatest amounts on lymphocytes. The antibody does not appear to react with other cell types.

In Table 32.6 the typical reaction patterns for undifferentiated tumours are shown using the above antibodies. Depending on the pattern detected by this primary screening a further characterization of the tumours can be performed by using supplementary antibodies as described below.

Table 32.6. Immunohistochemical screening panel for undifferentiated tumours.

	Keratin low/high molecular weight	Vimentin	Leukocyte common antibody	S-100
Carcinoma	+	–/(+)	–	–/(+)
Sarcoma	–/(+)	+	–	–/(+)
Lymphoma	–/(+)	+	+	–/(+)
Malignant melanoma	–/((+))	+	–	+

+ : always positive reaction; – : always negative reaction; (+): usually negative reaction, in a few cases positive; ((+)): predominantly negative reaction, very seldom positive.

32.5.2 Immunohistochemical Characterization of Carcinomas

Carcinomas are characterized by giving a positive reaction for *keratin* (Sect.32.5.1). Even though there is a tendency in the more well-differentiated tumours for carcinomas arising from more simple epithelia (e.g. liver and kidney) to contain predominantly low molecular keratin and for tumours arising from squamous epithelium to show more high molecular weight keratin, the keratin expression in more poorly differentiated tumours is usually so heterogeneous that it is impossible to arrive at any firm conclusions on this basis alone.

Some carcinomas also express *vimentin filaments* (Sect.32.5.1). These are carcinomas of the thyroid, lung, kidney, endometrium, and ovary. Furthermore, simultaneous expression of keratin and vimentin takes place in mesotheliomas, epithelioid sarcomas, and chordomas. A more specific characterization of carcinomas may be made on the basis of their secretory products.

Carcinoembryonic Antigen. Carcinoembryonic antigen (CEA) is a collective term for a group of glycoproteins (mol.wt. 150–220 kDa) primarily localized to foetal colonic mucosa. Polyclonal antibodies against CEA cross-react with a non-specific antigen (NCA) and with glycoprotein in the biliary tracts.

CEA can be demonstrated in epithelial tumours from the gastrointestinal tract, ductal carcinomas of the breast, and in carcinomas of the lungs and pancreas. On the other hand carcinoma of the liver, kidneys, prostate, and thyroid gland, with the exception of medullary carcinoma of the thyroid, are always negative. If the antibody cross-reacts with NCA, reaction is also seen in histiocytes, leukocytes, and several other kinds of tumours.

Epithelial Membrane Antigen. Epithelial membrane antigen (EMA) comprises a group of high molecular weight glycoproteins that have been isolated from milk fat globule membranes. EMA is particularly found in exocrine gland cells and can be demonstrated in most exocrine adenocarcinomas. Furthermore, it is present in squamous and transitional cell carcinomas, a number of mesotheliomas, and in small cell carcinomas of the lung.

Thyroglobulin. Carcinomas derived from follicle cells of the thyroid can be specifically characterized by the presence of thyroglobulin, a protein with a mol.wt. of 670 kDa. In addition to the more highly differentiated papillary and follicular thyroid carcinomas thyroglobulin can be demonstrated in a number of anaplastic thyroid carcinomas.

Prostate Specific Antigen. Carcinomas of the prostate can be specifically characterized by the presence of prostate specific antigen (PSA), a glycoprotein (mol.wt. 33 kDa) that is specific for glandular epithelium of the prostate. Other tissues, including epithelia of bladder, urethra, and seminal vesicles react negatively. The demonstration of PSA can thus be used to differentiate between carcinoma of the prostate and other carcinomas of the genito-urinary tract. In addition, metastases derived from the prostate may be identified.

Prostate Specific Acid Phosphatase. Similarly, prostate specific acid phosphatase (PSAP) is specific for prostatic glands and excretory ducts. A number of polyclonal antibodies to PSAP do, however, cross-react with other acid phosphatases. The application is otherwise quite similar to the use of prostate specific antigen.

Tumours Derived from Endocrine Glands. These can be subdivided on the basis of their production of hormones. Epithelial cells belonging to the diffuse neuroendocrine system (DNES) can be characterized, partly by staining with general neuroendocrine markers such as neuron specific enolase (NSE), chromogranin (CG-A), and synaptophysin (SY), and partly by demonstration of their polypeptide hormone production.

Neuron Specific Enolase. Neuron specific enolase (NSE) is an enzyme of glycolysis which comprises five isoenzymes that, according to their composition of subunits, are called $\alpha\alpha$, $\alpha\beta$, $\alpha\gamma$, $\beta\beta$, and $\gamma\gamma$. The latter is called neuron specific. NSE has been demonstrated in neural and neuroendocrine cells, and in addition in Schwann cells and a number of epithelial and mesenchymal cells. NSE can, for instance, be demonstrated in neuroendocrine tumours such as carcinoids and islet cell tumours in the pancreas, small cell bronchogenic carcinoma, phaeochromocytoma, and paraganglioma. Furthermore, a positive reaction is seen in a number of other malignant and benign tumours making the specificity rather questionable.

Chromogranin A. Chromogranin A (CG-A) is an acid calcium binding glycoprotein (mol.wt. 48 kDa) which is found in many neuroendocrine cells showing an argyrophil staining reaction. Related proteins are CG-B and CG-C. Chromogranin can be demonstrated in neuroendocrine epithelial tumours that are also argyrophilic (e.g. glucagonoma, gastrinoma, and carcinoid) but not in insulinoma, somatostatinoma, and small cell carcinoma of the lung.

Synaptophysin. Synaptophysin (SY) is a membrane associated calcium binding acid glycoprotein (mol.wt. 38 kDa) which can be demonstrated in presynaptic vesicles in neurons and in epithelial or neuroendocrine cells. It can be demonstrated in most epithelial or neuroendocrine tumours including small cell carcinoma of the lung, tumours derived from pancreatic islets, carcinoids, medullary carcinoma of the thyroid, and adenomas of the pituitary and parathyroid glands.

Serotonin. This can be demonstrated immunohistochemically in tumours derived from enterochromaffin cells, especially in the midgut, so-called EC-carcinoids.

32.5.3 Immunohistochemistry of Tumours of the Testis and Ovary

In the classification of tumours derived from ovary and testis immunohistochemical demonstration of epithelial markers and hormones can be of great value. A distinction is generally made between three types of tumours: Germ cell tumours, gonadal stromal tumours, and tumours derived from surface coelom epithelium.

Germ Cell Tumours. These comprise seminoma (in the ovary dysgerminoma) and non-seminomatous tumours. Seminoma/dysgerminoma are characterized by a positive reaction for placental alkaline phosphatase.

Placental Alkaline Phosphatase. Placental alkaline phosphatase (PLAP) is an isoenzyme of alkaline phosphatase and can in addition to germ cells be demonstrated in the syncytiotrophoblast, oviducts, endocervix, respiratory epithelium, and thymus.

Non-seminomatous tumours comprise embryonic carcinoma, yolk sac tumours, and choriocarcinoma. Embryonic carcinoma in addition to PLAP expresses α-foetoprotein, and keratin (Sect.32.5.1).

α-Foetoprotein. α-Foetoprotein (AFP) is a glycoprotein (mol.wt. 70 kDa) that in foetal life is produced in the yolk sac, liver, and gastrointestinal tract. In addition to embryonic carcinoma and endodermal sinus tumour a positive reaction is seen in a number of hepatocellular carcinomas.

Choriocarcinoma, which imitates the syncytiotrophoblast in the placenta, expresses partly PLAP (Sect.32.5.3) and keratin (Sect.32.5.1) and partly the hormones and proteins normally occurring in the placenta, i.e. *human chorionic gonadotrophin (HCG), human placental lactogen (HPL),* and *pregnancy-specific β-1-glycoprotein (SP1).*

Tumours Derived from Gonadal Stromal Cells. In the ovary and testis these tumours can be immunohistochemically classified by their hormone production. In the ovary *oestradiol* and *progesterone* can be demonstrated in granulosa and theca lutein cells. In the testis oestradiol and testosterone can be demonstrated in Sertoli cells and oestradiol (progesterone) and testosterone in Leydig cells.

Tumours Derived from the Surface Coelom Epithelium of the Ovary. These are characterized by positive reactions for keratin (Sect.32.5.1) and EMA (Sect.32.5.2), but only seldom by reactions for PLAP, HCG, and AFP (Table 32.7).

Table 32.7. Classification of germ cell neoplasms and their components.

	PLAP	HCG	AFP	KER	EMA
Seminoma/dysgerminoma	+	0[1]	0	0	0
Embryonic carcinoma	+/0	0[1]	+	+	0
Yolk sac tumour	+/0	0[1]	+	0/+	0
Choriocarcinoma	+	+	0	+	0
Non-germ cell carcinoma	0/+	0/+	0/+	+	+/0

[1] Positive in syncytiotrophoblast like cells.
+/0: usually positive; 0/+ : usually negative.
Antibodies to:
PLAP: placental alkaline phosphatase; HCG: human chorionic gonadotrophin; AFP: α-foetoprotein; KER: keratin; EMA: epithelial membrane antigen.

32.5.4 Immunohistochemical Characterization of Mesenchymal Tumours

Mesenchymal tumours comprise tumours derived from connective tissue cells, endothelial cells, muscle cells, cartilage and bone cells, as well as cells in the lymphatic and haemopoietic tissue. The majority of mesenchymal tumours are characterized by giving a positive immunohistochemical reaction with antibodies directed against *vimentin* (Sect.32.5.1). A number of tumours derived from muscle cells form an exception. Some mesenchymal tumours such as synovial sarcoma, epithelioid sarcoma, and chondroma express keratin filaments (Sect.32.5.1) in addition to vimentin. The preponderant part of tumours derived from muscle cells give a positive reaction for desmin.

Desmin. This is a polypeptide (mol.wt. 50–55 kDa) found in nearly all normal smooth muscle cells, striated muscle cells and in a few other kinds of cells. Some few smooth muscle cells, especially those in relationship to vessels, do not contain desmin, but only vimentin. Desmin can thus be demonstrated in benign and malignant tumours derived from smooth muscle (leiomyomas, leiomyosarcomas) as well as in tumours derived from striated muscle (rhabdomyoma and rhabdomyosarcoma) and also in tumours where a myogenic differentiation takes place, for example in mesodermal mixed tumour of the ovary or uterus. Desmin is the best general myogenic marker. More well differentiated myogenic tumours express actin in addition to desmin.

Actin. This is a group of contractile microfilaments (mol.wt. 43 kDa) which are found in nearly all cells (Sect.2.2.9). In muscle cells α- and γ-actin are found, whereas in other cells β- or γ-actin are present. Muscle specific antiactin can, just as desmin, be used for the identification of leio- and rhabdomyosarcomas as well as for the study of myofibroblast proliferation. Furthermore, the positive reaction in myoepithelial cells may be used for the differentiation between benign and malignant tumours of the breast, as malignant ductal elements are normally not surrounded by myoepithelial cells. A further characterization of especially well-differentiated tumours derived from striated muscle can be made by staining for myoglobin.

Myoglobin. This is an oxygen binding protein and can in tumours be demonstrated in rhabdomyosarcomas, rhabdomyoid differentiated cells in embryonic tumours, and mesodermal mixed tumours.

In addition to giving a positive reaction for vimentin (Sect.32.5.1), tumours derived from fibroblastic and fibrohistiocytic cells also in a number of cases react positively with *α-1-antitrypsin (AAT)*, *α-1-chymotrypsin (ACT)*, and ferritin.

Ferritin. Ferritin contains 20% iron which is surrounded by a shell of protein (mol.wt. 462 kDa). In addition to being present in cells derived from fibrohistiocytic cells (fibromatosis, malignant fibrous histiocytoma), it is found together with AAT and ACT in liposarcomas.

This positive reaction is, however, only of limited value in the differential diagnosis, as positive reaction also may be seen in many other cell types.

Lysozyme. The enzyme lysozyme (muramidase) cleaves muramic acid present in the cell wall of bacteria (Sect.2.3.1). Lysozyme can be demonstrated in neutrophils, monocytes, and macrophages as well as in the glandular epithelium in lachrymal and salivary glands, breast, stomach, intestine, and tubules of the kidney. In connection with neoplasia a positive reaction is seen in phagocytosing cells in fibrohistiocytic tumours in neoplasias derived from myeloid and monocytoid cells (Sect.32.5.5).

Von Willebrandt's Factor. Tumours derived from endothelium include haemangioma, haemangioblastoma, endotheliosarcoma, Kaposi's sarcoma, and cardiac myxoma. All, in addition to the reaction for vimentin (Sect.32.5.1), show a positive reaction for von Willebrandt's factor (VWF). VWF is also called factor VIII related antigen and is produced in normal endothelial cells, megakaryocytes, and platelets. The strongest reaction is seen in endothelial cells in small blood vessels, while the reaction in lymphatic vessels and sinusoids is frequently weak.

32.5.5 Immunohistochemical Characterization of Lymphoid and Myeloid Neoplasias

Lymphoid and myeloid neoplasia include neoplastic conditions derived from lymphoid cells (malignant lymphoma and leukaemias) and from myeloid cells (leukaemias and some localized tumours derived from early stages in the development of granular leukocytes, monocytes, red blood cells, and platelets).

Malignant lymphomas can be morphologically classified into two main groups: Hodgkin's lymphoma and non-Hodgkin lymphoma. Non-Hodgkin lymphomas can further be differentiated, depending on whether they are derived from T-cells or B-cells, into T- and B-cell lymphomas. The earliest B- and T-cells are characterized by a positive reaction for the enzyme *terminal deoxynucleotidyl transferase* (TDT) which may be used as a marker for lymphoblastic lymphomas and leukaemias (see Table 32.8).

Cluster of Differentiation System or CD-System. A considerable number of monoclonal antibodies reacting against leukocyte differentiation antigens have been isolated and characterized during recent years. The leukocyte differentiation antigens have been systematically described and arranged in a system called cluster of differentiation system or CD-system. A total of 45 antigens have hitherto been defined and characterized.

As previously described (Sect.26.1.2) all immunoglobulins (Ig) have a similar structure and are composed of four polypeptide chains—two identical heavy chains (H), and two identical light chains (L). There are five major types of heavy chains which give their name to the five major Ig classes: γ in IgG, α in IgA, μ in

IgM, δ in IgD, and ϵ in IgE. There are two types of light chains – κ and λ. An Ig-producing cell only produces one kind of light chain. The ratio between κ- and λ-chains in the organism is about 3:2. Immunoglobulins are produced by B-lymphocytes. The earliest chains that can be detected are μ-chains in the cytoplasm. Later immature and mature B-cell surface chains, that may be detected, are μ-, δ-, and L-chains. In the further development into plasma cells intracytoplasmatic Ig can be demonstrated.

Joining Chains. The joining chains (J) which link the subunits of IgM and IgA are produced in all B lymphocytes and plasma cells but are only secreted bound to IgA or IgM. The demonstration of J-chains can be used in the identification of B cells and in the classification of lymphomas. The phenotypic expression of T and B cell neoplasias with the reaction pattern for the most important CD-antigens, immunoglobulins, and TDT are given in Table 32.8.

CD 1–8. These are chiefly present in T lymphocytes, while CD 10, 19, and 22 are chiefly connected with B lymphocytes. While these antigens can only be satisfactorily demonstrated in cryostat sections the following antigens (CD 15, CD 30, CD 45, and L 26) can all be demonstrated in paraffin sections of formaldehyde fixed material.

CD 15. This is a carbohydrate, previously called hapten X, which is expressed by neutrophils, Hodgkin, and Reed-Sternberg cells, as well as by the cells in a few large-cell non-Hodgkin lymphomas. Demonstration can be used in the diagnosis of Hodgkin's lymphoma.

CD 30. CD 30 (105 kDa) is also called *χ-1 antigen*. It is found on a small, not more precisely defined, group of large lymphocytes. In malignant lymphomas CD 30 is found on Hodgkin and Reed-Sternberg cells and on most cells in pleomorphic T cell lymphoma as well as large cell anaplastic lymphomas.

CD 45. This antigen (200 kDa), also called *leukocyte common antigen* (LCA) (Sect.32.5.1), is – as previously described – an important marker in the differentiation of malignant lymphomas and other tumours. CD 45 R is a designation for a number of CD 45 proteins.

Antibodies can demonstrate either CD 45 R-B (205 and 220 kDa) which is found on most B lymphocytes and a smaller proportion of T lymphocytes or CD 45 R-T (185 kDa) found on most T lymphocytes. Both CD 45 R-B and CD 45 R-T react, as opposed to most other CD antibodies, on paraffin sections.

L 26. This is a monoclonal antibody which reacts with a, as yet not characterized, antigen found in the cytoplasm of most B lymphocytes. With L 26, 90–95% of all B lymphomas can be detected.

Table 32.9 shows the immunohistochemical classification of malignant lymphomas using antibodies which work on paraffin sections.

Table 32.8. Antigen patterns characterizing cell differentiation (and corresponding neoplasms) of the lymphoid system.

Cells	Neoplasms	T lymphocyte identification CD: 1	2	3	4	5	6/7	8	TdT
Early thymocyte	T-lymphoblastic lymphoma		+	+			+		+
Cortical thymocyte	Acute T-lymphoblastic leukaemia	+	+	+	+	+	+	+	+
Medullary thymocyte			+	+	+[1]	+	+	+[2]	+
Peripheral T cell	Peripheral T-lymphoma; T-CLL[3]			+	+[1]	+	+	+[2]	

[1] T helper cell; [2] T suppressor/cytotoxic cell; [3] Chronic lymphatic leukaemia.

Cells	Neoplasms	B lymphocyte identification CD: 10	19	22	μ	S-Ig[1]	C-Ig[2]	TdT
Pre-pre B	Common ALL[3]	+	+	+				+
Pre B	Common ALL[3]	+	+	+	+			
Immature B	B-lymphoblastic lymphoma; B-ALL[3]	+	+	+		+		
Small B	B-CLL[4]		+	+		+		
Plasma cell	Immunocytoma; myelomatosis; Waldenström's macroglobulinaemia		+	+			+	

[1] Surface immunoglobulin; [2] Cytoplasmic immunoglobulin; [3] Acute lymphatic leukaemia; [4] Chronic lymphatic leukaemia.
CD: cluster of differentiation system or CD-system; TdT: terminal deoxynucleotidyl transferase.

Table 32.9. Immunohistochemical classification of malignant lymphomas (paraffin sections).

	C-Ig[3]	J chain	L26	CD15	CD30	CD45	$CD45R_B$	$CD45R_T$
B-lymphoma	0/+	+	+	0/+[4]	0/+[4]	+/0	+/0[5]	0
T-lymphoma	0	0	0	0/+[4]	0/+[4]	+/0	0/+[6]	+/0
RS[1]	0/+	0	0	+	+	0	0	0
RS-LPNS[2]	0/+	0	0	0	+	+	?	?

+/0: usually positive; 0/+: usually negative.
[1] RS: Classical Reed-Sternberg cell in Hodgkin's lymphoma; [2] RS-PLNS: Reed-Sternberg cell in Hodgkin's lymphoma with lymphocyte predominance, nodular subtype; [3] Cytoplasmic Ig: only monoclonal occurrence in cells producing J chains; [4] Only certain large cell polymorphic lymphomas; [5] 90%; [6] Small group of cells.

In the classification of myeloid neoplasias, principally the leukaemias, the number of monoclonal antibodies that can be used as markers for the different lines of differentiation and differential stages is considerably less than the selection present for the description of lymphoid neoplasias.

As outlined in Table 32.10 antibodies against antigen CD 34 can be used to characterize the immature stages in myeloid differentiation. In the differentiation of neoplasias derived from megakaryocytes positive reactions are shown for CD 41 and CD 42. Antibodies against CD 13 are common for both myeloid and monocytoid leukaemias, while the monocytoid leukaemias are characterized by a positive reaction for CD 14. Leukaemias with granulocytic maturation are positive for CD 15. As regards characterization of cells from the erythroid line, no useful antibodies have so far been isolated.

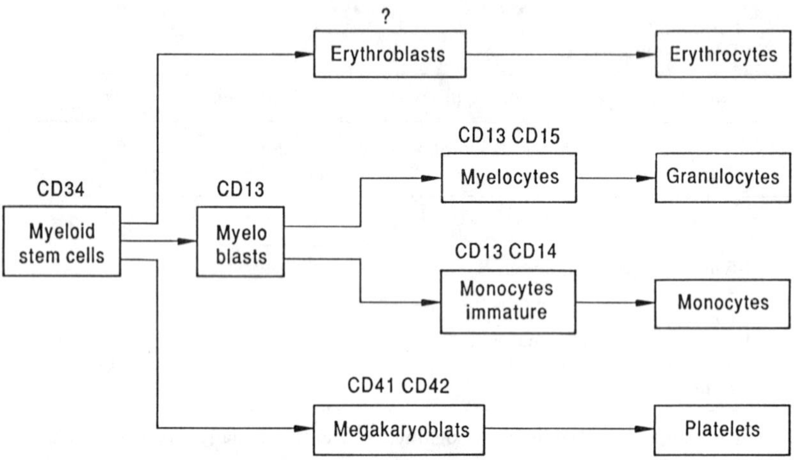

Table 32.10. Antigen patterns characterizing cell differentiation of the myeloid system.

32.5.6 Immunohistochemical Characterization of Neoplasias in the Nervous System

The neoplasias of the nervous system consist of tumours derived from the central and peripheral nervous system. Tumours of the central nervous system comprise in part tumours derived from neurons and in part tumours derived from glial cells and cells in the ependyma and meninges. The neurone derived tumours include, according to increasing maturity: neuroblastomas, ganglioneuroblastomas, ganglioneuromas, gangliogliomas, and gangliocytomas. These tumours, to increasing degrees, show positive reactions for the neuronal markers: neuron specific enolase (Sect.32.5.2), synaptophysin (Sect.32.5.2), and neurofilament.

Neurofilament. This comprises three polypeptide types (68, 150, and 200 kDa) which can be demonstrated in neurons and neuronal processes in the central nervous system and in peripheral nerves as well as ganglion cells and the suprarenal medulla. In addition to the neoplasias mentioned, positive reactions are also found in paragangliomas and phaeochromocytomas.

Gliafibrillary Acidic Protein. Astrocytomas are the largest group of tumours derived from the neuroglia. The most valuable marker for both normal and neoplastic astrocytes is gliafibrillary acidic protein (GFAP).

GFAP is a polypeptide (51 kDa) which can be demonstrated in astrocytes and a few ependymal cells. A positive reaction is also found in Schwann cells, folliculostellate cells in the pituitary gland, and satellite cells in ganglia. Positive reactions can be demonstrated in astrocytomas and tumours with an astrocyte component (glioblastoma, gliosarcoma) and in a few cases of ependymomas and oligodendrogliomas. Furthermore, in the central nervous system in a few non-glial tumours: medulloblastoma, neuroblastoma, retinoblastoma, and pinealoma as well as choroid plexus papilloma. Outside the central nervous system positive reaction has been demonstrated in schwannomas and pleomorphic adenomas.

S-100 Protein. S-100 protein (Sect.32.5.1) is in the nervous system mainly found in the same cells that express GFAP with the difference that the positive reaction in addition to being found in the cytoplasm is also often observed in the nuclei.

Tumours derived from oligodendroglia frequently show a positive reaction on staining with the antibody *anti-LEU-7* that reacts with a, as yet not characterized, membranous antigen found in NK cells (natural killer lymphocytes). A positive reaction can be shown in close to 90% of oligodendrogliomas but also in some astrocytomas, schwannomas, neurofibromas, a few malignant melanomas, and neuroendocrine tumours.

Meningeomas show a positive reaction for epithelial membranous antigen and vimentin (Sect.32.5.1).

Myelinic Basic Protein. Tumours of the peripheral nervous system comprise schwannomas and neurofibromas and their malignant counterparts. A positive reaction is seen on staining for S-100 and myelinic basic protein which is a protein (18 kDa) produced in Schwann cells as part in the formation of myelin sheaths.

32.5.7 Immunohistochemical Demonstration of Basement Membrane Antigens and Amyloid in Tumour Diagnosis

One of the most important characters of malignant tumours is their capacity for invasive growth. In tissues delimited by a basement membrane (epithelia) (Sect.2.4.3) infiltration of this boundary can be considered an early marker of invasive growth. The basement membrane consists of the basal lamina and the reticular lamina. The structure and antigenic composition of basement membranes of different tissues have been determined in considerable detail (Sect.2.4.3). Immunohistochemical investigations of tumour invasion depend on detecting the disruption of this ordered structure. The changes detected can be divided into those affecting the intrinsic and the extrinsic components, respectively.

Intrinsic Components. These are detected by a panel of antibodies to *collagen type 4, laminin, and heparan sulphate*. Examinations of epithelial tumours derived from the breast, skin, pancreas, and the prostate glands have shown that benign changes and in situ preinvasive malignant changes are characterized by an intact basement membrane with linear staining of intrinsic components. With invasion, thickening, fragmentation, and rupture of the intrinsic components has been described. The overall pattern remains somewhat unclear with a number of studies showing conservation of peripheral intrinsic components (particularly in highly differentiated tumours of the head and neck) while interrupted staining of intrinsic components has been described in dysplastic and in situ lesions.

These antigens have also been examined in sarcomas. Irregular pericellular staining has been described in highly differentiated leiomyosarcomas, neurofibrosarcomas, and liposarcomas. This feature is not present in malignant fibrous histiocytomas, osteosarcomas, haemangiopericytomas, and undifferentiated sarcomas.

Extrinsic Components. These are detected by antibodies to *fibronectin*. In carcinomas there is an overall reduction in fibronectin staining and the normal pattern is disrupted adjacent to neoplastic cells. The stromal reaction is frequently increased, particularly in tumours associated with the so-called desmoplastic reaction.

In sarcomas an irregular pericellular staining is seen in leiomyosarcomas, neurofibrosarcomas, and liposarcomas, while an irregular peri- and intercellular staining is seen in malignant, fibrous histiocytomas, osteosarcomas, haemangiopericytomas, and undifferentiated sarcomas.

Amyloid. The composition and demonstration of amyloid are described in Sect.21.7. The different types of amyloid may be identified with specific antibodies raised against the components detailed in Table 21.6.

In connection with tumour diagnosis, the demonstration of amyloid light chains (AL) is of interest in myelomatosis, while the demonstration of C-protein amyloid or APUD-amyloid (representing precipitation of polypeptide hormones) is of value in the characterization of tumours derived from neuroendocrine cells.

32.6 Immunohistochemical Identification of Microorganisms

The diagnosis of many infectious diseases depends on serologic tests (i.e. demonstration of serum antibodies against microorganisms) and culture of microorganisms from tissue or body fluids. While these diagnostic techniques are likely to remain important in the future, the possibility for identifying viruses and other parasites in situ combined with the increased frequency of opportunistic infections in immunocompromised patients has increased the need for reagents and methods for the in situ identification and localization of these microorganisms.

In recent years a number of antisera have been developed which make the immunohistochemical identification of microorganisms possible. These techniques have to be viewed in parallel with the potential of in situ hybridization particularly for viruses (Sect.20.6). Table 32.11 gives a survey of important microorganisms that can be demonstrated in routine treated tissue.

Table 32.11. Important microorganisms that can be demonstrated immunohistochemically in formaldehyde fixed, paraffin embedded material.

Campylobacter	Leishmania
Candida albicans	Morbilli virus
Chlamydia	Mycobacteria
Cytomegalovirus	Mycoplasma
Hepatitis virus B	Poliovirus
(Surface and core antigens)	Rotavirus
Herpes simplex I and II	Rubella virus
Human immunodeficiency virus	Toxoplasma
Influenza virus	Trichophyton
Klebsiella	Varicella-zoster virus
Legionella	

Appendix A: Standardization of Staining Methods

H. Lyon, D. Wittekind, E. Schulte

No detailed descriptions of staining methods are provided in this book. The reader is referred to one or more of the excellent texts covering this field (cf. Preface).

Nevertheless, we have felt it appropriate to include this appendix which gives some of our views on the technical aspects of procedures which should be given particular emphasis. It is our opinion that the need for standardization and quantitative methods in daily work is pressing. This appendix sets out the appropriate general considerations followed by a few selected methods in order to cover this area.

A.1 General Considerations

According to Boon and Wittekind (1986) the principle aim of standardizing staining methods is to render their application reproducible and therefore reliable. This is of the utmost importance when dyes and stains are used for automated cell pattern recognition (Wittekind, 1985; Wittekind and Schulte, 1987).

The theoretical background for standardization of cell and tissue preparation stems from the fact that any preparatory step – from cell sampling to mounting of the stained slide – will somehow affect the structure of the cell and ultimately lead to the production of a particular staining pattern which, in strict terms, is an artifact. What we eventually observe by microscopy is, from the perspective of a cell, the product of a rather violent procedure: In cytological preparations the cells have been isolated from their tissue, spread out on the surface of a glass slide and immersed in a liquid poison which abruptly arrests and – *sensu stricto* – "fixes" the cell in the very last moment of its life. During fixation some components of the cell might be removed by the fixative (alcohols remove lipids), and proteins may be precipitated or cross-linked. In histology the tissue is embedded in molten wax or in polymerizing plastic; the cell is cut into thin slices on a microtome. Finally – and this is true in both cytology and histology – dyes dissolved in aqueous or alcoholic solutions are bound by sometimes unknown mechanisms to one or the other substrate in the cell thus yielding "contrast" between stained and unstained – or less stained – components of the cell.

H. Lyon (Ed.)
Theory and Strategy in Histochemistry

What we finally look at has not much to do with the "true" structure of the living cell; this can be easily confirmed by the comparison of a living cell in phase contrast microscopy with the same cell after a conventional fixation and staining procedure in usual light microscopy.

Thus the aim of standardization in cell and tissue preparation is *to make* the staining pattern, i.e. the artifact, *reproducible*; in other words standardized preparatory techniques should guarantee *standard artifacts*.

Finally, we have to answer the question: "Which staining pattern should we choose as the standard". Variation of the preparatory technique will result in variation of the staining pattern: if for instance the pH of the staining solution is changed, hue and/or intensity of staining can change, and a whole palette of staining patterns will be found when *several* preparatory steps are changed simultaneously.

It seems logical both from a practical and a theoretical point of view to choose as standard the staining pattern which best fulfils the requirements of the observer, be it human or a computer; correspondingly, the preparatory technique yielding this staining pattern is defined as the standard technique. This is explained in detail below.

Standardization of a method for staining cytological or histological material requires consideration of all steps in the procedure.

A.1.1 Slide Preparation

Cytological material may primarily be air-dried and then fixed or fixation may be carried out on the wet material. The results achieved with a staining method may be quite different according to which of these procedures has been chosen. For instance with the Romanowsky-Giemsa method air drying is preferred, while for the Papanicolaou method air drying has a detrimental effect on the staining result. The choice of fixative also has a pronounced effect on the staining result; with the Papanicolaou method Boon and Drijver (1986) recommend the use of ethanol in concentrations between 50% and 70% and with added polyethylene glycol (300–1000), whereas this fixative cannot be recommended for use with the Romanowsky-Giemsa method.

Histological material may also primarily be treated in two different ways. These are either the preparation of cryostat sections (Sects. 11.2.2 and 11.3.1) or the use of a chemical fixative (Chap. 12) which is usually followed by dehydration (Sect. 14.2), clearing (Sect. 14.3), and embedding (Sect. 14.4) for light microscopy. All of these procedures have a profound effect on the material before the final sections are cut. Rigorous standardization is therefore of paramount importance to achieve reproducible results irrespective of the subsequent staining procedure.

A.1.2 Staining Methods

Concerning the staining methods themselves, assuming that the preparatory techniques have been standardized, the most important reagents are dyes (Sect.3.3).

Dyes. In general, dyes are coloured organic molecules with large systems of delocalized electrons (conjugated π-electronic systems). Dyes are available as crystals or as powders which on solution in a suitable solvent may bind by physico-chemical attractions to a substrate and impart colour to the latter. A *stain* is a solution of dye in a suitable solvent. Stains may be subdivided into stock solutions and working solutions, where a *stock solution* is a stable solution of one or more dyes at a concentration, which is higher than that usually employed for staining, while a *working solution* is a solution of one or more dyes in a suitable solvent at concentrations adapted to staining purposes. Finally, it should be mentioned that a *chromogenic reagent* is a colourless reagent which can react with suitable groups present or induced in the biological substrate with the formation of a dye in situ. Standardized dyes, stains, and chromogenic reagents are of course essential to a standardized method.

Standardization of Dyes. This can be carried out by specification of the physical and chemical characteristics of the dyes. Much work of this nature has been performed by the Biological Stain Commission (BSC) (Sect.3.3.10). However, exact specifications of completely pure dye samples are still lacking in what are probably the majority of cases. According to Boon and Wittekind (1986), this form for standardization is in theory sufficient. When a candidate sample of a dye complies with the specifications of the standard, the staining results would be reproducible if:

a. All other components and factors of the stain besides the constituent dyes are also standardized
b. The biological substrates to be stained are in reasonably comparable physico-chemical and technical states
c. The slide preparation is standardized

The difficulty with this approach is the achievement of sufficiently pure dye samples at the outset. This has led the BSC to base standardization not only on certain physico-chemical characteristics of the dye, but also according to its performance in the so-called "biological staining tests" (Clark, 1981).

Standardization of the Staining Solutions. This includes specification of the solvent to be used and the dye concentration. The latter should take into account possible precipitation and changes in concentration during use. Of further importance is the pH and content and concentration of ions. Here it must be remembered the addition of a buffer not only has a stabilizing effect on pH, but will also add ions to the solution and can thus give rise to "salting on/salting off" effects (Bennion and Horobin, 1974; Horobin and Goldstein, 1974).

Standardization of Technique of Staining. This includes the four factors of staining time, staining temperature, contact of stain with section, and rinses between or after staining.

Staining Time. The importance of this factor varies a great deal depending upon the complexity of the staining process. However, it should be ensured that the staining times are not too short. It should be appreciated that a staining equilibrium is very rarely achieved with most staining methods. Extended staining times nearly always result, therefore, in staining patterns which deviate considerably from the results obtained with "normal" staining times. Standardization of staining time should not, however, normally present any problem.

Staining Temperature. The temperature of the staining bath is very important and will of course effect the staining time. At elevated temperatures the transport of molecules and ions and reactions between these are accelerated. This means that increased staining temperature leads to shorter staining time. However, the increase in temperature may have deleterious effect on the staining result as the proteins of the tissue become denatured. It is usual to stain at room temperature which probably can vary between 18°C and 28°C. At the higher temperature chemical processes take place at double the speed of those at the lower. Control of staining temperature can therefore be quite important. The temperature of interest is that of the staining solution and not so much that of the surroundings. It should be remembered that buffers or other reagents taken directly from the refrigerator take a considerable amount of time to come into temperature equilibrium with the surroundings.

Contact of Stain with the Section or the Cells. This is a difficult factor to control as it is influenced by movement of the slides in the staining bath and also by the space between them in the staining rack. Standardization of this factor can, according to Zimmermann (1983), only be achieved when staining machines are employed.

Rinses between Baths for Differentiation, Blueing, and Dehydration. Rinsing procedures are usually critical. They are used to remove surplus dye from the substrate, or they are used to change the colour of the substrate-bound dye (as is the case with Haematoxylin), and normally the stained slides are rinsed in alcoholic baths to remove water prior to mounting.

The critical point with all these steps is that they might remove substrate-bound dye from the slide resulting in weak staining. Of course the rinsing procedure can be standardized (pH, temperature and ionic strength of the bath, rinsing time), but the final result depends substantially on the nature and form of the biological material. Thus thick sections or tissue imprints require longer times for sufficient differentiation than thinner ones, and section thickness is a factor which is difficult to control. There is also considerable variation in the sensitivity of different tissues to the rinsing agent; this means that despite strict standardization of rinsing procedures, some uncertainty about the result must remain. This is particularly the case when polychrome staining procedures are used; the different dyes may be differentially affected by the rinsing process.

A.2 Examples of Staining Methods

A.2.1 Methyl Green-Pyronin Y

The Methyl Green-Pyronin stain (Unna-Pappenheim; plasma cell stain) is a staining method characterized by the use of two cationic dyes, Methyl Green and Pyronin Y, in aqueous solution. The method was originally developed as a purely histological method (Sect.6.1.5), but work by Brachet (1940a; 1940b; 1942; 1953) showed that the method is valuable for the simultaneous demonstration of DNA (Methyl Green) and RNA (Pyronin Y).

Reagents.

Ethyl Green, C.I. 42590, or Methyl Green, C.I. 42585[1)]
Pyronin Y, C.I. 45005[2)]
Polyamide, e.g. Polyamide CG6, Machery-Nagel, cat. no. 81562
1-propanol
Formic acid
$NaBF_4$-solution
0.1 mol/l potassium hydrogen phthalate
0.1 mol/l HCl
DNase I
RNase A
99% v/v ethanol
95% v/v ethanol
70% v/v ethanol
Distilled water
1-butanol
Xylene
Hydrophobic synthetic resin

1) *Ethyl Green*, C.I. 42590, or *Methyl Green*, C.I. 42585, in pure form. If this is not available, commercial dyes may be purified by dissolving a quantity equivalent to 0.15 g pure Ethyl Green or Methyl Green in 50 ml warm distilled water. 5 g polyamide (Polyamide CC6, Macherey-Nagel, cat. no. 81562) is added. After stirring for 30 min with a magnetic stirrer, suction filtration is carried out through a glass filter (diameter 7 cm, pore size 3). The bottom of the glass filter is covered with a layer of polyamide (approximately 3 g) which has been flushed with water and dried by suction. This layer is covered by a piece of filter paper to prevent the polyamide from being suspended during filtration. The dye solution is poured over the filter without suction. When the funnel is filled up, suction is applied

gently. The residue in the flask is washed with a little water, and the filter cake is rinsed with water to obtain a total volume of 90 ml. The degree of purification is tested by performing thin layer chromatography with 1-propanol:formic acid, 4:1, as the eluting solvent. Crystal Violet is easily detected as it has a higher R_f value than Methyl Green and Ethyl Green as well as a characteristic colour. If proper purification has not been obtained, the solution is filtered once more through a layer of fresh polyamide on the filter. The solution is subsequently precipitated with a $NaBF_4$-solution and recrystallized from ethanol giving analytically pure Methyl or Ethyl Green tetrafluoroborate. (Andersen *et al.*, 1986).
2) *Pyronin Y*, C.I.45005. Certified samples containing at least 45%, and preferably between 80% and 100%, of the pure dye should be used.

Preparation of Ethyl Green-Pyronin Working Solution. Ethyl Green or Methyl Green in an amount corresponding to 0.15 g pure Ethyl Green or Methyl Green (calculated as the coloured cation) is dissolved in 90 ml warm distilled water. An amount corresponding to 0.03 g Pyronin Y (calculated as the coloured cation) is dissolved in 10 ml 0.1 mol/l phthalate buffer, pH 4.0. The latter solution is then added to the Methyl Green solution. The working solution is stable for at least one week when kept in a brown glass bottle at room temperature.

Material. Cryostat sections can be used or paraffin sections of material preferably fixed in Carnoy's mixture (99% v/v ethanol-chloroform-glacial acetic acid 6:3:1) at room temperature for 18 hours. However, material fixed in neutral buffered formaldehyde may also be used.

Staining Procedure.

1. The sections are hydrated
2. Stain for 5 min at room temperature (22°C) in the working solution
3. Rinse in two changes of distilled water, 2–3 seconds in each
4. Shake off surplus water
5. Agitate in three changes of 1-butanol until clear
6. Mount directly from 1-butanol or via two changes of xylene in a hydrophobic synthetic resin

Note that if glycol methacrylate sections are used instead of paraffin sections dehydration cannot be performed with 1-butanol. In these circumstances the sections should be dried over silica gel in a thermostat oven at 37°C.

Staining Results.

	Carnoy	Formaldehyde
chromatin	green	blue
nucleoli	red	(lilac)red
cytoplasmic RNA	red	(lilac)red
cartilage matrix	orange	orange
mast cells	orange	unstained
other structures	unstained	unstained

Specificity of Staining Results. The specificity of the Methyl/Ethyl Green staining is tested by subjecting an adjacent section to DNase I prior to staining with the working solution of Methyl/Ethyl Green and Pyronin Y. All green staining of chromatin (*DNA*) should be abolished. The specificity of the Pyronin Y staining is tested by subjecting an adjacent section to RNase A prior to staining with the working solution of Methyl/Ethyl Green and Pyronin Y. All red staining of nucleoli and cytoplasm (*RNA*) should be abolished by this procedure.

A.2.2 Azure B-Eosin Y

The Azure B-Eosin Y or Standard Romanowsky-Giemsa (RG) stain is a staining method characterized by the combined action of the cationic dye Azure B and the anionic dye Eosin Y in aqueous solution on biological material, namely, cytological films and histological sections.

Reagents.

Azure B-SCN, C.I. 52010, DIN 58981/1, in pure form[1)]
Eosinic acid, C.I. 45380, DIN 58981/2, in pure form[1)]
Dimethylsulphoxide (DMSO)
0.03 mol/l N-(2-hydroxyethyl) piperazine-N-2-ethanesulphoxide acid (HEPES) buffer, pH 6.5
Para-toluenesulphonic acid 0.1%
$NaHCO_3$ 0.1%
Distilled water
2-propanol
Toluene
Hydrophobic synthetic resin

1) Both dyes are easily soluble in dimethylsulphoxide (DMSO) at the appropriate concentration.

Preparation of Azure B-Eosin Y Stock Solution.

Azure B-SCN 7.5 g dissolved in 750 ml DMSO
Eosinic acid 1.2 g dissolved in 250 ml DMSO

The two solutions are prepared individually using a magnetic stirrer. Wait until the dyes are completely dissolved. Mix the two solutions ; use magnetic stirrer. The stock solution is ready for use. Note that the stock solution is now commercially available.

Preparation of the Buffer Solution.

a. HEPES (acid) 23.83 g in 1000 ml distilled water
b. HEPES (salt) 26.00 g in 1000 ml distilled water

Mix 900 ml solution a) and 100 ml solution b) to obtain a 0.1 mol/l buffer stock solution pH 6.5. The solution can be stored in the refrigerator for some months.

700 ml distilled water is added to 300 ml of this buffer stock solution to obtain the *buffer working solution* of 0.03 mol/l pH 6.5 (which is simply referred to as "buffer" in the following).

Material. Air dried cytological material (smears or imprints) may be fixed routinely in methanol p.a. 10 min. Stock solution diluted by methanol is strongly recommended as stain/fixative solution. The stain/fixative solution is prepared by dilution of 1 vol stock solution with 15 vol methanol analytical grade. Slides are immersed for three min in stain/fixative solution. Use of the stain/fixative solution avoids the loss of basophil granules which unavoidably occurs in methanol. Histological material can be fixed in 4% formaldehyde containing 1 g $ZnSO_4$ in 100 ml formaldehyde. Both paraffin and glycolmethacrylate (GMA) embedded material can be used.

Working Solution and Staining Procedure.

I. Cytology:

1. Blood films, tissue imprints and general cytology:
   Stock solution: 1 vol
   Buffer: 50 vol
   Staining time: blood films – 10–(20) min, general cytology – 20–(30) min
2. Bone marrow films:
   Stock solution: 1 vol
   Buffer: 25 vol
   Staining time: 25 min

Slides are immersed in the staining solution in vertical position. After staining they are rinsed in buffer for 1 min, followed by a 10 sec rinse in distilled water. Then the slides are placed in a vertical position for air drying.

II. Histology:

Stock solution: 1 vol
Buffer: 49 vol
Staining time: 4 hours (3–6 h, depending on material)

Slides are immersed in the staining solution in vertical position. After staining the slides are rinsed for 3 sec in distilled water, differentiated in para-toluenesulphonic acid 0.1% for 30 sec, washed in distilled water for 3 sec, followed by a rinse in $NaHCO_3$ 0.1% for 2 sec. After another 3 sec rinse in distilled water slides are dehydrated in 3 changes of isopropanol (30 sec each) and after clearing in toluene (2 changes, 2 min each) are mounted in synthetic resin.

Staining Results.

Cytology: Nuclear chromatin: purple; nucleoli: light blue; basophilic cytoplasm: blue; basophilic granules: purple-black; eosinophilic granules: red-orange; neutrophilic granules: purple; erythrocytes: pink-orange. *Note*: Staining has not been completed unless cell nuclei are purple.

Histology: Nuclear chromatin: purple; RNA-rich cytoplasm: blue; cytoplasm without or with low content of DNA: greenish-blue; intestinal goblet cells: purple/blue; eosinophil granules: orange; basophil/mast cell granules: red purple; cartilage: red purple; bone: violet; basement membranes: purple; elastic fibres: green; axons: purple.

Appendix B: Quantitative Morphological Methods in Microscopy

E. Hasselager

B.1 Definitions

An extensive range of methods have been developed in order to quantitate the dimensions and relative location of elements observed by microscopy. The terminology used to describe the different aspects of these methods is considered below.

Morphometry is the process by which morphological entities in the sample are measured and assigned numerical parameters such as numbers, distances, and areas, all of which refer to the observed field.

The objects that are to be measured are often within sections of tissue or cells but similar procedures are also applicable to monolayers such as blood smears, exfoliated cells, sediments, or cells on filters. The nature of the sample poses two important problems:

1. To what extent does it represent the original structure?
2. How can the two-dimensional parameters observed in the sample be transformed to three-dimensional parameters applicable to the object under study.

Stereology is a set of mathematical methods that provide quantitative information about three-dimensional structures as a result of observations made principally on sections. To avoid bias in producing estimates of "true" dimensions, certain well-defined requirements must be fulfilled concerning the sampling of tissue and the way sections are made.

A comprehensive introduction to stereology may be found in Weibel (1979). An excellent review and "state of the art" on stereological methods is given by Gundersen *et al.* (1988a).

B.2 Observations

Both light microscopy and transmission electron microscopy produce analogue images with a continuous grey scale. Light microscopy provides additional infor-

H. Lyon (Ed.)
Theory and Strategy in Histochemistry

mation based on colours. These may be a characteristic of the sample or result from a staining procedure.

Microscopists identify stuctures by comparing observed morphology with existing knowledge and experience. For image analysis, digital information is required and analogue images must be converted into digital information. No conversion is required for scanning electron microscopy as the primary image is already formed from digital signals displayed on a monitor.

B.2.1 Identification Based on Grey Level and Image Processing

Digital conversion of analogue images is achieved electronically by dividing the image into a two-dimensional array of picture elements known as pixels and reading the grey level in every element. For example an image may comprise 512 × 512 pixels, each with its own digital value of grey level intensity (Fig. B.2A–2B). Image conversion may be achieved using a scanner or more simply by a TV camera attached directly to the microscope. TV cameras based on a charge-coupled device produce pictures without the linear distorsions produced by ordinary thermoionic tube-based cameras. Section B.6 describes the further electronic processing of the digital information for recognition/identification.

B.3 Stereology

In the context of stereology, unbiased values are "without systematic deviation from the true value". To ensure this observations must be done on isotropic, uniform, random sections.

B.3.1 Sampling Procedures

Depending on the objectives of the study, tissue blocks must be selected either at random or systematically so that they are demonstrably representative of the structure or element under investigation. Fields for observation on individual sections must also be sampled systematically, independent of content and observer, and inside a well-defined region.

B.3.2 Two-dimensional Information

To determined the number of profiles (i.e. structures of any kind) per area, unbiased counting frames are used as discussed by Gundersen *et al.* (1988a).

The *area* of profiles per section is easily obtained by throwing a lattice of test points over the section and simple counting of points over profiles over total points on a section giving the area fraction of profiles.

The *boundary* of profiles per area can be measured by counting intersections of test lines and profile boundary. This is then related to area measurement as above.

B.3.3 Three-dimensional Information

As an estimate of volume density (V_V), meaning the volume of something as a fraction of a total reference volume, one simply uses the area density (A_A) or point density (P_P) obtained on sections.

Cavalieri's principle gives the volume of any object estimated from parallel sections separated by a known distance, by adding up the areas of all cross-sections of the object and multiplying this figure by the known distance.

B.3.4 Anisotropy

If anything in the tissue or cells has a preferred orientation, for example muscle fibres, capillaries, or mitochondria, the tissue is not isotropic. The contribution of these constituents to the section depends on the orientation of the section plane. Isotropic sections can be prepared by using the *"orientator"* where two consecutive sections at right angles to each other are made under random guidance (Gundersen *et al.*, 1988a).

In many biological tissues a particular orientation of sections is necessary to obtain information absent in other sections. Examples are skin covered with epithelium, intestinal crypts, contractile fibers in muscle cells, or the long axis of microvilli to identify their cellular origin.

A *vertical section* is a section plane perpendicular to a plane of reference for instance a basal lamina or the surface of an organ. Structures with preferred orientation can always be seen in one vertical section properly rotated. The use of test line systems for surface parameters must be weighted by the sine of the angle between test line and vertical plane axis, or the use of a special cycloid test line system (Baddeley *et al.*, 1986).

B.4 Special Stereological Tools

A set of new stereological methods for correct sampling and sizing of cells or particles has been reviewed by Gundersen *et al.* (1988b).

B.4.1 The Disector

This is a pair of parallel section planes a known distance apart. A particle is sampled if it hits one of the planes (reference plane) and *not* the other (look-up plane). By counting the number of particles only seen on reference planes the exact number of particles in the physical volume between the two planes is obtained.

The *optical disector* uses thin optical section planes produced inside one thick section (5–100 μm) of a block. The optical sections are produced by moving the focus plane down through the object. Objectives with high numerical aperture and small focal depth give well-defined thin focal planes. The confocal microscope gives excellent optical sectioning of thick blocks, but only in reflectance or epifluorescent microscopy (cf. Sects.28.4 and 28.3).

B.4.2 Other Stereological Tools

When measuring the distance l from any point to the boundary of a profile in any isotropic direction, and averaging the squared distances, then πl^2 is an unbiased estimate of the area. The area of a circle is a special case. This is the basis for the *nucleator* where measurements are carried out isotropically with respect to a fixed point, for example a nucleus in mononucleated cells, or more efficiently a nucleolus if only one is present in every cell. The nucleus in an oocyte can be used for follicular parameters.

With the *fractionator* it is possible to sample uniformly at random with a predetermined probability. Tissue or organs are cut into slabs, some of these are sampled uniformly, cut into bars, sampled again and cut into blocks for final sampling. The final blocks are embedded together as a conglomeration. Sections of the conglomeration will be effective unbiased samples of the original material with "concentrated" information.

B.5 Simple "Counting" Procedures

After proper sampling and sectioning something has to be measured on the sections. The simplest way of doing this is to have the section projected onto a test lattice with points, lines, etc., or to have the test system superimposed in the microscope. Areas and boundary lengths are estimated as number of points and intersections which in turn are transformed to volume and surface area respectively.

B.6 Manipulating Digital Images

B.6.1 Techniques for Digital Image Analysis

A digital image is a body of binary data from picture elements (pixels) each containing information of various local parameters describing intensity of light, colours, invisible parts of the spectrum, extinction, secondary electrons, x-rays or element analysis. The pixels are arranged in a 2-dimensional square, triangular, or hexagonal matrix consisting of for instance 512x512 elements. With grey levels from 0 to 255 (8 bits) in each element, one digital image needs approximately 0.25 Megabytes of memory. A colour image based on three colours in each pixel needs 0.75 Megabytes storage.

To handle *single stationary images* a limited amount of conventional computer power and memory is needed. If, however, the images arrive at *video rate*, 50 Hz, about 15 Megabytes information per second has to be put into a digital frame store, interfaced to and processed by a high performance computer, and, finally, the results have to be returned to the frame store for inspection on a monitor at video rate.

Conversion of continuous (analogue) images into a digital form acceptable to computers involves two important processes: sampling and quantitation, jointly referred to as "digitization". Sampling represents the image by measurements at spaced intervals traversing the image in a specific pattern. The distance between sample points determines spatial resolution. Quantitation transforms brightness values into a range of integer numbers. The number of bits used governs greyscale resolution. A diagram for obtaining and showing digital images is shown in Fig. B.1.

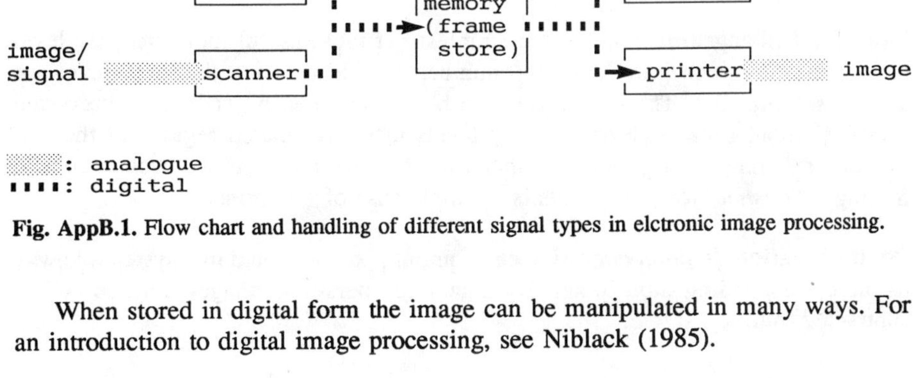

Fig. AppB.1. Flow chart and handling of different signal types in elctronic image processing.

When stored in digital form the image can be manipulated in many ways. For an introduction to digital image processing, see Niblack (1985).

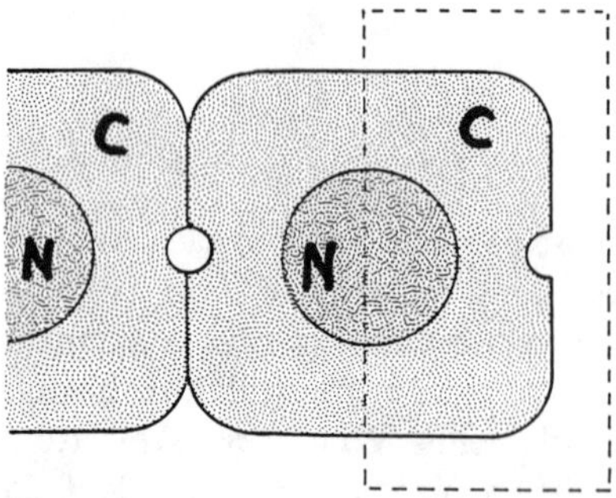

20	20	20	20	20	20
160	160	160	160	20	20
80	80	80	80	160	20
80	80	80	80	160	20
220	220	80	80	160	20
220	220	220	80	160	20
220	220	220	80	50	20
220	220	220	80	160	20
220	220	220	80	160	20
220	220	80	80	160	20
80	80	80	80	160	20
80	80	80	80	160	20
160	160	160	160	20	20
20	20	20	20	20	20

Fig. AppB.2. A: Section of 3 liver cells (only partly shown). Nuclei (N), cytoplasm (C), and background with discrete level of intensity. B: Corresponding digital array of numbers representing grey levels from rectangle indicated in A.

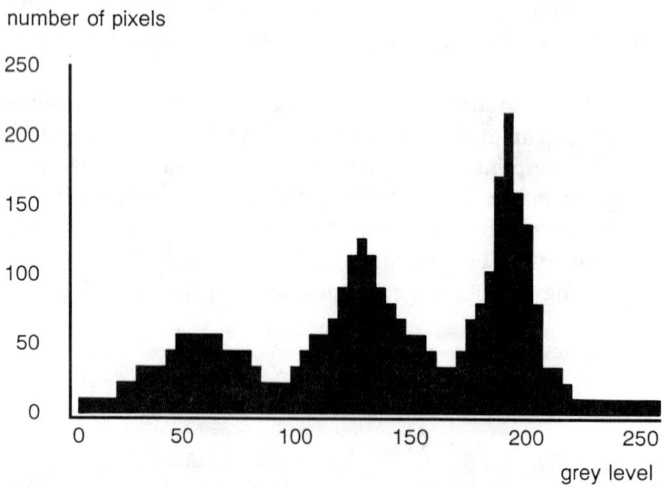

Fig. AppB.3. Grey level histogram of digitized image of Fig. AppB.2.A.

Grey level histograms. One of the simplest yet most useful tools are grey level histograms. For each grey level, the number of pixels having that grey level are shown, see Fig. B.3. The grey levels can be redistributed by non-linear greyscale transformation. By stretching the grey levels in the populated regions of the histogram and compressing them in others, the contrast may improve tremendously. Setting a *threshold* for grey levels is a simple case of this principle.

Point operation. In *point operation* each output pixel is related in a systematic way to the corresponding input pixel. By algebraic operations images may be added, subtracted, multiplied, etc.

By averaging M images of a stationary scene, the signal to noise ratio will improve by $\sqrt{M}$. *Temporal filtering* means that images from TV-cameras can have their signal to noise ratioes improved in real time, but with a delay of a few seconds for continuous averaging.

Local or neighbourhood operation. This means, that each output pixel may depend not only on the grey level of the corresponding input pixel, but also on the grey levels of certain of the neighbouring input pixels. A mask is applied to every single pixel in the input image, and the resulting value from the operation put into the corresponding output pixel. The masks shown here indicate the weights and disposition of pixels participating in the operation:

Mask	Weight	Characteristics, applications
0 1 0 1 0 1 0 1 0	4	Low pass filter, severe. Noise reduction and smoothing. Sharp edges will be blurred.
1 2 1 2 4 2 1 2 1	16	Low pass filter, mild. Effects of noise pixels are distributed over surrounding pixels.
−1 −1 −1 −1 9 −1 −1 −1 −1	1	High pass filter, isotropic. Produces a high output at points, wehre the grey level is changing rapidly, zero where grey levels remain constant.
−1 2 −1 −1 2 −1 −1 2 −1	0	Vertical line detector. Will amplify vertical lines but respond only weakly to lines or structures in other directions.

Frequency Methods, The Fourier Transform. In imaging, spatial frequency is the rate at which the brightness of an image changes with position in space. High spatial frequencies correspond to rapidly varying, fine detail; low spatial frequencies to slowly changing brightness. One of the best known techniques, the Fourier Transform, will break the signal down into a set of sine waves of different frequencies. In any image the spectrum can be deduced, and, correspondingly, given a spectrum, the original signal can be deduced. The spectrum contains the same information as the image, but in a different arrangement.

On an optical bench a Fourier plane can be observed in the back focal plane of a lens, also called the diffraction plane. From this *optical Fourier Transform* a real image may be reconstructed. In digital images a *discrete Fourier Transform* can be done mathematically along the same principles.

The real advantage is seen when larger masks or more elaborate filters than mentioned above have to be applied. When a mask (m) is intended for used on an image (i), only two Fourier Transforms (FT) have to be calculated as

$$FT(i) \times FT(m) = FT(i\&m)$$

and the inverse FT(i&m) gives the resulting image of (i) convoluted with (m). This approach is much faster for large masks than the use of direct spatial convolution. More information can be found in Pratt (1978).

Erosion and Dilation. Erosion strips off layers of pixels from the object. Dilation adds on layers of pixels. By combined sequences of erosion and dilation the connectivity of elements can be investigated and noise "defects" repaired. For example, mitotic figures can be retained, whereas insignificant particles are removed.

Image Restoration. When the output digital image is to be presented the single pixel values may be translated as they are sent to the display. This requires a hardware look-up table (LUT) adopting the grey values to ideal settings for the display system or transforming them to colours. False colours are useful both to improve grey level resolution as the human eye can not distinguish more than approximately 50 grey levels, and in order to display invisible bands or other information.

Three Dimensional Reconstruction. This can be done from serial sections a known distance apart. When sections first are stored in digital form, the whole object may be presented as seen from different angles and distances by appropriate computing.

Measurements of Height and Depth. These are made possible by viewing the object from different viewpoints. A parallax is generated by shifting the specimen, tilting the specimen, or by tilting the electron beam (TEM/SEM). This area is still developing.

B.6.1 Equipment for Digital Image Analysis

Software Programs for Digital Image Analysis. The essential elements of an image analysis system are shown in Fig. B.1. These are, conversion of the image to a digital set of information, some kind of a storage system, a computer to perform grey level operations or other operations at pixel level. Several programs for manipulating digital images are commercially available and may be used on standard computers. If only single or stationary images have to be analysed, fairly simple equipment can be used.

Integrated Systems for Digital Image Analysis. The single components of a digital image processing system must be set up with protocols ensuring compatibility.

With increasing demands on image resolution and the handling of images at video rate, larger frame stores and high performance computers are needed.

Again quite a number of different systems are offered by the commercial companies. Most of them are based on image registration by video cameras. All have facilities for thresholding, filtering and contrast enhancement. All are able to give parameters for areas, lengths, numbers, grey level histograms, etc. in the two-dimensional images. Many of them have special application programs for various disciplines. In biology for counting and classifying blood cells, finding mitotic figures, registering chromosomes, analyzing silver grains in autoradiographs or any kind of autometallographic technique, and contrast enhancement of radiographs. In biochemistry for analysis of chromatograms and electrophoresis patterns.

Other types of programmes have been developed for geology, carthography, aerial photography, and remote sensing data of any kind from satellites or space crafts.

All these techniques give useful results for two dimensions but the whole body of stereological methods mentioned above are required to transform the data to meaningful parameters for three-dimensional analysis. It is absolutely essential that the investigator should carefully consider shrinkage, anisotropy, uniform unbiased sampling, correction procedures, etc., irrespective of the elegance and ease by which two-dimensional results are obtained.

B.7 Applications of Stereology in Pathology

Stereology makes it possible to quantify morphology. In pathology, stereology increases diagnostic potential as description and grading can be done exactly without relying on crude semi-quantitative techniques. This makes it far easier to correlate morphology with other quantitative analyses such as biochemical analyses of tissue, blood, and other body fluids.

The content of collagen, degree of fibrosis, amount of lipid, ratio and diameter of different types of muscle cells are examples of easily obtained parameters.

The subjective assessment of malignancy of tumour cells is not easily reproducible among pathologists. Nielsen *et al.* (1986) found good correlation between mean nuclear volume in bladder tumours and prognosis, and later they also demonstrated correlation between the mean nuclear volume and the histological grading of the tumour in repeated biopsies during the course of the disease (Nielsen et al., 1989).

The total number of neurons in the human brain has been an intriguing problem for more than half a century. It has now been estimated in brain nuclei using the fractionator and the disector (Pakkenberg and Gundersen, 1988) and in neocortex using optical disectors (Braendgaard *et al.*, 1989).

References

Abe, M., Kramer, S.P., and Seligman, A.M.: The histochemical demonstration of pancreatic-like lipase and comparison with the distribution of esterase. J.Histochem.Cytochem. **12**:364–383 (1964).

Abrahart, E.N.: Dyes and their intermediates. 2nd edit. Edward Arnold, London (1977).

Adams, C.W.M.: A p-dimethylaminobenzaldehyde nitrite method for the histochemical demonstration of tryptophane and related compounds. J.Clin.Pathol. **10**:56–65 (1957).

Adams, C.W.M.: Histochemical mechanisms of the Marchi reaction for degenerating myelin. J.Neurochem. **2**:176–186 (1958).

Adams, C.W.M.: A histochemical method for the simultaneous demonstration of normal and degenerating myelin. J.Path.Bact. **77**:648–650 (1959).

Adams, C.W.M.: Osmium tetroxide and the Marchi method: Reactions with polar and non-polar lipids, protein and polysaccharide. J. Histochem.Cytochem. **8**:262–267 (1960).

Adams, C.W.M.: A perchloric acid-naphthoquinone method for the histochemical localisation of cholesterol. Nature (London) **192**:331–332 (1961).

Adams, C.W.M.: Neurohistochemistry. Elsevier, Amsterdam (1965).

Adams, C.W.M. and Bayliss, O.B.: The release of protein, lipid and polysaccharide components of the arterial elastica by proteolytic enzymes and lipid solvents. J.Histochem.Cytochem. **10**:222–226 (1962).

Adams, C.W.M. and Bayliss, O.B.: Histochemical observations on the localization and origin of sphingomyelin, cerebroside and cholesterol in normal and arteriosclerotic human artery. J.Path. Bact. **85**:113–119 (1963).

Adams, C.W.M. and Bayliss, O.B.: Lipid histochemistry. In: D. Glick and R.M. Rosenbaum, Eds.: Techniques of biochemical and biophysical morphology. Vol. 2. Wiley Interscience, New York (1974).

Adams, C.W.M. and Davison, A.N.: The histochemical identification of myelin phosphoglycerides by their ferric hydroxamates. J.Neurochem. **3**:347–351 (1959).

Adams, C.W.M., Abdullah, Y.H., and Bayliss, O.B.: Osmium tetroxide as a histochemical and histological reagent. Histochemie **9**:68–77 (1967).

Adams, C.W.M., Bayliss, O.B., and Ibrahim, M.Z.M.: Modifications to histochemical methods for phosphoglyceride and cerebroside. J. Histochem.Cytochem. **11**:560–561 (1963).

Adams, C.W.M., Abdullah, Y.H., Bayliss, O.B., and Weller, R.O.: Histochemical detection of triglyceride esters with specific lipases and a calcium lead sulphide technique. J.Histochem. Cytochem. **14**:385–395 (1966).

Aikens, R.S., Agard, D.A. and Sedat, J.W.: Solid state images for microscopy, pp.292–313. In: Y.-L. Wang and Taylor, D.L., Eds.: Methods in cell biology, Vol. 29. Fluorescence microscopy of living cells in culture. Part A. Fluorescent analogs, labeling cells, and basic microscopy. Academic Press, Boston (1989).

Albedi, F.M., Cassano, A.M., Ciarelli, F., Donelli, G., Giuliani, Mingazzini, P., and Marinozzi, V.: Influence of cetylpyridinium chloride on the ultrastructural appearance of sulphated glycosaminoglycans in human colonic mucosa. Histochemistry **89**:397–401 (1988).

Allison, R.T.: The effects of various fixatives on subsequent lectin binding to tissue sections. J.Histochem. **19**:65–74 (1987).

Altman, F.P.: Quantitative dehydrogenase histochemistry with special reference to the pentose shunt dehydrogenases. Progr.Histochem.Cytochem. **4/4**:225–273 (1972).

Altman, F.P.: Quantitation in histochemistry: a review of some commercially available microdensitometers. Histochem.J. **7:** 375–396 (1975).

Altman, F.P.: Tetrazolium salts and formazans. Progr.Histochem. Cytochem. **9/3:**1–56 (1976).

Altman, F.P.: The use of a heated stage for following histochemical reactions under a microdensitometer. Histochem.J. **10:**611–614 (1978).

Altman, F.P.: How thick is your section. Abstracts of the VIth International Histochemistry and Cytochemistry Congress. Royal Microscopical Society, Oxford, (1980a).

Altman, F.P.: Tissue stabilizer methods in histochemistry, pp. 81–102. In: Trends in enzyme histochemistry and cytochemistry. Ciba Foundation Symposium 73. Excerpta Medica, Amsterdam (1980b).

Altman, F.P. and Chayen, J.: Evidence for two types of hydrogen atom in reduced nicotinamide-adenine dinucleotide phosphate arising from glucose-6-phosphate oxidation, based on the inhibitory action of certain steroids. Biochem.J. **118/6:**6P–7P (1970).

Altman, F.P., Høyer, P.E., and Andersen, H.: Dehydrogenase histochemistry of lipid-rich tissues: A tetrazolium-metalchelation technique to improve localization. Histochem.J. **11:**485–488 (1979).

Altman, F.P., Moore, D.S. and Chayen, J.: The direct measurement of cytochrome P450 in unfixed tissue sections. Histochemistry **41:**227–232 (1975).

Andersen, A.P., Jakobsen, P., Lyon, H., and Høyer, P.E.: Purification of Methyl Green using polyamide. Histochem.J. **18:**461–462 (1986).

Andersen, H., and Høyer, P.E.: Studies in succinate dehydrogenase histochemistry. Histochemie **35:**173–188 (1973).

Andersen, H. and Høyer, P.E.: Simplified control experiments in the histochemical study of coenzyme-linked dehydrogenases. Histochemistry **38:**71–83 (1974).

Anderson, W.A., Kang, Y.-H., and Mohla, S.: Mammalian endogenous peroxidases as cellular markers and as biosynthetic endpoints of hormone-mediated activity: viewpoint from cytochemistry. Progr.Histochem.Cytochem. **11/4:**1–27 (1979).

Andrä, J. and van Duijn, P.: Quantitative aspects of cytochemical methods for acetylcholinesterase studied with a cytochemical model system. Histochemistry **83:**443–449 (1985).

Angermüller, S.: Peroxisomal oxidases: Cytochemical localization and biological relevance. Progr. Histochem.Cytochem. **20/1:**1–65 (1989).

Angermüller, S. and Fahimi, H.D.: Light microscopic visualization of the reaction product of cerium used for localization of peroxisomal oxidases. J. Histochem. Cytochem. **36:**23–28 (1988a).

Angermüller, S and Fahimi, H.D.: Heterogenous staining of D-amino acid oxidase in peroxisomes of rat liver and kidney. A light and electron microscopic study. Histochemistry **88:**227–285 (1988b).

Anthony, A., Colorso, G.J., Bocan, T.M.A. and Doebler, J.A.: Interferometric analysis of intrasection and intersection thickness variability associated with cryostat microtomy. Histochem.J. **16:**61–70 (1984).

Apte, S.S. and Puddle, B.: Bromodeoxyuridine (BrdUrd) immunohistochemistry in undecalcified plastic-embedded tissue. Elimination of the DNA denaturation step. Histochemistry **93:**631–635 (1990).

Araki, T., Chikamori, K., Sasaki, K., Kawata, S., Minami, S. and Yamada, M.: Topographic estimations by component spectroanalysis of two formazans of nitroblue tetrazolium in tissue sections. Histochemistry **86:**567–572 (1987).

Arndt-Jovin, D.J., Robert-Nicoud, M., Kaufman, S.J., and Jovin, T.M.: Fluorescence digital imaging microscopy in cell biology. Science **230:**247–256 (1985).

Atkin, N.: Principles and application of the Deeley integrating microdensitometer, pp.1–26. In: G.L. Wied and G.F. Bahr, Eds.: Introduction to quantitative cytochemistry-II. Academic Press, New York (1970).

Baddeley, A.J., Gundersen, H.J.G. and Cruz-Orive, L.M.: Estimation of surface area from vertical sections. J.Microsc. **142:**259–276 (1986).

Bader, C.A., Nasr, L.B., Monet, J.-D., Bachelet, M., Assailly, J. and Ulmann, A.: In situ biochemical studies of intestinal alkaline phosphatase in normal and phosphate-depleted rats by microdensitometry. J.Biol.Chem. **259:**11685–11661 (1984).

Baker, D.M. and Santer, R.M.; Development of a quantitative histochemical method for determination of succinate dehydrogenase activity in autonomic neurons and its application to the study of aging in the autonomic nervous system. J.Histochem.Cytochem. **38:** 525–531 (1990).

Baker, J.R. The histochemical recognition of lipine. Quart.J. Micr.Sci. **87:**441–470 (1946).

Baker, J.R.: Principles of biological microtechnique. Methuen, London, (1958).

Baker, J.R.: Cytological technique. The principles underlying routine methods. 5th edit. Science Paperbacks Methuen & Co. Ltd., London, (1966).

Bancroft, J.D. and Cook, C.: Manual of histological techniques. 2nd edit. Churchill-Livingstone, Edinburgh, (1984).

Bancroft, J.D., and Stevens, A. Eds.: Theory and practice of histological techniques. 3rd edit. Churchill-Livingstone, Edinburgh (1990).

Barka, T. and Anderson, P.J.: Histochemistry. Theory, practice, and bibliography. Hoeber, New York (1963).

Barer, M.R.: Development of a cell culture method for the detection and quantitation of bacterial enterotoxins. PhD Thesis. University of London (1987).

Barer, M.R., Lyon, H.O., and Drasar, B.S.: Quantitation of dye binding to cell monolayers in a microtiter system. Histochem.J. **18:**122-128 (1986a).

Barer, M.R., Mann, G.F., and Drasar, B.S.: A semi-automated system for the assessment of toxicity to cultured mammalian cells based on detection of changes in staining properties. Developments in Biological Standardization **64:**251-259 (1986b).

Barrett, A. and Heath, M.F.: Lysosomal enzymes, pp.19–145. In: J.T. Dingle, Ed.: Lysosomes: a laboratory handbook, 2nd edit., Elsevier, North Holland Biomedical Press (1977).

Barrett, A.J.: The many forms and functions of cellular proteinases. Fed.Proc. **39:**9–14 (1980).

Barrnett, R.J. and Seligman, A.M.: Demonstration of protein-bound sulfhydryl and disulfide groups by two new histochemical methods. J.Natl.Cancer Inst. **13:**215–216 (1952).

Bartholomew, J.W.: Stains for microorganisms in smears. In: G. Clark, Ed.: Staining procedures. 4th edit., pp.375–440. Williams & Wilkins, Baltimore (1981).

Bauman, J.G.J., van der Ploeg, M. and van Duijn, P.: Fluorescent hybridocytochemical procedures: DNA/RNA hybridization in situ, pp.41–87. In: J. Chayen and L. Bitensky, Eds.: Investigative microtechniques in medicine and biology. Vol. 1. Marcel Dekker, Inc., New York (1984).

Bayliss, O.B. and Adams, C.W.M.: Bromine Sudan Black (BSB). A general stain for tissue lipids including free cholesterol. Histochem.J. **4:**505–515 (1972).

Becher, S.: Untersuchungen über die Echtfärbung der Zellkerne mit künstlichen Beizenfarbstoffen und die Theorie des histologischen Farbprozesses mit gelösten Lacken. Gebr. Bornträger, Berlin (1921).

Belt, W.D. and Hayes, E.R.: An ultraviolet - Schiff reaction for unsaturated lipids. Stain Technol. **31:**117–122 (1956).

Bennion, P.J. and Horobin, R.W.: Some effects of salts on staining: Use of the Donnan equilibrium to describe staining of tissue sections with acid and basic dyes. Histochemistry **39:**71–82 (1974).

Benno, R.H., Tucker, L.W., Joh, T.H. and Reis, D.J.: Quantitative immunocytochemistry of tyrosine hydroxylase in rat brain. I. Development of a computer assisted method using the peroxidase-antiperoxidase technique. Brain Res. **246:**225–236 (1982a).

Benno, R.H., Tucker, L.W., Joh, T.H. and Reis, D.J.: Quantitative immunocytochemistry of tyrosine hydroxylase in rat brain. II. Variations in the amount of tyrosine hydroxylase among individual neurons of the locus coeruleus in relationship to neuronal morphology and topography. Brain Res. **246:**237–247 (1982b).

Bentley, S.A., Marshall, P.N., Wade, M.J. and Galbraith, W.: Azure B-eosin staining of blood cells. The effects of variation in stain formulation and staining technique on stain performance. Anal.Quant.Cytol. **1:**179–186 (1979).

Berrisford, R.G. Wells, M. and Dixon, M.F.: Gastric epithelial mucus—a densitometric histochemical study of aspirin-induced damage in the rat. Br.J.exp.Path. **66:**27–33 (1985).

Berube, G.R., Powers, M.M., Kerkay, J., and Clark, G.: The gallocyanin-chrome alum stain; influence of methods of preparation on its activity and separation of active staining compound. Stain Technol. **41:**73–81 (1966).

Bielschowsky, M.: Die Silberimprägnation der Neurofibrillen. J.Psychol.Neurol. **3:**169–189 (1904).

Bielschowsky, M.: Eine Modifikation meines Silberimprägnationsverfahrens zur Darstellung der Neurofibrillen. J.Psychol.Neurol. **12:**135–137 (1909).

Bigbee, J.W., Kosek, J.C. and Eng, L.F.: Effects of primary antiserum dilution on staining of "antigen-rich" tissues with the peroxidase antiperoxidase technique. J.Histochem.Cytochem. **25:**443–447 (1977).

Bitensky, L.: Cytochemical bioassays of polypeptide hormones. J.Reprod.Fert. **51:**287–294 (1977).

Bitensky, L.: Microdensitometry, pp.181–202. In: Trends in enzyme histochemistry and cytochemistry. Ciba Foundation Symposium 73. Excerpta Medica, Amsterdam (1980).

Bitensky, L. and Chayen, J.: Histochemical methods for the study of lysosomes, pp.209–243. In: J.T. Dingle, Ed.: Lysosomes. A laboratory handbook, 2nd edit. North Holland Publishing Co., Amsterdam (1977).

Bitensky, L. and Chayen, J.: Quantitative cytochemistry: The basis of sensitive bioassays, for comparison of bio- and immuno-reactive hormone values. Clin.Chem. **24:**1399–1407 (1978).

Bitensky, L., Alaghband-Zadeh, J. and Chayen, J.: Studies on thyroid stimulating hormone and the long-acting thyroid stimulating hormone. Clin.Endocrinol. **3:**363–374 (1974).

Björklund, A., Ehinger, B., and Falck, B.: A method for differentiating dopamine from noradrenaline in tissue sections by microspectrofluorometry. J.Histochem.Cytochem. **16:**263–270 (1968).

Björklund, A., Ehinger, B., and Falck, B.: Analysis of fluorescence excitation peak ratios for the cellular identification of noradrenaline, dopamine or their mixtures. J. Histochem.Cytochem. **20:**56–64 (1972).

Böck, G., Hilchenbach, M., Schauenstein, K., and Wick, G.: Photometric analysis of antifading reagents for immunofluorescence with laser and conventional illumination sources. J.Histochem.Cytochem. **33:**699–705 (1985).

Bodian, D.: A new method for staining nerve fibers and nerve endings in mounted paraffin sections. Anat.Rec. **65:**89–97 (1936).

Böhm, N.: Einfluß der Fixierung und der Säurekonzentration auf die Feulgen-Hydrolyse bei 28°C. Histochemie **14:**201–211 (1968).

Böhm, N., Sprenger, E., Schlüter, G., and Sandritter, W: Proportionalitätsfehler bei der Feulgen-Hydrolyse. Histochemie **15:**194–203 (1968).

Bohman, S.-O. and Maunsbach, A.B.: Effects on tissue fine structure of variations in colloid osmotic pressure of glutaraldehyde fixative. J.Ultrastruct.Res. **30:**195–208 (1970).

Booig, H.L.: Colloid chemical aspects of metachromasia. Acta Histochem. **Suppl.1:**37–54 (1958).

Boon, M.E. and Drijver, J.S.: Routine cytological staining techniques, theoretical background and practice. Macmillan Press, London (1986).

Boon, M.E. and Kok, L.P.: Microwave cookbook of pathology: The art of microscopic visualization. Coulomb Press Leyden, Leiden (1987).

Boon, M.E. and Kok, L.P.: Microwave cookbook of pathology: The art of microscopic visualization, 2nd edit. Coulomb Press Leyden, Leiden (1988).

Boon, M.E. and Wittekind, D.: Theoretical and practical aspects of standardization of staining in diagnostic cytology, pp.100–110. In: M.E. Boon and L.P. Kok, Eds.: Standardization and quantitation of diagnostic staining in cytology. Coulomb Press Leyden, Leiden (1986).

Boon, M.E., Hendrikse, F.C.J., Kok, P.G., Bolhuis, P. and Kok, L.P.: A practical approach to routine immunostaining of paraffin sections in the microwave oven. Histochem.J. **22:**347–352 (1990).

Boon, M.E., Schleicher, A., Wijsman-Grootendorst, A., Lyon, H. and Kok, L.P.: Staining of the nucleolus with protein and RNA stains for automatic measurement of nucleolar size in paraffin sections. Stain Technol. **63:**289–297 (1988).

Bourgeois, C. and Hubbard, B.: A method for simultaneous demonstration of choline-containing and neutral lipids in tissue sections. J.Histochem.Cytochem. **13:**571–578 (1965).

Brachet, J.: La détection histochimique des acides pentose nucleiques. C.R.Soc.Biol. **133:**88–90 (1940a).

Brachet, J.: La localisation des acides pentose nucleiques. C.R. Soc.Biol. **133:**90–91 (1940b).

Brachet, J.: La localisation des acides pentose nucleiques dans les tissus animaux et les oeufs d'amphibiens en voil de development. Arch.Biol. **53**:207–219 (1942).

Brachet, J.: The use of basic dyes and ribonuclease for the cytochemical detection of ribonucleic acid. Quart.J.Micr.Sci. **94**:1–10 (1953).

Braendgaard, H., Evans, S.M., Hovard, C.V. and Gundersen, H.J.G.: The total number of neurons in the human neocortex unbiasedly estimated using optical disectors. J. Microsc. **157**:285–304 (1989).

Bresser, J. and Evinger-Hodges, M.J.: Comparison and optimization of in situ hybridization procedures yielding rapid, sensitive mRNA detections. Gene Anal.Techn. **4**:89–104 (1987).

Brody, I.A. and Engel, W.K.: Isoenzyme histochemistry: the display of selective lactate dehydrogenase isoenzymes in sections of skeletal muscle. J.Histochem.Cytochem. **12**:928–929 (1964).

Buckingham, J.C. and Hodges, J.R.: The use of corticotrophin production by adenohypophyseal tissue in vitro for the detection and estimation of potential corticotrophin releasing factors. J.Endocrinol. **72**:187–193 (1977).

Buckingham, J.C., Chayen, J., Hodges, J.R., Robertson, W.R. and Weisz, J.: A cytochemical section assay method for the determination of luteinizing hormone. J.Endocrinol. **81**:160p (1979).

Buk, S.J.A. and Bayliss High, O.B.: Periodate oxidation of glycolipids: a borohydride-periodate-Schiff method for ganglioside demonstration in tissue sections. Histochem.J. **18**:228–232 (1986).

Bullock, G.R. and Petrusz, P., Eds.: Techniques in immunocytochemistry, vol.1. Academic Press, London (1982).

Bullock, G.R. and Petrusz, P., Eds.: Techniques in immunocytochemistry, vol.2. Academic Press, London (1983).

Bullock, G.R. and Petrusz, P., Eds.: Techniques in immunocytochemistry, vol.3. Academic Press, London (1985).

Burstone, M.S.: Histochemical methods for protein detection. In: W. Graumann and K. Neumann, Eds.: Handbuch der Histochemie, vol. III, 2nd edit. Gustav Fischer Verlag, Stuttgart (1959).

Burtner, H.J. and Lillie, R.D.: A five-hour variant of Gomori's methenamine silver method for argentaffine cells. Stain Technol. **24**:225–227 (1949).

Butcher, R.G.: The estimation of the nucleic acids of tissue sections and its use as a unit of comparison for quantitative histochemistry. Histochemie **13**:263–275 (1968).

Butcher, R.G.: The chemical determination of section thickness. Histochemie **28**:131–136 (1971).

Butcher, R.G.: The measurement in tissue sections of the two formazans derived from Nitroblue Tetrazolium in dehydrogenase reactions. Histochem.J. **10**:739–744 (1978).

Butcher, R.G.: The oxygen insensitivity phenomenon as a diagnostic aid in carcinoma of the bronchus, pp.241–251. In: J.R. Pattison, L. Bitensky and J. Chayen, Eds.: Quantitative cytochemistry and its applications. Academic Press, London (1979).

Butcher, R.G.: Grades of PVA and enzyme retention in tissue sections. Histochem.J. **14**:165–166 (1982).

Butcher, R.G.: Enzyme histochemistry of the myocardium. In: A.J. Drake-Holland and I.M. Noble, Eds.: Cardiac metabolism, pp.445–469. John Wiley & Sons, New York (1983).

Butcher, R.G.: Reaction rate studies of glucose-6-phosphate dehydrogenase in rat tracheal epithelium; the effect of section thickness. Histochemistry **81**:567–572 (1984).

Butcher, R.G. and Altman, F.P.: Studies on the reduction of tetrazolium salts. II. The measurement of the half reduced and fully reduced formazans of neotetrazolium chloride in tissue sections. Histochemie **37**:351–363 (1973).

Butcher, R.G. and Evans, A.W.: Diffusion during dehydrogenase reactions: the effects of intermediate electron acceptors. Histochem.J. **16**:885–895 (1984).

Butcher, R.G. and Papadoyannis, D.E.: Histochemical measurement of myocardial phosphofructokinase activity. J.Physiol. **289**:18P–19P (1979).

Butcher, R.G. and van Noorden, C.J.F.: Reaction rate studies of glucose-6-phosphate dehydrogenase activity in sections of rat liver using four tetrazolium salts. Histochem.J. **17**:993–1008 (1985).

Butcher, R.G., Bitensky, L., Cashman, B., and Chayen, J.: Differences in the redox balance in human rheumatoid and non-rheumatoid synovial lining cells. Beitr.Path. **148**:265–274 (1973).

Cabrini, R.L, Frasch, A.C.C. and Itoiz, M.E.: Quantitative histochemistry of esterases. Histochem.J. **9**:369–371 (1977).

Caffrey, J.L., Nett, T.M., Abel, J.H. and Niswender, G.D.: Activity of 3β-hydroxy-Δ^5-steroid dehydrogenase/Δ^5-Δ^4/isomerase in the ovine corpus luteum. Biol.Reprod. **20**:279–287 (1979).

Cain, A.J.: On the significance of the plasmal reaction. Quart. J.Micr.Sci. **90**:75–86 (1949a).

Cain, A.J.: A critique of the plasmal reaction, with remarks on recently proposed techniques. Quart.J.Micr.Sci. **90**:411–426 (1949b).

Cain, A.J.: The histochemistry of lipids in animals. Biol.Rev. (Cambridge) **25**:73–112 (1950).

Cajal, S. Ramon y: Sobre un neuvo proceder de impregnación de la neuroglia y sus resultados en los centros nerviosos del hombre y animales. Trab. Lab. Invest. Biol. **11**: Fasc. 3, Diciembre (1913) Cited in: G. Clark, Ed.: Staining Procedures, 4th edit., pp.134–135. Williams & Wilkins, Baltimore (1981).

Cajal, S. Ramon y: El proceder del oro-sublimado para la coloración de la neuroglia. Trab. Lab. Invest. Biol. **14**: Fasc. 3, (1916).

Caspersson, T.: Methods for the determination of the absorption spectra of cell structures. J.R. Microsc.Soc. **60**:8–25 (1940).

Casselman, W.G.B.: Acetylated Sudan Black B as a more specific histochemical reagent for lipids. Quart.J.Micr.Sci. **95**:321–322 (1954).

Chambers, D.J., Braimbridge, M.V., Frost, G.T.B., Nahir, A.M., and Chayen, J.: A quantitative cytochemical method for the measurement of β-hydroxyacyl CoA dehydrogenase activity in rat heart muscle. Histochemistry **75**:67–76 (1982).

Chambers, D.J., Dunham, J., Zanelli, J.M., Parsons, J.A., Bitensky, L. and Chayen, J.: A sensitive bioassay of parathyroid hormone in plasma. Clin.Endocrinol. **9**:375–379 (1978).

Chayen, J.: The cytochemical approach to hormone assay. Int.Rev. Cyt. **53**:333–396 (1978).

Chayen, J.: The cytochemical bioassay of polypeptide hormones. Springer-Verlag. Berlin (1980).

Chayen, J. and Denby, E.F.: Biophysical technique as applied to cell biology. Methuen, London (1968).

Chayen, J., Altman, F.P., and Butcher, R.G.: The effect of certain drugs on the production and possible utilization of reducing equivalents outside the mitochondria, pp.196–230. In: S.Dikstein, Ed.: Fundamentals of cell pharmacology. Thomas, Springfield (1973a).

Chayen, J., Bitensky, L., and Butcher, R.G.: Practical histochemistry. John Wiley & Sons, London (1973b).

Chayen, J., Loveridge, N. and Daly, J.R.: A sensitive bioassay for adrenocorticotrophic hormone (ACTH) in human plasma. Clin.Endocrinol. **1**:219–233 (1972).

Chayen, J., Daly, J.R., Loveridge, N. and Bitensky, L.: The cytochemical bioassay of hormones. Recent Progr.Horm.Res. **32**:33–79 (1976).

Chayen, J., Bitensky, L., Johnstone, J.J., Gooding, P.E. and Slater, T.F.: The application of microspectrophotometry to the measurement of cytochrome P-450, pp.129–137. In: J.R. Pattison, L. Bitensky and J. Chayen, Eds.: Quantitative cytochemistry and its applications. Academic Press, London (1979).

Chayen, J., Gilbert, D.M., Robertson, W.R., Bitensky, L. and Besser, G.M.: A cytochemical section-bioassay for thyrotropin. J.Immunoassay **1**:1–13 (1980).

Chayen, J., Frost, G.T.B., Dodds, R.A., Bitensky, L., Pitchfork, J., Baylis, P.H., and Barnett, R.J.: The use of a hidden metal-capture reagent for the measurement of Na^+-K^+-ATPase activity: A new concept in cytochemistry. Histochemistry **71**:533–541 (1981).

Cheng, K.K.: Experimental studies on the mechanism of the zonal distribution of beryllium liver necrosis. J.Path.Bact. **71**:265 (1956).

Chévremont, M. and Frédéric, J.: Une nouvelle méthode histochimique de mise en evidence des substances à fonction sulfhydrile. Arch.Biol. (Paris) **54**:589–605 (1943).

Chieco, P., Normanni, P., Moslen, M.T. and Maltoni, C.: Quantitative histochemistry of benzaldehyde dehydrogenase in hepatocellular carcinomas of vinyl chloride-treated rats. J.Histochem. Cytochem. **34**:151–158 (1986).

Clancy, M.J.: The formulation of buffers and media for enzyme histochemistry. Histochem.J. **19**:27–34 (1987).

Clark, G., Ed.: Staining procedures. 4th edit. Williams & Wilkins, Baltimore (1981).

Cobbold, P.H. and Rink, T.J.: Fluorescence and bioluminescence measurement of cytoplasmic free calcium. Biochem.J. **248:** 313–328 (1987).

Cole, L.C. and Nettleton, G.S.: A histochemical investigation of the mechanism of Aldehyde Fuchsin staining of pancreatic B-cell. Histochem.J. **20:**635–641 (1988).

Cornelese-ten Velde, I. and Prins, F.A.: New sensitive light microscopical detection of colloidal gold on ultrathin sections by reflection contrast microscopy. Combination of reflection contrast and electron microscopy in post-embedding immunogold histochemistry. Histochemistry **94:**61–71 (1990).

Cornelese-ten Velde, I., Bonnet, J., Tanke, H.J. and Ploem, J.S.: Reflection contrast microscopy. Visualization of (peroxidase-generated) diaminobenzidine polymer products and its underlying optical phenomena. Histochemistry **89:**141–150 (1988).

Cornelese-ten Velde, I., Wiegant, J., Tanke, H.J. and Ploem, J.S.: Improved detection and quantification of the (immuno)peroxidase product using reflection contrast microscopy. Histochemistry **92:**153–160 (1989).

Coulton, L.A., Henderson, B. and Chayen, J.: The assessment of DNA-synthetic activity. Histochemistry **72:**91–99 (1981).

Cowden, R.R. and Curtis, S.K.: Further comments on the selectivity of Mercury Orange for protein thiols. Histochem.J. **16:**1237–1240 (1984).

Cowen, T., Haven, A.J., and Burnstock, G.: Pontamine sky blue: A counterstain for background autofluorescence in fluorescence and immunofluorescence histochemistry. Histochemistry **82:**205–208 (1985).

Cremers, A.F.M., Jansen in de Wal, N., Wiegant, J., Dirks, R.W., Weisbeek, P., van der Ploeg, M. and Landegent, J.E.: Non-radioactive in situ hybridization. A comparison of several immunocytochemical detection systems using reflection-contrast and electron microscopy. Histochemistry **86:**609–615 (1987).

Croce, A.C., Bottiroli, G., Prosperi, E., Supino, R. and Stoward, P.J.: Limitations of the quantitative cytochemical assay of catechol oxidase in melanoma cells. Histochem.J. **20:** 595–602 (1988).

Crockard, A.D.: Cytochemistry of lymphoid cells: a review of findings in the normal and leukaemic state. Histochem.J. **16:**1027–1050 (1984).

Cross, S.A.M., Ewen, S.W.B., and Rost, F.W.D.: A study of the methods available for the cytochemical localization of histamine by fluorescence induced with o-phthalaldehyde or acetaldehyde. Histochem.J. **3:**471–476 (1971).

Culling, C.F.A., Reid, P.E., and Dunn, W.L.: A new histochemical method for the identification and visualization of both side chain acylated and non-acylated sialic acids. J.Histochem. Cytochem. **24:**1225–1230 (1976).

Culling, C.F.A., Reid, P.E., Clay, M.G., and Dunn, W.L.: The histochemical demonstration of O-acylated sialic acid in gastro-intestinal mucins. Their association with the potassium hydroxide-periodic acid-Schiff effect. J.Histochem.Cytochem. **22:** 826–831 (1974).

Danscher, G.: Autometallography. A new technique for light and electron microscopic visualization of metals in biological tissues (gold, silver, metal sulphides and metal selenides). Histochemistry **81:**331–335 (1984).

Deitch, A.D.: Microspectrophotometric study of the binding of the anionic dye Naphthol Yellow S by tissue sections and purified proteins. Lab.Invest. **4:**324–351 (1955).

Deitch, A.D., Wagner, D., and Richart, R.M.: Conditions influencing the intensity of the Feulgen reaction. J. Histochem. Cytochem **16:**371-379 (1968).

De Jong, A.S.H., van Kessel-van Vark, M., and Raap, A.K.: Sensitivity of various visualization methods for peroxidase and alkaline phosphatase activity in immunoenzyme histochemistry. Histochem.J. **17:**1119–1130 (1985).

De la Torre, J.C. and Surgeon, J.W.: A methodological approach to rapid and sensitive monoamine histofluorescence using a modified glyoxylic acid technique: The SPG method. Histochemistry **49:**81–93 (1976).

DeLellis, R.A., ed.: Diagnostic immunohistochemistry. Masson Publishing USA, New York (1981).

DeLellis, R.A., ed.: Advances in immunohistochemistry. Masson Publishing USA, New York (1984).

DeLellis, R.A., ed.: Advances in immunohistochemistry. Raven Press, New York (1988).
De Schepper, G.G., van Noorden, C.J.F. and Koperdraad, F.: A cytochemical method for measuring enzyme activity in individual preovulatory mouse oocytes. J.Reprod.Fert. **74**:709–716 (1985).
Desmet, V.J., Bullens, A.M., de Groote, J., and Heirwegh, K.P.M.: A new diazoreagent for specific staining of conjugated bilirubin in tissue sections. J.Histochem.Cytochem. **16**:419–427 (1968).
Dixon, K.C.: Hydrophobic lipids in injured cells. Histochem.J. **2**:151–187 (1970).
Dixon, K.C.: Solubilization and cellular fatty change. Histochem. J. **5**:363–387 (1973).
Dixon, M. and Webb, E.C.: Enzymes. 3rd edit. Academic Press, San Diego, CA (1979).
Dodds, R.A., Pitsillides, A.A. and Frost, G.T.B.: A quantitative cytochemical method for ornithine decarboxylase activity. J.Histochem.Cytochem. **38**:123–127 (1990).
Doebler, J.A., Wickersham, E.W. and Anthony, A.: Quantitative cytochemical analyses of ovarian luteal and interstitial cell RNA in relation to plasma progestins during pseudopregnancy in the rat. Cell Biochem.Funct. **1**:173–178 (1983).
Drexhage, H.A., Wilders, M.M., Lens, J.W. and Boyages, S.: Immunoglobulins die das Wachstum von Thyreozyten stimulieren. Aktuel.Endokrinol.Stoffwechsel **10,suppl.2**:180–184 (1989).
Dubeau, L., Chandler, L.A., Gralow, J.R., Nichols, P.W. and Jones, P.A.: Southern blot analysis of DNA extracted from formalin-fixed pathology specimens. Cancer Res. **46**:2964–2969 (1986).
Dubowitz, V. and Brooke, M.H.: Muscle biopsy: A modern approach. W. B. Saunders, London (1973).
Duijndam, W.A.L. and van Duijn, P.: The influence of chromatin compactness on the stoichiometry of the Feulgen-Schiff procedure studied in model films. J.Histochem.Cytochem. **23**:882-890 (1975).
Duijndam, W.A.L., van Duijn, P. and Riddersma, S.H.: Optical errors in scanning stage absorbance cytophotometry. II. Application of correction factors for residual distributional error, glare and diffraction error in practical cytophotometry. J.Histochem.Cytochem. **28**:395–400 (1980a).
Duijndam, W.A.L., Smeulders, A.W.M., van Duijn, P. and Verweij, A.C.: Optical errors in scanning stage absorbance cytophotometry. I. Procedures for correcting apparent integrated absorbance values for distributional, glare, and diffraction errors. J.Histochem.Cytochem. **28**:388–394 (1980b).
Dunham, J.: Metabolism of the fracture callus, pp.213–220. In: J.R. Pattison, L. Bitensky and J. Chayen, Eds.: Quantitative cytochemistry and its application. Academic Press, London (1979).
Dunnigan, H.G.: Chromatographic separation and photometric analysis of the components of Nile Blue Sulphate. Stain Technol. **43**:242–248 (1968a).
Dunnigan, H.G.: The use of Nile Blue Sulphate in the histochemical identification of phospholipids. Stain Technol. **43**:249–256 (1968b).
Duve, C. de, Pressman, B.C., Gianetto, R., Wattiaux, R. and Appelmans, F.: Tissue fractionation studies. VI. Intracellular distribution patterns of enzymes in rat liver tissue. Biochem.J. **60**:604–617 (1955).
Ealey, P.A., Henderson, B., and Loveridge, N.: A quantitative study of peroxidase activity in unfixed tissue sections of the guinea-pig thyroid gland. Histochem.J. **16**:111–122 (1984).
Ebel, J.P., Colas, J., and Muller, S.: II. Mise au point de methodes de détection cytochimiques specifiques des polyphosphates. Exp.Cell Res. **15**:21–28 (1958).
Edgar, G.W.F. and Donker, C.H.M.: Influence of lipid solvents on sphingolipids (sphingomyelins, cerebrosides, gangliosides) in tissue sections. Acta neurol.Psychiat.Belg. **5**:451–461 (1957).
Ehrlich, P.: Referate aus den Verein für innere Medicin zu Berlin. Deutsche Med.Wochenschr. **8**:269 (1882). Cited in: Lison p.273 (1960).
Einarson, L.: A method for progressive selective staining of Nissl and nuclear substance in nerve cells. Amer.J.Pathol. **8**:295–307 (1932).
Einarson, L.: On the theory of gallocyanin-chrome alum staining and its application for quantitative estimation of basophilia. A selective staining of exquisite progressivity. Acta Pathol. Microbiol.Scand. **28**:82–102 (1951).
Elftman, H.: Controlled chromation. J.Histochem.Cytochem. **2**:1–8 (1954).
Elleder, M. and Lojda, Z.: Studies in lipid histochemistry IX. The specificity of the Holczinger's reaction for fatty acids. Histochemie **32**:301–305 (1972).

Ellwart, J. and Dörmer, P.: Effect of 5-fluoro-2'-deoxyuridine (FdUrd) on 5-bromo-2'-deoxyuridine (BrdUrd) incorporation into DNA measured with a monoclonal BrdUrd antibody and by the BrdUrd/Hoechst quenching effect. Cytometry **6**:513–520 (1985).

El-Sherif, G., van Noorden, C.J.F. and Bácsy, E.: Specificity of the metal-salt method for the localization of Ca^{+}-ATPase activity studied in rat hypophysis tissue in a model system of polyacrylamide gel films. Histochem.J. **22**:51–56 (1990).

Engbæk, K., Johansen, K.S., and Jensen, M.E.: A new technique for Gram staining paraffin-embedded tissue. J.Clin.Path. **32**:187–190 (1979).

Ericsson, J.L.E. and Biberfeld, P.: Studies of aldehyde fixation. Fixation rates and their relation to fine structure and some histochemical reactions in liver. Lab.Invest. **17**:281–298 (1967).

Erlich, H.A.: PCR technology. Macmillan Press, London (1989).

Erlich, H.A., Gelfand, D.H. and Saiki, R.K.: Specific DNA amplification. Nature **331**:461–462 (1988).

Ernst, S.A.: Transport adenosine triphosphatase cytochemistry. I. Biochemical characterization of a cytochemical medium for the ultrastructural localization of ouabain-sensitive, potassium-dependent phosphatase activity in the avian salt gland. J.Histochem.Cytochem. **20**:13–22 (1972a).

Ernst, S.A.: Transport adenosine triphosphatase cytochemistry. II. Cytochemical localization of ouabain-sensitive, potassium-dependent phosphatase activity in the secretory epithelium of the avian salt gland. J.Histochem.Cytochem. **20**:23–38 (1972b).

Eränkö, O.: Distribution of fluorescing islets, adrenaline and noradrenaline in the adrenal medulla of the hamster. Acta Endocrinol. **18**:174–179 (1955).

Eränkö, O.: The practical histochemical demonstration of catecholamines by formaldehyde-induced fluorescence. J.Roy. Micr.Soc. **87**:259–276 (1967).

Eränkö, O. and Eränkö, L.: Differentiation of dopamine from noradrenaline in tissue sections by microspectrofluorimetry. J. Histochem.Cytochem. **19**:131–132 (1971).

Esmann, V.: The glycogen content of leukocytes from diabetic and non-diabetic subjects. Scand.J. clin.Lab.Invest. **13**:134–139 (1961).

Evans, A.W., Johnson, N.W., and Butcher, R.G.: A quantitative cytochemical study of three oxidative enzymes during experimental oral carcinogenesis in the hamster. Brit.J.Oral Surg. **18**:3–16 (1980).

Fahimi, H.D.: Fine-structural cytochemical localization of peroxidatic activity of catalase. Techniques of biochemical and biophysical morphology **2**:197–245 (1975).

Fahimi, H.D.: Qualitative cytological criteria for the validation of enzyme histochemical techniques, pp.33–51. In: Trends in enzyme histochemistry and cytochemistry. Ciba Foundation Symposium 73. Excerpta Medica, Amsterdam (1980).

Falck, B. and Torp, A.: A fluorescence method for histochemical demonstration of noradrenalin in the adrenal medulla. Med.Exp. **5**:429–432 (1961).

Falck, B., Hillarp, N.-Å., Thieme, G., and Torp, A.: Fluorescence of catechol amines and related compounds condensed with formaldehyde. J.Histochem.Cytochem. **10**:348–354 (1962).

Faulkner, R.R. and Lillie, R.D.: A buffer modification of the Warthin-Starry silver method for spirochaetes in single paraffin sections. Stain Technol. **20**:81–82 (1945).

Feigl, F.: Spot tests in organic analysis. 5th edit. Elsevier, Amsterdam, (1958).

Felgenhauer, K. and Glenner, G.G.: Quantitation of tissue-bound renal amino-peptidase by a microdensitometric technique. J.Histochem.Cytochem. **14**:53–63 (1966).

Fenton, S., Clarkson, E., MacGregor, G., Alaghband-Zadeh, J. and De Wardener, H.E.: An assay of the capacity of biological fluids to stimulate renal glucose-6-phosphate dehydrogenase activity in vitro as a marker of their ability to inhibit sodium potassium-dependent adenosine triphosphatase activity. J.Endocrinol. **94**:99–110 (1982).

Ferguson, M.M. and MacPhee, G.B.: Kinetic study of 11β-hydroxysteroid dehydrogenase in rat submandibular salivary gland. Archs. oral Biol. **20**:241–245 (1975).

Feulgen, R. and Rossenbeck, H.: Mikroskopisch-chemischer Nachweis einer Nucleinsäure von Typus der Thymonucleinsäure und die darauf beruhende elektive Färbung von Zellkernen in mikroskopischen Präparaten. Z.Physiol.Chem. **135**:203–248 (1924).

Filipe, M.I.: Value of histochemical reactions for mucosubstances in the diagnosis of certain pathological conditions in the colon and rectum. Gut **10**:577–586 (1969).

Filipe, M.I.: Malignant and inflammatory diseases of the gastrointestinal tract, pp.126–135. In: M.I. Filipe and B.D. Lake, Eds.: Histochemistry in pathology. Churchill Livingstone, Edinburgh (1983).

Filipe, M.I., Abbas, S., and Greaves, P.: Mucin composition and CEA content in assessing malignant potential of colonic adenomas. VIIIth European Congress of Pathology, Helsinki, p.22 (abstract) (1981).

Firth, J.A.: Quantitative assessment of Na^+,K^+-ATPase localization by direct and indirect p-nitrophenyl phosphatase methods. J.Histochem.Cytochem. **35:**507–513 (1987).

Flanders, A., Galbraith, W. and Marshall, P.N.: Microspectrophotometric studies of Romanowsky stained blood cells. IV. Maturation of myeloid and erythroid cell lines in bone marrow. Stain Technol. **59:**91–103 (1984).

Flitney, F.W.: The time course of the fixation of albumin by formaldehyde, glutaraldehyde, acrolein and other aldehydes. J. Roy.Micr.Soc. **85:**353–364 (1966).

Fontana, A.: Verfahren zur intensiven und raschen Färbung des Treponema pallidum und anderer Spirochäten. Dermatol.Wochenschr. **55:**1003–1004 (1912).

Francis, R.J.: Amyloid, pp.155–175. In: J.D. Bancroft, and A. Stevens, Eds.: Theory and practice of histological techniques. 3rd edit. Churchill-Livingstone, Edinburgh (1990).

Franks, F.: Biological freezing and cryofixation. In: Low temperature biological microscopy and microanalysis. Royal Microscopical Society, Oxford, (1978).

Frederiks, W.M.: Some aspects of the value of Sudan black B in lipid histochemistry. Histochemistry **54:**27–37 (1977).

Frederiks, W.M. and Marx, F.: Quantitative aspects of the histochemical tetrazolium salt reaction on monoamine oxidase in rat liver. Histochem.J. **17:**707–715 (1985).

Frederiks, W.M. and Marx, F.: A quantitative histochemical study of 5'-nucleotidase activity in rat liver using the lead salt method and polyvinyl alcohol. Histochem.J. **20:**207–214 (1988).

Frederiks, W.M., Marx, F., and Bosch, K.S.: Diurnal variation in glycogen phosphorylase activity in rat liver. A quantitative histochemical study. Eur.J.Cell Biol. **43:**339–341 (1987a).

Frederiks, W.M., Marx, F., and van Noorden, C.J.F.: Quantitative histochemical assessment of the heterogeneity of glycogen phosphorylase activity in liver parenchyma of fasted rats using the semipermeable membrane technique and the PAS reaction. Histochem.J.: **19:**150–156 (1987b).

Frederiks, W.M., Marx, F., and van Noorden, C.J.F.: Histochemical demonstration of creatine kinase activity using polyvinyl alcohol and auxiliary enzymes. Histochem.J. **19:**529–532 (1987c).

Frederiks, W.M., Marx, F., Chamuleau, R.A.F.M., van Noorden, C.J.F. and James, J.: Immunocytochemical determination of ploidy class-dependent bromodeoxyuridine incorporation in rat liver parenchymal cells after partial hepatectomy. Histochemistry **93:**627–630 (1990).

Freifelder, D.: Physical biochemistry. Applications to biochemistry and molecular biology. W.H. Freemann, San Francisco (1976).

Fritz, P., Hönes, J., Rautenstrauch, H., Lutz, D., Tuczek, H.V., Mischlinski, A., Saal, H.J., Klein, C. and Laschner, W.: Immunophotometric quantification of extravascular immunoglobulin deposits in the synovial membrane of patients with osteoarthritis and rheumatoid arthritis. Bas.Appl.Histochem. **32:**455–469 (1988).

Fukuda, M.: Application of the microcomputer in multiparametric fluorescence cytophotometry. J.Histochem.Cytochem. **31:**241–243 (1983).

Fukuda, M., Böhm, N. and Fujita, S.: Cytophotometry and its biological application. Prog.Histochem.Cytochem. **11/1:**1–119 (1978).

Gahrton, G. and Yataganas, X.: Quantitative cytochemistry of glycogen in blood cells. Methods and clinical application. Progr.Histochem.Cytochem. **9/1:**1–30 (1976).

Galbraith, W. and Marshall, P.N.: Studies on Papanicolaou staining. III. Quantitative investigations of orangeophilia and cyanophilia. Stain Technol. **59:**133–142 (1984).

Galbraith, W., Marshall, P.N. and Bacus, J.W.: Microspectro-photometric studies of Romanowsky stained blood cells. I. Subtraction analysis of a standardized procedure. J.Microsc. **119:**313–330 (1980).

Galbraith, W., Marshall, P.N., Lee, E.S. and Bacus, J.W.: Studies on Papanicolaou stain. I. Visible-light spectra of stained cervical cells. Anal.Quant.Cytol. **1:**160–168 (1979).

Gall, J.G. and Pardue, M.L.: Formation and detection of RNA-DNA hybrid molecules in cytological preparations. Proc.Natl.Acad.Sci. **63**:378–383 (1969).

Garrett, J.R., Kidd, A., Kyriacou, K. and Smith, R.E.: Use of different derivatives of D-val-leu-arg for studying kallikrein activities in cat submandibular glands and saliva. Histochem.J. **17**:805–818 (1985).

Garvey, W., Fathi, A., Bigelow, F., Carpenter, B., and Jimenez, C.: A reliable method for demonstrating Gram-positive and Gram-negative bacteria. Stain Technol. **61**:251–253 (1986).

Geerts, A. and Roels, F.: Quantitation of catalase activity by microspectrophotometry after diaminobenzidine staining. Histochemistry **72**:357–367 (1981).

Gilbert, D.M, Besser, G.M. Bitensky, L and Chayen, J.: Development of a cytochemical section-assay for thyroid stimulators. J.Endocrinol. **75**:40P (1977).

Gill, J.E. and Jotz, M.M.: Further observations on the chemistry of pararosaniline-Feulgen staining. Histochemistry **46**:147–160 (1976).

Gilliland, G., Perrin, S. and Bunn, H. Franklin: Quantitative amplification of mRNA using polymerase chain reaction. J.Cell.Biochem. Suppl. V, **13**, WH:001 (1989)

Glauert, A.M., Ed.: Practical methods in electron microscopy, Vol. 5. North-Holland Publishing Company, Amsterdam (1977).

Gleich, G.L.: The eosinophil: New aspects and structure and function. J.Allergy Clin.Immunol. **60**:73–82 (1977).

Glenner, G.G.: The histochemical demonstration of indole derivatives by the rosindole reaction of E. Fischer. J.Histochem. Cytochem. **5**:297–304 (1957).

Glenner, G.G.: The bases of the staining of amyloid fibers: Their physico-chemical nature and the mechanism of their dye-substrate interaction. Progr.Histochem.Cytochem. **13/3** (1981).

Glenner, G.G. and Lillie, R.D.: The histochemical demonstration of indole derivatives by the post-coupled p-dimethylaminobenzylidene reaction. J.Histochem.Cytochem. **5**:279–296 (1957).

Glenner, G.G. and Lillie, R.D.: Observations on the diazotization-coupling reaction for the histochemical demonstration of tyrosine: Metal chelation and formazan variants. J.Histochem.Cytochem. **7**:416–422 (1959).

Glick, D.: Trends in quantification in histochemistry and cytochemistry. Histochem.J. **13**:227–240 (1981).

Glick, D., Engstrom, A. and Malmstrom, B.G.: A critical evaluation of quantitative histo- and cytochemical microscopic techniques. Science **114**:253–258 (1951).

Goldstein, D.J.: Ionic and non-ionic bonds in staining, with special reference to the action of urea and sodium chloride on the staining of elastic fibres and glycogen. Quart.J.Micr.Sci. **103**:477–492 (1962).

Goldstein, D.J.: A microdensitometric method for the analysis of staining kinetics. J.Microsc. **113**:331–343 (1980).

Goldstein, D.J.: Errors in microdensitometry. Histochem.J. **13**: 251–267 (1981).

Golgi, C.: Sulla fina struttura dei bulbi olfattorii. Riv.Sper. Freniatr.Med.Legal. **1**:405–425 (1875).

Golodetz, L. and Unna P.G.: Zur Chemie der Haut, III. Das Reduktions-vermögen der histologischen Elemente der Haut. Monatschr.Prakt.Dermatol. **48**:149–166 (1909).

Goltzman, D., Henderson, B. and Loveridge, N.: Cytochemical bioassays of parathyroid hormone. Characteristics of the assay and analysis of circulating hormonal forms. J.Clin.Invest. **65**:1309–1317 (1980).

Gomori, G.: Microtechnical demonstration of phosphatase in tissue sections. Proc.Soc.Exp.Biol. Med. **42**:23–26 (1939).

Gomori, G.: A new histochemical test for glycogen and mucin. Amer.J.Clin.Pathol. **16**:177–179 (1946).

Gomori, G.: Aldehyde Fuchsin: A new stain for elastic tissue. Amer.J.Clin.Pathol. **20**:665–666 (1950a).

Gomori, G.: An improved technic for acid phosphatase. Stain Technol. **25**:81–85 (1950b).

Gonchoroff, N.J., Greipp, P.R., Kyle, R.A. and Katzmann, J.A.: A monoclonal antibody reactive with 5-bromo-2-deoxyuridine that does not require DNA denaturation. Cytometry **6**:506–512 (1985).

Gooding, P.E., Chayen, J., Sawyer, B. and Slater, T.F.: Cytochrome P-450 distribution in rat liver and the effect of sodium phenobarbitone administration. Chem.-Biol.Interact. **20:** 299–310 (1978).

Gordon, H. and Sweets, H.H.,Jr.: A simple method for the silver impregnation of reticulum. Amer.J.Pathol. **15:**545–552 (1936).

Gordon, M. and Robertson, W.R.: The application of continuous monitoring microdensitometry to an analysis of NAD^+ binding and 3β-hydroxy-Δ^5-steroid dehydrogenase activity in the regressing corpus luteum of the pro-oestrous rat ovary. Histochem.J. **18:**41–44 (1986).

Gossrau, R.: Tetrazoliummethoden zum histochemischen Hydrolasennachweis. Histochemistry **58:**203–218 (1978).

Gossrau, R.: Cytochemistry of membrane proteases. Histochem.J. **17:**737–771 (1985).

Gossrau, R. and Lojda, Z.: Study on dipeptidylpeptidase II (DPP II). Histochemistry **70:**53–76 (1980).

Gossrau, R., van Noorden, C.J.F. and Frederiks, W.M.: Enhanced light microscopic visualization of oxidase activity with the cerium capture method. Histochemistry **92:**349–353 (1989).

Gossrau, R., Graf, R., Ruhnke, M. and Hauski, C.: Proteases in the human full-term placenta. Histochemistry **86:**405–413 (1987).

Graf, M., Baici, A. and Strauli, P.: Histochemical localization of cathepsin B at the invasion front of the rabbit V2 carcinoma. Lab.Invest. **45:**587–596 (1981).

Graham, R.C. and Karnovsky, M.J.: The early stages of absorption of injected horseradish peroxidase in the proximal tubules of mouse kidney: Ultrastructural cytochemistry by a new technique. J.Histochem.Cytochem. **14:**291–302 (1966).

Gram, C.: Ueber die isolierte Färbung der Schizomyceten in Schnitt- und Trockenpräparaten. Forschr.Med. **2:**185 (1884).

Grange, J.M.: The mycobacteria. In: G. Wilson, A. Miles, and M.T. Parker, Eds.: Topley and Wilson's Principles of bacteriology, virology and immunity, vol. II: Systematic bacteriology, 7th edit. Edward Arnold Ltd., London (1983).

Gratzner, H.G.: Monoclonal antibody to 5-bromo- and 5-iododeoxyuridine: A new reagent for detection of DNA replication. Science **218:**474–475 (1982).

Gratzner; H.G., Leif, R.C., Ingram, D.J. and Castro, A.: The use of antibody specific for bromodeoxyuridine for the immunofluorescent determination of DNA replication in single cells and chromosomes. Exptl.Cell Res. **95:**88–94 (1975).

Gray, P., Day, M.W., Hayweiser, L.J., and Nevsimal, C.: Oxazine dyes. 2. Celestine blue B, gallocyanin and gallamin blue with mordants other than ferric alum. Stain Technol. **32:**161-165 (1957).

Grimelius, L.: A silver nitrate stain for α_2 cells in human pancreatic islets. Acta Soc.Med.Ups. **73:**243–270 (1968).

Grocott, R.G.: A stain for fungi in tissue sections and smears. Amer.J.Clin.Pathol. **25:**975–979 (1955).

Grynkiewicz, G., Poenie, M., and Tsien, R.Y.: A new generation of Ca^{2+} indicators with greatly improved fluorescence properties. J. Biol.Chem. **260:**3440–3450 (1985).

Gundersen, H.J.G., Bendtsen, T.F., Korbo, L., Marcussen, N., Møller, A., Nielsen, K., Nyengaard, J.R., Pakkenberg, B., Sørensen, F.B., and West, M.J.: Some new, simple and efficient stereological methods and their use in pathological research and diagnosis. Acta Pathol.Microbiol.Immunol.Scand. **96:**379–394 (1988a).

Gundersen, H.J.G., Bagger, P., Bendtsen, T.F., Evans, S.M., Korbo, L., Marcussen, N., Møller, A., Nielsen, K., Nyengaard, J.R., Pakkenberg, B., Sørensen, F.B., Vesterby, A. and West, M.J.: The new stereological tools: Disector, fractionator, nucleator and point sampled intercepts and their use in pathological research and diagnosis. Acta Pathol.Microbiol. Immunol.Scand. **96:**857–881 (1988b).

Gutschmidt, S.: "In situ" determinations of apparent K_M and V_{max} of brush border disaccharidases along the villi of normal human jejunal biopsy specimens. A quantitative histochemical study. Histochemistry **71:**451–462 (1981).

Gutschmidt, S., Emde, C.,and Riecken, E.O.: Quantification of α-glucosidases along the villus of the small intestine in man. Introduction of a computerized histochemical method. Histochemistry **67:**85–97 (1980a).

Gutschmidt, S., Kaul, W., and Riecken, E.O.: A quantitative histochemical technique for the characterisation of α-glucosidases in the brush-border membrane of rat jejunum. Histochemistry **63:**81–101 (1979).

Gutschmidt, S., Lange, U., and Riecken, E.O.: Kinetic characterization of unspecific alkaline phosphatases at different villus sites of rat jejunum. Histochemistry **69:**189–202 (1980b).

Gutschmidt, S., Lange, U. and Riecken, E.O.: "In situ"-measurements of protein contents in the brush border region along rat jejunal villi and their correlations with four enzyme activities. Histochemistry **72:**467–479 (1981).

Haase, A.T.: Analysis of viral infections by in situ hybridization. J.Histochem.Cytochem. **34:**27–32 (1986).

Håkanson, R., Juhlin, L., Owman, C., and Sporrong, B.: Histochemistry of histamine: microspectrofluorometric characterization of the fluorophores induced by o-phthaldialdehyde. J.Histochem.Cytochem. **18:**93–99 (1970).

Halkjær-Kristensen, J. and Ingemann-Hansen, T.: Quantitation of the PAS reaction in skeletal muscle of man. Histochem.J. **11:**737–746 (1979).

Hamada, S.: A double labeling technique combining ^{3}H-thymidine autoradiography with BrdU immunocytochemistry. Acta Histochem.Cytochem. **18:**267–269 (1985).

Hanker, J.S.: Oxidoreductases, pp.1–139. In: M.A. Hayat, Ed.: Electron microscopy of enzymes. Principles and methods. Vol. 4. Van Nostrand Reinhold Company, New York (1975).

Harada, K.: The nature of mycobacterial acid-fastness. Stain Technol. **51:**255–260 (1976).

Hardonk, M.J.: The use of phenazine methosulphate as an electron carrier in the histochemical demonstration of dehydrogenases. Histochemie **4:**563–568 (1965).

Harkema, J.R., Plopper, C.G., Hyde, D.M. and St. George, J.A.: Regional differences in quantities of histochemically detectable mucosubstances in nasal, paranasal, and nasopharyngeal epithelium of the bonnet monkey. J.Histochem.Cytochem. **35:**279–286 (1987).

Hartmann, F. and Fleck, U.: Vergleichende chemische und histologische Analyse der Leberverfettung. Klin.Wochenschr. **30:**652–654 (1952).

Haugland, R.P.: Molecular Probes. Handbook of Fluorescent probes and research chemicals. Molecular Probes, Inc. Eugene, Oregon, USA (1989).

Hayat, M.A.: Principles and techniques of electron microscopy: Biological applications. Vol. I. Van Nostrand-Reinhold, New York, (1970).

Hayes, E.R.: A rigorous re-definition of the plasmal reaction. Stain Technol. **24:**19–23 (1949).

Hayhoe, F.G.J. and Quaglino, D.: Haematological cytochemistry. Churchill-Livingstone, Edinburgh (1980).

Henderson, B.: Quantitative cytochemical measurement of glyceraldehyde 3-phosphate dehydrogenase activity. Histochemistry **48:**191–204 (1976).

Henderson, B.: Quantitative dehydrogenase cytochemistry in the study of cellular metabolism with special reference to rheumatoid arthritis, pp.83–98. In: J.R. Pattison, L. Bitensky and J. Chayen, Eds.: Quantitative cytochemistry and its applications. Academic Press, London (1979).

Henderson, B.: Increase in the activity of lysosomal acid hydrolases in the chondrocytes of arthritic joints of rabbits with experimental allergic arthritis. Histochem.J. **16:**287–293 (1984).

Henderson, B. and Loveridge, N.: Intermediate electron-acceptors in quantitative cytochemistry. Comparison of phenazine methosulphate and Meldola Blue. Histochemistry **72:**617–623 (1981).

Henderson, B., Bitensky, L. and Chayen, J.: Altered phospholipids in human rheumatoid synoviocytes. Ann.Rheum.Dis. **37:**24–29 (1978a).

Henderson, B., Bitensky, L. and Chayen, J.: Mitochondrial oxidative activity in human rheumatoid synovial lining cells. Ann.Rheum.Dis. **37:**548–551 (1978b).

Henderson, B., Johnstone, J.J. and Chayen, J.: 5' Nucleotidase in the human synovial lining in rheumatoid arthritis. Ann.Rheum.Dis. **39:**248–252 (1980).

Henderson, B., Loveridge, N., and Robertson, W.R.: A quantitative study of the effects of different grades of polyvinyl alcohol on the activities of certain enzymes in unfixed tissue sections. Histochem.J. **10:**453–463 (1978c).

Henderson, B., Glynn, L.E., Bitensky, L. and Chayen, J.: Evidence for cell division in acutely inflamed rabbit joints. Ann.Rheum.Dis. **40:**177–181 (1981).

Herzog, R.E.: Cytophotometric determinations of DNA, histone, arginine, lysine, and their concentrations in eu- and heterochromatin of the cell nucleus of dysplasias, carcinoma in situ, and carcinoma of the human cervix uteri. Arch.Gynecol. **231**:91–98 (1982).

Hess, J., Jensen, C.V. and Diemer, N.H.: Calcium-imaging with Fura-2 in isolated cerebral microvessels. Acta Histochem. **87**:107–114 (1989).

Hildebrand, R. and Schleicher, A.: Image analysis of the histochemical demonstration of glucose-6-phosphatase activity in rat liver. Histochemistry **86**:181–190 (1986).

Hillarp, N.-Å. and Hökfelt, B.: Evidence of adrenaline and noradrenaline in separate adrenal medullary cells. Acta Physiol. Scand. **30**:55–68 (1953).

Holczinger, L.: Histochemischer Nachweis freier Fettsaüren. Acta Histochem. (Jena) **8**:167–175 (1959).

Hooghwinkel, G.J.M. and Smits, G.: The specificity of the periodic acid-Schiff technique studied by a quantitative test tube method. J.Histochem.Cytochem. **5**:120–126 (1957).

Hopwood, D.: Some aspects of fixation with glutaraldehyde. A biochemical and histochemical comparison of the effects of formaldehyde and glutaraldehyde fixation on various enzymes and glycogen with a note on penetration of glutaraldehyde into liver. J.Anat. **101**:83–92 (1967).

Hopwood, D.: Fixatives and fixation: A review. Histochem.J. **1**:323–360 (1969).

Horobin, R.W.: Histochemistry. Gustav Fischer, Stuttgart and Butterworths, London (1982a).

Horobin, R.W.: Selection of optimum tetrazolium salts for use in histochemistry: the value of structure-staining correlations. Histochem.J. **14**:301–310 (1982b).

Horobin, R.W.: Thoughts arising from Cowden and Curtis's letter. Histochem.J. **16**:1240–1241 (1984).

Horobin, R.W. and Bennion, P.J.: The interrelation of the size and substantivity of dyes; the role of van der Waals attractions and hydrophobic bonding in biological staining. Histochemie **33**:191–204 (1973).

Horobin, R.W. and Flemming, L.: Is Mercury Orange a selective stain for thiols? Histochem.J. **14**:1003–1006 (1982).

Horobin, R.W. and Flemming, L.: One-bath trichrome staining: investigation of a general mechanism based on a structure-staining correlation analysis. Histochem.J. **20**:29–34 (1988).

Horobin, R.W. and Flemming, L.: 'Trouble-shooting' microwave accelerated procedures in histology and histochemistry: understanding and dealing with artefacts, errors and hazards. Histochem.J. **22**:371–376 (1990).

Horobin, R.W. and Goldstein, D.J.: The influence of salt on the staining of tissue sections with basic dyes: An investigation into the general applicability of the critical electrolyte concentration theory. Histochem.J. **6**:599–609 (1974).

Horobin, R.W. and Murgatroyd, L.G.: The composition and properties of gallocyanin-chrome alum stains. Histochem.J. **1**:36–54 (1968).

Hougaard, D.M. and Larsson, L.-I.: Towards microfluorometric quantitation of polyamines in situ. Relationship between cellular polyamine concentration and fluorescence yield of the formaldehyde fluorescamine method. Histochemistry **93**:359–362 (1990).

Howie, A.J., Burnett, D. and Crocker, J.: The distribution of cathepsin B in human tissues. J.Path. **145**:307–314 (1985).

Høyer, P.E.: Histoenzymology of the human ovary: dehydrogenases directly involved in steroidogenesis, pp.52–67. In: P.M. Motta and E.S.E. Hafez, Eds.: Biology of the ovary. Martinus Nijhoff, The Hague (1980).

Høyer, P.E.: A quantitative cytochemical study of estradiol 17β-dehydrogenase in the human term placenta. J.Histochem.Cytochem. **36**:875 (1988).

Høyer, P.E. and Andersen, H.: Specificity in steroid histochemistry, with special reference to the use of steroid solvents. Distribution of 11β-hydroxysteroid dehydrogenase in kidney and thymus from the mouse. Histochemie **24**:292–306 (1970).

Høyer, P.E. and Andersen, H.: Histochemistry of 3β-hydroxysteroid dehydrogenase in rat ovary. I. A methodological study. Histochemistry **51**:167–193 (1977).

Høyer, P.E. and Byskov, A.G.: A quantitative study of Δ^5,3β-hydroxysteroid dehydrogenase activity in the rete system of the immature mouse ovary. In: A.G. Byskov and H. Peters, Eds.: International Congress Series No. 559. Development and function of reproductive organs, pp.216-224. Excerpta Medica, Amsterdam-Oxford-Princeton (1981).

Høyer, P.E, Lyon, H., Jakobsen, P., and Andersen, A.P.: Standardized Methyl Green-Pyronin Y procedures using pure dyes. Histochem.J. **18**:90–94 (1986).

Høyer, P.E., Wohlrab, F. and Stoward, P.J.: Dehydrogenases involved in steroid and prostaglandin metabolism. Chapter 23. In: P.J. Stoward, Ed.: Pearse, A.G.E.: Histochemistry: Theoretical and applied. 4th edit. Vol. III. Churchill Livingstone, Edinburgh In press (1991).

Hukill, P. and Putt, F.A.: A specific stain for iron using 4,7-diphenyl-1,10-phenanthroline (bathophenanthroline). J.Histochem. Cytochem. **10**:490–494 (1962).

Husain, O.A.N.: Cytology: preparative techniques, pp.296–300. In: Filipe, M.I. and Lake, B.D., Eds.: Histochemistry in pathology. Churchill Livingstone, Edinburgh (1983).

Husain, O.A.N. and Watts, K.C.: Rapid demonstration of nucleic acids using "oxidized" gallocyanin and chromic potassium sulphate: methods and applications. J.Clin.Pathol. **37**:99–101 (1984).

Hussein, K.A., Milne, G., and Hopwood, D.:Glycosaminoglycans in human gallbladder basement membrane: nature and quantitative changes in chronic cholecystitis. Histochem.J. **20**:449–454 (1988).

Ibrahim, K.S., Husain, O.A.N., Bitensky, L. and Chayen, J.: A modified tetrazolium reaction for identifying malignant cells from gastric and colonic cancer. J.clin.Pathol. **36**:133–136 (1983).

Illsley, N.P. and Verkman, A.S.: Membrane chloride transport measured using a chloride-sensitive fluorescent probe. Biochemistry **26**:1215–1219 (1987).

Imes, N.K.,Jr., Sanders, D.C., Crane, C.R., and Clark, G.: Assaying actual hematein content of commercial hematoxylins and hemateins. Stain Technol. **44**:167–172 (1969).

Impraim, C.D., Saiki, R.K., Erlich, H.A. and Teplitz, R.L.: Analysis of DNA extracted from formalin-fixed, paraffin-embedded tissue by enzymatic amplification and hybridization with sequence-specific oligonucleotides. Biochem.Biophys.Res.Comm. **142**:710–716 (1987).

Innis, M.A., Myambo, K.B., Gelfand, D.H. and Brow, M.A.D.: DNA sequencing with Thermus aquaticus DNA polymerase and direct sequencing of polymerase chain reaction-amplified DNA. Proc.Nat.Acad.Sci. **85**:9436–9440 (1988).

Innis, M.A., Gelfand, D.H., Sninsky, J.J. and White, T.J.: PCR protocols. Academic Press, New York (1989).

Inoue, S.: Video microscopy. Plenum Press, New York (1986).

IUB: Enzyme nomenclature. Academic Press, San Diego, California, (1984).

Jacobsen, N.O.: The histochemical localization of lactic dehydrogenase isoenzymes in the rat nephron by means of an improved polyvinyl alcohol method. Histochemie **20**:250–265 (1969).

Jawetz, E., Melnick, J.L., Adelberg, E.A. Brooks, G.F., Butel, J.S and Ornston, L.N.: Medical microbiology, 18th ed. Prentice-Hall International Inc., London (1989).

Jedrzejewski, K. and Kugler, P.: Peptidases in the kidney and urine of rats after castration. Histochemistry **74**:63–84 (1982).

Jeffrey, I.J.M., Mosley, S.M., Jones, C.J.P., and Stoddart, R.W.: Proteolysis and lectin histochemistry. Histochem.J. **19**:269–275 (1987).

Jöbsis, A.C.: Prostate, pp.263–267. In: Filipe, M.I. and Lake, B.D., Eds.: Histochemistry in pathology. Churchill Livingstone, Edinburgh (1983).

Jobst, K. and Sandritter, W.: Über den quantitativen histochemischen Nachweis von basischen Kernproteinen mit Gallocyaninchromalaun. Histochemie **4**:277-285 (1964).

Johannes, M.-L. and Klessen, C.:Alcianblue/PAS or PAS/Alcianblue? Remarks on a classical technic used in carbohydrate histochemistry. Histochemistry **80**:129–132 (1984).

John, H.A., Birnstiel, M.L. and Jones, K.W.: RNA-DNA hybrids at the cytological level. Nature **223**:582–587 (1969).

Johnson, M.A.: Skeletal muscle, pp.89–113. In: Filipe, M.I. and Lake, B.D., Eds.: Histochemistry in pathology. Churchill Livingstone, Edinburgh (1983).

Johnston, K.I. and Ashford, A.E.: A simultaneous-coupling azo dye method for the quantitative assay of esterase using α-naphthyl acetate as substrate. Histochem.J. **12**:221–234 (1980).

Jones, D.: The reaction of formaldehyde with unsaturated fatty acids during histological fixation. Histochem.J. **1**:459–491 (1969).

Jones, D.: Reactions of aldehydes with unsaturated fatty acids during histological fixation, pp.1–45. In: P.J. Stoward, Ed.: Fixation in histochemistry. Chapman and Hall, London (1973).

Jonges, G.N. and van Noorden, C.J.F.: In situ kinetic parameters of glucose-6-phosphate dehydrogenase and phosphogluconate dehydrogenase in different areas of the rat liver acinus. Histochem.J. **21**:585–594 (1989).

Jonsson, G.: Fluorescence methods for the histochemical demonstration of monoamines. VII. Fluorescence studies on biogenic monoamines and related compounds condensed with formaldehyde. Histochemie **8**:288–296 (1967).

Juarranz, A. Ferrer, J.M., Tato, A., Cañete, M., and Stockert, J.C.: Metachromatic staining and electron dense reaction of glycosaminoglycans by means of Cuprolinic Blue. Histochem.J. **19**:1–6 (1987).

Junqueira, L.C.U., Bignolas, G., and Brentani, R.R.: Picrosirius staining plus polarization microscopy, a specific method for collagen detection in tissue sections. Histochem.J. **11**:447–455 (1979).

Kasten, F.H.: The chemistry of Schiff's reagent. Int.Rev.Cytol. **10**:1–100 (1960).

Kashiwa, H.K. and Atkinson, W.B.: The applicability of a new Schiff base, glyoxal bis (2-hydroxyanil), for the cytochemical localisation of ionic calcium. J.Histochem.Cytochem. **11**:258 (1963).

Kayser, K. and Bubenzer, J.: Microwave-assisted staining procedures in routine histopathology. Histochem.J. **22**:365–370 (1990).

Kiefer, G.: Recent developments in Gallocyanine-Chrome Alum staining, pp.199–208. In: G.L. Wied and G.F. Bahr, Eds.: Introduction to Quantitative Cytochemistry. Academic Press, New York, London (1970).

Kiefer, R., Kiefer, G., Salm, R., Rossner, R., and Sandritter, W.: A method for the quantitative evaluation of eu- and heterochromatin in interphase nuclei using cytophotometry and pattern analysis. Beitr.Path. **150**:163-173 (1973).

Kiernan, J.A.: Histological & histochemical methods: Theory & Practice. Pergamon Press, Oxford (1981).

Kiviranta, I., Jurvelin, J., Tammi, M., Säämänen, A.-M., and Helminen, H.J.: Microspectrophotometric quantitation of glycosaminoglycans in articular cartilage sections stained with Safranin O. Histochemistry **82**:249–255 (1985a).

Kiviranta, I., Tammi, M., Jurvelin, J., Säämänen, A.-M., and Helminen, H.J.: Demonstration of chondroitin sulphate and glycoproteins in articular cartilage matrix using periodic acid-Schiff (PAS) method. Histochemistry **83**:303–306 (1985b).

Kjellstrand, P.T.T.: Control of extraction of deoxyribonucleic acid and apurinic acid by polyethylene glycol in Feulgen hydrolysis. J.Histochem.Cytochem. **25**:371-375 (1977).

Kjellstrand, P.T.T. and Lamm, C.J.: A model of the breakdown and removal of the chromatin components during Feulgen acid hydrolysis. Histochem.J. **8**:419-430 (1976).

Klunk, W.E., Pettegrew, J.W. and Abraham, D.J.: Two simple methods for quantifying low-affinity dye-substrate binding. J.Histochem.Cytochem. **37**:1293–1297 (1989a).

Klunk, W.E., Pettegrew, J.W. and Abraham, D.J.: Quantitative evaluation of Congo red binding to amyloid-like proteins with a beta-pleated sheet conformation. J.Histochem.Cytochem. **37**:1273–1281 (1989b).

Koch, J.E., Kolvraa, S., Petersen, K.B., Gregersen, N. and Bolund, L.: Oligonucleotide-priming methods for the chromosome-specific labeling of alpha satelite DNA in situ. Chromosoma **98**: 259–265 (1989).

Koch, R.: Die Ätiologie der Tuberkulose. Berl.Klin.Wochenschr. **19**:221 (1882).

Kok, L.P. and Boon, M.E.: Microwave in microscopy. J.Microsc. **158**:291–322 (1990).

Konradi, C., Kornhuber, J., Froelich, L., Fritze, J., Heinsen, H., Beckmann, H., Schulz, E., and Riederer, P.: Demonstration of monoamine oxidase-A and -B in the human brain stem by a histochemical technique. Neuroscience **33**:383–400 (1989).

Krapf, R., Illsley, N.P., Tseng, H.C., and Verkman, A.S.: Structure-activity relationships of chloride-sensitive fluorescent indicators for biological application. Anal.Biochem. **169**:142–150 (1988).

Krawetz, S.A., Pon, R.T. and Dixon, G.H.: Increased efficiency of the Taq polymerase catalyzed chain reaction. Nucl.Acid.Res. **17**:819 (1989).

Krikelis, H. and Smith, A.A.: Palladium toning of silver-impregnated reticular fibers. Stain Technol. **63**:97–100 (1988).

Kugler, P.: Aminopeptidase A is angiotensinase A. I. Quantitative histochemical studies in the kidney glomerulus. Histochemistry **74**:229–245 (1982a).

Kugler, P.: Quantitative dehydrogenase histochemistry with exogenous electron carriers (PMS, MPMS, MB). Histochemistry **75**:99–112 (1982b).

Kugler, P.: Quantitative histochemistry of the lysosomal dipeptidyl aminopeptidase II in the proximal tubule of the rat kidney. Histochemistry **76**:557–566 (1982c).

Kugler, P.: Microphotometric determination of enzymes in brain sections. I. Hexokinase. Histochemistry **93**:295–298 (1990a).

Kugler, P.: Microphotometric determination of enzymes in brain sections. III. Glutamate dehydrogenase. Histochemistry **93**:537–540 (1990b).

Kugler, P. and Baier, G.: Microphotometric determination of enzymes in brain sections. II. GABA transaminase. Histochemistry **93**:501–505 (1990).

Kugler, P. and Wrobel, K.-H.: Meldola Blue: a new electron carrier for the histochemical demonstration of dehydrogenases (SDH, LDH, G-6-PDH). Histochemistry **59**:97–109 (1978).

Kuhlmann, W.D.: Ultrastructural immunoperoxidase cytochemistry. Progr.Histochem.Cytochem. **10/1**:1–57 1977.

Kurauchi, O., Mizutani, S., Nomura, S., Hotta, R., Ito, Y., Asada, Y., Furuhashi, M., Kasugai, M. and Tomoda, Y.: Effects of progesterone and estradiol on aminopeptidase A (AAP) and leucine aminopeptidase (LAP) activities in the pregnancy serum and placenta of the rat. Exp.Clin.Endocrinol. **93**:90–94 (1989).

Lake, B.D.: Alcian Blue 8GX. Histochem.J. **12**:717 (1980).

Lake, B.D.: Metabolic disorders of the central and peripheral nervous system. In: Filipe, M.I. and Lake, B.D., Eds.: Histochemistry in pathology, pp.53–69. Churchill Livingstone, Edinburgh (1983).

Lake, B.D., Puri, P., Nixon, H.H. and Claireaux, A.E.: Hirschsprung's disease. An appraisal of histochemically demonstrated acetyl-cholinesterase activity in suction rectal biopsies as an aid to diagnosis. Arch.Pathol.Lab.Med. **102**:244–247 (1978).

Landegent, J.E., Jansen in de Wal, N, Ploem, J.S. and van der Ploeg, M.: Sensitive detection of hybridocytochemical results by means of reflection-contrast microscopy. J.Histochem.Cytochem. **33**:1241–1246 (1985).

Larsson, L.-I.: Immunocytochemistry: theory and practice. CRC Press, Inc., Boca Raton, Florida (1988).

Larsson, L.-I. and Hougaard, D.M.: Optimization of non-radioactive in situ hybridization: image analysis of varying pretreatment hybridization and probe labelling conditions. Histochemistry **93**:347–354 (1990).

Lawrence, G.M., Beesley, A.C.H. and Matthews, J.B.: The use of continuous monitoring and computer-assisted image analysis for the histochemical quantification of hexokinase activity. Histochem.J. **21**:557–564 (1989).

Lautrop, H., Højby, N., Bremmelgaard, A. and Korsager, B.: Bakteriologiske undersøgelsesmetoder. FADL's forlag, Copenhagen (1979).

LaVelle, A.: Some introductory comments on silver staining. Stain Technol. **60**:271–273 (1985).

Leeflang-de Pijper, A.M. and Hülsmann, W.C.: Pitfalls in histochemical localization studies of NADPH generating enzymes of enzyme systems in rat small intestine. Histochemistry **39**:143–153 (1974).

LeHir, M., Herzog, V., and Fahimi, H.P.: Cytochemical detection of catalase with 3,3′-diaminobenzidine. A quantitative reinvestigation of the optimal conditions. Histochemistry **64**:51–66 (1979).

Lillie, R.D.: Phenolic oxidative activities of the skin: Some reactions of keratohyalin and trichohyalin. J.Histochem.Cytochem. **4**:318–330 (1956).

Lillie, R.D.: Acetylation and nitrosation of tissue amines in histochemistry. J.Histochem.Cytochem. **6**:352–362 (1958).

Lillie, R.D. (Ed.): Conn's biological stains. 9th edit. Williams & Wilkins, Baltimore, (1977).

Lillie, R.D. and Donaldson, P.T.: Histochemical azocoupling of protein histidine. Brunswick's nitration method. J.Histochem. Cytochem. **20**:929–937 (1972).

Lillie, R.D. and Donaldson, P.T.: The mechanism of the ferric ferricyanide reduction reaction. Histochem.J. **6**:679–684 (1974).

Lillie, R.D. and Fullmer, H.M.: Histopathologic technic and practical histochemistry. 3rd edit. McGraw-Hill Book Co., New York, (1976).

Lillie, R.D. and Pizzolato, P.: Histochemical azo coupling reactions of the pigments of obstructive icterus and of hematoidin. J.Histochem.Cytochem. **17**:738–748 (1969).

Lillie, R.D. and Pizzolato, P.: Histochemical azocoupling reactions of the pigments of obstructive icterus and of hematoidin. II. Effect of blockade, bleaching and extraction procedures on the pigments and their azo derivatives. J.Histochem.Cytochem. **18**:75–79 (1970).

Lillie, R.D., Pizzolato, P., Dessauer, H.C., and Donaldson, P.T.: Histochemical reactions at tissue arginine sites with alkaline solutions of β-naphthoquinone-4-sodium sulfonate and other o-quinones and oxidized o-diphenols. J.Histochem.Cytochem. **18**:487–497 (1971).

Lillie, R.D., Pizzolato, P., Henderson, R., and Donaldson, P.T.: The influence of formaldehyde fixation time on the rate of deamination of erythrocytes and spinal cord proteins. Histochemie **24**:156–166 (1970).

Lillie, R.D., Pizzolato, P., Vacca, L.L., Catalano, R.A., and Donaldson, P.T.: Use of the diazo-sulfanilic acid, pH 1 Azure A sequence on adrenal medulla and the effects of prior oxidation and reduction on the reaction in several species. J.Histochem. Cytochem. **21**:448–454 (1973).

Lison, L.: Histochemie et Cytochemie Animales. 3rd edit. Gauthier Villars, Paris, (1960).

Lojda, Z.: Malabsorption. In: Filipe, M.I. and Lake, B.D., Eds.: Histochemistry in pathology, pp.136–144. Churchill Livingstone, Edinburgh (1983).

Lojda, Z.: The importance of protease histochemistry in pathology. Histochem.J. **17**:1063–1089 (1985).

Lojda, Z. and Gossrau, R.: Study on aminopeptidase A. Histochemistry **67**: 267–290 (1980).

Lojda, Z., Gossrau, R., and Schiebler, T.H.: Enzyme histochemistry. A laboratory manual. Springer Verlag, Berlin (1979).

Lojda, Z, van der Ploeg, M. and van Duijn, P.: Phosphates of the naphthol AS series in the quantitative determination of alkaline and acid phosphatase activities "in situ" studies in polyacrylamide membrane model systems and by cytospectro-photometry. Histochemie **11**:13–32 (1967).

Lomax. R.B. and Robertson, W.R.: Mitochondrial α-glycerol phosphate dehydrogenase activity in IIA fibres of the rat lateral gastrocnemius muscle; the effect of Ca^{2+} and ATP. Histochem.J. **22**:119–124 (1990).

Lomax, R.B., Daglish, A., Taylor, R.J., Gordon, M.T. and Robertson, W.R.: Rapid continuous monitoring of enzyme activity in tissue sections: experience with the M85A and Zeiss UMSP-30 systems. Histochem.J.**21**:595–599 (1989).

Lorén, I., Björklund, A., Falck, B., and Lindvall, O.: An improved histofluorescence procedure for freeze-dried paraffin-embedded tissue based on combined formaldehyde-glyoxylic acid perfusion with high magnesium content and acid pH. Histochemistry **49**:177–192 (1976).

Loveridge, N.: A quantitative cytochemical method for measuring carbonic anhydrase activity. Histochem.J. **10**:361–372 (1978).

Loveridge, N. and Robertson, W.R.: Stimulation of adrenal 5-ene, 3β-hydroxysteroid dehydrogenase by corticotrophin in vitro. J.Endocrinol. **78**:457–458 (1978).

Loveridge, N., Alaghband-Zadeh, J., Daly, J.R. and Chayen, J.: The nature of the redox change measured in the cytochemical bioassay of corticotrophin. J.Endocrinol. **67**:28P (1975).

Loveridge, N., Bloom, S.R., Welbourn, R.B. and Chayen, J.: Quantitative cytochemical estimation of the effect of pentagastrin (0.005–5 pg/ml) and of plasma gastrin on the Guinea-pig fundus in vitro. Clin.Endocrinol. **3**:389–396 (1974).

Loveridge, N., Hoile, R., Johnson, A.G. and Chayen, J.: The cytochemical measurement of gastrin-like activity. J.Endocrinol. **77**:40P–41P (1978).

Loveridge, N., Hoile, R.W., Johnson, A.G., Gardner,J.D. and Chayen, J.: The cytochemical section-bioassay of gastrin-like activity. J.Immunoassay **1**:195–209 (1980).

Lowry, O.H., Rosebrough, N.J., Farr, A.L. and Randall, R.J.: Protein measurement with the Folin phenol reagent. J.Biol.Chem. **193**:265–275 (1951).

Lukaszyk, A.: The use of tris (hydroxymethyl) aminomethane as a copper complexing agent in the histochemical method for α-glycerophosphate-ferricyanide oxidoreductase activity detection. Histochemie **27**:366–369 (1971).

Lund-Hansen, T., Høyer, P.E., and Andersen, H.: A quantitative cytochemical assay of β-galactosidase in single cultured human skin fibroblasts. Histochemistry **81:**321–330 (1984).

Lyon, H. and Prentø, P.: Alcian blue-silver impregnation. A method differentiating between acid carbohydrates, reticulin fibres and collagen fibres. Acta Path.Microbiol.Scand. Sect.A **81:**6–8 (1973).

Lyon, H. and Prentø, P.: Aldehyde Fuchsin staining of pancreatic B cells. Reproducible high-contrast staining of formalin-fixed and paraffin-embedded material. Histochem.J. **12:**97–105 (1980).

Lyon, H., Bæksted, M., and Møller, M.: Method for direct induction of serotonin fluorescence in a carcinoid tumor after embedding in hydroxyethyl methacrylate. Lab.Invest. **45:**316–320 (1981).

Lyon, H., Bæksted, M., and Møller, M.: Induced fluorescence of serotonin in paraffin sections from carcinoid tumors. Brain Res. Bull. **9:**751–756 (1982).

Lyon, H., Schulte, E. and Høyer, P.E.: The correlation between uptake of Methyl Green and Feulgen staining intensity of cell nuclei. An image analysis study. Histochem.J. **21:**508–513 (1989).

Lyon, H., Jakobsen, P., Clausen, P.P., and Andersen, A.P.: Methyl Green-Pyronin staining with pure Pyronin Y. Histochem.J. **15:**605–606 (1983).

Mager, M., McNary, W.F.,Jr., and Lionetti, F.: The histochemical detection of zinc. J.Histochem. Cytochem. **1:**493–504 (1953).

Magun, B.E. and Kelly, J.W.: A new fluorescent method with phenanthrenequinone for the histochemical demonstration of arginine residues in tissues. J.Histochem.Cytochem. **17:**821–827 (1969).

Mai, J.K., Schmidt-Kastner, R., and Tefett, H.-B.: Use of acridine orange for histologic analysis of the central nervous system. J.Histochem.Cytochem. **32:**97–104 (1984).

Maniatis, T., Fritsch, E.F. and Sambrook, J.: Molecular cloning, a laboratory manual. Cold Spring Harbour Lab. 2nd edit. (1989).

Marshall, P.N.: Thin-layer chromatography of Sudan dyes. J.Chromatogr. **136:**353–357 (1977).

Marshall, P.N. and Galbraith, W.: Microspectrophotometric studies of Romanowsky stained blood cells. III. The action of methylene blue and azure B. Stain Technol. **59:**17–36 (1984a).

Marshall, P.N. and Galbraith, W.: The calculation of differences in the colours of histological objects. Histochem.J. **16:**211–218 (1984b).

Marshall, P.N. and Horobin, R.W.: The chemical nature of the gallocyanin-chrome alum staining complex. Stain Technol. **47:** 155–161 (1972).

Marshall, P.N., Galbraith, W. and Bacus, J.W.: Studies on Papanicolaou staining. II. Quantitation of dye components bound to cervical cells. Anal.Quant.Cytol. **1:**169–178 (1979).

Marshall, P.N., Galbraith, W., Navarro, E.F. and Bacus, J.W.: Microspectrophotometric studies of Romanowsky stained blood cells. II. Comparison of the performance of two standardized stains. J.Microsc. **124:**197–210 (1981).

Masson, M P.: La glande endocrine de l'intestin chez l'homme. C.R.Acad.Sci. **158:**59–61 (1914).

Mayall, B.H.: Deoxyribonucleic acid cytophotometry of stained human leukocytes. I. Differences among cell types. J.Histochem. Cytochem. **17:**249–257 (1969).

Mayall, B.H. and Mendelsohn, M.L.: Errors in absorption cytophotometry: Some theoretical and practical considerations, p.89. In: G.L. Wied and G.F. Bahr, Eds.: Introduction to quantitative cytochemistry-II. Academic Press, New York (1970).

Mayer, P.: Über das Färben mit Hämatoxylin. Mitt.Zool.Stat.Neapel **10:**170–186 (1891).

McAllister, H.A. and Rock, D.L.: Comparative usefulness of tissue fixatives for in situ viral nucleic acid hybridization. J.Histochem.Cytochem. **33:**1026–1032 (1985).

McAuliffe. W.G.: A note on the purification of Alcian blue. Stain Technol. **58:**374–376 (1983).

McDonald, J.K.: An overview of protease specificity and catalytic mechanisms: aspects related to nomenclature and classification. Histochem.J. **17:**773–785 (1985).

McGadey, J.: A tetrazolium method for non-specific alkaline phosphatase. Histochemie **23:**180–184 (1970).

McGee-Russell, S.M.: Histochemical methods for calcium. J. Histochem.Cytochem. **6:**22–42 (1958).

McKenzie, J.M. and Zakarija, M.: LATS in Grave's disease. Recent Progr.Horm.Res. **33**:29–57 (1977).

McManus, J.F.A.: Histological demonstration of mucin after periodic acid. Nature **158**:202 (1946).

McMillan, P.J.: Differential demonstration of muscle and heart type lactic dehydrogenase of rat muscle and kidney. J.Histochem.Cytochem. **15**:21–31 (1967).

Medawar, P.B.: The rate of penetration of fixatives. J.Roy.Micr. Soc. **61**:46–57 (1941).

Meijer, A.E.F.H.: Semipermeable membrane techniques in quantitative enzyme histochemistry, pp.103–120. In: Trends in enzyme histochemistry and cytochemistry. Ciba Foundation Symposium 73. Excerpta Medica, Amsterdam (1980).

Mendelson, D., Tas, J., and James, J.: Cuprolinic Blue: a specific dye for single-stranded RNA in the presence of magnesium chloride. II. Practical applications for light microscopy. Histochem.J. **15**:1113–1121 (1983).

Menten, M.L., Junge, J., and Green, M.H.: A coupling histochemical azo dye test for alkaline phosphatase in kidney. J.Biol.Chem. **153**:471–477 (1944).

Mihály, A., Király, E., Nógrádi, A., and Bencsik, K.: A modified semipermeable membrane technique for carbonic anhydrase histochemistry of the nervous system. Histochemistry **88**:485–487 (1988).

Mizutani, S.; Akiyama, H., Kurauchi, O., Taira, H., Narita, O. and Tomoda, Y.: In vitro degradation of angiotensin II (A-II) by human placental subcellular fractions, pregnancy sera and purified placental aminopeptidases. Acta Endocrinol. **110**:135–139 (1985a).

Mizutani, S., Sumi, S., Oka, K., Yamada, R., Kurauchi, O., Taira, H., Narita, O. and Tomoda, Y.: In vitro degradation of oxytocin by pregnancy serum, placental subcellular fractions and purified placental aminopeptidases Exp.Clin.Endocrinol. **86**:310–316 (1985b).

Montes, G.S., Krisztán, R.M., Shigihara, K.M., Tokoro, R., Mourao, and Junqueira, L.C.U.: Histochemical and morphological characterization of reticular fibers. Histochemistry **65**:131–141 (1980).

Morselt, A.F.W.: Application and possibilities of cytophotometry for red cell hematology. Progr. Histochem. Cytochem. **11/3**:1–42 (1978).

Morstyn, G., Pyke, K., Gardner, J., Ashcroft, R., De Fazio, A. and Bhathal, P.: Immunohistochemical identification of proliferating cells in organ culture using bromodeoxyuridine and a monoclonal antibody. J.Histochem.Cytochem. **34**:697–701 (1986).

Mowry, R.W.: Aldehyde fuchsin staining, direct or after oxidation: Problems and remedies, with special reference to human pancreatic B cells, pituitaries and elastic fibres. Stain Technol. **53**:141–154 (1978).

Mowry, R.W. and Kent, S.P.: Aldehyde pararosanilin (true aldehyde fuchsin) stains purified insulin, oxidized or not, following adequate fixation with either formalin or Bouin's fluid. Stain Technol. **63**:311–323 (1988).

Mullis, K.B.: The unusual origin of the polymerase chain reaction. Sci.Amer. **262/4**:36–43 (1990).

Murray, G.I., Burke, M.D., and Ewen, S.W.B.: Dehydrogenase enzyme histochemistry on freeze-dried or fixed resin-embedded tissue. Histochem.J. **20**:491–498 (1988).

Nadji, M. and Morales, A.R.: Immunoperoxidase techniques. A practical approach to tumor diagnosis. Raven Press, New York (1987).

Nauta, W.J.H. and Gygax, P.A.: Silver impregnation of degenerating axons terminals in the central nervous system: 1) Technic, 2) Chemical notes. Stain Technol. **26**:5–11 (1951).

Neelsen, F.: Ein casuistischer Beitrag zur Lehre von der Tuberkulose. Centralbl.Med.Wiss. **21**:497–501 (1883).

News and Views. Histochem.J. **17**:841–844 (1985).

News and Views. Histochem.J. **18**:59–60 (1986).

Newsholme, E.A. and Start, C.: Regulation in metabolism, pp.1–349. John Wiley & Sons, Chichester (1976).

Newsholme, E.A., Crabtree, B. and Zammit, V.A.: Use of enzyme activities as indices of maximum rates of fuel utilization. pp.245–258. In: Trends in enzyme histochemistry and cytochemistry. Ciba Foundation Symposium 73. Excerpta Medica, Amsterdam (1980).

Nibbering, P.H., Marijnen, J.G.J., Raap, A.K., Leijh, P.C.J. and van Furth, R.: Quantitative study of enzyme immunocytochemical reactions performed with enzyme conjugates immobilized on nitrocellulose. Histochemistry **84**:538–543 (1986).

Niblack, W.: An introduction to digital image processing. Strandberg Publishing Company, Birkerød, Denmark (1985).

Nielsen, K., Colstrup, H., Nielsson, T. and Gundersen, H.J.G.: Stereological estimates of nuclear volume correlated with histopathological grading and prognosis of bladder tumours. Virchows Arch. (Cell. Pathol.)**52:**41–54 (1986).

Nielsen, K., Berild, G.H., Bruun, E., Jørgensen, P.E. and Weis, N.: Stereological estimation of mean nuclear volume in prostatic cancer, the reproducibility and the possible value of estimations on repeated biopsies in the course of disease. J. Microsc. **154:**63–69 (1989).

Nöhammer, G.: A modification of the amidoblack-TCA-staining for quantitative microspectrometrical determination of proteins in tissue sections. Histochemistry **80:**395–400 (1984).

Norén, O., Dabelsteen, E., Høyer, P.E., Olsen, J., Sjöström, H. and Hansen, G.H.: Onset of transcription of the aminopeptidase N (leukemia antigen CD 13) gene at the crypt/villus transition zone during rabbit enterocyte differentiation. FEBS Lett. **259:**107–112 (1989).

Novikoff, A.B.: The endoplasmic reticulum: A cytochemist's view. Proc.Nat.Acad.Sci.U.S.-Biol. Sci. **73:**2781–2787 (1976).

Nuovo, G.J. and Silverstein, S.J.: Methods in laboratory investigation. Comparison of formalin, buffered formalin, and Bouin's fixation on the detection of human papilloma virus deoxyribonucleic acid from genital lesions. Lab.Invest. **59:**720–724 (1988).

Ogawa, H., Taneda, A., Kanaoka, Y. and Sekine, T.: The histochemical distribution of protein bound sulfhydryl groups in human epidermis by the new staining method. J.Histochem.Cytochem. **27:**942–946 (1979).

Old, S.L. and Johnson, M.A.: methods of microphotometric assay of succinate dehydrogenase and cytochrome c oxidase activities for use on human skeletal muscle. Histochem.J. **21:**545–555 (1989).

Olenev, S.N.: Cytophotometrical histochemical method of simultaneous measurement of two ingredients (DNA, proteins, enzymes) in the same cell. Acta histochem. **73:**151–161 (1983).

Olsson, I. and Dahlqvist, A.: The kinetics of the periodate-Schiff reaction of human leukocyte homogenates. J.Histochem.Cytochem. **15:**646–651 (1967).

Olsson, I., Venge, P., Spitznagel, J.K., and Lehrer, R.I.: Arginine-rich cationic proteins and human eosinophil granules. Comparison of the constituents of eosinophilic and neutrophilic leukocytes. Lab.Invest. **36:**493–500 (1977).

Ormerod, M.G. and Sloane, J.P.: Breast tumours, pp.236–244. In: Filipe, M.I. and Lake, B.D., Eds.: Histochemistry in pathology. Churchill Livingstone, Edinburgh (1983).

Ornstein, L.: The distributional error in microspectro-photometry. Lab.Invest. **1:**250–262 (1952).

Oud, P.S.: Protein staining for quantitative analysis, pp.52–54; pp.121–122. In: M.E. Boon and L.P. Kok, Eds.: Standardization and quantitation of diagnostic staining in cytology. Coulomb Press Leyden, Leiden (1986).

Oud, P.S., Henderik, J.B.J., Huysmans, A.C.L.M., Pahlplatz, M.M.M., Hermkens, H.G., Tas, J., James, J. and Vooijs, G.P.: The use of Light Green and Orange II as quantitative protein stains, and their combination with the Feulgen method for the simultaneous determination of protein and DNA. Histochemistry **80:**49–57 (1984).

Ovadia, M. and Stoward, P.J.: Some quantitative histochemical studies of the periodic acid oxidation of glycogen in situ. Histochem.J. **3:**233–240 (1971).

Pabst, M.A.: Staining of different endocrine cells with hydrochloric acid-toluidine blue in Epon embedded rat tissue. Stain Technol. **60:**93–98 (1985).

Pakkenberg, B. and Gundersen, H.J.G.: Total number of neurons and glia cells in human brain nuclei estimated by the disector and fractionator. J. Microsc. **150:**1–20 (1988).

Pal, M.K. and Biswas, M.: Small molecules as chromotropes in metachromasia. Histochemie **27:**36–42 (1971).

Papadimitriou, J.M. and van Duijn, P.: Effects of fixation and substrate protection on the isoenzymes of aspartate aminotransferase studied in a quantitative cytochemical model system. J.Cell Biol. **47:**71–83 (1970).

Park, C.M., Reid, P.E., Owen, D.A., Dunn, W.L., and Volz, D.: Histochemical procedures for the simultaneous visualization of neutral sugars and either sialic acid and its O-acyl variants or O-sulphate ester. II. Methods based upon the periodic acid-phenylhydrazine-Schiff reaction. Histochem.J. **19:**257–263 (1987).

Paschen, W.: Imaging of energy metabolites (ATP, glucose and lactate) in tissue sections: A bioluminescent technique. Progr. Histochem.Cytochem. **20/4:**1–122 (1990).

Patau, K.: Absorption microphotometry of irregular-shaped objects. Chromosome **5:**341–362 (1952).

Patzelt, W.J.: Reflexionskontrast, eine neue lichtmikroscopische Technik. Mikrokosmos **3:**78–81 (1977).

Pearse, A.D. and Marks, R.: Measurement of section thickness in quantitative microscopy with special reference to enzyme histochemistry. J.Clin.Pathol. **27:**615–618 (1974).

Pearse, A.G.E.: Common cytochemical and ultrastructural characteristics of cells producing polypeptide hormones (the APUD series) and their relevance to thyroid and ultimobranchial C cells and calcitonin. Proc.Roy.Soc. **170:**71–80 (1968).

Pearse, A.G.E.: Histochemistry: Theoretical and applied. 3rd edit. Vol. II. Churchill-Livingstone, Edinburgh (1972).

Pearse, A.G.E.: Histochemistry: Theoretical and applied. 4th edit. Vol. I. Churchill Livingstone, Edinburgh (1980).

Pearse, A.G.E.: Histochemistry: Theoretical and applied. 4th edit. Vol. II. Churchill Livingstone, Edinburgh (1985).

Pedersen, K.J.: Vævsbiologi. Fra celler til væv. FADL's Forlag, Copenhagen (1982).

Pellicciari, C. and Fraschini, A.: Methods of denaturation and renaturation of DNA in interphasic chromatin: cytochemical quantitative analysis by Methyl Green staining. Histochem.J. **10:**213–222 (1978).

Pentz, S. and Schulle, H.: Darstellung des Adhäsionsverhaltens kultivierter Leberzellen an Glasoberflächen während der Mitose durch Reflexions-Kontrast-Mikroskopie. Zeiss Inf., Oberkochen **25:**41–43 (1980).

Perrild, H., Frost, J., and Robertson, W.R.: The cytochemical assay of NAD^+ kinase activity in the follicular cells of guinea-pig thyroid gland. Histochem.J. **16:**71–83 (1984).

Perrild, H., Loveridge, N., Reader, S.C.J. and Robertson, W.R. Acute stimulation of thyroidal NAD^+ kinase, NADPH reoxidation, and peroxidase activities by physiological concentrations of thyroid stimulating hormone acting in vitro: A quantitative cytochemical study. Endocrinology **123:**2499–2505 (1988).

Perrild, H., Høyer, P.E., Loveridge, N., Reader, S.C.J. and Robertson, W.R.: Acute in vitro thyrotropin regulation of lysosomal enzyme activity in the thyroid follicular cell. Mol.Cell.Endocrinol. **65:**75–80 (1989).

Peters, A.: Staining of nervous tissue by protein-silver mixtures. Stain Technol. **33:**47–53 (1958).

Petersen, O.W., Høyer, P.E., and van Deurs, B.: Effect of oxygen on the tetrazolium reaction for glucose 6-phosphate dehydrogenase in cryosections of human breast carcinoma, fibrocystic disease and normal breast tissue. Virchows Arch. [Cell Pathol.] **50:**13–25 (1985).

Petersen, O.W., Høyer, P.E., and van Deurs, B.: Frequency and distribution of estrogen receptor-positive cells in normal, nonlactating human breast tissue. Cancer Res. **47:**5748–5751 (1987).

Pette, D. and Brandau, H.: Intracellular localization of glycolytic enzymes in cross-striated muscles of Locusta migratoria. Biochem.Biophys.Res.Commun. **9:**367–370 (1962).

Pette, D. and Wimmer, M.: Microphotometric determination of enzyme activities in cryostat sections by the gel film technique, pp.121–134. In: Trends in Enzyme Histochemistry and Cytochemistry, Ciba Foundation Symposium 73. Excerpta Medica, Amsterdam (1980).

Pfüller, Y., Franz, H., and Preiss, A.: Sudan black B: Chemical structure and histochemistry of the blue main components. Histochemistry **54:**237–250 (1977).

Pilgrim, C. and Stumpf, W.E.: Applications of autoradiography in neurobiological research. J.Histochem.Cytochem. **35:**917–928 (1987).

Piller, H.: Microscope photometry. Springer Verlag, Berlin (1977).

Ploem, J.S.: Standards for fluorescence microscopy, pp.137–153. In: E.J. Holborrow, Ed.: Standardization in immunofluorescence. Blackwell Scientific Publications, Oxford (1970).

Ploem, J.S.: Reflection-contrast microscopy as a tool for investigation of the attachment of living cells to a glass surface, pp.405–421. In: R. van Furth, Ed.: Mononuclear phagocytes in immunity, infection and pathology. Blackwell, Oxford (1975).

Ploem, J.S.: Appropriate technology for the quantitative assessment of the final reaction product of histochemical techniques, pp.275–303. In: Trends in enzyme histochemistry and cytochemistry. Ciba Foundation Symposium 73. Excerpta Medica, Amsterdam (1980).

Ploem, J.S., de Sterke, J.A., Bonnet, J. and Wasmund, H.: A microspectrofluorometer with epi-illumination operated under computer control. J.Histochem.Cytochem. **22**:668–677 (1974).

Poenie, M., Alderton, J., Steinhardt, R., and Tsien, R.: Calcium rises abruptly and briefly throughout the cell at the onset of anaphase. Science **233**:886–889 (1986).

Polak, J.M. and van Noorden, S., Eds.: Immunocytochemistry. Practical applications in pathology and biology. John Wright & Sons, Bristol (1983).

Polak, J.M. and van Noorden, S.: An introduction to immunocytochemistry: Current techniques and problems. Oxford University Press, Oxford (1984).

Polak,J.M. and Varndell, I.M., Eds.: Immunolabelling for electron microscopy. Elsevier, Amsterdam (1984).

Polasaky, Z., Kiefer, G., and Sandritter, W.: Quantitative histochemische Untersuchungen über die säurefeste Anfärbung des Spermienkopfes. Histochemie **4**:312–321 (1964).

Pollister, A.W. and Ornstein, L.: The photometric chemical analysis of cells, pp.431–518. In: R.C. Mellors, Ed.: Analytical cytology, 2nd edit., McGraw-Hill Book Company, New York (1959).

Poulter, L.W., Campbell, D.A., Munro, C. and Butcher, R.G.: The quantitation of HLA-DR expression on human cells using immunocytochemistry. J.Immunol.Method. **98**:227–234 (1987).

Pratt, W.K.: Digital image processing. Wiley, New York (1978).

Prentø, P.: Metals and phosphate in the chloragosomes of Lumbricus terrestris and their possible physiological significance. Cell Tiss.Res. **196**:123–134 (1979).

Prentø, P.: The effect of histochemical methylation on the phosphate groups of nucleic acids: Interpretation of the absence of nuclear basophilia. Histochem.J. **12**:661–668 (1980).

Proctor, G.B. and Horobin, R.W.: The aging of Gomori's aldehyde-fuchsin: the nature of the chemical changes and the chemical structures of the coloured components. Histochemistry **77**:255–267 (1983).

Proctor, G.B. and Horobin, R.W.: Chemical structures and staining mechanisms of Weigert's resorcin-fuchsin and related elastic fiber stains. Stain Technol. **63**:101–111 (1988).

Proescher, F. and Arkush, A.S.: Metallic lakes of the oxazines (gallamin blue, gallocyanin and celestin blue) as nuclear stain substitutes for haematoxylin. Stain Technol. **3**:28-38 (1928).

Puchtler, H. and Meloan, S.N.: On the chemistry of formaldehyde fixation and its effects on immunohistochemical reactions. Histochemistry **82**:201–204 (1985).

Puchtler, H., Meloan, S.N., and Brewton, B.R.: On the binding of Schiff's reagent in the PAS reaction: effects of molecular configurations in models. Histochemistry **40**:291–299 (1974).

Puchtler, H., Meloan, S.N., and Spencer, M.: Current chemical concepts of acids and bases and their application to anionic ("acid") and cationic ("basic") dyes. Histochemistry **82**:301–306 (1985).

Puchtler, H., Meloan, S.N., and Waldrop, F.S.: Application of current chemical concepts to metal-hematein and -brazilein stains. Histochemistry **85**:353–364 (1986).

Puchtler, H., Meloan, S.N., and Waldrop, F.S.: Are picro-dye reactions for collagen quantitative? Chemical and histochemical considerations. Histochemistry **88**:243–256 (1988).

Raap, A.K.: Localization properties of fluorescence cytochemical enzyme procedures. Histochemistry **84**:317–321 (1986).

Raap, A.K. and van Duijn, P.: Studies on the phenazine methosulphate tetrazolium salt capture reaction in NAD(P)-dependent dehydrogenase cytochemistry. II. A novel hypothesis for the mode of action of PMS and the study of the properties of reduced PMS. Histochem.J. **15**:881–893 (1983).

Raja, S.: Histochemical demonstration of conjugated and unconjugated bilirubin using a modified diazo-reagent. Nature (London). **205**:304–305 (1965).

Raju, B., Murphy, E., Levy, L.A., Hall, R.D., and London,R.E.: A fluorescent indicator for measuring cytosolic free magnesium. Amer.J.Physiol. **256**:C540–548 (1989).

Raspail, F.-V.: Nouveau systéme de chimie organique. Paris (1833). Cited in: Pearse, p.3 (1980).

Rees, L.H., Holdaway, I.M., Kramer, R., McNeilly, A.S. and Chard, T.: New bioassay for luteinizing hormone. Nature **244**: 232–234 (1973).

Reid, P.E., Culling, C.F.A., and Dunn, W.L.: Saponification-induced increase in the periodic acid-Schiff reaction in the gastrointestinal tract. Mechanism and distribution of the reactive substance. J.Histochem.Cytochem. **21**:473–482 (1973).

Reid, P.E., Volz, D., Cho, K.Y., and Owen, D.A.: A new method for the histochemical demonstration of O-acyl sugars in human colonic epithelial glycoproteins. Histochem.J. **20**:510–518 (1988).

Reid, P.E., Dunn, W.L., Ramey, C.W., Coret, E., Trueman, L., and Clay, M.G.: Histochemical studies of the mechanism of the periodic acid-phenylhydrazine-Schiff (PAPS) procedure. Histochem.J. **16**:641–649 (1984).

Reid, P.E., Owen, D.A., Fletcher. K., Rowan, R.E., Reimer, C.L., Rouse, G.J., and Park, C.M.: The histochemical specificity of High Iron Diamine-Alcian Blue. Histochem.J. **21**:501–504 (1989).

Rietveld, W.J.: The reliability of color perception in man, pp.73–81. In: M.E. Boon and L.P. Kok, Eds.: Standardization and quantitation of diagnostic staining in cytology. Coulomb Press Leyden, Leiden (1986).

Rink, T.J., Tsien, R.Y., and Pozzan, T.: Cytoplasmic pH and free Mg^{2+} in lymphocytes. J.Cell Biol. **95**:189–196 (1982).

Rio-Hortega, P. del: Noticia de un neuvo y fácil método para la coloración de la neuroglia y del tejido conjuntivo. Trab.Lab. Invest.Biol. **15**:367–378 (1917).

Robertson, W.R.: A quantitative cytochemical method for the demonstration of Δ^5,3β-hydroxysteroid dehydrogenase activity in unfixed tissue sections of rat ovary. Histochemistry **59**:271–285 (1979).

Robertson, W.R.: A quantitative study of N-acetyl-β-glucosaminidase activity in unfixed tissue sections of the guinea-pig thyroid gland. Histochem.J. **12**:87–96 (1980).

Robertson, W.R., Earnshaw, R.J. and Lambert, A.: The determination by microdensitometry of the initial maximum velocity rate of rat ovarian 3β-hydroxysteroid dehydrogenase activity. Histochemistry **74**:435–442 (1982a).

Robertson, W.R., Frost, J., Høyer, P.E., and Weinkove, C.: 20α-Hydroxysteroid dehydrogenase activity in the rat corpus luteum; a quantitative cytochemical study. J.Steroid Biochem. **17**:237–243 (1982b).

Roelfzema, W.H., Tohyama, C., Nishimura, Nishimura, N. and Morselt, A,F.W.: Quantitative immunohistochemistry of metallothionein in rat placenta. Histochemistry **90**:365–369 (1989).

Rogers, A.W.: Techniques of autoradiography. 3rd edit. Elsevier and North-Holland, Amsterdam (1979).

Roozemond, R.C.: The effect of calcium chloride and formaldehyde on the release and composition of phospholipids from cryostat sections of rat hypothalamus. J.Histochem.Cytochem. **17**:273–279 (1969a).

Roozemond, R.C.: The effect of fixation with formaldehyde and glutaraldehyde on the composition of phospholipids extractable from rat hypothalamus. J.Histochem.Cytochem. **17**:482–486 (1969b).

Ruhnke, M. and Gossrau, R.: Kinetic end-point microdensitometry (section biochemistry) of γ-glutamyl transpeptidase and dipeptidyl peptidase IV in the mature mouse decidua and visceral yolk sac. Acta histochem. **87**:23–31 (1989a).

Ruhnke, M. and Gossrau, R.: Reaction rate measurements of proteases and glycosidases with chromogenic methods. Histochem.J. **21**:535–544 (1989b).

Ruijgrok, J.M., Boon, M.E. and de Wijn, J.R.: The effect of heating by microwave irradiation and by conventional heating on the aldehyde concentration in aqueous glutaraldehyde solutions. Histochem.J. **22**:389–393 (1990).

Rüter, A., Wittekind, D., Gunzer, U., Aus, H.M. and Harms, H.: Comparative colorimetry in cytophotometric measurements of azure B-eosin Y-stained and Giemsa-stained blood cell smears. Anal.Quant.Cytol. **4**:128–139 (1982).

Rys, P. and Zollinger, H.: The theory of coloration of textiles. The Dyers Company Publications Trust (1975).

Saiki, R.K., Gelfand, D.H., Stoffel, S., Scharf, S.J., Higuchi, R., Horn, G.T., Mullis, K.B. and Erlich, H.A.: Primer-directed enzymatic amplification of DNA with a thermostable DNA polymerase. Science **239**:487–491 (1988).

Salthouse, T.N.: Luxol fast blue ARN: A new solvent azo dye with improved staining qualities for myelin and phospholipids. Stain Technol. **37**:313–316 (1962).

Sandritter, W., Kiefer, G., and Rick, W.: Über Stöchiometrie von Gallocyanin-chromalaun mit Desoxyribonukleinsäure. Histochemie **3**: 314-340 (1963).

Sannes, P.L.: The histochemical and cytochemical localization of proteases. Progr.Histochem. Cytochem. **18/1**:1–48 (1988).

Sarkar, G. and Sommer, S.S.: Shedding light on PCR contamination. Nature **343**:27 (1990).

Schauenstein, E., Desoye, G. and Nöhammer, G.:Quantitative Proteinbestimmung in Einzelzellen mit Amidschwarz. Histochemistry **68**:75–90 (1980).

Schenk, E.: Note from the Biological Stain Commission. A newly certified dye-Alcian Blue 8GX. Stain Technol. **56**:129–131 (1981).

Scherini, E., Mares, V., Bernocchi, G. and Barni, S.: Intranuclear differences in the response of Purkinje cell DNA of the rat cerebellum to bleomycin. A microphotometric and autoradiographic study. Histochemistry **89**:227–230 (1988).

Schieck, R., Taubert, G. and Krug, H.: Dry mass, DNA and non-histone protein determinations in lung cancer cells. Histochem.J. **19**:504–508 (1987).

Schmidt, H.: Phenol oxidase (E.C. 1.14.18.1) a marker enzyme for defence cells. Progr.Histochem. Cytochem. **17/3**:1–194 (1988).

Schmorl, G.: Die Pathologisch-Histologische Untersuchungsmethoden. 15th edit. Vogel, Leipzig (1928).

Schofield, K.P., Stone, P.C.W., Beddall, A.C., and Stuart, J.: Quantitative cytochemistry of the toxic granulation blood neutrophil. Brit.J.Haematol. **53**:15–22 (1983).

Schubert, M. and Hamerman, B.: Metachromasia; chemical theory and histochemical use. J.Histochem.Cytochem. **4**:159–189 (1956).

Schulte,E., Wittekind, D., and Kretschmer, V.: Victoria Blue B—a nuclear stain for cytology. A cytophotometric study. Histochemistry **88**:427–433 (1988).

Scopsi, L. and Larsson, L.-I.: Increased sensitivity in peroxidase immunocytochemistry. Histochemistry **84**:221–230 (1986a).

Scopsi, L. and Larsson, L.-I.: Bodian's silver impregnation of endocrine cells. A tentative explanation to the staining mechanism. Histochemistry **86**:59–62 (1986b).

Scott, J.E.: On the mechanism of the Methyl Green-Pyronin stain for nucleic acids. Histochemie **9**:30–47 (1967).

Scott, J.E.: Histochemistry and Alcian Blue. I. Metachromasia of Alcian Blue, Astrablau and other cationic phthalocyanin dyes. Histochemie **21**:227–285 (1970).

Scott, J.E.: The histochemistry of Alcian Blue. Note on the presence and removal of boric acid as the major diluent in Alcian blue 8GX. Histochemie **29**:129–133 (1972a).

Scott, J.E.: Amplification and staining by Alcian blue and similar ingrain dyes. J.Histochem.Cytochem. **20**:750–752 (1972b).

Scott, J.E.: Histochemistry of Alcian Blue. III. The molecular biological basis of staining by Alcian Blue 8GX and analogous phthalocyanins. Histochemie **32**:191–212 (1972c).

Scott, J.E.: Alcian dyes: I.C.I. cease manufacture and release details of composition. Histochemie **37**:379–380 (1973a).

Scott, J.E.: The second BDH-lecture. Trans.Biochem.Soc. **1**:787–806 (1973b).

Scott, J.E.: Collagen-proteoglycan interactions. Localization of proteoglycans in tendon by electron microscopy. Biochem.J. **187**:887–891 (1980).

Scott, J.E. and Dorling, J.: Differential staining of acid solutions. Histochemie **5**:221–233 (1965).

Scott, J.E. and Mowry, R.W.: Alcian Blue—a consumer's guide. J. Histochem.Cytochem. **18**:842 (1970).

Seidler, E.: New nitro-monotetrazolium salts and their use in histochemistry. Histochem.J. **12**:619–630 (1980).

Seifert, G., ed.: Morphological tumor markers. General aspects and diagnostic relevance. Springer-Verlag, New York (1987).

Seligman, A.M., Karnovsky, M.J., Wasserkrug, H.I. and Hanker, J.S.: Non-droplet ultrastructural demonstration of cytochrome oxidase activity with a polymerising osmiophilic reagent, diaminobenzidine (DAB). J.Cell Biol. **38**:1–14 (1968).

Shea, J.R., Jr.: A method for in situ cytophotometric estimation of absolute amount of ribonucleic acid using azure B. J.Histochem.Cytochem. **18:**143–152 (1970).

Shennawy, I.E., Gee, D.J., Horobin, R.W., and Aparicio, S.R.: Novel application of aniline blue. Fluorescent staining of glycogen and some protein structures. Histochemistry **81:**93–98 (1984).

Shibata, D.K., Arnheim, N. and Martin, W.J.: Detection of human papilloma virus in paraffin-embedded tissue using the polymerase chain reaction. J.Exp.Med. **167:**225–230 (1988).

Shires, T.K.: Tests for nucleic acid selectivity of Einarson's gallocyanin-chrome alum stain on monolayers of strain L-929 mouse fibroblasts. Stain Technol. **42:**67-74 (1967).

Sinapius, D.: Über Chromierungsmethoden zur Lipoiddarstellung. Histochemie **2:**217–233 (1961).

Singer-Sam, J., Tanguay, R.L. and Riggs, A.D.: Use of chelex to improve the PCR signal from a small number of cells. Amplifications—A forum for PCR users, Perkin Elmer Cetus, **3:**11 (1989).

Sippel, T.O.: The histochemistry of thiols and disulphides. I. The use of N-(4-aminophenyl)maleimide for demonstrating thiol groups. Histochem.J. **5:**413–423 (1973).

Sippel, T.O.: New fluorochromes for thiols: maleimide and iodoacetamide derivatives of a 3-phenylcoumarin fluorophore. J.Histochem.Cytochem. **29:**314–316 (1981).

Sisken, J.E.: Fluorescent standards, pp.113–126. In: D.Lansing Taylor and Yu-Li Wang, Eds.: Methods in cell biology, Vol. 30: Fluorescence microscopy of living cells in culture, Part B. Quantitative fluorescence microscopy—Imaging and spectroscopy. Academic Press, Inc., San Diego (1989).

Sjögren, S.: Lactate dehydrogenase isozymes in developing rat oral mucosa. A comparative study of LDH biochemistry and histochemistry. J.Histochem.Cytochem. **32:**958–964 (1984).

Smith, H.: The leukaemias, pp.206–214. In: M.I. Filipe and B.D. Lake, Eds.: Histochemistry in pathology. Churchill Livingstone, Edinburgh (1983).

Smith, M.T., Redick, J.A. and Baron, J.: Quantitative immunohistochemistry: A comparison of microdensitometric analysis of unlabeled antibody peroxidase-antiperoxidase staining and of microfluorometric analysis of indirect fluorescent antibody staining for nicotinamide adenosine dinucleotide phosphate (NADPH)-cytochrome-c (P-450) reductase in rat liver. J.Histochem.Cytochem. **31:**1183–1189 (1983).

Smith, M.T., Loveridge, N., Wills, E.D. and Chayen, J.: The distribution of glutathione in the rat liver lobule. Biochem.J. **182:**103–108 (1979).

Society of Dyers and Colourists: The colour index. 3rd edit. Bradford (England) (1971).

Solcia, E., Vassallo, G., and Capella, C.: Selective staining of endocrine cells by basic dyes after acid hydrolysis. Stain Technol. **43:**257–263)1968).

Spicer, S.S.: The use of various cationic reagents in histochemical differentiation of mucopolysaccharides. Amer.J. Clin.Pathol. **36:**393–409 (1961).

Spicer, S.S.: Diamine methods for differentiating mucosubstances histochemically. J.Histochem. Cytochem. **13:**211–234 (1965).

Stellmach, J. and Severin, E.: A fluorescent redox dye. Influence of several substrates and electron carriers on the tetrazolium salt-formazan reaction of Ehrlich ascites tumour cells. Histochem.J. **19:**21–26 (1987).

Sternberger, L.A. and Sternberger, N.H.: The unlabeled antibody method: Comparison of peroxidase-antiperoxidase with avidin-biotin complex by a new method of quantification. J.Histochem.Cytochem. **34:**599–605 (1986).

Sternberger, N.H., Itoyama, Y., Kies, M.W. and Webster, H. de F.: Myelin basic protein demonstrated immunocytochemically in oligodendroglia prior to myelin sheath formation. Proc. Natl.Acad.Sci.USA **75:**2521–2524 (1978).

Stoward, P.J.: Criteria for the validation of quantitative histochemical enzyme techniques, pp.11–31. In: Trends in enzyme histochemistry and cytochemistry. Ciba Foundation Symposium 73. Excerpta Medica, Amsterdam (1980).

Stoward, P.J. and Al-Sarraj, B.: Quantitative histochemical investigations of semipermeable membrane techniques for the assay of acid phosphatase in skeletal muscle. I. Meijer's method. Histochemistry **71:**405–431 (1981a).

Stoward, P.J. and Al-Sarraj, B.: Quantitative histochemical investigations of semipermeable membrane techniques for the assay of acid phosphatase in skeletal muscle. III. A modified simultaneous coupling technique. Histochemistry **71:**599–607 (1981b).

Stoward, P.J. and Nakae, Y.: Simultaneous histochemical assay of two dehydrogenases in the same cell. Histochem.J. **20**:610–616 (1988).

Stoward, P.J., Al-Sarraj, B., and Ireland, H.: Quantitative histochemical investigations of semipermeable membrane techniques for the assay of acid phosphatase in skeletal muscle. II. Nonspecific reactions in control sections. Histochemistry **71**:585–598 (1981).

Stoward, P.J., Campbell, J.B., and Al-Sarraj, B.: Quantitative histochemical investigations of semipermeable membrane techniques for the assay of acid phosphatase in skeletal muscle. IV. A post-coupling technique. Histochemistry **74**:367–377 (1982).

Stoward, P.J., Meijer, A.E.F.H., Seidler, E. and Wohlrab, F.: Dehydrogenases. Chapter 22. In: P.J. Stoward, Ed.: Pearse, A.G.E.: Histochemistry: Theoretical and applied. 4th edit. Vol. III. Churchill Livingstone, Edinburgh In press (1991).

Sumi, Y.: Histochemical staining of cadmium with thiazolylazophenol derivatives. Histochemistry **67**:1–6 (1980).

Sumi, Y., Muraki, T., and Suzuki, T.: Histochemical staining of cadmium with benzothiazolylazonaphthol derivatives. Histochemistry **73**:481–486 (1982).

Suurmeijer, A.J.H., Boon, M.E. and Kok, L.P.: Notes on the application of microwaves in histopathology. Histochem.J. **22**:341–346 (1990).

Swift, H.H. and Rasch, E.: Microphotometry with visible light. In: G. Oster and A.W. Pollister, Eds.: Physical techniques in biological research, Vol 3. Academic Press, New York (1956).

Takamatsu, H.: Histologische und biochemische Studien über die Phosphatase. I. Mitteilung. Histochemische Untersuchungsmethodik Phosphatase und deren Verteilung in verschiedenen Organen und Geweben. Trans.Soc.Path.Jap. **29**:492–498 (1939).

Tanasugarn, L., McNeil, P., Reynolds, G.T., and Lansing Taylor, D.: Microspectrofluorometry by digital image processing: Measurement of cytoplasmic pH. J.Cell Biol. **98**:717–724 (1984).

Tas, J.: The Alcian Blue and combined Alcian Blue-Safranin O staining of glycosaminoglycans studied in a model system and in mast cells. Histochem.J. **9**:205–230 (1977).

Tas, J. Mendelson, D., and Noorden, C.J.F.: Cuprolinic Blue: a specific dye for single-stranded RNA in the presence of magnesium chloride. I. Fundamental aspects. Histochem.J. **15**:801–814 (1983).

Taylor, C.R.: Immunomicroscopy. W.B. Saunders Company, Philadelphia (1986).

Taylor, D.L. and Salmon, E.D.: Basic fluorescence microscopy, pp.207–237. In: Y.-L. Wang and Taylor, D.L., Eds.: Methods in cell biology, Vol. 29. Fluorescence microscopy of living cells in culture. Part A. Fluorescent analogs, labeling cells, and basic microscopy. Academic Press, Inc., Boston (1989).

Thomas, R.C.: Bicarbonate and pH_i response. Nature **337**:601 (1989).

Troyer, H., Fisher, O.T. and Rosenquist, T.H.: Bone alkaline phosphatase kinetics studied by a new method. Histochemistry **50:** 251–259 (1977).

Tsien, R.Y.: New calcium indicators and buffers with high selectivity against magnesium and protons: Design, synthesis, and properties of prototype structures. Biochemistry **19**:2396–2404 (1980).

Tsien, R.Y.: A non-disruptive technique for loading calcium buffers and indicators into cells. Nature **290**:527–528 (1981).

Tsien, R.Y.: Fluorescence measurement and photochemical manipulation of cytosolic free calcium. TINS **11**:419–424 (1988).

Tsien, R.Y.: Fluorescent indicators of ion concentrations. In: D Lansing Taylor and Y.-L. Wang, Eds: Methods in cell biology, vol 30 part B. Fluorescence microscopy of living cells in culture. pp.127–156. Academic Press, Inc. San Diego, California (1989).

Tsien, Y., Pozzan, T., and Rink, T.J.: Calcium homeostasis in intact lymphocytes: Cytoplasmic free calcium monitored with a new, intracellularly trapped fluorescent indicator. J.Cell Biol. **94**:325–334 (1982).

Tsien, R.Y., Rink, T.J., and Poenie, M.: Measurement of cytosolic free Ca^{2+} in individual small cells using fluorescence microscopy with dual excitation wavelengths. Cell Calcium **6**:145–157 (1985).

Tuckett, F. and Morriss-Kay, G.: Alcian Blue staining of glycosaminoglycans in embryonic material: effect of different fixatives. Histochem.J. **20**:174–182 (1988).

Tuqon, N.A. and Adams, C.W.M.: Histochemistry of myelin. I. Proteins and lipid-protein complexes in the normal sheath. J. Neurochem. **6**:327–333 (1961).

Umezawa, H., Aoyagi, T., Ogawa, K., Naganawa, H., Hamada, M. and Takeuchi, T.: Diprotins A and B, inhibitors of dipeptidyl aminopeptidase IV, produced by bacteria. J.Antibiotics **37**:422–425 (1984).

Uzman, L.L.: Histochemical localization of copper with rubeanic acid. Lab.Invest. **5**:299–305 (1956).

Van Bogaert, L.J., Quinones, J.A., and van Craynert, M.P.: Difficulties involved in diaminobenzidine histochemistry of endogenous peroxidase. Acta Histochem. **67**:180–194 (1980).

Van de Kant, H.J.G., van Pelt, A.M.M., Vergouwen, R.P.F.A. and de Rooij, D.G.: A rapid immunogold-silver staining procedure for detection of bromodeoxyuridine in large numbers of plastic sections, using microwave irradiation. Histochem.J. **22**:321–326 (1990).

Van den Brink, W.J., Zijlmans, H.J.M.A.A., Kok, L.P., Bolhuis, P., Volkers, H.H., Boon, M.E. and Houthoff, H.J.: Microwave irradiation in label-detection for diagnostic DNA-in situ hybridization. Histochem.J. **22**:327–334 (1990).

Van der Ploeg, M. and van Duijn, P.: 5,6-dihydroxy indole as a substrate in a histochemical peroxidase reaction. J.R.microsc.Soc. **83**:415–423 (1964).

Van der Ploeg, M. and van Duijn, P.: Cytophotometric determination of alkaline phosphatase activity of individual neutrophilic leukocytes with a biochemical calibrated model system. J.Histochem.Cytochem. **16**:693–706 (1968).

Van der Ploeg, M. and van Duijn, P.: Reflection versus fluorescence. A note on the physical backgrounds of two types of light microscopy. Histochem. **62**:227–232 (1979).

Van der Ploeg, M., van den Broek, K. and Mitchell, J.P.: Dual wavelength scanning cytophotometry (Bicoscan). Histochemistry **62**:29–43 (1979).

Van Duijn, P.: Fundamental aspects of enzyme cytochemistry, pp.3–23. In: E. Wisse, W.Th. Daems, I. Molenaar and P. van Duijn, Eds.: Electron microscopy and cytochemistry. North-Holland, Amsterdam (1974).

Van Duijn, P.: Prospects for microscopical cytochemistry. Histochem.J. **8**:653–676 (1976).

Van Duijn, P. and van der Ploeg, M.: Microscopic cytochemistry as matrix chemistry, pp.209–229. In: Ciba Foundation symposium 73. Excerpta Medica, Amsterdam (1980).

Van Duijn, P. and van Noorden, C.J.F.: Plateau absorbance measurements: an alternative approach to enzyme activity determination illustrated by the example of alkaline phosphatase. Histochem.J. **21**:619–624 (1989).

Van Holde, K.E.: Physical biochemistry. Prentice-Hall, Englewood Cliffs, New Jersey (1971).

Van Kuppevelt, T.H.M.S.M., Cremers, F.P.M., Domen, J.G.W., and Kuyper, C.M.A.: Staining of proteoglycans in mouse lung alveoli. II. Characterization of the Cuprolinic Blue-positive anionic sites. Histochem.J. **16**:671–686 (1984a).

Van Kuppevelt, T.H.M.S.M., Domen, J.G.W., Cremers, F.P.M., and Kuyper, C.M.A.: Staining of proteoglycans in mouse lung alveoli. I. Ultrastructural localization of anionic sites. Histochem.J. **16**:657–669 (1984b).

Van Noorden, C.J.F.: Histochemistry and cytochemistry of glucose-6-phosphate dehydrogenase. Progr.Histochem.Cytochem. **15/4**:1–85 (1984).

Van Noorden, C.J.F.: Enzyme reaction rate studies in tissue sections. Proceedings Royal Microscopical Society **23**:93–97 (1988a).

Van Noorden, C.J.F.: On the role of oxygen in dehydrogenase reactions using tetrazolium salts. Histochem.J. **20**:587–593 (1988b).

Van Noorden, C.J.F.: Principles of cytophotometry in enzyme histochemistry and validity of the reactions. Acta Histochem. **37 Suppl.**:21–35 (1989).

Van Noorden, C.J.F. and Butcher, R.G.: Histochemical localization of NAD(P)-dependent dehydrogenase activity with four different tetrazolium salts. J.Histochem.Cytochem. **32**: 998–1004 (1984).

Van Noorden, C.J.F. and Butcher, R.G.: A quantitative histochemical study of NADPH-ferrihemoprotein reductase activity. Histochem.J. **18**:364–370 (1986a).

Van Noorden, C.J.F. and Butcher, R.G.: The out-of-range error in microdensitometry. Histochem.J. **18**:397–398 (1986b).

Van Noorden, C.J.F. and Butcher, R.G.: Quantitative enzyme histochemistry. Chapter 33. In: P.J. Stoward, Ed.: Pearse, A.G.E.: Histochemistry: Theoretical and applied. 4th edit. Vol. III. Churchill Livingstone, Edinburgh In press (1991).

Van Noorden, C.J.F. and Jonges, G.N.: Quantification of the histochemical reaction for alkaline phosphatase activity using the indoxyl-tetranitro BT method. Histochem.J. **19:**94–102 (1987).

Van Noorden, C.F.J. and Tas. J.: Advantages of 1-methoxyPMS as electron carrier in dehydrogenase cytochemistry. Proc.R.microsc.Soc. **16:**248 (1981a).

Van Noorden, C.F.J. and Tas, J.: Model film studies in enzyme histochemistry with special reference to glucose-6-phosphate dehydrogenase. Histochem.J. **13:**187–206 (1981b).

Van Noorden, C.J.F. and Tas, J.: Model film studies in enzyme histochemistry with special reference to glucose-6-phosphate dehydrogenase. Histochem.J. **13:**187–206 (1981c).

Van Noorden, C.F.J. and Tas, J.: Advantages of 1-methoxyPMS as electron carrier in dehydrogenase cytochemistry. Histochem.J. **14:**837–845 (1982a).

Van Noorden, C.F.J. and Tas, J.: The role of exogenous electron carriers in NAD(P)-dependent dehydrogenase cytochemistry studied in vitro and with a model system of polyacrylamide films. J.Histochem.Cytochem. **30:**12–20 (1982b).

Van Noorden, C.J.F. and Vogels, I.M.C.: Enzyme histochemical reactions in unfixed and undecalcified cryostat sections of mouse knee joints with special reference to arthritic lesions. Histochemistry **86:**127–133 (1986).

Van Noorden, C.J.F. and Vogels, I.M.C.: Polyvinyl alcohol and other tissue protectants in enzyme histochemistry: a consumer's guide. Histochem.J. **21:**373–379 (1989a).

Van Noorden, C.J.F. and Vogels, I.M.C.: Cytophotometric analysis of succinate and lactate dehydrogenase activity in rat liver, heart muscle and tracheal epithelium. Histochem.J. **21:**575–583 (1989b).

Van Noorden, C.J.F., Vogels, I.M.C. and van Wering, E.R.: Enzyme cytochemistry of unfixed leukocytes and bone marrow cells using polyvinyl alcohol for the diagnosis of leukemia. Histochemistry **92:**313–318 (1989).

Van Noorden, C.J.F., Kooij, A., Vogels, I.M.C., and Frederiks, W.M.: On the nature of the 'nothing dehydrogenase' reaction. Histochem.J. **17:**1111–1118 (1985).

Van Noorden, C.J.F., Vogels, I.M.C., Everts, V. and Beertsen, W.: Localization of cathepsin B activity in fibroblasts and chondrocytes by continuous monitoring of the formation of a final fluorescent reaction product using 5-nitrosalicylaldehyde. Histochem.J. **19:**483–487 (1987).

Van Oostveldt, P. and Bauwens, S.: Quantitative fluorescence in confocal microscopy. The effect of the detection pinhole aperture on the re-absorption and inner filter phenomena. J.Microsc. **158:**121–132 (1990).

Van Prooijen-Knegt, A.C., Raap, A.K., van der Burg, M.J.M., Vrolijk, J. and van der Ploeg, M.: Spreading and staining of human metaphase chromosomes on aminoalkylsilane-treated glass slides. Histochem.J. **14:**333–344 (1982).

Vejlsted, H.: The influence of light on gallocyanin-chrome alum stained sections. J.Histochem. Cytochem. **17:**545–546 (1972).

Volz, D., Reid, P.E., Park, C.M., Owen, D.A., Dunn, W.L., and Ramey, C.W.: Can "mild" periodate oxidation be used for the specific histochemical identification of sialic acid residues? Histochem.J. **18:**579–582 (1986).

Volz, D., Reid, P.E., Park, C.M., Owen, D.A., and Dunn, W.L.: Histochemical procedures for the simultaneous visualization of neutral sugars and either sialic acid and its side chain O-acyl variants or O-sulphate ester. I. methods based upon the selective periodate oxidation of sialic acids. Histochem.J. **19:**249–256 (1987a).

Volz, D., Reid, P.E., Park, C.M., Owen, D.A., and Dunn, W.L.: A new histochemical method for the selective periodate oxidation of total tissue sialic acids. Histochem.J. **19:**311–318 (1987b).

Von Bülow, F.A. and Høyer, P.E.:Use of epoxy resin slides for handling unfixed cryostat sections intended for histochemistry at the ultrastructural level. Histochem.J. **15:**825–827 (1983).

Wachsmuth, E.D.: Assessment of immunocytochemical techniques with particular reference to the mixed-aggregation immuno-cytochemical technique, pp.135–159. Trends in enzyme histochemistry and cytochemistry. Ciba Foundation Symposium 73, Excerpta Medica, Amsterdam (1980).

Wachsmuth, E.D.: A comparison of the highly selective fluorescence staining of fungi in tissue sections with Uvitex 2B and Calcofluor White M2R. Histochem.J. **20:**215–221 (1988).

Wachsmuth, E.D., Thöner, M., and Pfleiderer, G.: The cellular distribution of aldolase isozymes in rat kidney and brain determined in tissue sections by the immuno-histochemical method. Histochemistry **45:**143–161 (1975).

Wachstein, M. and Meisel, E.: Histochemistry of substrate specific phosphatases at a physiological pH. J.Histochem.Cytochem. **4:**424–425 (1956).

Warford, A.: In situ hybridisation: a new tool in pathology. Med.Lab.Sci. **45:**381–394 (1988).

Waterhouse, D.F.: Histochemical detection of barium and strontium. Nature, London **167:**358 (1951).

Waters,S.E. and Butcher, R.G.: Studies on the Gomori acid phosphatase reaction: the preparation of the incubation medium. Histochem.J. **12:**191–200 (1980).

Weakley, B.S.: A beginner's handbook in biological transmission electron microscopy. 2nd edit. Churchill-Livingstone, New York (1981).

Weber, P., Harrison, F.W., and Hof, L.: Histochemical application of dansylhydrazine as a fluorescent labelling reagent for sialic acid in cellular glycoconjugates. Histochemie **45:**271–277 (1975).

Weibel, E.R.: Stereological Methods: Practical methods for biological morphometry. Vol.1. Academic Press, London, (1979).

Weibel, E.R.: Stereological Methods: Theoretical foundations. Vol.2. Academic Press, London, (1980).

Weier, H.-U.G., Segraves, R., Pinkel, D. and Gray, J.W.: Synthesis of Y chromosome-specific labeled DNA probes by in vitro DNA amplification. J.Histochem.Cytochem. **38:**421–426 (1990).

Welch, J.P., Mitchell, R.B. and Davis, J.C.: Cytophotometric measurements of DNA, RNA and basic protein in FSH treated rat Sertoli cells in culture. Biol.Reprod. **21:**69–74 (1979).

White, T.J., Arnheim, N. and Erlich, H.A.: The polymerase chain reaction. Trends Genetics **5:**185–188 (1989).

Whittaker, J.R.: Quantitative measurement by microdensitometry of tyrosinase (dopa oxidase) development in whole small ascidian embryos. Histochemistry **71:**349–359 (1981).

Wieland, H. and Scheuing, G.: Die Fuchsin-Schweflige Säure und ihre Farbreaktion mit Aldehyden. Ber.Dtsch.Chem.Ges. **54:**2527–2555 (1921).

Williams, G. and Jackson, D.S.: Two organic fixatives for acid mucopolysaccharides. Stain Technol. **31:**189–191 (1956).

Williams, M.A.: Autoradiography and immunocytochemistry (Part 1). Quantitative methods in biology (Part 2). In: Glauert, A.M., Ed.: Practical methods in electron microscopy. Vol. 6. North-Holland Publishing Company, Amsterdam (1977).

Winzek, C., Plieninger, P. and Baumgärtel, H.: An improved method to investigate staining kinetics in single cells. Histochemistry **86:**421–426 (1987).

Wittekind, D.: Standardization of dyes and stains for automated cell pattern recognition. Analyt.Quant.Cytol.Histol. **7:**6–30 (1985).

Wittekind, D. and Schulte. E.: Die Bedeutung der Standardisierung der Zell- und Gewebspräparation für bildanalytische Operationen. In: Eins, S., Ed.: Quantitative und strukturelle Bildanalyse in der Medizin. GIT-Verlag, Darmstadt. (1987).

Wittekind, D., Hilgarth, M., and Kretschmer, V.: Die einfache und reproduzierbare Papanicolaou-Färbung. Geburtshilfe Frauenheilkd. **39:**969–972 (1979).

Wittekind, D., Hilgarth, M., and Kretschmer, V.: The new and reproducible Papanicolaou stain. Morphologic and spectrophotometric observations on the influence and stain composition on staining results. Anal.Quant.Cytol. **4:**286–294 (1982).

Wittekind, D.H.: On the nature of Romanowsky-Giemsa staining and its significance for cytochemistry and histochemistry: an overall view. Histochem.J. **15:**1029–1047 (1983).

Wittekind, D.H., Reinhardt, E.R., Kretschmer, V., and Zipfel, E.: Influence of staining on fast automated cell segmentation, feature extraction and cell image analysis. Anal.Quant.Cytol. **5:**55–60 (1983).

Wohlrab, F., Seidler, E. and Kunze, K.D.: Histo- und Zytochemie dehydrierender Enzyme. Grundlagen und Problematik. J.A. Barth, Leipzig (1979).

Wolman, M. and Kasten, F.H.: Polarized light microscopy in the study of the molecular structure of collagen and reticulin. Histochemistry **85**:41–49 (1986).

Wolters, G.H.J., Pasma, A., Wiegman, J.B. and Konijnendijk, W.: Changes in histochemically detectable calcium and zinc during tolbutamide-induced degranulation and subsequent regranulation of rat pancreatic islets. Histochemistry **78**:325–338 (1983).

World Health Organization, Expert Committee Biological Standardization, 26th Report. WHO Techn.Rep.Ser. **565**:5–72 (1975).

Wright, J.R., Calkins, E., and Humphrey, R.L.: Potassium permanganate reaction in amyloidosis. A histologic method to assist in differentiating forms of this disease. Lab.Invest. **36**:274–281 (1977).

Wolman, M. and Kasten, F.H.: Polarized light microscopy in the study of the molecular structure of collagen and reticulin. Histochemistry **85**:41–49 (1986).

Yamabayashi, S.: Periodic acid-Schiff-Alcian Blue: a method for the differential staining of glycoproteins. Histochem.J. **19**:565–571 (1987).

Yasuma, A. and Itchikawa, T.:Ninhydrin-Schiff and alloxan-Schiff staining. A new histochemical staining method for protein. J.Lab. Clin.Med. **41**:296–299 (1953).

Yoffe, J.R.: A comparison of two cytochemical methods used to determine lysosomal function in cultured endothelial cells. Histochem.J. **12**:537–542 (1980).

Ziehl, F.: Zur Färbung des Tuberkelbacillus. Dtsch.Med. Wochenschr. **8**:451 (1882).

Ziehl, F.: Ueber die Färbung des Tuberkelbacillus. Dtsch.Med. Wschr. **9**:247–249 (1883).

Zimmermann, H.: Physikalisch-chemische Grundlagen der Färbung für manuelle und apparative Zytodiagnostik. In: Wittekind, D., Ed.: Manual and automated diagnosis in cytology. Microsc.Acta [Suppl]**6**:45–58 (1983).

Zipfel, E., Grezes, J.-R., Naujok, A., Seifert, W., Wittekind, D.H. and Zimmermann, H.W.: Über Romanowsky-Farbstoffe und den Romanowsky-Giemsa-Effekt. 3. Mitteilung: Mikrospektralphotometrische Untersuchung der Romanowsky-Giemsa-Färbung. Spektroskopischer Nachweis eines DNA-Azur B-Eosin Y-Komplexes, der den Romanowsky-Giemsa-Effekt verursacht. Histochemistry **81**:337–351 (1984).

Zoller, L.C., Mulhern, P. and Hulik, D.: Quantitative and qualitative cytochemical analysis of changes in polysaccharides-proteins in rat ovulable and atretic ovarian follicles during the estrous cycle. Histochemistry **73**:339–351 (1981).

[illegible] and [illegible] [illegible] Scand J Haematol [illegible] (1967).

Subject Index

Please note: the numbers in the indices refer to section numbers in the text

Index of Constituents (Including Chemical Groups, Cells, and Tissues)

Please note: the numbers in the indices refer to section numbers in the text

Index of Dyes (Including Pigments, Chromogenic Reagents, and Stains)

Please note: the numbers in the indices refer to section numbers in the text

Index of Histochemical and Histological Methods

Please note: the numbers in the indices refer to section numbers in the text